T. Addison Richards

Appleton's Companion Hand-Book of Travel

T. Addison Richards

Appleton's Companion Hand-Book of Travel

ISBN/EAN: 9783337211257

Printed in Europe, USA, Canada, Australia, Japan

Cover: Foto ©Andreas Hilbeck / pixelio.de

More available books at **www.hansebooks.com**

COMPANION HAND-BOOK

OF

TRAVEL:

CONTAINING

A FULL DESCRIPTION

OF THE PRINCIPAL

CITIES, TOWNS, AND PLACES OF INTEREST,

TOGETHER WITH

HOTELS AND ROUTES OF TRAVEL

THROUGH THE

UNITED STATES AND THE CANADAS.

WITH COLORED MAPS.

EDITED BY

T. ADDISON RICHARDS.

NEW YORK:
D. APPLETON & COMPANY, 443 & 455 BROADWAY.
LONDON: TRÜBNER & CO.
1865.

NOTICE.

No expense or labor will be spared to make the Companion Hand-Book attractive, comprehensive, concise, thorough, and every way *reliable*.

As we shall frequently print new editions, any information in regard to errors and omissions which those who use the work may detect, or any facts of interest and value—particularly in respect to new routes and accommodations—will be gratefully received and considered. Such communications should be addressed to the Publishers.

The population of Cities and Towns mentioned in this Work are those of the last Census (1860), except when otherwise stated.

WE ALSO PUBLISH

APPLETONS' ILLUSTRATED RAILWAY GUIDE, Containing Seventy Maps, and the latest Time Tables, corrected to date. Published semi-monthly, under the supervision of the Railway Companies. One Volume, 288 pages. 25 Cents. Subscription Price, $3 per annum.

APPLETONS' ILLUSTRATED HAND-BOOK OF AMERICAN TRAVEL, a Full and Reliable Guide, by Railway, Steamboat, and Stage, to the Cities, Towns, Waterfalls, Battle-fields, Mountains, Rivers, Lakes, Hunting and Fishing Grounds, Watering Places, Summer Resorts, and all Scenes and Objects of Importance and Interest in the United States and British Provinces. By T. ADDISON RICHARDS. With Careful Maps of all Parts of the Country, and 100 Pictures of Famous Places and Scenes, from Original Drawings by the Author and other Artists.

PART I., containing the NORTHERN AND EASTERN STATES, $1.
PART II., " SOUTHERN AND WESTERN STATES, $1.

OR THE TWO PARTS BOUND IN ONE VOLUME, $1 50.

ENTERED, according to Act of Congress, in the year 1860, by

D. APPLETON & COMPANY,

In the Clerk's Office of the District Court of the United States for the Southern District of New York.

TO THE TRAVELLER:

SOME PARTING WORDS OF EXPLANATION AND ADVICE.

---•◆•---

IN a journey over so vast a country as the United States, occupying nearly half a Continent, and measuring its length and breadth by thousands, and its routes of travel by tens of thousands of miles, one may very readily be pardoned if he sometimes stumbles by the way. May *we* not beg the benefit of this consideration, if, in our present laborious *itinéraire*, we have occasionally chanced, despite all our watchfulness, to only half look at points of interest or to overlook them altogether; or if, amidst the intricate reticulation of the roads, we may have momently lost our way? We hope, however, that we have not been thus unlucky in any considerable degree, for we have made very honest effort to guide our traveller truly and surely; to show him —hastily, to be sure, as needs must be, yet intelligently—the past and the present, the physique and the *morale*, of the great country through which we have led him; its differing peoples and places, from the mountains to the prairies—from the cities and palaces of the East to the wildernesses and wigwams of the West.

Though we have thus done our best for the present, we hope to do still better hereafter, as we revise and extend our volume year after year, with the benefit of enlarged personal observation and of the good counsels of others: for we trust that those who follow our guidance will do us the kindness to advise us of any and all errors and omissions they may discover in our pages. To assist them in rendering us this generous service, we have placed some blank leaves for memoranda, at the end of our book.

THE PLAN OF THIS BOOK.

We have thought it best to follow the familiar geographical order of the various divisions of the country, and thus to begin at Canada on the extreme north-east, and, continuing along the shores of the Atlantic and the Gulf of Mexico, and upon the Pacific, westward. With rare exceptions, we have, instead of selecting a particular route and seeing all it offers of attraction, jumped at once to our especial destination, and then intimated the way by which it is reached. Thus, if the traveller happens to be in New York or Boston, and desires to go to New Orleans, he will, by turning to

"New Orleans," find the routes thither. The chief cities are taken as starting points for all other and lesser places in their neighborhood. It has not, of course, been possible to mention every village or town in the Union, in the narrow limits of a pocket volume, like this.

GENERAL REMARKS.

The foreign tourist will soon observe, to his satisfaction, (and the *citizen* might remember it oftener, with thanks to his stars,) the great convenience of the total absence in the United States, of all annoying demands for passports—of scowling fortifications and draw-bridges, of jealous gates, closed at a fixed hour of the evening and not to be re-opened before another fixed hour of the morning; of custom-houses between the several States, and of all rummaging of baggage by gens d'armes for the octroi; and yet, nevertheless, of as perfect a feeling of security, everywhere, as in the most vigilantly policed kingdoms of Europe.

He may or may not like the *table d'hôte* system of our hotels—the uniform fare and the unvarying price; that, excepting in the few metropolitan cities, where the habits of all nations obtain, he must submit to.

From the social equality everywhere and without exception, he will not suffer, however high his rank at home; and if it be not the highest, he will surely gain in consideration. To win attention and care, both the lofty and the lowly have, and have only, to dispense good will and kind manners as they pass along.

MONEY.

"Greenbacks," in place of gold and silver, are current all through the Union, while bank-notes, and especially of distant States, very often are not. Change, too, will save trouble; especially half-dollars, generally the fare of omnibuses and hacks, and invariably the price of meals. Twenty-five cent. notes, too, are useful, as fees for little services by the way. In travelling through the settled districts by railway and steamboats, and at the best hotels, the daily expenses should be estimated at not less than five or six dollars per day for each person.

BAGGAGE.

As little baggage as possible is always a good rule, though a very liberal supply is permitted on the railways and almost any quantity on the steamboats. On the stages, the prescribed limit of sixty or eighty pounds cannot be exceeded without extra charge.

The regular carriages of his hotel will convey the traveller securely and in season, to the railway station or the steamboat landing, where his first care must be to deposit his trunks in the keeping of the baggage-master, and receive a check for each one—corresponding marks will be attached to the baggage, and it will be delivered at the end of the route only to the holder of the checks. It is best to get baggage checked for the entire journey, or for the longest possible stage thereof, and thus save one's self the trouble of looking out for it more frequently than is necessary.

Before arriving at his destination, the traveller will, on the principal routes, receive a call from an express agent, to whom he may safely resign his check and address, confident that his baggage will be duly delivered, and at a charge of about thirty-five cents for each piece or trunk. On arriving at the end of his journey, he should put himself in one of the carriages marked as in the particular service of the hotel to which he is going. If he employs other vehicles, it will be well to learn the fare beforehand, particularly in the city of New York, where hackmen pay but little attention, when they can help it, to the law in the case.

TICKETS.

Tickets on the railways should be purchased at the office before starting, otherwise a small additional charge will be made. If a long journey over various roads is intended, it is cheaper and more convenient to buy a *through ticket* to the end of the route, or for as long a distance as possible. On the steamboats, the tickets for passage and for meals will be purchased at leisure, after starting, at the captain's office.

HOTELS.

The hotels of the United States are famous all the world over, for their extent, convenience, comfort, and elegance. They are often truly palatial in their sumptuousness, with means and appliances for the prompt gratification of every want and whim. The universal price of board, from one end of the country to the other, is $2 50 to $4 per day at the most fashionable, and indeed at all the principal houses. Private parlors and extra rooms involve an additional charge, according to their position. Wines are always extra and always dear enough.

WAITERS OR SERVANTS.

It is not the general custom in America, as in Europe, to fee waiters at the hotels, though it may very properly be done for especial personal service. It is often done by those who like hot dinners better than cold, or who may have a fancy for some rare dish when it unluckily happens to be " all out."

COSTUME.

At the watering places, the same resources of toilette are needed as in the city *salon;* but though you be thus provided, do not be unprovided with a travelling suit equal to rude usage. If the color be a gray or a brown, so much the better in the dust of railway and stage routes. Don a felt hat,—it does not crush itself on your head in car or carriage, or blow overboard on steamboats. Leave thin boots (this especially to the ladies) at home, and be well, and comfortably, and safely shod, in *stout calf-skin*. It is a pity to be kept in doors by the fear of spoiling one's gaiters or wetting one's feet, when the meadows and hills and brooks are waiting to be explored. In mountain tramps, a generous sized flask, filled with most excellent brandy, may be swung over the shoulder with very picturesque effect.

Now that we have told our traveller *how* to go, it only remains to us, before starting, to add a word of suggestion as to

WHERE TO GO.

If you are in New York, with one, or two, or three, or more summer days to spare, run up to one or other of the many delightful places on the Hudson River,—to West Point, or Newburgh, for example; or to the Catskill Mountains; or run down to Rockaway, or Long Branch, or any of the many healthful and inviting resorts along the coast of Long Island and New Jersey.

If a week is at your command, go to Lake George, or to Trenton Falls, or Niagara; explore the varied route of the Erie Railway, or seek some one of the innumerable Springs of the State.

If a fortnight or a month can be spared, make a trip to Canada. See Montreal and the Ottawa River, then go to Quebec and the Saguenay, returning through Maine; or from Montreal go up the St. Lawrence to Toronto, and thence to the great Lakes; or spend a part or all of your time among the wonderful White Hills of New Hampshire.

If the whole summer is waiting to be disposed of, visit the beauties of the Upper Mississippi, or explore the marvels of the vast region lying west of the great Father of waters, where Territories and States are springing up as by magic, and the wilderness may be literally seen blossoming as the rose.

In winter leisure, make a voyage to Havana, or elsewhere in the neighboring islands, and find the summer airs again which you have lost in more northern latitudes.

There is no lack of inviting resorts for a day, or week, or month, or forever. Look in this respect at our Skeleton Tours, and at the detailed descriptions and routes in the pages which follow. *Go somewhere, if you can*, all of you; and wherever and whenever you go, God speed you on your way and send you duly back wiser, and better, and healthier, and happier men and women.

SKELETON TOURS

From New York to Various Parts of the United States and the Canadas.

WITH AN APPROXIMATE STATEMENT OF THE TIME REQUIRED TO TRAVEL FROM PLACE TO PLACE, AND OF THE DURATION OF THE HALTS TO BE MADE AT THE MOST REMARKABLE SPOTS.

See Description of Routes, Hotels, Places, and Scenes in the following pages.*

A TOUR OF SIX DAYS,

Visiting West Point, Newburgh and the Catskills.

MONDAY. New York to West Point (52 miles), by morning steamboat up the Hudson River, through the Highlands, or by an early train on the Hudson River Railway, stopping at Garrison's, and crossing by steam-ferry to the West Point Hotel or to Cozzen's, just below. Arrive in three hours, by or before noon. Visit the Military Academy, the ruins of Fort Putnam, Kosciusko's Garden, Weir's Studio, etc.

TUESDAY. Morning steamboat or early train to Newburgh (9 miles, crossing ferries included, one hour), stop at the Orange Hotel on the Main street, or at the Powelton, an elegant summer establishment in the suburbs; visit Washington's Head Quarters in the village. After dinner, take a carriage for "Idlewild," the charming home of N. P. Willis, four miles down the river. Explore the grounds and the beautiful mountain brook and glen. Visit "Cedarlawn," the residence of the author Headley, on the way, a mile below Newburgh.

WEDNESDAY. Morning steamboat or by railway from Fishkill, opposite Newburgh, to Oakhill, opposite Catskill; 51 miles, 2 hours, *besides ferries*. From Catskill village, in good coaches, 12 miles, through a most picturesque hill and valley region, to the Mountain House.

THURSDAY. Look out for the grand spectacle from this point, of the sun-rise. After breakfast walk to the North Mountain, overlooking the hotel and the two lakes; next, join the usual morning party in the two-mile ride to the High Falls: back to dinner.

FRIDAY. Ride from the Mountain House through the great Kauterskill Clove, westward to the village of Palenville, returning by valley and mountain road eastward; or explore the ravines and cascades of the Clove, better *on foot* —a good day's tramp.

SATURDAY. Return to New York, via Catskill Village and the Hudson River.

*** If more time is at command, devote a day to a visit to High Peak, another to the Stony Clove, and another to the Plauterkill Clove and Creek.

A TOUR OF SIX DAYS,

Visiting Albany and Troy (via the Hudson River), Saratoga Springs, Lake George, Fort Ticonderoga, and Whitehall, on Lake Champlain.

MONDAY. From New York by morning boat or cars, via Hudson River, 146 miles, 5 or 6 hours, to Albany (see Albany and Troy.)

TUESDAY. Railway, time about two hours, from Albany, through the city of Troy, to Saratoga Springs. Stop at the United States Hotel, at the Union Hall, at the Clarendon, or at the Congress Hall.

WEDNESDAY. To Lake George by railway 15 miles, to Moreau Station, and thence by plank road, an hour or two, via Glen's Falls to Caldwell. Stop over night at the Lake House, or at the Fort William Henry Hotel, close by.

THURSDAY. Spend the day on the Lake boating and fishing, or sketching.

FRIDAY. Make the voyage of the lake in the favorite little steamer, the "Minnehaha," a few hours' sail to the village of Ticonderoga, at the foot or north end of the lake; thence 3 or 4 miles by coach to the ruins of Fort Ticonderoga.

* For Railway Time-Tables consult Appletons' Monthly Railway and Steam Navigation Guide.

SATURDAY. Return home by the Lake Champlain steamers to Whitehall, and thence by Railway via Troy and the Hudson.

∗ Same tour (except on the Hudson), within the same time from Boston, taking the Western Railway, thence (*Monday*) 200 miles to Albany.

A TOUR OF SIX DAYS,

Visiting Trenton and Niagara Falls, via the Central Railway, and Returning by the New York and Erie Road.

MONDAY. From New York to Trenton Falls, via Hudson River, 146 miles to Albany, Central Railway, 95 miles to Utica, thence 15 miles to the Falls.
TUESDAY. Explore the Falls.
WEDNESDAY. Return to Utica and resume journey on the Central Road, via Syracuse and Rochester (Falls of the Genesee) to Niagara.
THURSDAY. At Niagara.
FRIDAY. To Buffalo, and thence by the Erie Railway, passing the night at Binghamton.
SATURDAY. Erie Road from Binghamton to New York.

∗ If more time is at command, remain over Sunday at Niagara, and follow the picturesque route of the Erie Road more leisurely, seeing the cascades and ravines of the Genesee, and the great Railway Bridge at Portage, 61 miles from Buffalo. Elmira, 273 miles from New York; Owego, 236 miles; Great Bend, 200 miles; and Port Jervis, 88 miles, are pleasant stopping places on the way.

TOUR OF A WEEK,

Visiting Philadelphia, Baltimore and Washington City.

MONDAY. From New York by morning line to Philadelphia on the New Jersey Railway, 87 miles, or by the Camden and Amboy route. Arrive in the early afternoon.
TUESDAY. At Philadelphia.
WEDNESDAY. Morning train to Baltimore, 97 miles; arrive in the early afternoon.
THURSDAY. Spend the day in Baltimore, and take the evening train, 40 miles, 2 hours, to Washington.
FRIDAY. At Washington
SATURDAY. Return to New York by Baltimore and Philadelphia, 224 miles.

∗ If more time can be spared, remain in Washington Saturday and Sunday, visiting Mount Vernon, Georgetown, Alexandria, etc. Return on Monday to Philadelphia, and next day leisurely to New York.

TOUR OF A WEEK,

Visiting the Valley of Wyoming and the Delaware Water Gap.

MONDAY. From New York by the Erie Railway, 200 miles, to Great Bend.
TUESDAY. By the Delaware, Lackawanna and Western Railway to Scranton, an interesting place; thence to Wilkesbarre, on the Susquehanna, and in the Valley of Wyoming.
WEDNESDAY. Explore the valley, visiting Prospect Rock, 3 miles from the village and Nanticoke, in the beautiful passage of the Susquehanna, at the Southern extremity of Wyoming.
THURSDAY. Returning, take the cars via Mauch Chunk, in the coal region, to Easton and the Water Gap.
FRIDAY. At the Water Gap.
SATURDAY. Reach home by the Delaware, Lackawanna, and Western Railway, and other routes across New Jersey.

∗ With more time, it would be agreeable to spend a day at Scranton, two or three in and below the Valley of Wyoming; to stop at Mauch Chunk, and see the coal mines and the bold landscape of the Lehigh River.

TOUR OF TWO WEEKS,

Visiting the White Mountains and the Lake Scenery of New Hampshire, via Boston.

FIRST WEEK.

MONDAY. From New York to Boston, journey occupying the night, by the Stonington and Providence, the Fall River or the Norwich routes (morning and evening), or by the Boston Express; or the Shore Line via New London.
TUESDAY. Boston.
WEDNESDAY. Boston to Centre Harbor, on Lake Winnipiseogee; arrive at dinner time; spend the afternoon on the lake or lake shores.
THURSDAY. Visit Red Hill (on horseback), a few miles distant, and overlook the beautiful lake region.
FRIDAY. Proceed by the White Mountain stages to North Conway, one of the most charming valleys in the world; stop over night at Thompson's.
SATURDAY. Continue journey by stage 24 miles to Crawford House, in the Great White Mountain Notch—traversing the valleys of Conway, Bartlett, etc.
SUNDAY. Crawford House

SECOND WEEK.

MONDAY. Ascend Mount Washington.

TUESDAY. Visit the Silver Cascade and other scenes in the neighborhood of the Notch.
WEDNESDAY. Continue journey, by stage, 27 miles, to the Profile House, in the Franconia group of the White Hills, following the course of the Ammonoosuc.
THURSDAY. Profile House; visit Echo Lake and Profile Lake, and see the Old Man of the Mountain, Eagle Cliff, Cannon Mountain, and other sights of the vicinage.
FRIDAY. Ride 5 miles, from the Profile to the Flume House; visit the Flume and its neighboring marvels.
SATURDAY. Returning, take stage to Littleton, 12 miles; thence by Railway 20 miles to Wells' River; thence through the valley of the Connecticut to Bellows Falls, Brattleboro, or Northampton.
SUNDAY. At Bellows Falls, Brattleboro, or Northampton.
THIRD MONDAY. Home, from Northampton, by Springfield, Hartford, and New Haven, or (from Bellows Falls) by Albany and the Hudson.

A TOUR OF TWO WEEKS,

From New York to the White Mountains, via Boston and Portland, returning by the Connecticut Valley Routes.

FIRST WEEK.

MONDAY. New York to Boston.
TUESDAY. At Boston.
WEDNESDAY. Boston to Portland, Maine.
THURSDAY. At Portland.
FRIDAY. From Portland by the Grand Trunk Railway, 91 miles, to Gorham, N. H., White Mountain Station: continue journey in coaches 8 miles to the Glen House.
SATURDAY. Journey by stage from the Glen House, 34 miles, to the Crawford House, White Mountain Notch.
SUNDAY. At the Crawford House.

SECOND WEEK.

Explore the White Mountains and return home, as in preceding tour.

A TOUR OF TWO WEEKS,

Visiting the New England Cities, New Haven, Hartford, Springfield, Boston, Providence, and Newport.

FIRST WEEK.

MONDAY. From New York to New Haven, Ct., 76 miles, by the New Haven Railway; visit Yale College, the Trumbull Gallery, etc.
TUESDAY. Continue journey, 36 miles, to Hartford, Ct.
WEDNESDAY. To Springfield, Mass., 26 miles; visit the United States Armory.
THURSDAY. To Boston, 98 miles.
FRIDAY. At Boston.
SATURDAY. At Boston.
SUNDAY. At Boston.

SECOND WEEK.

MONDAY. Morning train from Boston, 43 miles, to Providence; see the Library of Brown University and the Athenæum; visit the Seekonk River and "What Cheer Rock," on the edge of the city, the village and Falls of Pawtucket, near by, etc.
TUESDAY. At Providence; take a sail down the Narragansett Bay and back, in one of the numerous excursion steamers.
WEDNESDAY. Take the steamboat down the Narragansett Bay, from Providence to Newport; a charming voyage of some two hours.
THURSDAY. At Newport.
FRIDAY. At Newport.
SATURDAY. At Newport.
SUNDAY. At Newport.
MONDAY. Home, by Fall River Steamers direct.

TOUR OF TWO WEEKS,

From New York up the Valley of the Housatonic to Great Barrington, Stockbridge, etc., in Berkshire, Mass.; Lebanon Springs and Shaker Village, N. Y. Returning via the Hudson River and West Point.

FIRST WEEK.

MONDAY. From New York, via New Haven Railway, to Bridgeport, Ct.; thence without stopping, by the Housatonic Railway up the valley and river of the Housatonic to Great Barrington, in Berkshire, Mass.
TUESDAY. At Great Barrington.
WEDNESDAY. From Great Barrington, Railway 26 miles to Old Stockbridge.
THURSDAY. At Old Stockbridge.
FRIDAY. Lebanon Springs.
SATURDAY. Lebanon Springs.
SUNDAY. Lebanon Springs. Visit Shaker village, near by.

SECOND WEEK.

MONDAY. Visit Pittsfield, Williamstown, Lenox, Adams, etc.
TUESDAY. Visit Pittsfield, Williamstown, Lenox, Adams, etc.
WEDNESDAY. Visit Pittsfield, Williamstown, Lenox, Adams, etc.
THURSDAY. Return via Western Railway to

Albany, or by the Hudson and Boston Road to Hudson, and thence down the Hudson River to West Point.
FRIDAY. At West Point.
SATURDAY. Back in New York.

TOUR OF TWO WEEKS,

Visiting the Valley of the Connecticut.

MONDAY. By Railway from New York via New Haven and Hartford, Ct., to Springfield, Mass. 138 miles; dine, visit the U. S. Armory, etc.
TUESDAY. To Northampton, 17 miles, by Railway, near the banks of the Connecticut.
WEDNESDAY. At Northampton, visiting Mount Holyoke, and other scenes of great interest in the immediate neighborhood.
THURSDAY. Continue on the Railways up the valley and river 19 miles to Greenfield, Mass.; walk in the evening to the high ridge called Poet's Seat, finely overlooking all the country round.
FRIDAY. Resume the journey (by Railway always), up the valley, 24 miles further, to Brattleboro, in Vermont. This is one of the most agreeable resting places on the route; one of the most attractive in scenery, society, hotel comforts, etc.
SATURDAY. Visit the grounds of the Insane Asylum, West River, the Cemetery, and other charming localities in the vicinage of Brattleboro.
SUNDAY. Still at Brattleboro; a pleasant place for a Sunday halt, all travel being suspended on that day hereabouts.
MONDAY. Resume journey 24 miles further up the river to Bellows Falls. At this point the traveller may turn back if he pleases by railway via Rutland, Vt., Whitehall, on Lake Champlain, Saratoga Springs, Albany or Troy, and the Hudson River; going on Tuesday to Saratoga, and on Wednesday to New York; or he may continue on with us yet further up the valley of the Connecticut.
TUESDAY. From Bellows Falls 26 miles to Windsor, Vt., a very quiet, picturesque, and pleasant place.
WEDNESDAY. Ascend Mount Ascutney, near Windsor
THURSDAY. From Windsor (returning) by the Vermont Central Road, through the charming valley of the Winooski to Burlington, on Lake Champlain.
FRIDAY. Cross the Lake from Burlington to Port Kent, and visit the bold ravine called the Walled Banks of the Ausable.
SATURDAY. Home by Whitehall, Troy, Albany, and the Hudson.
*** At Windsor (Second Tuesday of this tour), the traveller being on one of the most agreeable routes thence, may continue his journey eastward to the White Mountain Region.

TOUR OF THREE WEEKS,

Visiting the Hudson River, Saratoga Springs, Lake George, Lake Champlain, Montreal, Quebec, and the Saguenay River, the St. Lawrence River, Niagara Falls, and the Scenery of the Erie Railway.

FIRST WEEK.

MONDAY. From New York to Albany, by steamboat or railway (Hudson River), thence by railway to Saratoga.
TUESDAY. Saratoga Springs.
WEDNESDAY. To Caldwell on Lake George.
THURSDAY. Down Lake George to Fort Ticonderoga on Lake Champlain.
FRIDAY. Steamer on Lake Champlain (a pleasant voyage) and onward by railway to Montreal.
SATURDAY. Montreal.
SUNDAY. Montreal.

SECOND WEEK.

MONDAY. Railway or St. Lawrence River to Quebec.
TUESDAY. Quebec.
WEDNESDAY. Down the St. Lawrence to the mouth of the Saguenay
THURSDAY. Voyage up the Saguenay.
FRIDAY. Back to Quebec.
SATURDAY. Grand Trunk Railway to Montreal.
SUNDAY. Montreal.

THIRD WEEK.

MONDAY. Up the St. Lawrence to Kingston.
TUESDAY. Grand Trunk Railway via Toronto to Hamilton; thence, by the Great Western Road to Suspension Bridge, Niagara.
WEDNESDAY. Niagara Falls.
THURSDAY. Niagara.
FRIDAY. Erie Railway (returning) to Owego or Binghamton, or to Utica on Central Route.
SATURDAY. Home.
*** Omit the détour from Montreal to Quebec, and back, and make this tour within two weeks instead of three.

HUNTING TOUR OF THREE WEEKS,

To the Saranac Lakes, in the Wilderness of Northern New York.

FIRST WEEK.

MONDAY. From New York to Port Kent, opposite Burlington, on Lake Champlain, via Hudson River, Saratoga Springs, and Whitehall. From Port Kent, by omnibus or stage, five miles back, to Keeseville. Stop at the Ausable House.

TUESDAY. Visit the remarkable ravines and cascades near Keeseville, called the Walled Banks of the Ausable.

WEDNESDAY. Take the tri-weekly mail wagon or private carriage, for the banks of the Lower Saranac, stopping at Baker's, a mile distant, or at Martin's on the shore.

THURSDAY. Secure the services of a guide and hunter, with his boat, dogs, tent, and all necessary equipments and provisions for camp life, all the journey hence being by water.

FRIDAY. On the Lower Saranac, crossing the "carrying place" in the afternoon to the Middle Saranac on the shore of which camp for the night, after a supper of trout readily taken, with venison, perchance, to boot.

SATURDAY and SUNDAY. Camp on the Upper Saranac, one of the most beautiful of these lakes, and a fine hunting and fishing ground.

SECOND WEEK.

MONDAY and TUESDAY. Visit the St. Regis Lake.

WEDNESDAY. Return to the Middle Saranac (or Round Lake), make a short portage to the Stony Creek Pond; and thence reach "the Racquette River," by a pull of three miles on the Stony Creek. Camp for the night.

THURSDAY. Voyage on the Racquette River of 20 miles to Tupper's Lake. The tourist is here at the last and most charming portion of the region comprised in our present tour; and here, be he artist or hunter, he will be very willing to pass the remainder of the time which his furlough grants to him. Lough Neah is a continuation of the picturesque waters of Tupper's Lake.

FRIDAY. Tupper's Lake.
SATURDAY. Tupper's Lake.
SUNDAY. Tupper's Lake.

THIRD WEEK.

MONDAY. Tupper's Lake.
TUESDAY. Returning; retraverse the Racquette River.
WEDNESDAY. Arrive at the Middle Saranac Lake.
THURSDAY. Back to the starting point on the Lower Saranac.
FRIDAY. Regain Lake Champlain at Port Kent, or at Westport.
SATURDAY. Home.

*** If the traveller in this wonderful region be addicted to the rifle, the rod, or the pencil, he may extend his visit with pleasure from three weeks to three months. The Adirondack hills and lakes—another portion of this marvellous wilderness—are not far removed from the Saranac: and one, two, or more weeks might be spent there with great satisfaction.

TOUR OF FOUR WEEKS,

To the Grand Lakes, via Quebec, Montreal, the St. Lawrence Niagara Falls, &c.

FIRST WEEK.

MONDAY. From New York, via Albany and Troy, to Saratoga.
TUESDAY. Saratoga Springs.
WEDNESDAY. To Montreal, by Railway or Steamer on Lake Champlain.
THURSDAY. Montreal.
FRIDAY. To Quebec.
SATURDAY. Quebec and visiting the Montmorenci, the Chaudiere, &c.
SUNDAY. Quebec.

SECOND WEEK.

MONDAY. Great Trunk Railway, by Montreal, to Toronto, on Lake Ontario.
TUESDAY. Take the Northern Railway of Canada, 95 miles, to Collingwood, on the Georgian Bay, an arm of Huron.
WEDNESDAY. By Steamer, on Lake Huron, to the Straits of Mackinac.
THURSDAY. Mackinac.
FRIDAY. Mackinac.
SATURDAY. Steamer to the Sault St. Marie—the connecting link of the waters of Huron and Lake Superior.
SUNDAY. At the Sault de St. Marie, or the "Soo," as it is familiarly called.

THIRD WEEK.

Voyage on Lake Superior.

FOURTH WEEK.

MONDAY. From the Sault de St. Marie (returning) (Steamer on Lake Huron) to Detroit, Michigan.
TUESDAY. Great Western Railway to Suspension Bridge, Niagara Falls.
WEDNESDAY. Niagara Falls.
THURSDAY. Niagara Falls.
FRIDAY. To Utica Central Railway, or to Binghamton, Erie Route.
SATURDAY. To New York.

TOUR OF FOUR WEEKS,

To the Virginia Springs, Weir's Cave, the Natural Bridge, the Peaks of Otter, &c.

FIRST WEEK.

MONDAY. From New York to Philadelphia.
TUESDAY. Philadelphia to Baltimore.
WEDNESDAY. Baltimore to Washington City.
THURSDAY. At Washington City—visit Mount Vernon.
FRIDAY. To Alexandria; and thence, by the Alexandria Railway, 88 miles; and from Gordonsville, on the Virginia Central Road, 64 miles to Staunton.
SATURDAY. Stage or Carriage, 17 miles, to Weir's Cave.
SUNDAY. At Weir's Cave.

SECOND WEEK.

MONDAY. At Weir's Cave, returning in the afternoon to Staunton.
TUESDAY. Continue journey on the Central Road, to Jackson's River, thence to the Sulphur Springs by Stage. (For Natural Bridge, take Stage at Millboro'.)
WEDNESDAY. *En Route.*
THURSDAY. White Sulphur Springs.
FRIDAY. White Sulphur Springs.
SATURDAY. White Sulphur Springs.
SUNDAY. White Sulphur Springs.

THIRD WEEK

May be devoted to the other Springs of this Region.

FOURTH WEEK.

Visit the Natural Bridge, 63 miles from the White Sulphur Springs; 12 miles from Lexington; 36 miles from Lynchburg, on the Virginia and Tennessee Railway, from Richmond, west; next, see the Peaks of Otter, in the same region. Return home by the Virginia and Tennessee Road, from Lynchburg to Richmond; thence, by the Great Southern Mail Route to Washington; or, more agreeably, by the James River and the Chesapeake Bay, to Baltimore; from Baltimore to Philadelphia; from Philadelphia to New York.

TOUR OF FOUR WEEKS,

From New York, via Boston and Portland, to Quebec and the Saguenay, Montreal, the Ottawa, and the St. Lawrence, returning by Niagara and Trenton Falls, Saratoga Springs, and the Hudson River. Détour of ten days (extra) to the White Mountains.

FIRST WEEK.

MONDAY. New York to Boston.
TUESDAY. At Boston.
WEDNESDAY. Boston to Portland, Maine.
THURSDAY. At Portland.
FRIDAY. From Portland to Quebec, by the Grand Trunk Railway.

Détour of Ten Days to White Mountains.

[The White Mountains may be pleasantly visited from this part of our present Route (in ten extra days), stopping at Gorham, N. H., 91 miles on the way from Portland, reaching Glen House, 8 miles from Gorham, same day; Crawford House, White Mountain Notch, on Saturday; and so on, as per programme of SECOND WEEK, in previous Tours, returning to the Glen House by the Second Sunday, and resuming journey (from Gorham to Quebec) on Monday following.]

SATURDAY. At Quebec.
SUNDAY. At Quebec.

SECOND WEEK.

MONDAY. At Quebec, visiting Falls of Montmorenci, of the Chaudiere, of St. Anne, &c.
TUESDAY. Excursion to Saguenay River and back to Quebec.
WEDNESDAY. Excursion to Saguenay River and back to Quebec.
THURSDAY. Excursion to Saguenay River and back to Quebec.
FRIDAY. From Quebec, by grand Trunk Railway, or St. Lawrence River, to Montreal.
SATURDAY. Montreal.
SUNDAY. Montreal.
MONDAY. Excursion up the Ottawa River from Montreal and back.
TUESDAY. Excursion up the Ottawa River from Montreal and back.
WEDNESDAY. Excursion up the Ottawa River from Montreal and back.
THURSDAY. Up the St. Lawrence and Lake Ontario (or by Grand Trunk Railway) to Niagara Falls.
FRIDAY. Up the St. Lawrence and Lake Ontario (or by Grand Trunk Railway) to Niagara Falls.
SATURDAY. At Niagara Falls.
SUNDAY. At Niagara Falls.

FOURTH WEEK.

MONDAY. Still at Niagara.
TUESDAY. By Central Railway to Utica.
WEDNESDAY. From Utica, 15 miles, to Trenton Falls
THURSDAY. At Trenton Falls, returning to Utica in the evening.
FRIDAY. Journey to and stay at Saratoga Springs.
SATURDAY. Back to New York, via Troy, Albany, and the Hudson River.

SKELETON TOURS. 15

TOUR OF FOUR WEEKS.

To the Upper Mississippi, via Niagara, Detroit, Chicago, Milwaukee, St. Paul, St. Louis, Louisville, Cincinnati, etc.

FIRST WEEK.

MONDAY. From New York to Niagara by the Erie Railway, 444 miles, or by the Central route, 466 miles—a journey more comfortably made in two days than one, if time serves. By Canandaigua, direct, 439 miles.
TUESDAY. Niagara.
WEDNESDAY. By the Great Western Railway, 229 miles, to Detroit.
THURSDAY. By the Michigan Central road, 284 miles, to Chicago.
FRIDAY. Chicago, Ill.
SATURDAY. To Milwaukee by steamer on Lake Michigan, or by railway along shore, 85 miles.
SUNDAY. At Milwaukee, Wis.

SECOND WEEK.

Visit to St. Paul, Minnesota, leaving Milwaukee on Monday for Madison, Wis., and thence (circuitously) by railway to Dubuque on the Mississippi, or returning, to Chicago, and thence to Dubuque direct, by the Galena and Chicago route. From Dubuque by steamer up the Mississippi River to St. Paul and the Falls of St. Anthony. Returning by the end of the week (second of the tour) via the river, to St. Louis.

THIRD WEEK.

MONDAY. At St. Louis.
TUESDAY. By the Ohio and Mississippi Railway, and the New Albany and Salem road to Louisville.
WEDNESDAY. At Louisville, Ky.
THURSDAY. At Louisville.
(Another week would permit the traveller to visit the Mammoth Cave very agreeably from this the chief point of détour thither.)
FRIDAY. By railway or steamer on the Ohio River to Cincinnati.
SATURDAY. At Cincinnati, Ohio.
SUNDAY. At Cincinnati

FOURTH WEEK.

MONDAY. By railway to Columbus, Ohio.
TUESDAY. Railway to Zanesville, Ohio.
WEDNESDAY. To Wheeling, Va.
THURSDAY AND FRIDAY. By the Baltimore and Ohio road to Baltimore, or by the Pennsylvania Railway to Philadelphia. Both these noble routes are as magnificent in their pictorial attractions as in their grand extent—each traversing a wide extent of country, replete with every variety of natural beauty.
SATURDAY. To New York.

A WINTER TOUR OF SIX WEEKS,

Visiting the Invalid Resorts of Florida, Savannah and Augusta, Geo., Charleston and Columbia, S. C., Richmond, Va., and Washington City.

FIRST WEEK.

SATURDAY. Leave New York by the steamer of Saturday afternoon, and arrive in Savannah Tuesday morning. Spend the rest of the week in Savannah at the Pulaski House, the Scriven, or the City Hotel.

SECOND WEEK.

SATURDAY. Leave Savannah in the steamer for Jacksonville, Pilatka and other places on the St. John's river. Spend the week hereabouts.

THIRD WEEK.

At St. Augustine, on the coast, below the mouth of the St. John's. St. Augustine, or the "Ancient City," as it is sometimes called, from its venerable age, which exceeds that of any other place in the Union, will tempt the visitor to a long tarry with the social attractions which its fame as an invalid resort has secured. The peculiar natural features of the city and the neighborhood, will also win his particular interest.

FOURTH WEEK.

At St. Augustine.

FIFTH WEEK.

Return to Savannah and take the Georgia Central railway to Augusta, thence by the South Carolina road to Charleston.

SIXTH WEEK.

MONDAY. By South Carolina Railway from Charleston to Columbia.
TUESDAY. At Columbia, resuming journey in the afternoon.
WEDNESDAY. En route.
THURSDAY. At Richmond, Va.
FRIDAY. Arrive at Washington City.
SATURDAY. To Baltimore in the evening.
SUNDAY. At Baltimore.
MONDAY. To New York.

THROUGH ROUTES.

From ALBANY to

ROUTES from ALBANY to the following Alphabetically arranged List of Towns.

The numerical notation of Routes does not indicate our preference of any one over the others.

For Time of Departure see Appletons' Railway Guide.

Baltimore............	Via Hudson R. R., Harlem R. R., or North River Steamboats to New York, N. J. R. R., or Camden and Amboy to Philadelphia, and Phila., Wil'n and Balt. R. R. to Baltimore.
Boston...............	Via Western R. R.
Buffalo..............	Via N. Y. Central R. R.
Chicago.,...........	Via N. Y. Central to Suspension Bridge, Great Western to Detroit and Michigan Central R. R. to Chicago ; or via N. Y. Central to Buffalo, Lake Shore Route to Cleveland, Clevel'd and Toledo R. R. to Toledo and Mich'n, Southern to Chicago.
Cincinnati..........	Via N. Y. Central to Buffalo, Lake Shore Route to Cleveland, Cleveland and Columbus to Columbus and Little Miami to Cincinnati.
Montreal, C. E......	Via Rensselaer and Sar. R. R. to Saratoga Springs, to Whitehall by Saratoga and Whitehall R. R., Steamers on Lake Champlain to Rouse's Point, and R. R. to Montreal.
New York...........	Via Hudson R. R., Harlem R. R., or the North River Boats.
Niagara Falls.......	Via N. Y. Central.
Philadelphia........	See Route to Baltimore.
St. Louis............	ROUTE No. 1.—Via N.Y. Central to Buffalo, Lake Shore Route to Cleveland, Bellefontaine Route to Indianapolis, Terre Haute and Richmond to Terre Haute, and Terre Haute, Alton and St. Louis to St. Louis.
"	ROUTE No. 2.—Via same route to Cleveland, Cleve. and Tol. to Toledo, and Mich. Southern to Chicago ; St. Louis, Alton and Chi. R. R. to St. Louis.
"	ROUTE No. 3.—Via N. Y. Central, Great Western and Mich. Central to Chicago, and Chicago, Alton and St. Louis to St. Louis.
"	ROUTE No. 4.—Via the route to Cleveland ; Cleveland and Col. R. R., Little Miami R. R. to Cincinnati, and Ohio and Miss., or Ind. and Cin. R. R., Terre Haute and Rich'nd, and Terre Haute, Alton and St. Louis R. R.'s to St. Louis.
"	ROUTE No. 5.—Via the Route to Cleveland and Toledo ; via Toledo and Wabash R. R. at Toledo for St. Louis.

From BALTIMORE to

ROUTES from BALTIMORE to the following Alphabetically arranged List of Towns.

For Time of Departure see Appletons' Railway Guide.

Albany..............	See Albany to Baltimore.
Boston...............	Via Phila., Wil. and Balt. R. R. to Phila.; Camden and Amboy to New York, and, at the traveller's choice, by the Boston Express Line, Shore R. R. Line, or the Steamers via Norwich and Worcester, Fall River, or the Stonington Route, to Boston.
Buffalo..............	Via Northern Central to Williamsport, Williamsport and Elmira to Elmira, and Erie R. R. to Buffalo.
Charleston, S. C. ...	ROUTE No. 1.—Via Balt. and Ohio to Washington ; Rich'd, Fred. and Potomac to Richmond ; Rich'd and Petersburg F. R. to Petersburg ; Petersburg and Weldon R. R. to Weldon ; Wilmington and Weldon R. R. to Wilmington ; Wilmington and Manchester R. R. to Florence ; and North Eastern R. R. to Charleston.
"	ROUTE No. 2.—Via Steamboat to Norfolk and Seaboard and Roanoke R. R. to Weldon ; thence to Charleston via R. R.'s in Route No. 1.
Chicago, Ill.........	ROUTE No. 1.—Via B. and Ohio route to Wheeling, Cl. Ohio to Newark ; Newark, Mansfield and Sandusky R. R. to Munroeville ; Cleve. and Tol. and Mich. South'n R. R. to Toledo and Chicago.
"	ROUTE No. 2.—Same to Wheeling ; Central Ohio to Columbus ; Little Miami to Xenia ; Dayton, Xenia and Belpre R. R. to Dayton ; Bellefontaine Route to Indianapolis ; Lafayette and Indianapolis R. R. to Lafayette ; Louisville, N. A. and Chicago to Michigan City ; Mich. Central to Chicago.

THROUGH ROUTES. 17

From BALTIMORE to	ROUTES from BALTIMORE to the following Alphabetically arranged List of Towns—CONTINUED. For Time of Departure see Appletons' Railway Guide.
Chicago, Ill.	ROUTE NO. 3.—Baltimore and Ohio R.R. and N. West. Va. R.R. to Parkersburg; Mar. and Cin. R.R. to Cincinnati; Indianapolis and Cin. R.R. to Indianapolis; Lafayette and Ind. R.R. to Lafayette; Louisville, N.A. R.R. and Chicago R.R. to Mich. City; Mich. Central to Chicago.
Cincinnati	Via Baltimore and Ohio R.R. and N. West'n R.R. to Parkersburg; Marietta and Cincinnati R.R. to Cincinnati; or via Balt. and Ohio R.R. to Wheeling; Ohio Central to Columbus, and Little Miami to Cincinnati.
Mobile	ROUTE NO. 1.—Via Balt. and Ohio R.R. to Washington; Washington to Alexandria by Steamboat; by the Orange and Alex. R.R. to Lynchburg; Va. and Tenn. R.R. to Bristol; East Tenn. and Va. R.R. to Knoxville; E. Tenn. and Ga. R.R. to Cleveland and Dalton R.R. to Dalton; Western and Atlantic to Atlanta, Ga.; Atlanta and West Point R.R. to West Point; Montg. and W. Point R.R. to Montgomery; and Steamboat to Mobile.
"	ROUTE NO. 2.—Via Charleston Route to Florence; Wil. and Manchester R.R. to Kingsville; So. Ca. R.R. to Augusta; Georgia R.R. to Atlanta; and from Atlanta as in the preceding route.
New Orleans	ROUTE NO. 1.—Washington Branch R.R. to Washington; thence via Steamboat to Alexandria; Orange and Alexandria R.R. to Lynchburg; thence via Virginia and Tennessee R.R. to Bristol; thence via East Tennessee and Virginia R.R. to Knoxville; thence via East Tennessee and Georgia R.R. to Chattanooga; thence via Nashville and Chattanooga R.R. to Stevenson; thence via Memphis and Charleston R.R. to Grand Junction; thence via Mississippi Central R.R. to Canton; thence via New Orleans, Jackson and Great Northern R.R. to New Orleans.
"	ROUTE NO. 2.—See Route to Mobile No. 2; by steamer from Mobile to New Orleans.
"	ROUTE NO. 3.—Route No. 1 to Cincinnati; thence by Ohio and Miss. R.R. to Odin, on the Ill. Central; Illinois Central to Cairo; by Steamboat to Columbus; Mobile and Ohio to Junction; Miss. Central to Jackson and N.O.; Jackson R.R. to New Orleans.
New York	See Route from New York to Baltimore.
Niagara Falls	See Route Balt. to Buffalo; thence by N.Y. Central to Niagara Falls.
Philadelphia	Via Phila., Wilmington and Balt. R.R.
St. Louis	ROUTE NO. 1.—Via Route 1 to Cincinnati, and Ohio and Miss. R.R. to St. Louis, or from Cincinnati via Ind. and Cincinnati R.R., Terre Haute and Rich., and Terre Haute, Alton and St. Louis R.R. to St. Louis.
"	ROUTE NO. 2.—Via Balt. and Ohio R.R. to Wheeling; Ohio Central R.R. to Columbus; Columbus and Indianapolis and Bellefontaine R.R. to Indianapolis; Terre Haute and Rich'd, and Terre Haute, Alton and St. Louis R.R. to St. Louis.

From BOSTON to	ROUTES from BOSTON to the following Alphabetically arranged List of Cities. For Time of Departure see Appletons' Railway Guide.
Albany	Via Boston and Wor. R.R. and Western R.R.
Baltimore	See Routes from Boston to New York, and Baltimore to New York.
Charleston	See Routes from Boston to New York, and New York to Charleston.
Cincinnati	See Routes from Boston to Albany, and Albany to Cincinnati.
"	Also, Routes from Boston to New York, and routes thence to Cincinnati.
Chicago	See Routes from Boston to Albany, and Albany to Chicago.
"	Or see Routes from Boston to New York, and Routes from New York to Chicago.
Montreal	ROUTE NO. 1.—Via Eastern R.R. or Boston and Maine to Portland, and Grand Trunk to Montreal.
"	ROUTE NO. 2.—Via Boston and Lowell R.R., Cheshire R.R., Ver. Central to Rouse's Point; thence by Montreal and Champlain R.R.
Memphis	See Routes to New Orleans.
New Orleans	See Routes from Boston to New York, and Routes thence to New Orleans.
New York	Five Routes, viz., Boston Express; the Shore Line via Providence; or via either the Fall River, Stonington or Norwich Lines.

THROUGH ROUTES.

From BOSTON to	ROUTES from BOSTON to the following Alphabetically arranged List of Cities—CONTINUED. *For Time of Departure see Appletons' Railway Guide.*
Philadelphia	See Routes from Boston to New York, and New York to Philadelphia.
Savannah	See Routes from Boston to New York, New York to Balt., and Routes from Balt. to Charleston, and Steamer to Savannah. A steamer leaves New York twice a week for Savannah direct.
St. Louis	See Routes from Boston to Albany, and Routes thence to St. Louis.
"	Or Routes from Boston to New York, and Routes thence to St. Louis.

From CINCINNATI to	ROUTES from CINCINNATI to the following Cities Alphabetically arranged. *For full Time Tables see Appletons' Railway Guide.*
Albany	See Routes from Albany to Cincinnati.
Baltimore	See Routes from Baltimore to Cincinnati.
Boston	See Routes from Boston to Cincinnati.
Chicago	Via Indianapolis and Cin. R. R. to Indianapolis; Lafayette and Indianapolis to Lafayette; Louisville, N. Albany and Chicago R. R. to Mich. City; Mich. Central to Chicago.
Cleveland	ROUTE NO. 1.—Via Little Miami R. R. to Columbus, and Cleve. and Col. R. R. to Cleveland.
"	ROUTE NO. 2.—Via Cin., Ham. and Dayton, Springfield, Mt. Vernon and Pitts'g R. R., and Cleveland, Columbus and Cin. R. R.
Charleston	See Routes from Baltimore to Cincinnati, and Baltimore to Charleston.
Detroit	Via Cin., Ham. and Dayton, and Dayton and Michigan R. R.
Montreal	ROUTE NO. 1.—Via Cin., Ham. and Dayton R. R., and Dayton and Michigan R. R. to Detroit; Grand Trunk Railway to Montreal.
"	ROUTE NO. 2.—Via Niagara Falls; via Cin., Ham. and Dayton, and Dayton and Michigan R.R. to Detroit; Great West'n R. R. to Niagara Falls, or via Hamilton to Toronto; thence via Grand Trunk R. R.; or from Niagara Falls via Steamer down the St. Lawrence.
"	ROUTE NO. 3.—To Cleveland by either Cleveland Routes; Lake Shore to Buffalo, and N. Y. Central to Niagara Falls; thence by Gt. West. R. R. and Grand Trunk R. R., or by Steamer on the Lakes and St. Lawrence to Montreal.
New Orleans	ROUTE NO. 1.—Via Ohio and Miss. and Illinois Central to Cairo; Steamboat to Columbus; Mobile and Ohio to Junction; Miss. Central to Canton; and N. O., Jackson and Gt. Northern to New Orleans.
New Orleans, via Nashville & Mammoth Cave	ROUTE NO. 2.—Via Ohio and Miss. R. R. to Seymour; Jeffersonville R. R. to Louisville; Louisville and Lexington R. R. via Mammoth Cave, to Nashville; Nashville and Chattanooga R. R. to Stevenson; Memphis and Charleston R. R. to Junction. Here the traveller may continue on to Memphis, or journey direct to New Orleans, via Miss. Central and N. O., Jackson and Gt. Northern R. R.
New York	See Routes from New York to Cincinnati.
Philadelphia	See Routes from Phila. to Cincinnati.
St. Louis	ROUTE NO. 1.—Via Ohio and Miss. R. R.
"	ROUTE NO. 2.—Via Indianapolis and Cincinnati R. R. to Indianapolis; and Terre Haute and Rich'd and Terre Haute, Alton and St. Louis R. R. to St. Louis.
Washington, D. C.	See Route from Baltimore to Cincinnati, and Baltimore to Washington.

From CHICAGO to	ROUTES from CHICAGO to the following Cities Alphabetically arranged. *For full Time Tables see Appletons' Railway Guide.*
Albany	See Albany to Chicago.
Baltimore	See Baltimore to Chicago.
Boston	ROUTE NO. 1.—Via Mich. Central, Gt. Western R. R., N. Y. Central to Albany; and Western R. R. to Boston.

THROUGH ROUTES. 19

From CHICAGO to	ROUTES from CHICAGO to the following Cities Alphabetically arranged—CONTINUED. For full Time Tables see Appletons' Railway Guide.
Boston............	ROUTE No. 2.—Via Mich. Southern, Cleveland and Toledo, Lake Shore and N. Y. Central to Albany; and Western R. R. to Boston.
Charleston.........	See Routes from Balt. to Chicago, and Balt. to Charleston.
Cincinnati.........	See Routes from Cincinnati to Chicago.
Cleveland.........	Via Mich. Southern, and Cleveland and Toledo R. R.
Kansas............	Chicago, Burlington and Quincy R. R. to Quincy and Hannibal; Hannibal and St. Jo. R. R. to St. Josephs; and Steamer to Kansas City and Leavenworth.
Montreal..........	Via Mich. Central to Detroit; and Grand Trunk R. R. to Montreal.
New Orleans.......	ROUTE No. 1.—Via Illinois Central to Cairo; to Columbus by Steamboat; Mobile and Ohio R. R. to Junction; Miss. Central to Canton; N. O. Jackson R. R. to New Orleans.
"	ROUTE No. 2.—Via St. Louis, Alton and Chicago R. R. to St. Louis; and Steamboat to New Orleans.
New York..........	See Routes from New York to Chicago.
Philadelphia.......	See Routes from Philadelphia to Chicago.
St. Louis..........	Via St. Louis, Alton and Chicago R. R.

From MONTREAL to	ROUTES from MONTREAL to the following Alphabetically arranged Cities. For full Time Tables see Appletons' Railway Guide.
Albany............	ROUTE No. 1.—Via Montreal and Champlain R. R. to Rouse's Point; Champlain Steamer to Whitehall; Saratoga and Whitehall and Rensselaer and Saratoga R. R.'s to Albany.
"	ROUTE No. 2.—Via Mont. and Champlain R.R. to Rouse's Point; Ver't Central to Burlington; Rutland and Burlington to Rutland; Troy and Boston to Troy; and Hudson River to Albany.
Baltimore..........	See Routes to Albany, Albany to New York, and New York to Baltimore.
Boston............	See Routes from Boston to Montreal.
Charleston.........	Via Routes to Boston and New York, or Routes to New York; thence by Routes from New York to Charleston; or by Steamer from New York direct to Charleston.
Cincinnati.........	See Routes from Cincinnati to Montreal.
Chicago...........	See Routes from Chicago to Montreal.
New Orleans.......	ROUTE No. 1.—Via Grand Trunk R. R. to Detroit; Mich. Central to Chicago; Ill. Central R. R. to Cairo; Steamboat to Columbus; Mobile and Ohio R. R. to Junction; Miss. Central to Canton, and No. Jackson to Gt. N.
"	ROUTE No. 2.—Via Grand Trunk to Detroit; Dayton and Michigan to Toledo; Toledo and Wabash to Tolono; and via Ill. Central, and as in Route No. I.
"	ROUTE No. 3.—Via Grand Trunk to Detroit; Dayton and Michigan to Sydney; Bellefontaine Route to Indianapolis, Terre Haute and Rich'd, and Terre Haute, Alton and St. Louis to Matoon; and Illinois Central, and as in Route No. 1 to New Orleans.
New York..........	ROUTE No. 1.—Either of the Routes to Albany, and Route from Albany to New York.
"	ROUTE No. 2.—Via Montreal and Champlain R. R. to Rouse's Point; via Ver. Central to Windsor; Cheshire R. R. and Conn. R. R. to Springfield; N. H., Hartford and Springfield Route to New Haven; and New Haven R. R. to New York.
Philadelphia.......	See Routes to New York, and Route from New York to Philadelphia.
Quebec............	Via Grand Trunk R. R., or by Steamer down the St. Lawrence.
St. Louis..........	ROUTE No. 1.—Via Route to Chicago; and St. Louis, Alton and Chicago to St. Louis.
"	ROUTE No. 2.—Via same Route to Detroit; and Detroit and Michigan to Toledo; Toledo, Wabash and Gt. West'n to St. Louis.
Saguenay River....	Via Steamer twice a week in Summer.
St. Paul and St. Anthony's Falls......	Via Grand Trunk R. R. to Detroit; Detroit and Milwaukee R. R. and Steamer to Milwaukee; La Crosse and Milwaukee R. R. to La Crosse; Steamer to St. Paul.
Washington, D. C...	Via Routes to New York; Routes from New York to Balt., and Baltimore to Washington, to Charleston.

From NEW YORK to	ROUTES from NEW YORK to the following Alphabetically arranged List of Towns. *For full Time Tables see Appletons' Railway Guide.*
Albany	Via Hudson or Harlem R. R., or Steamboats.
Buffalo	ROUTE No. 1.—Via Routes to Albany, and N. Y. Central to Buffalo.
"	ROUTE No. 2.—Via New York and Erie, and Buffalo, New York and Erie.
Boston	ROUTE No. 1.—Via Boston Express, through New Haven, Hartford, Springfield and Worcester.
"	ROUTE No. 2.—Via the Shore Line, passing through New Haven, New London and Stonington.
"	ROUTE No. 3.—Via Steamboat to Stonington, and R. R. to Boston.
"	ROUTE No. 4.—Via Steamboat to Norwich, and R. R. to Boston.
"	ROUTE No. 5.—Via Fall River Steamboats, stopping at Newport, and Fall R. R. to Boston.
Baltimore	Via N. Jersey R. R. to Phila., and Phil., Wil. and Balt. R. R.
Chicago	ROUTE No. 1.—Via Hudson R. R. or Harlem R. R. to Albany; via N. Y. Central to Suspension Bridge; Great Western to Detroit; and Michigan Central to Chicago.
"	ROUTE No. 2.—Via Hudson River or Harlem R. R. to Albany; N. Y. Central to Buffalo; Lake Shore R. R. to Cleveland; Cleveland and Toledo R. R. to Cleveland; thence via Mich. Southern to Chicago.
"	ROUTE No. 3.—Via N. Y. and Erie R. R. to Dunkirk; Lake Shore R. R. to Cleveland; Cleveland and Toledo R. R. to Toledo; and Mich. Southern R. R. to Chicago.
"	ROUTE No. 4.—Via N. J. R. R. or Camden and Amboy to Philadelphia; Pennsylvania R. R. to Pittsburg; and Pittsburg, Fort Wayne and Chicago R. R. to Chicago.
Charleston	Via N. J. R. R. or Camden and Amboy to Phila.; Philadelphia, Wilmington and Balt. R. R. to Baltimore. See Baltimore to Charleston.
Cincinnati	ROUTE No. 1.—Via New Jersey R. R. or Camden and Amboy R. R. to Philadelphia; Phil., Wil. and Balt. R. R. to Baltimore. See Routes from Balt. to Cincinnati.
"	ROUTE No. 2.—Via Erie R. R. to Dunkirk; Lake Shore to Cleveland; Cleve., Col. and Cin. R. R. to Columbus, and Little Miami to Cincinnati.
"	ROUTE No. 3.—Via Hudson River or Harlem R. R. to Albany; N. Y. Central to Buffalo; Lake Shore to Cleveland; Cleve., Col. and Cin. R. R. to Columbus, and Little Miami to Cincinnati.
"	ROUTE No. 4.—Via Camden and Amboy or N. Jersey R. R. to Philadelphia; Penn. R. R. to Pittsburg; Cleveland and Pittsburg R. R. to Steubenville Junction; Steubenville and Indiana R. R. to Newark; Central Ohio R. R. to Columbus; Little Miami R. R. to Cincinnati.
Dubuque	Via any of the Routes to Chicago; via Galena and Chicago R. R. to Freeport; Illinois Central R. R. to Dubuque.
Kansas	ROUTE No. 1.—Via any of the Routes to Chicago; Chicago, Burlington and Quincy R. R. to Quincy and Hannibal; Hannibal and St. Jo. R. R. to St. Jo. and Kansas City.
"	ROUTE No. 2.—Via any of the Routes to St. Louis; Pacific R. R. to Jefferson City, and Steamboat to Kansas City.
Milwaukee	ROUTE No. 1.—Via Hudson River or Harlem R. R. to Albany; N. Y. Central to Suspension Bridge; Great Western to Detroit; Detroit and Milwaukee R. R. to Grand Haven; thence by Steamboat on Lake to Milwaukee.
"	ROUTE No. 2.—Via any of the Routes to Chicago; and Chic. and Milwaukee R.R. to Milwaukee.
Montreal	ROUTE No. 1.—Via Hudson R. R. or Steamers to Troy; Rensselaer and Sar. R. R. to Saratoga Springs; Saratoga and Whitehall R. R. to Whitehall; Steamers to Rouse's Point; Mont. and Champ. R. R. to Montreal.
"	ROUTE No. 2.—Via Hudson R. R. or Steamboat to Troy; Troy and Boston, and Rutland and Washington R. R. to Rutland; thence via Rutland and Burlington R. R. to Burlington; Vermont and Canada R. R. to Rouse's Point; and Montreal and Champlain R. R. to Montreal.
"	ROUTE No. 3.—See Route No. 2 Montreal to New York.
Nebraska City	Via any of the Routes to Chicago; Chicago, Burlington and Quincy to Quincy and Hannibal; Hannibal and St. Jo. R. R. to St. Josephs; Steamboat to Nebraska and Omaha Cities.
Newport	ROUTE No. 1.—Via Fall River Steamers, Pier 3 N. R., to Newport.
"	ROUTE No. 2.—Via Shore Line to New Haven, New London and Greenwich; Steamboat from Greenwich.
Niagara Falls	ROUTE No. 1.—Via Hudson River, Harlem R. R. or Steamers to Albany; and N. Y. Central to the Falls.

THROUGH ROUTES 21

From NEW YORK to	ROUTES from NEW YORK to the following Alphabetically arranged List of Towns—CONTINUED. *For full Time Tables see Appletons' Railway Guide.*
Niagara Falls........	ROUTE No. 2.—Via Erie R. R. to Buffalo, and N. Y. Central to the Falls.
New Orleans.........	ROUTE No. 1.—See Route to Baltimore, and Routes from thence to New Orleans.
"	ROUTE No. 2.—Via any of the Routes to Chicago; Illinois Central to Cairo; by Steamboat to Columbus; Mobile and Ohio R. R. to Junction; Miss. Central R. R. to Jackson, N. O. Jackson and Great Northern to New Orleans.
"	ROUTE No. 3.—Via Erie R. R. to Dunkirk; or via Hudson River or Harlem R. R. to Albany, and thence by N. Y. Central to Buffalo; from Dunkirk or Buffalo, by Lake Shore Line, to Cleveland; Cleveland and Columbus to Crestline; Bellefontaine Route to Indianapolis; Terre Haute, Alton and St. Louis to Mattoon; Illinois Central to Cairo; Steamboat to Columbus; Mobile and Ohio R.R. to Junction; Miss. Central to Jackson, N. O.; Jackson R. R. to New Orleans.
"	ROUTE No. 4.—Via N. J. R. R. to Phila.; Penn. R. R. to Pittsburg; Pittsburg, Fort Wayne and Chicago to Crestline, and as in Route No. 3 for the remaining portion of the route.
Pike's Peak.........	Via any of the Routes to Chicago; Chicago, Burlington and Quincy R. R. to Quincy and Hannibal; Hannibal and St. Jo. R. R. to St. Josephs, or Leavenworth City; and Cal. Central Overland and Pike's Peak Express.
St. Louis	ROUTE No. 1.—Via Hudson River or Harlem R. R. to Albany; N. Y. Central to Suspension Bridge or Buffalo, connecting at Suspension Bridge with Great West'n and Mich. Central for Chicago, and at Buffalo with Lake Shore; Cleveland and Toledo and Mich. Southern for Chicago, via St. Louis, Alton and Chicago R. R. to St. Louis.
"	ROUTE No. 2.—Via any of the Routes to Crestline; Bellefontaine to Indianapolis; Terre Haute and Richmond, and Terre Haute, Alton and St. Louis to St. Louis.
"	ROUTE No. 3.—Via any of the Routes to Cincinnati; thence via Ohio and Miss., or Indianapolis and Cin., and Terre Haute and Richmond, and Terre Haute, Alton and St. Louis to St. Louis.
"	ROUTE No. 4.—Via any of the Routes to Philadelphia; thence via Pennsylvania R. R. to Pittsburg; thence via Cleveland and Pittsburg R. R. to Steubenville; thence via Pittsburg, Columbus and Cincinnati R. R. to Newark; thence via Central Ohio R. R. to Columbus; thence via Dayton and Western and Indiana Central R. Rs. to Indianapolis; thence via Terre Haute and Richmond, and Terre Haute, Alton and St. Louis R. Rs. to St. Louis.
"	ROUTE No. 5.—Via Erie R. R., Lake Shore, and Cleveland and Toledo R. Rs. to Toledo, or via Hudson River, N. Y. Central, Lake Shore, and Cleveland and Toledo R. Rs. to Toledo; and thence by Toledo and Wabash, and Gt. Western R. Rs. to Springfield; and St. Louis, Alton and Chicago R. R. to St. Louis.

From PHILADELPHIA to	ROUTES from PHILADELPHIA to the following Alphabetically arranged Cities. *For full Time Tables see Appletons' Railway Guide.*
Albany	Via Camden and Amboy R.R. to New York; and Hudson River or Harlem R. R., or Boat on the Hudson to Albany.
Baltimore............	Via Phil., Wil. and Baltimore R. R.
Boston	Via Camden and Amboy to New York, and either one of the 2 railway or 3 steamboat routes.
Charleston...........	Via Phil., Wil. and Balt. R. R. to Balt., and Route from Baltimore to Charleston.
Cincinnati	ROUTE No. 1.—Via Penn. R. R. to Pittsburg; Pittsburg, Fort Wayne and Chic. to Crestline; Cleveland and Columbus to Columbus; Little Miami to Cincinnati.
"	ROUTE No. 2.—Via Penn. R. R. to Pittsburg; Pittsburg, Col. and Cin. R. R. to Columbus; and Little Miami to Cincinnati.
"	ROUTE No. 3.—Via Phil., Wil. and Baltimore R. R., and Routes from Balt. to Cincinnati.
Chicago...............	Via Penn. R. R. to Pittsburg; and Pittsburg, Fort Wayne and Chicago R. R. to Chicago.
Montreal.............	See Route to New York, and Routes from New York to Montreal.

THROUGH ROUTES.

From PHILADELPHIA to	ROUTES from PHILADELPHIA to the following Alphabetically arranged Cities—CONTINUED. *For full Time Tables see Appletons' Railway Guide.*
New Orleans	ROUTE NO. 1.—Via Penn. R. R. to Pittsburg; thence by Pittsburg, Fort Wayne and Chi. R. R. to Crestline; thence via Bellefontaine R. R. to Indianapolis; Terre Haute and Richmond R. R. to Indianapolis; Terre Haute, Alton and St. Louis to Matoon; Illinois Central to Cairo; Steamboat to Columbus; Mobile and Ohio R. R. to Junction; Miss. Central to Jackson; and N. O., Jackson and Gt. Northern to New Orleans.
"	ROUTE NO. 2.—The same Route to Crestline; Cleve. and Col. R. R., and Little Miami to Cincinnati; thence by Ohio and Miss. R. R. to Odin; Ill. Central to Cairo; and the remainder of the route as in Route No. 1.
"	ROUTE NO. 3.—Via Phil., Wil. and Balt. to Baltimore; and either of the Routes from thence to New Orleans.
New York	Via Camden and Amboy R. R.
Savannah	Via Phil., Wil. and Balt. R. R., and Routes to Charleston; thence by Steamer to Savannah.
St. Louis	ROUTE NO. 1.—Via Penn. R. R. to Pittsburg; thence via Pittsburg, Fort Wayne and Chi. to Crestline; thence by Bellefontaine to Indianapolis; Terre Haute and Richmond to Indianapolis; Terre Haute, Alton and St. Louis to St. Louis.
"	ROUTE NO. 2.—Via Penn. R. R. to Pittsburg; via Pittsburg, Col. and Cin. to Newark; thence via Central Ohio to Columbus; thence via Dayton and Western and Ind. Central to Indianapolis; Terre Haute and Rich'd to Terre Haute; and Terre Haute, Alton and St. Louis to St. Louis.
"	ROUTE NO. 3.—Same Route to Columbus; Little Miami to Cincinnati; thence via Ohio and Miss., or Indianapolis and Cin. R. R. to Ind.; and Terre Haute and Rich'd, and Terre Haute, Alton and St. Louis to St. Louis.
"	ROUTE NO. 4.—Via Phil., Wil. and Balt., and any of the Routes from Balt. to St. Louis.

THE TRAVELLER'S MEMORANDUM.

THE TRAVELLER'S MEMORANDUM.

BRITISH AMERICA.

The possessions of the British Crown in North America, occupy nearly all the upper half of the Continent; a vast territory, reaching from the Arctic seas to the domains of the United States, and from the Atlantic to the Pacific Oceans. Of this great region, our present explorations will refer only to the lower and settled portions, known as the British Provinces—the Canadas, New Brunswick, and Nova Scotia. The rest is for the most part yet a wilderness.

CANADA.

GEOGRAPHY AND AREA. Canada, the largest and most important of the settled portions of the British territory in North America, lies upon all the northern border of the United States, from the Atlantic coast to the waters of Lake Superior and the Mississippi. The two provinces into which it is divided, were formerly known as Upper and Lower Canada, or Canada East and Canada West; and thus, indeed, their differing manners, habits and laws, still virtually divide and distinguish them, though they are now nominally and politically united. The entire length of the Canadian domain, from east to west, is between twelve hundred and thirteen hundred miles, with a breadth varying from two to three hundred miles.

DISCOVERY, SETTLEMENT, AND RULERS. The earliest discovery of the Canadas is ascribed to Sebastian Cabot, 1497; Jacques Cartier, a French adventurer, spent the winter of 1541 at St. Croix, now the River St. Charles, upon which Quebec is partly built. The first permanent settlement, however, was at Tadousac, at the confluence of the Saguenay and the St. Lawrence. From that time (about 1608) until 1759, the country continued under the rule of France; and then came the capture of Quebec by the English, under General Wolfe, and the transfer, within a year thereafter, of all the territory of New France, as the country was at that time called, to the British power, under which it has ever since remained. The mutual disagreement which naturally arose from the conflicting interests and prejudices of the two opposing nationalities, threatened internal trouble from time to time, and finally displayed itself in the overt acts recorded in history as the rebellion of 1837. It was after these incidents, and as a consequence thereof, that the two sections of the territory were formed into one. This happened in 1840.

GOVERNMENT. Canada is ruled by an executive, holding the title of Governor-General, received from the crown of Great Britain, and by a legislature called the Provincial Parliament. This body consists of an Upper and a Lower House; the members of the one were formerly appointed by the Queen, but now (as fast as those thus placed die) this body is, like the other branch, chosen by the people; each for a term of eight years.

RELIGION. The dominant religious faith in Lower Canada or Canada East, is that of the Romish Church; while in the upper province the creed of the English Establishment prevails.

LANDSCAPE. The general topography of Upper or Western Canada, is that of a level country, with but few variations excepting the passage of some table heights, extending south-westerly. It is the most fertile division of the territory, and thus, to the tourist in search of the picturesque, the least attractive.

The Lower Province, or Canada East, is extremely varied and beautiful in its physical aspect; presenting to the delighted eye a magnificent gallery of charming pictures of forest wilds, vast prairies, hill and rock-bound rivers, rushing waters, bold mountain heights, and all, every where intermingled, and their attractions embellished by intervening stretches of cultivated fields, and rural villages, and villa homes.

MOUNTAINS. The hill ranges of Canada are confined entirely to the lower or eastern province. The chief lines, called the Green Mountains, follow a parallel course south-westerly. They lie along the St. Lawrence River, on its southern side, extending from the latitude of Quebec to the Gulf of St. Lawrence. There is another and corresponding range on the north side of the river, with a varying elevation of about 1,000 feet. The Mealy Mountains, which extend to Sandwich Bay, rise in snow-capped peaks to the height of 1,500 feet. The Wotchish Mountains, a short, crescent-shaped group, lie between the Gulf of the St. Lawrence and Hudson's Bay.

RIVERS. Canada has many noble and beautiful rivers, as the St. Lawrence, one of the great waters of the world; the wild, mountain-shored floods of the Ottawa, and the Saguenay; and the lesser waters of the Sorel or Richelieu, the St. Francis, the Chaudière and other streams.

The St. Lawrence. This grand river, which drains the vast inland seas of America, extends from Lake Ontario, 750 miles to the Gulf of St. Lawrence, and thence to the sea. Its entire length, including the great chain of lakes by which it is fed, is not less than 2,200 miles. Ships of the largest size ascend the river as far as Montreal. Its chief affluents are the Saguenay, eastward, and the Ottawa on the west. The width of the St. Lawrence varies from about a mile to four miles; at its mouth it is 100 miles across. It abounds in beautiful islands, of which there is a vast group, near its egress from Lake Ontario, known and admired by all the world as the "Thousand Isles."

The Thousand Islands. It is a curious speculation to the voyager always, how his steamer is to find its way through the labyrinth of the thousand islands, which stud the broad waters like the countless tents of an encamped army, and ever and anon his interest is aroused up to the highest pitch at the prospective danger of the passage of some angry rapid. All the journey east, from lake to lake of the great waters, past islands now miles in circuit, and now large enough only for the cottage of Lilliputian lovers, is replete with ever-changing pleasure.

Montreal and Quebec, the chief cities of Canada, are upon the St. Lawrence, while Toronto lies on the shores of Lake Ontario, the continuing waters westward.

The Ottawa River flows 800 miles and enters the St. Lawrence on both sides of the Island of Montreal, traversing in its way Lake Temiscaming, Grand Lake, and others. Rapids and falls greatly impede the navigation of its waters, but lend to them wonderful beauty. It is a wild forest region; that of the Ottawa, but little occupied heretofore by others than the rude lumbermen, though numerous settlements are now springing up, and its agricultural capacities are being developed.

The Committee on Railways of the House of Assembly of the Province, in its report, thus speaks of this river:—

"At the head of the lake the Blanche River falls in, coming about 90 miles from the north. Thirty-four miles farther down the lake it receives the Montreal River, coming 120 from the north-west. Six miles lower down, on the east or Lower Canada bank, it receives the Keepawasippi, a large river which has its origin in a lake of great size, hitherto but partially explored, and known as Lake Keepawa. This lake is connected with another chain of irregularly shaped lakes, from one of which proceeds the river Du Moine, which enters the Ottawa about 100 miles below the mouth of the Keepawa-sippi; the double discharge from the same chain of lakes in opposite directions presents a phenomenon similar to the connection between the Orinoco and Rio Negro in South America. The Keepawa-sippi has never been surveyed, but on a partial survey of the lake from which it proceeds, it was found flowing out with a slow and noiseless current, very deep, and about 800 feet in width; its middle course is unknown, but some rafts of timber have been taken out a few miles above the mouth. It is stated in the report from which we quote, that there is a cascade at its mouth 120 feet in height; this is a fable; the total descent from the lake to the Ottawa may be 120 feet, but there is no fall at the mouth of the river.

"From the Long Sault at the foot of Lake Temiscaming, 233 miles above Bytown, and 360 miles from the mouth of the Ottawa, down to Deux Joachim Rapids, at the head of the Deep River, that is for 89 miles, the Ottawa, with the exception of 17 miles below the Long Sault, and some other intervals, is not at present navigable except for canoes. Besides other tributaries in the interval, at 197 miles from Bytown, now called Ottawa, it receives on the west side the Mattawan, which is the highway for canoes going to Lake Huron by Lake Nipissing. From the Mattawan the Ottawa flows east by south to the head of Deep River reach, nine miles above which it receives the river Du Moine from the north.

"From the head of Deep River, as this part of the Ottawa is called, to the foot of Upper Allumettes Lake, two miles below the village of Pembroke, is an uninterrupted reach of navigable water, 43 miles in length. The general direction of the river in this part is south-east. The mountains along the north side of Deep River are upwards of 1,000 feet in height, and the many wooded islands of Allumettes lake render the scenery of this part of the Ottawa magnificent and exceedingly picturesque—far surpassing the celebrated lake of the Thousand Islands on the St. Lawrence.

"Passing the short rapid of Allumettes, and turning northward, round the lower end of Allumettes Island, which is 14 miles long and 8 at its greatest width, and turning down south-east through Coulonge Lake, and passing behind the nearly similar islands of Calumet, to the head of the Calumet Falls, the Ottawa presents, with the exception of one slight rapid, a reach of 50 miles of navigable water. The mountains on the north side of Coulonge lake, which rise apparently to the height of 1,500 feet, add a degree of grandeur to the scenery, which is in other respects beautiful and varied. In the Upper Allumettes Lake, 115 miles from Ottawa, the river receives from the west the Petawawee, one of its largest tributaries. This river is 140 miles in length, and drains an area of 2,200 square miles. At Pembroke, 9 miles lower down on the same side, an inferior stream, the Indian River, also empties itself into the Ottawa.

"At the head of Lake Coulonge, the Ottawa receives from the north the Black River, 130 miles in length, draining an area of 1,120 miles; and 9 miles lower, on the same side, the river Coulonge, which is probably 160 miles in length, with a valley of 1,800 square miles.

"From the head of the Calumet Falls to Portage du Fort, the head of steamboat navigation, a distance of 8 miles, are impassable rapids. Fifty miles above the city, the Ottawa receives on the west the Bonnechère, 110 miles in length, draining an area of 980 miles. Eleven miles lower, it receives the Madawaska, one of its greatest feeders, a river 210 miles in length, and draining 4,100 square miles.

"Thirty-seven miles above Ottawa there is an interruption to the navigation, caused by three miles of rapids and falls, to pass which a railroad has been made. At the foot of the rapids, the Ottawa divides among islands into numerous channels, presenting a most imposing array of separate falls.

"Six miles above Ottawa begin the rapids terminating in the Ottawa Chaudière Falls, which, inferior in impressive grandeur to the Falls of Niagara, are perhaps more permanently interesting, as presenting greater variety.

"The greatest height of the Chaudière Falls is about 40 feet. Arrayed in every imaginable variety of form in vast dark masses, in graceful cascades, or in tumbling spray, they have been well described as a hundred rivers struggling for a passage. Not the least interesting feature which they present is the lost Chaudière, where a body of water greater in volume than the Thames at London, is quietly sucked down, and disappears under ground.

"At the city of Ottawa the river receives the Rideau from the west, running a course of 116 miles, and draining an area of 1,350 square miles."

The city of Ottawa, on the banks of the river, is thought to be excelled in the beauty of its position, only by Quebec, on the St. Lawrence. From Barrack Hill here, the wide panorama includes the Falls of the Chaudière, the Suspension Bridge, which connects the upper and lower provinces, the islanded stretch of the river above, and of the far-away mountain ranges.

The Rideau Falls, near the mouth of the Rideau, just below the city of Ottawa, is a charming scene.

"A mile lower it receives, from the north, its greatest tributary, the Gatineau, which, with a course-probably of four hundred and twenty miles, drains an area of twelve thousand square miles. For about two hundred miles the upper course of this river is in the unknown northern country. At the farthest point surveyed, two hundred and seventeen miles from its mouth, the Gatineau is still a noble stream, a thousand feet wide, diminished in depth but not in width.

"Eighteen miles lower down, the Rivière au Lièvre enters from the north, after running a course of two hundred and sixty miles in length, and draining an area of four thousand one hundred miles. Fifteen miles below it, the Ottawa receives the North and South Nation Rivers on either side, the former ninety-five and the latter a hundred miles in length. Twenty-two miles further, the River Rouge, ninety miles long, enters from the north. Twenty-one miles lower, the Rivière du Nord, a hundred and sixty miles in length, comes in on the same side; and lastly, just above its mouth, it receives the River Assumption, which has a course of a hundred and thirty miles.

"From Ottawa the river is navigable to Grenville, a distance of fifty-eight miles, where the rapids that occur for twelve miles are avoided by a succession of canals. Twenty-three miles lower, at one of the mouths of the Ottawa, a single lock, to avoid a slight rapid, gives a passage into Lake St. Louis, an expansion of the St. Lawrence above Montreal.

"The remaining half of the Ottawa's waters find their way to the St. Lawrence, by passing in two channels, behind the Island of Montreal and the Isle Jesus, in a course of 31 miles. They are interrupted with rapids; still it is by one of them that all the Ottawa lumber passes to market. At Bout de l' Isle, therefore, the Ottawa is finally merged in the St. Lawrence, a hundred and thirty miles below, from the city of Ottawa.

Routes from Montreal up the Ottawa.—Steamers run daily, during the summer months, between Montreal and Ottawa, and Kingston and Ottawa, via the Rideau Canal. Above Ottawa the traveller may proceed by carriage or by stage, nine miles, to the village of Aylmer, and thence by steamer to the Chats; thence by railway, two miles; then again by steamer to the Portage du Fort: now, wagons for awhile, and then again a steamer to Pembroke, and yet another from thence to Deux Joachims; afterwards he must canoe it. The Ottawa may also be reached by railway direct, from Prescott on the St. Lawrence, to Ottawa City.

The Saguenay. The journey up this beautiful river may be made semi-weekly, by steamer from Quebec, or by the Grand Trunk railway, 101 miles to St. Paschal Rivière du Loup, opposite the mouth of the Saguenay, and thence by steamer. The course of the Saguenay—between lofty and precipitous heights; and in its upper part, amid rushing cataracts, is 120 miles from Lake St. John to the St. Lawrence, which it enters 140 miles below Quebec. Large ships ascend 60 miles.

In the trip from Quebec to the Saguenay beauties, there are many interesting points to be noted in the preceding journey of 120 miles down the St. Lawrence; the ancient-looking settlements on its banks, and the not less picturesque *habitans* of the country. A day's sail lands the voyager at Rivière du Loup, where he passes the night on board his steamer, waiting for the following morning to resume his journey.

The Saguenay is a perfectly straight river, with grand precipices on either side. It has neither windings nor projecting bluffs, nor sloping banks nor sandy shores like other rivers, nor is its stern, strange aspect varied by either village or villa.

"It is," says a voyager thither, "as if the mountain range had been cleft asunder, leaving a horrid gulf of 60 miles in length and 4,000 feet in depth, through the gray mica schist, and still looking fresh and new. One thousand five hundred feet of this is perpendicular cliff, often too steep and solid for the hemlock or dwarf-oak to find root; in which case, being covered with colored lichens and moss, their fresh-looking fractures often appear, in shape and color, like painted fans, and are called the pictured rocks. But those parts more slanting are thickly covered with stunted trees, spruce and maple, and birch growing wherever they can find crevices to extract nourishment; and the bare roots of the oak, grasping the rock, have a resemblance to gigantic claws. The bases of these cliffs lie far under water, to an unknown depth. For many miles from its mouth no soundings have been obtained with two thousand feet of line; and, for the entire distance of 60 miles, until you reach Ha Ha Bay, the largest ships can sail, without obstruction from banks or shoals, and, on reaching the extremity of the bay, can drop their anchor in 30 fathoms. The view up this river is singular in many respects; hour after hour, as you sail along, precipice after precipice unfolds itself to view, as in a moving panorama; and you sometimes forget the size and height of the objects you are contemplating, until reminded by seeing a ship of one thousand tons lying like a small pinnace under the towering cliff to which she is moored; for, even in these remote and desolate regions, industry is at work, and, although you cannot much discern it, saw-mills have been built on some of the tributary streams which fall into the Saguenay. But what strikes one most, is the absence of beach or strand, for except in a few places where mountain torrents, rushing through gloomy ravines, have washed down the *detritus* of the hills, and formed some alluvial land at the mouth, no coves, nor creeks, nor projecting rocks are seen in which a boat could find shelter, or any footing be obtained. The characteristic is a steep wall of rock rising abruptly from the water; a dark and desolate region, where all is cold and gloomy; the mountains hidden with driving mist, the water black as ink, and cold as ice. No ducks nor sea-gulls sitting on the water, or screaming for their prey. No hawks nor eagles soaring overhead, although there is an abundance of what might be called ' Eagle Cliffs.' No deer coming down to drink at the streams, no squirrels nor birds to be seen among the trees. No fly on the water, nor swallows skimming over the surface. It reminds you of

'That lake whose gloomy shore
Sky-lark never warbled o'er.'

Two living things you may see, but these are cold-blooded animals; you may see the cold seal, spreading himself upon his clammy rock, watching for his prey. You may see him make his sullen plunge into the water, like to the Styx for blackness. You may see him emerge again, shaking his smooth oily sides, and holding a huge living salmon writhing in his teeth; and you may envy the fellow faring so sumptuously, until you recollect that you have just had a hearty breakfast of fresh grilled salmon yourself, and that you enjoyed it as much as your fellow creature is now enjoying his raw morsel. And this is all you see for the first twenty miles, save the ancient settlement of Tadousac at the entrance, and the pretty cove of L'Anse a l'Eau, which is a fishing station.

"Now you reach Cape Eternity, Cape Trinity, and many other overhanging cliffs, remarkable for having such clean fractures, seldom equalled for boldness and effect, which create constant apprehensions of danger, even in a calm; but if you happen to be caught in a thunder-storm, the roar, and darkness, and flashes of lightning are perfectly frightful. At last you terminate your voyage at Ha Ha Bay; that is, smiling or laughing bay, in the Indian tongue, for you are perfectly charmed and relieved to arrive at a beautiful spot, where you have sloping banks, a pebbly shore, boats, and wherries, and vessels riding at anchor; birds and animals, a village, a church, French Canadians, and Scottish Highlanders."

After duly enjoying the pleasant "let down" from the high tragic tone of the landscape you have been so long gazing upon and wondering at, formed in the comparatively pastoral character of this upper region of the Ottawa, you return to your steamer, and descending the stern and solemn river, come again, at nightfall, to the Rivière du Loup, from whence you started in the morning. This is the second day of your journey, and on the third you are back once more in Quebec.

SPRINGS.

The Caledonia Springs.—HOTELS:—

The Caledonia Springs, a place of much resort, are at the village of Caledonia, 72 miles from Montreal. Leave Montreal by the Lachine railway, and take the steamer to Carillon. At Point Fortune, opposite Carillon, on the other side of the Ottawa, take stage to the Springs, arriving the same evening.

Plantagenet Springs. From Montreal to Point Fortune, as in the route to the Caledonia Springs; and thence by stage, arriving same evening. Distance 88 miles. The consumption of the "Plantagenet water" is said to be very great.

The St. Leon Springs are at the village of St. Leon, on the Rivière du loup, "en haut," between Montreal and Quebec; 26 miles by stage from Three Rivers, a landing of the St. Lawrence steamers.

St. Catharine's.—HOTELS:—

St. Catharine's, Canada West, on the Great Western Railway, 11 miles from Niagara Falls, and 32 miles from Hamilton. See St. Catharine's in route from Montreal to Niagara via the St. Lawrence.

WATERFALLS IN CANADA.

Niagara. See chapter on the state of New York.

Falls of Montmorenci. See Quebec.

The Chaudiere Falls on the Ottawa. See Ottawa River.

The Chaudiere Falls, Quebec. See City of Quebec.

The Rideau Falls. See Ottawa river.

The Falls of Shawanegan are on the River St. Maurice, 25 miles from Three Rivers, on the St. Lawrence river, between Montreal and Quebec. The St. Maurice, 186 feet in breadth at this point, makes a perpendicular descent of about 200 feet. The imposing character of this scene is, as yet, but little known. Between the Falls and the town of Three Rivers, the St. Maurice affords excellent fishing.

St. Anne's Falls are 24 miles below Quebec. See Quebec.

RAILWAYS.

The Grand Trunk connects Montreal with Quebec, and each with Portland in Maine. From Montreal it follows the upper shore of the St. Lawrence and of Lake Ontario to Toronto, and thence continues westward, across the peninsula of Canada West, via Port Sarnia, on the southern extremity of Lake Huron to the city of Detroit in Michigan. The whole length of the road, with its present branches, is 1050 miles. It connects with routes to Niagara Falls, with the line of the Great Western Railway, and with the routes Mississippiwards.

The Great Western Railway extends from Niagara Falls, 229 miles west to Detroit, Michigan, connecting with the Michigan Central route for Chicago, &c.

The Montreal and New York road extends from Montreal 67 miles to Plattsburg, and is a part of a route from Montreal to New York.

The Champlain and St. Lawrence extends

from Montreal, 44 miles, to Rouse's Point on Lake Champlain, thence to New York, Boston, &c.

The Northern Railway of Canada extends 94 miles from Toronto on Lake Ontario to Collingwood on the Georgian Bay, Lake Huron. A part of a pleasant route from New York to Lake Superior.

The Ottawa and Prescott Railway extends from Prescott (opposite Ogdensburg) on the St. Lawrence, 54 miles to Ottawa, on the Ottawa river.

The Hamilton and Toronto road extends 38 miles from Toronto to Hamilton, connecting the Grand Trunk and the Great Western routes.

The Coburg and Peterboro' Railway, 28 miles from Peterboro' to Coburg, on the line of the Grand Trunk, between Montreal and Toronto.

Many other routes are either in progress or in contemplation—Canada vying with the "States" in this field of enterprise.

MONTREAL.

Hotels. The Donegana, Notre-Dame street; the St. Lawrence, Great St. James street, a fine house, centrally located; the Ottawa, Great St. James street; and the Montreal House, Custom House square, and opposite the Custom House. Besides these leading establishments, there are many other comfortable houses and cafés, where travellers of all ranks and classes may be lodged and regaled according to the varied humors of their palates and their purses. We forbear to name too many, lest the bewildered stranger should, in the very abundance of the good things placed before him, starve while choosing.

Montreal may be reached daily from New York in from 15 to 18 hours, by the Hudson River or Harlem railway to Troy; rail to Whitehall, and steamer on Lake Champlain, or by rail through Vermont via Rutland, Burlington and St. Alban's to Rouse's Point, or via Plattsburg on Lake Champlain. From Boston via Albany, or other routes to Lake Champlain, &c.; or, via Portland and the Grand Trunk railway.

Montreal, the most populous city in British North America, is picturesquely situated at the foot of the Royal Mountain, from which it takes its name, upon a large island at the confluence of the Ottawa and St. Lawrence, which, both in fertility and cultivation, is considered the garden of Canada East. The main branch of the Ottawa, which is the timber highway to Quebec, passes north of Montreal Island, and enters the St. Lawrence about 18 miles below the city; about one-third of its waters is, however, discharged into Lake St. Louis, and joining but not mingling at Caughnawaga, the two distinct bodies pass over the Sault St. Louis and the Lachine Rapids—the dark waters of the Ottawa washing the quays of Montreal, while the blue St. Lawrence occupies the other shore. Nor do they merge their distinctive character until they are several miles below Montreal. The quays of Montreal are unsurpassed by those of any city in America; built of solid limestone, and uniting with the locks and cut stone wharves of the Lachine Canal, they present for several miles a display of continuous masonry, which has few parallels. Unlike the levees of the Ohio and the Mississippi, no unsightly warehouses disfigure the river side. A broad terrace, faced with gray limestone, the parapets of which are surmounted with a substantial iron railing, divides the city from the river throughout its whole extent.

The people in Montreal number over 75,000, and the population is steadily increasing. The houses in the suburbs are handsomely built in the modern style, and mostly inhabited by the principal merchants. Including its suburbs, of which it has several, the city stretches along the river for two miles from s. w. to N. E., and, for some distance, extends between one and two miles inland. It was formerly surrounded by a battlemented wall; but this having fallen into decay, it is now entirely open. St. Paul st., the chief commercial thoroughfare, extends along the river the whole length of the city. Great St. James and Notre-Dame streets are the fashionable promenades.

The Victoria Bridge which spans the great St. Lawrence at Montreal, is one of the noblest structures which we shall see in the whole long course of our American journeyings. Its length is 10,284 feet, or nearly two miles. It rests, in this splendid transit, upon 24 piers and two abutments of solid masonry, the central span being 330 feet in length. The heavy iron tubes through which the railway track is laid is, in its largest dimensions 22 feet high and 16 feet wide. The total cost of this bridge was over six millions of dollars. It has been recently completed, and was formally opened, with high pomp and ceremony, amidst great popular rejoicings, by the young heir to the British Crown, the Prince of Wales, during his visit, in the summer of 1860, to America.

The French Cathedral. Of the public buildings, the most remarkable is the Roman Catholic Cathedral, Place d'Armes, constructed in the Gothic style, with a length of 255½ feet, and a breadth of 134½ feet. It has two towers, each of which has a noble elevation of two hundred and twenty-five feet. The view from these towers—embracing the city and its suburbs, the river, and the surrounding country—is exceedingly beautiful. The principal window of the Cathedral is 64 feet high and 32 broad. Of the vastness of the interior of this edifice an idea may be formed from the fact that it is capable of accommodating eight or nine thousand persons. This immense assembly may, by numerous outlets, disperse in five or six minutes.

The Seminary of St. Sulpice, adjoining the Cathedral, is 132 feet long, and 20 deep, and is surrounded by spacious gardens and court yard.

The Bank of Montreal and the City Bank, the first a fine example of Corinthian architecture, stand side by side on the square called the Place d'Armes.

St. Patrick's Church (Catholic) occupies a commanding position at the west end of Lagauchetiere street.

The Bishop's Church (Catholic) is a very elegant structure in St. Denis street.

The remaining Catholic churches are the Recollet, in Notre Dame street, the Bonsecours, near the large market, and the St. Mary's in Griffintown. There are also chapels attached to all the Nunneries, in some of which excellent pictures may be seen.

Nunneries. The Grey Nuns, in Foundling street, was founded in 1692, for the care of lunatics and children. The Hotel Dieu was established in 1644, for the sick generally. The Black, or congregational nunnery, in Notre Dame street, dates from 1659. The Sisterhood, at this third and last of the conventual establishments of Montreal, devote themselves to the education of young persons of their own sex.

The stranger desirous of visiting either of the nunneries should apply to the Lady Superior for admission, which is seldom refused.

The Protestant churches worthy of notice are St. Andrew's Church, a beautiful specimen of Gothic architecture, being a close imitation of Salisbury Cathedral, in England, though of course on a greatly reduced scale. This, with St. Paul's Church, in St. Helen street, are in connection with the Established Church of Scotland. The Episcopalian churches are, the beautiful new edifice Christ Church Cathedral, St. George's Church, in St. Joseph street, St. Stephen's, in Griffintown, Trinity in St. Paul street, and St. Thomas's, in St. Mary street. Various other denominations of Christians have churches—the Wesleyans, a large and very handsome building, in St. James street, and also others in Griffintown and Montcalm street; the Independents formerly had two houses, but now only the one in Radegonde street. This last was the scene of the sad riot and loss of life on the occasion of Gavazzi's lecture in 1852. The Free Church has also two places of worship, one in Coté street; and one in St. Gabriel street; besides these, there are the American and the United Presbyterian, the Baptist, and the Unitarian Churches; a small Jewish Synagogue, the last named being classical in design.

Directly opposite the city is the wharf of the New York and St. Lawrence Railway Company. Below Nun's Island are seen the gigantic piers of the Tubular Bridge, a wonderful structure, which spans the great St. Lawrence.

The Bonsecours Market is an imposing Doric edifice, erected at a cost of $280,000. In one of the upper stories are the Offices of the Corporation and Council Chamber, and a concert or ball-room capable of seating 4,000 people. The view from the dome of this structure, overlooking the river and St. Helen's isle, are well worth the seeing.

At the head of Place Jacques Cartier there is a column erected to the memory of the naval hero, Lord Nelson.

The Court House is one of the most striking of the architectural specialties of the city.

The Post Office is in Great James street.

The Custom House is a neat building on the site of an old market-place, between St. Paul street and the river.

The Merchants' Exchange and Reading-Room are in St. Sacrament street. The latter is a large and comfortable room, well supplied with newspapers and periodicals, English and American, all at the service of the stranger when properly introduced.

The General Hospital and St. Patrick's are in Dorchester street; the latter, however, at the west end of the town.

M'Gill's College is beautifully situated at the base of the mountain. The High School department of the college is in Belmont street.

The city also possesses, besides the University of McGill's College, many excellent institutions for the promotion of learning—French and English seminaries, a royal grammar-school, with parochial, union, national, Sunday and other public schools. It has numerous societies for the advancement of religion, science, and industry; and several public libraries.

The Water-Works, a mile or so from the city, are extremely interesting for their own sake, and for the fine view of the neighborhood to be seen thence.

The Mount Royal Cemetery is two miles from the city, on the northern slope of the mountain. From the high road round its base, a broad avenue through the shaded hill-side gradually ascends to this pleasant spot.

There are other romantic burying-grounds, both of the Catholic and the Protestant population, in the vicinity of Montreal, and other scenes which the visitor should enjoy—pleasant rides all about, around the mountain and by the river, before he bids good-bye to the Queen City of Canada.

QUEBEC.

Hotels. The leading hotels are Russell's, and the Clarendon. Russell's is in Palace street, Upper Town, and is the favorite head-quarters of American tourists, while the English establish themselves generally at the Clarendon, on St. Louis street.

Quebec may be pleasantly reached from New York, via Boston to Portland, Maine, and thence 317 miles by the Grand Trunk Railway, total distance, by this route, from New York to Quebec, 650 miles; or, from New York by the Hudson River Railway or steamboats; or by the Harlem Railway to Albany, thence to Whitehall, thence on Lake Champlain to Plattsburg, thence by the Montreal and New York Railway to Montreal, and from Montreal by steamer down the St. Lawrence, or by the Grand Trunk Railway. Distance by railway, from Montreal to Quebec, 168 miles. There are other railway routes from Boston to Quebec, via Albany, or by the Vermont Central and Vermont and Canada lines through St. Albans to Montreal.

Quebec is the capital of United Canada, and, after Montreal, the most populous city in British North America. It is upon the left bank of the St. Lawrence river, and some 340 miles from the Ocean.

The city was founded in 1608, by the geographer, Champlain. It fell into the possession of the British in 1619, but was restored three years later. The English made an unsuccessful attempt to regain possession in 1690, but it did not finally come into their hands until taken by General Wolfe in 1759.

The city is divided into two sections, called the Upper and the Lower Towns; the Upper Town occupying the highest part of the promontory, which is surrounded by strong walls and other fortifications; and the Lower Town, being built around the base of Cape Diamond. The latter is the business quarter.

The Citadel, a massive defence, crowning the summit of Cape Diamond, covers about 40 acres with its numerous buildings. Its impregnable position makes it perhaps the strongest fortress on this continent; and the name of the "Gibraltar of America" has been often given to it not inaptly. The access to the Citadel is from the Upper Town, the walls of which are entered by five gates. Near the Palace gate is the Hospital and a large Guard House. By St. Louis gate, on the south-west, the tourist will reach the memorable Plains of Abraham, the scene of Wolfe's victory and death, in the year 1759.

The Prescott Gate is the only entrance on the St. Lawrence side of the fortress.

The view from the Citadel is remarkably fine, taking in, as it does, the opposite banks of the great river through many picturesque miles up and down. The promenade here, on the ramparts above the esplanade, is charming. In the public garden, on Des Carrierres street, there is an obelisk to the memory of Wolfe and Montcalm.

The Parliament House. Among the chief public edifices of Quebec is the New Parliament House, which supplies the place of the building destroyed by fire in 1854.

The Roman Catholic Cathedral was erected under the auspices of the first Bishop of Quebec, and was consecrated in 1666. It is 216 feet long, and 180 feet in breadth.

The Ursuline Convent and the Church of St. Ursula are agreeable buildings, encompassed by pleasant gardens. This establishment was founded in 1639, and holds a high position in the public esteem. It contains a Superior, fifty nuns, and six novices, who give instruction in reading, writing, and needle work. The convent was destroyed by fire in 1650, and again in 1686. The remains of the Marquis de Montcalm are buried here.

The Artillery Barracks form a range of stone buildings 5,000 feet in length.

Durham Terrace is the site of the old castle of St. Louis, which was entirely consumed by fire in 1834.

The English Protestant Cathedral, consecrated in 1804, is one of the finest modern edifices of the city.

St. Andrew's Church, in St. Anne street, is in connection with the Scotch Establishment. The Methodists have a chapel in St. Stanislaus' street, and another in St. Louis suburb, called the Centenary Chapel.

The Lower Town. The passage from the Upper to the Lower Town is by the Prescott gate. It is in this portion of the city that the traveller will find the Exchange, the Post Office, the Banks, and other commercial establishments.

The Plains of Abraham may be reached via the St. Louis Gate, and the counterscarp on the left, leading to the glacis of the citadel; hence towards the right; approaching one of the Mertello Towers, where a fine view of the St. Lawrence opens. A little beyond, up the right bank, is the spot where General Wolfe fell on the famous historic ground of the Plains of Abraham. It is the highest ground, and is surrounded by wooden fences. Within an enclosure lower down is a stone well, from which water was brought to the dying hero.

CANADA. 33

Wolfe's Cove, the spot where Montgomery was killed, and other scenes, telling tales of the memorable past, will be pointed out to the traveller in this neighborhood.

The Mount Hermon Cemetery is about three miles from the city, on the south side of the St. Louis Road. The grounds are 32 acres in extent, sloping irregularly but beautifully down the precipices which overhang the St. Lawrence. They were laid out by the late Major Douglass, of the U. S. Engineers, who had previously displayed his skill and taste in the arrangements of the Greenwood Cemetery, near New York.

Lorette. To see Lorette may be made the motive of an agreeable excursion from Quebec, following the banks of the St. Charles.

Lake St. Charles is four miles long, and one broad. It is divided by projecting ledges into two parts. It is a delightful spot, in its natural attractions, and in the fine sport it affords to the angler.

The Falls of Montmorenci, eight miles distant, are among the chief delights of the vicinage of Quebec. The river here is 60 feet wide, and the descent of the torrent 250 feet.

"The effect of the view of these falls upon the beholder is most delightful. The river at some distance, seems suspended in the air, in a sheet of billowy foam, and contrasted, as it is, with the black frowning abyss into which it falls, it is an object of the highest interest. The sheet of foam which first breaks over the ridge, is more and more divided as it plunges and is dashed against the successive layers of rock, which it almost completely veils from view; the spray becomes very delicate and abundant, from top to bottom, hanging over, and revolving around the torrent, till it becomes lighter and more evanescent than the whitest fleecy clouds of summer, than the finest attenuated web, than the lightest gossamer, constituting the most airy and sumptuous drapery that can be imagined. Yet, like the drapery of some of the Grecian statues, which, while it veils, exhibits more forcibly the form beneath, this does not hide but exalts the effect produced by this noble cataract.

"Those who visit the falls in the winter, see one fine feature added to the scene, although they may lose some others. The spray freezes, and forms a regular cone, of 100 feet and upwards in height, standing immediately at the bottom of the cataract, like some huge giant of fabulous notoriety."

The extraordinary formation called the Natural Steps, will not fail to interest the visitor at Montmorenci.

The Falls of St. Anne, in the river St. Anne, 24 miles below Quebec, are in a neighborhood of great picturesque beauty. Starting from the city in the morning betimes, one may visit Montmorenci nicely, and proceed thence the same evening to St. Anne. Next morning after a leisurely survey of these cascades, there will be most of the day left to get back, with any *détours* that may seem desirable, to Quebec.

The Falls of the Chaudiere are reached via Point Levi. The rapid river plunges over a precipice of 130 feet, presenting very much the look of boiling water, from whence its name of chaudiere or caldron. The cataract is broken into three separate parts by the intervention of huge projecting rocks, but it is reunited before it reaches the basin beneath.

We take our leave of this venerable city, its unique natural beauties, and its winning stories, with the remembrance of some of the impressions it made upon Professor Silliman, when he visited it years ago :—"Quebec," he writes, "at least for an American city, is certainly a very peculiar place. A military town—containing about 20,000 inhabitants—most compactly and permanently built—environed, as to its most important parts, by walls and gates—and defended by numerous heavy cannon—garrisoned by troops having the arms, the costume, the music, the discipline of Europe—foreign in language, features, and origin, from most of those whom they are sent to defend—founded upon a rock, and in its highest parts overlooking a great extent of country—between 300 and 400 miles from the ocean—in the midst of a great continent, and yet, displaying fleets of foreign merchantmen in its fine capacious bay—and showing all the bustle of a crowded seaport—its streets narrow, populous, and winding up and down almost mountainous declivities—situated in the latitude of the finest parts of Europe—exhibiting in its environs the beauty of an European capital—and yet in winter smarting with the cold of Siberia—governed by a people of different language and habits from the mass of the population—opposed in religion, and yet leaving that population without taxes, and in the full enjoyment of every privilege, civil and religious."

Toronto, on Lake Ontario, and the line of the Grand Trunk Railway; from Quebec, 501 miles; from Montreal, 333 miles; from Hamilton, 38 miles; from Niagara Falls, 81 miles. For description, see *Toronto*, in route from Montreal to Niagara, via the St. Lawrence.

Kingston, at the foot of Lake Ontario, on the St. Lawrence, and on the line of the Grand Trunk Railway; from Quebec, 341 miles; from Montreal, 173 miles; from Toronto, 160 miles. See *Kingston* in route from Montreal to Niagara.

Hamilton, near the eastern terminus of the Great Western Railway, at the head of Lake

Ontario; from Quebec, 539 miles; from Montreal, 371 miles; from Toronto, 38; from Niagara, 43. See route from Montreal to Niagara.

London is a prosperous town, midway on the line of the Great Western Railway, in its traverse of the peninsula of Canada West from Lake Ontario to Lake Huron. Distant from Niagara Falls, 119 miles, west; from Hamilton, 76 miles; from Toronto, 114 miles; from Montreal, 447 miles; from Quebec, 615 miles. In 1820, the present site of London was a wilderness, occupied by the savages and the wild deer; now its population exceeds 12,000. Like Hamilton, Toronto, and all the growing towns of Canada, it is well built, upon wide streets, and with elegant and substantial architecture.

MONTREAL TO NIAGARA FALLS; UP THE ST. LAWRENCE RIVER AND LAKE ONTARIO.

The traveller may go from Montreal to Niagara, either by steamer on the St. Lawrence, or by the Grand Trunk Railway, 333 miles to Toronto on Lake Ontario. At Toronto he may cross the western end of the lake to the town of Niagara, and thence reach the Falls by the Erie and Ontario Railway 14 miles long; or he may go less directly, by water or by rail to Hamilton, and thence by rail again to the Falls.

UP THE ST. LAWRENCE FROM MONTREAL. See rivers of Canada for general mention of the St. Lawrence.

Lachine. From Montreal the traveller will proceed nine miles to Lachine by railway, avoiding the rapids which the steamers sometimes descend. At Lachine is the residence of Sir George Simpson, Governor of the Hudson's Bay Company and of the officers of this, the chief post of that corporation. It is from this point that the orders from head-quarters in London are sent to all the many posts throughout the vast territory of the Company; and near the end of April each year a body of trained *voyageurs* set out hence in large canoes, called *maîtres canots*, with packages and goods for the various posts in the wilderness. Two centuries ago, the companions of the explorer Cartier on arriving here, thought they had discovered a route to China, and expressed their joy in the exclamation of La Chine! Hence, the present name, or so at least says tradition. A costly canal overcomes the obstruction of the rapids at Lachine.

The Village of the Rapids; or, Caughnawaga. An Iroquois settlement lies opposite Lachine, at the outlet of the expansion of the river called Lake St. Louis. The Indians at Caughnawaga, subsist chiefly by navigating barges and rafts down to Montreal, and in winter, by a trade in moccasins; snow-shoes, &c. They are mostly Roman Catholics, and possess an elegant church.

Lake St. Louis. The brown floods of the Ottawa assist in forming this great expanse of the St. Lawrence. They roll unmixed through the clearer water of the great river. On the northern shore of Lake St. Louis is the island of Montreal, 30 miles long. At the western extremity is *Isle Perrot*. The *Cascade Rapids* separate the expanse just passed from Lake St. Francis. The Beauharnois Canal here is 11¼ miles in length, and has nine locks.

Lake St. Francis, into which the voyager now enters, extends 40 miles. Midway on the right, is the village of LANCASTER, where a pile of stones or *cairn* has been thrown up in honor of Sir John Colborne, formerly Governor-General of Canada, now Lord Seaton. Leaving Lake St. Francis, we pass the passage of the celebrated *Long Sault* rapids. Here, too, is the *Cornwall Canal*, 11¼ miles in length, with 7 locks of noble size.

Cornwall is a pleasant town, formerly called "Pointe Maline," in memento of the labor of ascending the river at this point.

The Village of St. Regis lies across from Cornwall. It forms the boundary between Canada and the State of New York, and also intersects the tract of land occupied by the 1,000 Iroquois, American and British, who dwell here.

Dickenson Landing is at the head of the Cornwall Canal, within the space of the 38 miles which follow to Prescott; the villages of MOULINETTE, MARIA TOWN, and MATILDA, are successively passed.

The Battle Field of Chryseler's Farm, where the Americans met a defeat in the last war, lies a little above Maria Town.

Prescott is rapidly recovering its prestige lost when the construction of the Rideau Canal won its trade away to Kingston; for now a railway from New York approaches it at Ogdensburg, and another connects it with Ottawa city, on the Ottawa river. Besides which advantages, it is on the line of the Grand Trunk route. From Prescott may be seen the windmill and the ruined houses, mementoes of the attempt at invasion by Schultz and his band in 1838.

Ogdensburg, New York, the western terminus of the Northern Railway from Lake Champlain, is opposite Prescott.

Maitland, built upon the site of an old French fort, is seven miles above Prescott.

Brockville is yet five miles more, onward. It is one of the best built towns in Canada West.

Gananoque is 32 miles above Brockville.

At Kingston, 20 miles yet beyond Gananoque, we leave the St. Lawrence, and approach the waters of Lake Ontario. In descending the river, the wonderful labyrinth of the Thousand Isles is passed just east of Kingston. Wolfe's Island, a well-cultivated spot, is opposite Kingston.

Kingston.—HOTELS :—*Kent's British American; Iron's Hotel.*

The city of Kingston, modern as it appears, looks far back for its history, as its advantageous locale did not fail to attract the notice of the early French discoverers. It was once occupied as a small fort called Cataraqui, otherwise known as Frontenac, and was the scene of various sieges and exploits before it passed with all the territory of the Canadas from French to British rule. It was from this point that murderous expeditions were made by the Indians in the olden times against Albany and other English settlements of New York; which in turn sent back here its retributive blows. The present city was founded in 1783. It has now a population of about 16,000. Among its objects of interest are the fortifications of Fort Henry, on a hill upon the eastern side of the harbor; four fine Martello Towers off the town; and other defensive works; the University of Queen's College; the Roman Catholic College of Regiopolis; and the Provincial Penitentiary a little to the west of the city.

As the navigation of the St. Lawrence ends at Kingston, the river boats are exchanged here for others more suited to the lake voyages.

Lake Ontario—American shore.—Let us, before we enter the great waters of Ontario, say a word to the traveller who may prefer to make the voyage along the *American* or lower shore of the lake. From the boundary line 45° the entire *littoral* is in the State of New York.

French Creek comes into the St. Lawrence as we leave it. It was here that General Wilkinson embarked (November 1813) with 7,000 men, with the purpose of descending the river and attacking Montreal. A week subsequently, an engagement took place near Williamsburg, on the Canadian side, when the Americans came off but poorly. General Wilkinson being disappointed in his expectation of reinforcements from Plattsburg, retired to French Mills, and there went into winter-quarters. This place was afterwards named Fort Covington, in memory of General Covington, who fell at the battle of Williamsburg.

Sackett's Harbor, (N. Y.,) lies 20 miles below the mouth of French Creek. It is on the eastern extremity of the lake, on the south side of Black River Bay. This was the naval station of the United States during the English and American War of 1812. In May, 1813, Sir George Prevost made a landing with 1,000 troops, but re-embarked without accomplishing any thing. The Navy Yard here is a prominent object as we land.

Oswego.—HOTELS—*The American.*

Oswego, (N. Y.,) is the chief commercial port on the American shore of Ontario. It is very agreeably situated at the mouth of the Oswego river. The Oswego Canal comes in here (38 miles) from Syracuse, and the railway, also, from the same place.

Charlotte, the port of the city of Rochester, (N. Y.,) is at the approach to Lake Ontario, of the beautiful Genesee river. (See "Rochester.")

From the mouth of the Genesee to Fort Niagara, a distance of 85 miles, the coast now presents a monotonous and forest-covered level, with a clearing only here and there.

Having now peeped at the American, or southern shore, we will go back to Kingston, and start again on the upper side of the lake, making first for Toronto, 165 miles distant; from Montreal, 333 miles.

Coburg, with a population of about 5,000, is 70 miles from Toronto, and 90 miles from Kingston. It has many and varied manufactories. A railway from Peterboro' (30 miles distant) comes in here. In the vicinage is the *Victoria College*, founded by Act of the Provincial Legislature in 1842.

Port Hope is seven miles above Coburg. From this point, or from Coburg, the journey to Kingston is often charmingly made *overland*, through a beautiful country at the head of the *Bay of Quinté*, a singular arm of the St. Lawrence.

Toronto.—HOTELS.

Toronto is the largest and most populous city in Canada West. Sixty years ago the site of the present busy mart was occupied by two Indian families only. In 1793, Governor Simcoe began the settlement under the name of York, changed, when it was incorporated in 1834, to Toronto—meaning, in the Indian tongue, "The place of meeting." The population, in 1817, numbered only 1,200; in 1850, it had reached 25,000; and now, it is, perhaps, 50,000, or upwards.

The Provincial Legislature meets at Toronto and Quebec, alternately, every four years—an arrangement made since the disturbances of 1849, which resulted in the burning of the Parliament Houses at Montreal.

Among the public buildings of Toronto, the traveller will perhaps please himself with a peep at the Catholic Church of St. Michael, the St. James' Cathedral (English), the University of

Toronto, the St. Lawrence Hall and Market, the Parliament House, Osgoode Hall, the Post Office, the Court-House, the Exchange, the Mechanics' Institute, Knox's Church, Trinity College, Upper Canada College, the Lunatic Asylum, the Jail, and the Normal and Model Schools. At Toronto, the traveller may if he pleases, reach Niagara direct, without touching at Hamilton, as we propose to do in our present journey.

Hamilton.—HOTELS.—*Anglo-American.*

Hamilton is among the most beautiful and most prosperous cities of Canada. It aspires, even to run a race with Toronto, one of the "2:40" nags of the province. Many advantages promise it a brave future. It is at the head of the western extremity of Lake Ontario, connected with the eastern capitals of the United States, and with Quebec, Montreal, and Toronto, by the Grand Trunk, and the Hamilton and Toronto Railways; and with Lake Huron and the Mississippi States, by the Great Western Railway, which traverses the garden lands of Canada; and, via the Suspension Bridge at Niagara, with the whole railway system of New York. The distance from Toronto to Hamilton, by the steamer, is 45 miles—time, two and a half hours; by railway, 38 miles—time (express), 1 hour 24 minutes. The population of Hamilton, in 1845, was 6,500; at this time it much exceeds 20,000.

From Hamilton to the Falls. Distance, by the Great Western Railway, from Hamilton to the Suspension Bridge, 43 miles—time, 1 hour, 35 minutes. Stations, Ontario, Grimsby, Beamsville, Jordan, St. Catharine's, Thorold, Niagara Falls.

St. Catharine's is the chief point of interest on this part of our route. Its pleasant topography, and, more particularly, its *mineral waters*, is making it a place of great summer resort. Here we leave the reader to establish himself at Niagara, and to see all its marvels, having elsewhere pointed out where he should go, and what should be his *itinéraire* while there. See *Niagara Falls* (New York.)

THE GREAT LAKES.

A delightful tour of a few weeks, may be made, in the heat of the summer, among the natural wonders of the region of the Great Lakes, to Mackinac, the Sault de St. Marie, and the shores of Lake Superior, returning, perhaps, by some one of the lower routes to the Atlantic, from the head waters of the Mississippi.

At Toronto, on Lake Ontario, which may be easily and speedily reached by routes which we shall hereafter travel—from New York, by the Hudson River and Lake Champlain to Montreal, and thence by the Grand Trunk Railway; or by the Central road from Albany to Buffalo, and by Niagara; or, by Niagara, via the New York and Erie railway; or, from Portland or Boston, by railroad to Montreal, &c. By steamboat daily, from Buffalo or from Chicago, &c., to Mackinac or *Mackinaw*, as the word is pronounced.

At Toronto, the traveller will take the Collingwood route, by the Ontario, Simcoe, and Huron Railway, 94 miles to Collingwood at the head of Georgian Bay or Manitoulin Lake, the north-east part of Lake Huron. Huron is the third in size of the five great inland seas, which pour their floods into the St. Lawrence. It lies between 43° and 46° 15' north latitude, having the State of Michigan on the south-south-west, and Canada West upon all other points, excepting where the Straits of Mackinac and the Falls, or Sault de Ste. Marie enter it from Lakes Michigan and Superior, and at its outlet in the St. Clair river. It is divided by the peninsula of Cabot's head, and the Manitouline Islands, the upper portions being the north channel and the Georgian Bay, which we reach at Collingwood. The length of Lake Huron, following its crescent shape, is about 280 miles, and its greatest breadth, not including the Georgian Bay, is 105 miles; its average width is 70 miles. Lake Huron is 352 feet above Lake Ontario, and 600 feet above the level of the sea. The depth is 1,000 feet—greater than that of any other in the grand chain of which it is a link. Off Saginaw, leads, it is said, have been dropped to a depth of 18,000 feet, which is 12,000 feet below the level of the Atlantic, and yet without finding bottom. The waters here are so pure and clear that objects may be distinctly seen from 50 to 100 feet below the surface. In these noble waters there are said to be more than 3,000 islands.

From Collingwood, the route is by suitable steamers to Mackinac, or the Straits of Mackinac, which are the connecting links between Lake Michigan and Lake Huron. The Island of Mackinac has a circumference of about nine miles, and its shores and vicinage are picturesque and romantic in the highest degree. The Arched Rock, facing the water, and rising to the elevation of some 200 feet, makes a bold and striking picture from all points on the lake, and especially as you look through its rude arches from the summit. Robinson's Folly is an attractive bluff on the north shore—years ago a Mr. Robinson, after whom the bluff is named, erected a summer-house upon its crest. Here he passed his days, and oftentimes his nights, despite the cautions of the people about him, until, in an unlucky tempest, he and his eyrie nest were swept away together.

The Cave of Skulls is upon the western

shore of the island. Once upon a time, it is said, a party of Sioux Indians were pursued hither by the Ottawas, who imprisoned and destroyed their foes in this cavern, by building fires at its mouth. The traveller, Henry, was one night secreted here, by a friendly Indian, when, to his surprise and horror, the morning light showed that he had been sleeping soundly upon a bed of human bones.

The Needles, another natural wonder of Mackinac, is a bold rock, in form not unlike a light-house. This elevation commands a panorama of the entire island, and a fine view of the crumbling and weed-covered ruins of Fort Holmes. Days of delight may be passed amidst the natural beauties of land and water at Mackinac, made doubly picturesque by the wild frontier life yet found here, and mingled, too, with the still existing homes and presence of the Red men.

Fort Mackinac stands upon a rocky height, 150 feet above the village, which it overlooks. An agency for Indian affairs is established here, which is, from time to time, the resort of deputations and bands of the wild dwellers of the surrounding wilderness. Immense quantities of fish are sent from Mackinac. Steamboats from Detroit, Chicago, and other places, stop here continually.

Sault Ste. Marie. Passing on towards Lake Superior, a voyage of eight pleasant hours, in a steamer, will bring us to the famous Falls of St. Mary, in the Strait of St. Mary, which connects the waters of Lake Superior and Lake Huron, and separates Canada West from the upper part of Michigan. The strait extends 63 miles from the south-east extremity of Lake Superior until it reaches Lake Huron. Its course is sometimes narrow, and broken into angry rapids; again, it widens into beautiful lakelets, and winds amid enchanting islands. It is navigable for vessels drawing eight feet of water, up to within a mile of Lake Superior, where the passage is interrupted by the great "Sault" or Falls. The Sault is a series of turbulent rapids, with a total descent of 22 feet in the course of three quarters of a mile. The passage of these falls, or the "running the rapids," as it is called, is most exhilarating sport.

The rapids are broken up into several different channels, and among them are scattered little islands, such as you see at Niagara, and, like them, bristling with cedars in all possible attitudes.

At this point, on the American side, is the little village of the Sault—an old settlement in the State of Michigan, founded by the Jesuits about two centuries ago. It has evidently seen and felt nothing of the great progress which has been building up cities and states. Here is to be seen the native owner of the soil and the half-breed (a cross of the French and Indian blood); and many other objects of interest.

These rapids are not unlike those of Niagara, excepting that, instead of ending upon the brink of a terrible precipice, they decline with the steady flow of a wide river; and steamers and canoes may fearlessly enter them. They run in different channels, everywhere dodging the numerous little cedar-covered islands in their way. The Sault yields abundant supplies of finny inhabitants; for the excellence of its white fish it is particularly renowned.

The village of the Sault on the Michigan shore, was founded by the Jesuits 200 years ago, but so little progress has it made, that the Aboriginal owner of the soil is still found in possession. Upon the British side of the river, there is an ancient-looking establishment, occupied as an agency of the Hudson's Bay Company.

The St. Mary's Ship Canal, a noble work, now overcomes the obstruction made by these rapids in the passage from Lake Huron to Lake Superior. Heretofore, merchandise from Chicago, Detroit, Buffalo, and other places, had to be discharged and conveyed over a railroad to the upper end of the Sault, and then hauled down to the waters at the opposite extremity; and the locks in this massive canal are, perhaps, the largest in the world.

The Chippewa Hotel is a good house on the American side of the rapids; and Pine's Hotel is a well-kept establishment on the British shore.

Steamers leave the Sault, daily, for all places on Lake Superior, and the neighboring waters.

Lake Superior. We enter Lake Superior after the passage of the *Sault Ste. Marie*, between two bold promontories, rising to the height of 200 to 300 feet, called Cape Gro and Cape Iroquois.

This grand inland sea is the largest body of fresh water on the globe. Its greatest length is 420 miles, its extreme breadth is 160 miles, and its circuit, 1,750 miles. On its west and north-west shore is Minnesota, on the southern border are Wisconsin and Michigan, while British America lies on all other sides. The waters, which are wonderfully transparent, come by more than two hundred streams, from a basin covering an area of 100,000 square miles. The north, south, and western parts are full of islands, while in the central portions of the lake there are few or none. In the north, these islands are many of them large enough to afford ample shelter for vessels. The picturesque regions of the lake are along the northern shore. In this direction the scenery is of a very bold and striking character. For many miles here there are continuous ranges of cliffs, which reach sometimes an elevation of 1,500 feet; on the south the banks are low and sandy, except where they are broken by occasional limestone ridges. These ridges rise near the eastern extremity, upon this side, 300 feet, in unique and surprising perpendicular walls and cliffs, broken into the oddest forms, indented with grotesque caverns, and jutting out into ghostly headlands. It is these strange formations which are famous under the name of the "Pictured Rocks." This range is on the east of Point Keweenaw. The rocks have been colored by continual mineral drippings. A similar rocky group lies to the west of the Apostle Islands. It is some hundred feet high, and is broken by numberless arches and caves of the most picturesque character. On the summit of these bluffs, there is everywhere a stunted growth of Alpine trees.

The Porcupine Mountains upon the southern shore of the lake, appear, says a voyager, to be about as extensive (though not so lofty) as the Catskils.

Of the islands of Lake Superior, the largest, which is some 40 miles in length, and from seven to ten broad, is called Royal Isle. Its hills rise to the altitude of 400 feet, with fine bold shores, on the north, and many fine bays on the south. It is, like all this region, a famous fishing-ground. Near the western extremity of the lake, there is a group known as the Apostle Islands. They form a trio of forest-covered heights, adding greatly to the beauty of the landscape around; on the extreme end of the largest, is the trading post called *La Pointe*, inhabited by Indians and white adventurers. It is a great place of annual rendezvous for the red man and the trader, and a starting point for tramps to the regions of the Mississippi.

The shores of Lake Superior have long been extensively explored for their abundant copper wealth; and mines have been opened at all points.

Fond du Lac is in Minnesota, on the Saint Louis river, 22 miles from its entrance into Lake Superior. It is accessible by steamboat; and its wonderfully wild and romantic hills, and rocks, and glens, are well worth a visit from the tourist of the Great Lakes.

We shall come back to this region, when we visit the head waters of the Mississippi, by-and-by.

NEW BRUNSWICK.

NEW BRUNSWICK, a Province of Great Britain, lies upon the eastern boundary of the State of Maine. The Landscape is of great variety and of most picturesque beauty; the whole Province (excepting the dozen miles lying directly on the sea) being broken into attractive valleys and hills, which northward assume a very marked, and sometimes a very rugged aspect. Much of its area of 230 miles in length, and 130 in breadth, is covered with magnificent forests, which, as in the neighboring State of Maine, constitute its chief source of industry and wealth.

The hills are nowhere of a very wonderful height, but they often rise in precipitous and sharp acclivities, which give them almost an Alpine aspect; all the more striking in contrast with the peaceful plains and vales which they protect from the tempests of the sea.

Like the neighboring Province of Nova Scotia, New Brunswick so abounds in lakes and rivers, that ready water access may be had with the help of a short portage, now and then, over its entire area. Thus a canoe may easily be floated from the interior to the Bay de Chaleaur, the Gulf of St. Lawrence, and the ocean on the north, or to the St. John River, and thence to the Bay of Fundy on the south.

The St. John River is the largest in New Brunswick, and one of the most remarkable and beautiful in America. It rises in the Highlands which separate Maine from Canada, not very far from the sources of the Connecticut. For 150 miles it flows in a north-east direction, to the junction of the St. Francis. From the mouth of the St. Francis, the course of the St. John is irregularly E. S. E. to the Grand Falls; at which point it makes a descent of from 70 to 80 feet, presenting a splendid picture for the gratification of the tourist. The leap of the Grand Falls passed, the river makes its way almost southward for some distance, after which it turns abruptly to the eastward, and so continues its way for 100 miles, passing Fredericton, to the outlet of the Grand Lake, in the southern central part of the Province. From Grand Lake its passage is in a wide channel, due south to Kingston, and thence south-west to

NEW BRUNSWICK.   39

St. John, at its mouth in St. John Harbor, on the Bay of Fundy.

The entire length of this beautiful river is about 600 miles, and from the Grand Falls to the sea, 225 miles, its course is within the British territory. The river and its affluents are thought to afford 1,800 miles of navigable waters. Very much of the shores of the St. John is wild forest land. In some parts, the banks rise in grand rocky hills, forming in their lines and interlacings, pictures of wonderful delight.

The chief tributaries of the St. John, besides the St. Francis and other waters already mentioned, are Aroostook, the Oromocto, and the Eel, on the west; and the Salmon, the Nashwaak, the Tobique, the Kennebecasis, and the Washedemoak, on the east.

The coast, and bays, and lakes, and rivers, of New Brunswick, abound with fish of almost every variety, and in immense supplies. The fisheries of the Bay of Fundy are of great value, and employ vast numbers of the population. In the harbor of St. John alone, there have been, at one time, two hundred boats, with five hundred men taking salmon, shad, and other fish. Nearly six hundred fishermen have been seen at one period at the Island of Grand Manan; while at the West Isles, about seven hundred men have been thus employed at one moment; and so on, at many other of the countless fishing grounds and stations of the New Brunswick and the Nova Scotia coasts.

The climate here is healthful, but subject to great extremes of heat and cold; the mercury rising sometimes to 100° in the day time, and falling to 50° at night.

INTERNAL COMMUNICATION. Besides the steamers and stages which connect the various towns and cities of New Brunswick and Nova Scotia, lines of railway are in active progress, which will unite the two Provinces, and both to the Canadas and the States.' A portion of the European and North American Railway is now open (Aug. 1, 1860) from St. John to Shediac, 108 miles; from whence steamers connect with Charlottetown, P. E. Island, Pictou, N. S., the northern ports of New Brunswick and Quebec, in Canada. This line will open up new and pleasant ground to the tourist. Another road is to extend from St. Andrews to Woodstock, and thence to Quebec. The magnetic telegraph already connects New Brunswick, Nova Scotia, and Prince Edward's Island with the States. The connection between Nova Scotia and Prince Edward Island is by a submarine cable, nine miles from Cape Tormentine to Cape Traverse.

St. John.—HOTELS :—*Waverley House.*

Routes.—From Boston, Mass., every Monday and Thursday, at 9 A. M., by steamer. From Halifax via Windsor, N. S., 45 miles by rail, and thence by steamer, 110 miles, to St. John.

St. John, at the mouth of the St. John River, is the principal city of New Brunswick, with a population of over 30,000. It is superbly situated upon a bold, rocky peninsula, and is seen very imposingly from the sea. The scenery of the St. John River is very striking in the passage immediately preceding its entrance into the harbor, and a mile and a half above the city. It makes its impetuous way here in a chain of grand rapids, through rugged gaps, 240 feet wide, and 1,200 feet long. This passage is navigable only during the very brief time of high and equal tides in the harbor and the river; for at low water the river is about 12 feet higher than the harbor, while at high water, the harbor is five feet above the river. It is thus, only, when the waters of the harbor and of the river are on a level, that vessels can pass; and this occurs only during a space of from fifteen to twenty minutes, at each ebb and flow of the tide. Immense quantities of timber are rafted down from the forests of the river above, to St. John. It is the entrepôt also of the agricultural and mineral products of a wide region of country.

Fredericton.—HOTELS :—*Barker House.*

Routes.—From Boston via St. John.

Fredericton, the capital of New Brunswick, stands upon a flat tongue of land, in a bend of the St. John River, 80 miles from its mouth. This sandy plain is about three miles long, sometimes reaching a breadth of half a mile. The river, which is navigable up to this point, is here three-quarters of a mile wide. Small steamers ascend 60 miles yet above to Woodstock, and sometimes to the foot of the Great Falls.

The view both up and down the valley is most interesting—to the north and uncleared range of highlands, with detached cones and broken hills thrown out in bold relief upon the landscape. Villas enclosed in the woods, and farms upon the clearings, are the chief objects it presents; while to the South the river is seen widening, like a silver cord, through the dark woodlands, until it disappears among the islands in the distance.

St. Andrews, with a population of about 8,000, is at the north-east extremity of Passamaquoddy Bay, three miles from the shores of the United States, near Eastport, in Maine, and 60 miles from St. John. A railway will connect St. Andrews with Woodstock, 80 miles distant, and will be continued to Canada.

NOVA SCOTIA.

Nova Scotia, the ancient Acadia, including the Island of Cape Breton and Sable Island, lies south-east of New Brunswick, from which it is separated by the Bay of Fundy, except only at the narrow Isthmus of Chignecto. It may be reached at Halifax, its capital, by the British steamers from New York and Boston. The railways now in progress within its limits will soon more conveniently unite it to the cities of the Canadas and the United States. The area of the Province is 18,746 square miles, including the 3,000 of Cape Breton, and the 69 of Sable Island. The southern shores are often very rugged. The interior is diversified with hills and valleys, though not of very bold character, as the highest land is but 1,200 feet above the sea. The numerous lakes cover much of the southern part of the Province. The agricultural capabilities vary much for the area of the country. On the Atlantic Coast much of the soil is rocky and barren. The richest soils are in that section of the country bordering upon the Bay of Fundy and the Gulf of St. Lawrence, and the streams emptying into them; and, generally speaking, this is the most thickly settled region. Nova Scotia has become so much denuded of its valuable timber, that its lumber trade is now neither very large nor productive, compared with that of New Brunswick or Canada. Farming, however, especially in the finer agricultural districts just named, is extensively carried on, and is very remunerative.

The extensive mineral deposits of Coal, Iron, and Copper, have become, of late years, an object of great and constantly increasing attention to the inhabitants; and Gypsum, Grindstones, and Building Stone of various kinds, have long been important articles of export. Upon the Atlantic coast, too, many people are occupied in the extensive fishing trade, which has been prosecuted here more actively than upon any of the British American shores, excepting only that of Newfoundland.

The Coast of Nova Scotia.—The greatest length of Nova Scotia is 356 miles, and the greatest breadth 120 miles. The south-east coast, in a distance of 110 miles only from Cape Canso to Halifax, has no less than 12 ports capacious enough to receive ships of the line, and 14 deep enough for merchantmen. A belt of rugged broken land, of which the greatest height is 500 feet, formed of granite and primary rock, extends along all the Atlantic shore, from Cape Canso to Cape Sable. This belt varies in breadth from 10 to 50 miles, and covers about one-third of the whole Province. From Briar's Island, off Digby Neck, 130 miles to Capes Split and Blomidon, along the northern coast on the Bay of Fundy, there is a ridge of wooded frowning precipices of trap rock, which overhang the waves at an elevation of from 100 to 600 feet. These magnificent cliffs are picturesque and grand in the extreme. They are, too, (which is something in this utilitarian age,) not only ornamental, but useful, for they serve to protect the interior from the terrible fogs of the bay.

The Rivers and Lakes and Bays of Nova Scotia.—The lakes here, though generally small, are almost countless in number, covering the southern portions of the peninsula as with a net-work of smiling waters. In some instances, no less than a hundred are grouped within a space of twenty square miles. Lake Rosignol, the largest of the region, is 30 miles long. It is near the western end of the peninsula. Grand Lake comes next, then College Lake eastward. Minas Bay on the north coast, the eastern arm of the Bay of Fundy, penetrating 60 miles inland, is very remarkable for the tremendous tides which rush in here, sometimes to the height of 60 to 70 feet, while they do not reach more than from 6 to 9 feet in the harbor of Halifax, directly opposite; these are the spring-tides. They form what is called the *bore*. The bays of St. Marys, the Gut of Canso, Townsend Bay, George Bay, and Chedabucto Bay, in the eastern part of the Province, and St. Margarets and Mahone Bays on the south, are all large and most interesting waters.

The Annapolis River flows into the Bay of Fundy, 100 miles from the Garden of Acadia. Besides this principal river there are many others navigable for a greater or less distance from their mouths, as the Shubenacadie, which, by the help of a canal, connects Cobequid Bay, from the Bay of Fundy on the north side of the peninsula, with Halifax Harbor on the south; the Tusket and the Clyde in the south-west extremity of the Province, the Mersey, the Musquodoboit, and the St. Marys. Indeed, rivers pour their waters into all the many bays and harbors which so thickly stud the whole line of these remarkable coasts.

Halifax.—HOTELS :—

ROUTES. From New York direct, by the British Mail Steamers. From St. John, N. B., by steamer, 110 miles to Windsor, thence by rail, 45 miles to Halifax.

Halifax, the capital of Nova Scotia, is upon the south coast of the peninsula, on the declivity of a hill, about 250 feet high, rising from one of the finest harbors on the continent. The streets are generally broad, and for the most part macadamized. Viewed from the water, or from the opposite shore, the city is prepossessing and animated. In front, the town is lined with wharves, which, from the number of vessels constantly loading and discharging, always exhibit a spectacle of great commercial activity. Warehouses rise over the wharves, or tower aloft in different parts of the town, and dwelling-houses and public buildings roar their heads over each other, as they stretch along and up the sides of the hill. The spires of the different churches, the building above the town in which the town-clock is fixed, a rotunda-built church, the signal-posts on Citadel Hill, the different batteries, the variety of style in which the houses are built (some of which are painted white, some blue, and some red); rows of trees showing themselves in different parts of the town; the ships moored opposite the dockyard, with the establishments and tall shears of the latter; the merchant vessels under sail, at anchor, or along the wharves; the wooded and rocky scenery of the background, with the islands and the small town of Dartmouth on the east shore,—are all objects most agreeable to see.

Of the public buildings, the chief is a handsome edifice of stone, called the Province Building, 140 feet long by 70 broad, and ornamented with a collonade of the Ionic order. It comprises suitable chambers for the accommodation of the Council and Legislative Assembly, and also for various Government offices. The Government House, in the southern part of the town, is a solid, but gloomy-looking structure, near which is the residence of the military commandant. The Admiral's residence, on the north side of the town, is a plain building of stone. The north and south barracks are capable of accommodating three regiments. The Wellington Barracks (in the northern part of the town), which comprises two long ranges of substantial stone and brick buildings, is the most extensive and costly establishment of the kind in North America. There is also a Military Hospital, erected by the late Duke of Kent. Dalhousie College is a handsome edifice of freestone. Among the churches of various denominations are several of the English establishment, and of the Presbyterian order, and two of the Roman Catholic faith. The Court House is a spacious freestone structure, in the southern part of the town. In the suburbs is a new Hospital. The banking establishments are four in number. The hotels and boarding-houses are not of the highest order. The inhabitants of Halifax are intelligent and social, and travellers will remark a tone of society here more decidedly English than in most of the other colonial cities.

The harbor opposite the town is more than a mile wide, and has, at medium tides, a depth of 12 fathoms. About a mile above the upper end of the town it narrows to one-fourth of a mile, and then expands into Bedford Basin, which has a surface of ten square miles, and is completely land-locked. On an island opposite the town are some strong mounted batteries. The harbor is also defended by some other minor fortifications. The Citadel occupies the summit of the heights commanding the town, and is a mile in circumference. It is a costly work, and, after that of Quebec, is the strongest fortress in the British North American Colonies.

Halifax, ever since its settlement in 1749, has been the seat of a profitable fishery. Its trade, which is in a very prosperous condition, is principally with the West Indies and other British colonies, with the United States, and the mother country. It is also the chief rendezvous and naval depot for the British navy on the North American station. The British Government having made Halifax one of the stopping-places of the Cunard line of steamers, in their trips either way across the Atlantic, has added greatly to its importance as a maritime city, as well as advanced its commercial prosperity.

THE UNITED STATES.

The grand territory of the United States, through which we propose to travel in our present volume, occupies no meaner area than that of 2,936,166 square miles, scarcely less than that of the continent of Europe. In form, it is nearly a parallelogram, with an average length of 2,400 miles, from east to west, and a mean breadth, from north to south, of 1,300 miles. Its extreme length and breadth are, respectively, 2,700 and 1,600 miles; reaching from the Atlantic on the east, to the Pacific on the west; from British America on the north, to the Gulf of Mexico and the Mexican Republic on the south. Its present division is into thirty-four States and nine Territories, besides the District of Columbia. The States have been popularly grouped into 4 classes, according to their geographical position; as the Eastern Group, or "New England," embracing Maine, New Hampshire, Vermont, Massachusetts, Rhode Island, and Connecticut; the "Middle" group, of New York, New Jersey, Pennsylvania, Delaware, and Maryland; the "Southern States," Virginia, North Carolina, South Carolina, Georgia, Florida, Alabama, Mississippi, Louisiana, and Texas; and the "Western States," of Tennessee, Kentucky, Ohio, Indiana, Illinois, Michigan, Iowa, Wisconsin, Missouri, Arkansas, California, Oregon, Minnesota, and Kansas.

All the Territories—Washington, New Mexico, Utah, Nebraska, Colorado, Dakotah, Nevada, Idaho, and Arizona, are included in this division of the country.

POPULATION OF THE UNITED STATES.

The District of Columbia, (D. C.,†) 1860.......................... 75,076

THE EASTERN OR NEW ENGLAND STATES.

Connecticut, (Conn.,) 1860............	460,151	New Hampshire, (N. H.,) 1860..........	326,072
Rhode Island, (R. I.,) 1860............	174,621	Vermont, (Vt.,).......................	315,116
Massachusetts, (Mass.,) 1860...........	1,231,065	Maine, (Me.,) 1860....................	628,276

THE MIDDLE STATES.

New York, (N. Y.,) 1860................	3,887,542	Pennsylvania, (Pa.,) 1860...............	2,906,370
New Jersey, (N. J.,) 1860...............	672,031	*Delaware, (Del.,) 1860...............	112,218

THE SOUTHERN STATES.

*Maryland, (Md.,) 1860................	687,034	*Florida, (Fla.,) 1860..................	140,439
*Virginia, (Va.,) 1860.................	1,596,083	*Alabama, (Ala.,) 1860.................	964,296
*North Carolina, (N. C.,) 1860.........	992,667	*Louisiana, (La.,) 1860................	709,433
*South Carolina, (S. C.,) 1860..........	703,812	*Texas, (Tex.,) 1860...................	601,039
*Georgia, (Ga.,) 1860..................	1,057,327	*Mississippi, (Miss.,) 1860.............	791,395

* Slave States.   † Abbreviations used in the address of letters, etc.

THE WESTERN STATES.

*Arkansas, (Ark.,) 1860	435,427	Michigan, (Mich.,) 1860	749,112
*Tennessee, (Tenn.,) 1860	1,109,847	Wisconsin, (Wis.,) 1860	775,873
*Kentucky, (Ky.,) 1860	1,155,713	Iowa, (Io.,) 1860	674,948
Ohio, (O.,) 1860	2,339,599	*Missouri, (Mo.,) 1860	1,173,317
Indiana, (Ia.,) 1860	1,350,479	Oregon, (Or.,) 1860	52,464
Minnesota, (Min.,) 1860	162,022	California, (Cal.,) 1860	380,015
Illinois, (Ill.,) 1860	1,711,753	Kansas, (Ka.,) 1860	107,110

TERRITORIES.

New Mexico, (N. M. Ty.,)	93,541	Dakotah, 1860	4,839
Washington, (Was. Ty.,) 1860	11,578	Nevada, 1860	6,857
Utah, (Ut. Ty.,)	40,295	Idaho	
Nebraska, (Na. Ty.,) 1860	28,842	Arizona	
Colorado 1860	34,197		
Total			31,429,891

MAINE.

MAINE is the extreme eastern portion of New England, and the border State of the Union in that direction, with the British provinces of New Brunswick on the north and north-east, and Canada on the north-west. It has three distinct topographical aspects—in the comparatively level, and somewhat sandy and marshy character of the southern portion, lying back 20 miles from the Atlantic coasts—in the pleasant hill and valley features of the interior, and in the rugged, mountainous, and wilderness regions of the north.

A great portion of the State is yet covered by dense forests, the utilization of which is the chief occupation and support of its inhabitants. The most fertile lands lie in the central southern regions, between the Penobscot river on the east, and the Kennebec on the west, and in the valley borders of other waters. The mountain ranges are often very bold and imposing—one summit, that of Katahdin, having an elevation of 5,385 feet above the level of the sea. The lakes are very numerous, sometimes of great extent, and often very beautiful, all over the State; and more especially among the mountains in the north. Indeed, it is estimated, that one-tenth part of the whole area of Maine is covered by water. The rivers are numerous and large, and present everywhere scenes of great and varied beauty. The Atlantic coast, which occupies the whole southern line of the State, is the finest in the Union, in its remarkably bold, rocky character, and in its beautiful harbors, bays, islands, and beaches. The sea-islands of Maine are over 400 in number; and many of them are very large, and covered by fertile and inhabited lands. The climate, though marked by extremes, both of heat and cold, is yet everywhere most healthful; and its rigor is much modified by the proximity of the ocean.

The Mountains and Lakes. The most interesting route for the tourist here, is perhaps a journey through the hills, lakes, and forests of the north; but we warn him, beforehand, that it will not be one of ease. Rugged roads and scant physical comforts will not be his most severe trial: for, in many places, he will not find road or inn at all, but must trudge along painfully on foot, or by rude skiff over the lakes, and trust to his rifle and his rod to supply his larder. In these wildest regions the exploration may be made with great satisfaction by a party well provided with all needed tent equipage, and with all the paraphernalia of the chase; for deer, and the moose, and the wild fowl are abundant in the woods; and the finest fish may be freely taken in the water. Still he may traverse most of the mountain lands and lakes by the roads and paths of the lumbermen, who have invaded all the region; and he may bivouac as comfortably as should content an orthodox forester, in these humble shanties.

MAINE.

The mountains of Maine are broken and distinct peaks. A range, which seems to be an irregular continuation of the White Hills of New Hampshire, extends along the western side of the State for many miles; and, verging towards the north-east, terminates in Mars Hill. This chain divides the waters which flow north into the St. John's River from those which pass southward to the Atlantic. Many beautiful lakes lie within this territory.

Mount Katahdin, with its peaks 5,385 feet above the sea, is the loftiest summit in the State, and is the *ultima thule*, at present, of general travel in this direction. The ordinary access is in stages from Bangor over the Aroostook Road, starting in tolerable coaches on a tolerable road, and changing always, in both, from bad to worse. A pleasant route *for the adventurer* is down the West Branch of the Penobscot, in a canoe, from Moosehead Lake. Guides and birches, as the boats are called, may be procured at the foot of Moosehead, or at the Kinco House, near the centre of the lake. By this approach Katahdin is seen in much finer outlines than from the eastward.

Sugar Loaf Mountain, upon the Sebools River, north-east of Mount Katahdin, is nearly 2,000 feet high, and from its summit a magnificent view is commanded, which embraces some fifty mountain peaks, and nearly a score of wonderful lakes. Then there are Bigelow, Saddleback, Squaw, Bald, Gilead, the Speckled Mountain, the Blue Mountain, and other heights more or less noble, amidst which are brooks, and lakelets, and waterfalls of most romantic character.

Moosehead Lake, the largest in Maine, is among the northern hills. It is 35 miles long, and, at one point, is 10 miles in breadth, though near the centre there is a pass not over a mile across. Its waters are deep, and furnish ample occupation to the angler, in their stores of trout and other fish. This lake may be traversed in the steamboats employed in towing lumber to the Kennebec. A summer hotel occupies a very picturesque site upon the shore. The Kinco House, midway, is the usual stopping-place. There are numerous islands on the Moosehead Lake, some of which are of great interest. On the west side, Mount Kinco overhangs the water, at an elevation of 600 feet. Its summit reveals a picture of forest beauty well worth the climbing to see. Moosehead is 15 miles north of the village of Monson, and 60 north-west of Bangor. The roads thither, lying through forest land, are necessarily somewhat rough and lonely. This lake is the source of the great Kennebec river, by whose channels its waters reach the sea.

Lake Umbagog lies partly in Maine and yet more in New Hampshire. Its length is about 12 miles, and its breadth varies from 1 to 5 miles. The outlet of Umbagog and the Margallaway river form the Androscoggin.

Androscoggin and Moosetocknoguntic Lakes are in the vicinity of Umbagog.

Sebago Pond, a beautiful lake 12 miles long and from 7 to 8 miles broad, is about 20 miles from Portland, on a route thence to Conway and the White Mountains. It is connected with Portland by a canal.

The Penobscot, the largest and most beautiful of the rivers of Maine, may be reached daily from Boston and Portland, by steamer, as far up as Bangor, and also by railway from Portland to Bangor. It is formed by two branches, the east and the west, which unite near the centre of the State, and flow in a general south-west course to Bangor, 60 miles from the sea, and at the head of navigation. Large vessels can ascend to Bangor, and small steamboats navigate the river yet above. At Bangor the tide rises to the great height of 17 feet, an elevation which is supposed to be produced by the wedge-shaped form of the bay, and by the current from the Gulf Stream. The length of the Penobscot, from the junction of the east and west branch, is 135 miles, or measuring from the source of the west branch, it is 300 miles; though, as far as the tourist is concerned, it is only 60 miles—being that portion between Bangor and the ocean. This part, then, the Penobscot proper, ranks, in its pictorial attractions, among the finest river scenery of the United States. In all its course there are continual points of great beauty, and very often the shores rise in striking and even grand lines and proportions. We have met tourists who have been hardly less impressed with the landscape of this fine river than with that of the Hudson even, though we do not admit such a comparison.

Bangor.—HOTELS :—*The Bangor House; the Hatch House.*

Bangor, at the head of tide water and of navigation on the Penobscot River, 60 miles from its mouth, is one of the largest cities of Maine, having a population of more than 17,000. Steamboats connect it daily with Portland and Boston; and it is reached also by the Androscoggin and Kennebec, and Penobscot and Kennebec railways, via Waterville, on the Kennebec. The distance from Bangor to Portland, by railway, is 138 miles. Bangor is connected with Old Town (12 miles) by railway, and another road is contemplated to Lincoln, 50 miles up the Penobscot valley. The Bangor Theological Seminary, founded 1816, occupies a fine site in the higher portion of the city. The "specialty" of Bangor is lumber, of which it is the greatest depot in the world. All the vast country above, drained by the Penobscot and its affluents, is covered with dense forests of pine, and hemlock, and spruce, and cedar, from which immense quantities of lumber are continually cut and sent from the marvellous saw-mills, down the river to market at Bangor. During the eight or nine months of the year through which the navigation of the river is open, some 2,000 vessels are employed in the transportation of this freight. The whole industry of Bangor is not, however, in the lumber line, as she is also engaged in ship-building, and has a large coasting trade, and a considerable foreign commerce.

Belfast and Castine are some 30 miles below Bangor, where the Penobscot enters its name-sake bay. Belfast on the west, and Castine on the east shore, are nine miles apart. They are both small ship-building and fishing towns.

The Kennebec River is in the western part of the State, extending from Mooschead Lake, 150 miles to the sea. It makes a descent in its passage of a thousand feet, thus affording a great and valuable water-power. The scenery of the Kennebec, though pleasant, is far less striking than that of the Penobscot. Its shores are thickly lined with towns and villages, among which is Augusta, the capital of the State.

Augusta.—HOTELS :—*The Stanley House.*

Augusta is at the head of sloop navigation on the Kennebec, 43 miles from its mouth. It is 60 miles N.N.E. of Portland by railway, and 69 S.W. of Bangor. Steamboats run hence to Portland and Boston, calling at the river landings. The city is chiefly upon the right bank of the river, which is crossed here by a bridge 520 feet long; and a quarter of a mile above, by a railroad bridge, 900 feet in length. The private residences, and some of the hotels, are upon a terrace, a short distance west of the river, while the business parts of the town lie along shore. The State House is an elegant structure of white granite. Its site, in the southern part of the city, is lofty and very picturesque ; in front is a large and well-cared-for park. The United States Arsenal, surrounded with extensive and elegant grounds, is upon the east side of the river. Here, too, is the Hospital for the Insane, built upon a commanding and most beautiful eminence. The principal hotels here are the Stanley House, the Augusta House, and the Mansion House. Augusta is upon the railway route from Portland to Bangor. Population, 8,000.

Hallowell is a pretty village, two miles below Augusta, on the river, and on the line of the Kennebec and Portland Railway.

Gardiner on the Kennebec, at the mouth of the Cobbesseeontee River. This point is the head of ship navigation on the Kennebec. The city is seven miles below Augusta, and 53 miles from Portland by the Kennebec and Portland R. R.

Waterville is on the Kennebec, at the Ticonic Falls, and at the northern terminus of the Androscoggin and Kennebec Railway, connecting with the Kennebec and Penobscot line. It is the seat of Waterville College, a prosperous establishment, controlled by the Baptists.

Bath.—HOTELS :— *The Sagadahock House.*

Bath, a flourishing city of over 8,000 people, is on the Kennebec, 12 miles from the sea ; 30 miles south of Augusta ; and 36 north-east of Portland. It is the terminus of a branch road from Brunswick, on the Kennebec and Portland Railway ; it is to be united at Lewiston with the Androscoggin and Kennebec route from Portland to Bangor.

The Androscoggin River is a fine stream, flowing from Lake Umbagog, partly in New Hampshire, but chiefly through the south-western corner of Maine, into the Kennebec, 20 miles from the ocean.

Brunswick, on the Androscoggin, is 27 miles from Portland by railway. It is the seat of Bowdoin College, which is beautifully located on a high terrace, near the edge of the village. This popular institution was founded in 1802. The Medical School of Maine, which is connected with Bowdoin College, has a very valuable library, and anatomical cabinet. The Androscoggin here falls 50 feet within the reach of half a mile.

Mount Desert Island. A summer trip to Mount Desert Island has of late years been a pleasant treat to American landscape painters, and a visit thither might be equally grateful to the general tourist. The vigorous and varied rock-bound coast of New England can be nowhere seen to greater advantage. Mount Desert Island is an out-of-the-way nook of beauty in Frenchman's Bay, east of the mouth of the Penobscot River. It is 40 miles from Bangor, and may be reached from Boston by boat, via Rockville, and thence by another steamer, on to Bucksport (on the Penobscot), and thence by stage via Ellsworth, or from Castine on the Penobscot Bay, hard by. If the visitor here cannot sketch the bold, rocky cliffs, he can beguile the fish to his heart's content.

Eastport, upon the waters of Passamaquoddy Bay, at the extreme eastern point of the territory of the United States is well-deserving of a visit from the tourist in quest of the beautiful in nature; for more charming scenes on land and on sea, than are here, may rarely be found.

The traveller may see Eastport and its vicinage and then go home, if he pleases; for it is the jumping-off place—the veritable Lands-End—the latitude and longitude beyond which the stars and stripes give place to the red cross of England.

Eastport is 234 miles N.E. of Portland, and is reached thence and from Boston by regular steamboat communication. Steamboats run also to Calais and places *en route*, 30 miles above, at the head of navigation on the St. Croix River. The town is charmingly built on Moose Island, and is connected to the mainland of Perry by a bridge; and by ferries with Pembroke, Lubec, and the adjoining British Islands. It is not a very ponderous place, the population of the township scarcely exceeding 5,000. Fort Sullivan is its shield and buckler against any possible foes from without.

The *Passamaquoddy Bay* extends inland some 15 miles, and is, perhaps, 10 miles in breadth. Its shores are wonderfully irregular and picturesque, and the many islands which stud its deep waters, help much in the composition of pictures to be enjoyed and remembered.

TO PORTLAND, MAINE.

From Boston, 107 miles by the Eastern Railway, via Lynn, Salem, Newburyport, Portsmouth, N. H., &c.; or by the Boston and Maine route, 111 miles through Reading, Lawrence, Andover, Haverhill, Exeter, Dover, &c.; or by steamer daily.

From Montreal, by the Grand Trunk Railway.

Portland.—Hotels :—*The Preble House; the United States; the American; the Elms.*

Portland, the commercial metropolis of Maine, is handsomely situated on a peninsula, occupying the ridge and side of a high point of land, in the S. W. extremity of Casco Bay, and on approaching it from the ocean, is seen to great advantage. The harbor is one of the best on the Atlantic coast, the anchorage being protected on every side by land, whilst the water is deep, and communication with the ocean direct and convenient. It is defended by Forts Preble, Scammel, and Gorges. On the highest point of the peninsula is an observatory 70 feet in height, commanding a fine view of the city, harbor, and islands in the bay. The misty forms of the White Mountains, 60 miles distant, are discernible in clear weather.

This city is elegantly built, and the streets are beautifully shaded and embellished with trees; and so profusely, that there are said to be here no less than 3,000 of these rural delights. Congress street, the main highway, follows the ridge of the peninsula through its entire extent. Among the public buildings of Portland, the City Hall, the Court House, and some of the churches, are worthy of particular attention. The Society of Natural History possesses a fine cabinet, containing specimens of the ornithology of the State, more than 4,000 species of shells, and a rich collection of mineralogical and geological examples, and of fishes and reptiles. The Athenæum has a library of 10,000 volumes, and the Mercantile library possesses, also, many valuable books. The Portland Sacred Music Society is an interesting Association here.

The long line of the Grand Trunk Railway connects the city with Montreal and Quebec, and thence with all the region of the St. Lawrence and the Saguenay Rivers. Two lines of railway unite it with Boston and the western cities, and with the interior of Maine, at Augusta on the Kennebec, and at Bangor, on the Penobscot.

The **Grand Trunk Railway, Route from Portland, North.** This great thoroughfare connects the navigable waters of Portland harbor with the great commercial capital of Canada. Its route passes through a fertile and productive country, generally under fine cultivation, the streams in its vicinity abounding in water privileges of the first importance. From Portland it passes onward to the valley of Royal's River, and follows up the valley of the Little Androscoggin. It strikes and crosses that river at Mechanic Falls, 43 miles from Portland, at which place the Buckfield Branch Railroad connects with it. Pursuing its course upward, it passes in the vicinity of the "Mills" on its way to Paris Cape, in the neighborhood of Norway and Paris, drawing in upon it the travel and business of that rich and populous region. Still following up the valley of the Little Androscoggin, passing on the way two important falls, it reaches Bryant's Pond, the source of that river. This point is 15 miles from Rumford Falls, on the Great Androscoggin, one of the greatest and most available water-powers in the State. Passing hence into the valley of Alder stream, the route strikes the Great Androscoggin, near Bethel, a distance of 70 miles from Portland. Crossing that stream, it follows up its picturesque and romantic valley, bordered by the highest mountains in New England, till, in its course of about 20 miles from Bethel, it reaches Gorham in New Hampshire, distant from the base of Mount Washington a few miles only. From this point that celebrated shrine may be approached and ascended with more ease, in a shorter distance, and less time, than from any other accessible quarter in the vicinity of the White Hills.

MAINE.—MASSACHUSETTS. 47

(See routes to White Mountains.) This point also is only five miles distant from Berlin Falls, the greatest waterfall in New England, where the waters of the Great Androscoggin, larger in volume than the waters of the Connecticut, descend nearly 200 feet in a distance of about two miles. From the valley of the Androscoggin the road passes into the valley of the Connecticut, reaching the banks of that river at North Stratford in New Hampshire. Following up this rich and highly productive valley about 35 miles, the road reaches the parallel of 45° N. Lat. ; at the boundary between the United States and Canada; continuing thence to Quebec, and up the St. Lawrence via Montreal, to Toronto on Lake Ontario, where it connects with other routes for Lake Superior and all parts of the great West.

Lewiston is a flourishing manufacturing village, containing about 7,000 inhabitants situated upon the Androscoggin River between Portland and Waterville on the Androscoggin and Kennebec Railway, 33 miles north of Portland. The waterfall here is one of exceeding beauty. The entire volume of the Androscoggin is precipitated some 50 feet over a broken ledge, forming in its fall a splendid specimen of natural scenery. The river immediately below the fall, subsides into almost a uniform tranquillity, and moves slowly and gracefully along its course, in strange though pleasing contrast with its wild and turbid appearance at and above the cataract. The Androscoggin and Kennebec road communicates with the Grand Trunk Railway at Danville, six miles below Lewiston, and with the Androscoggin road at Leeds, 11 miles above.

MASSACHUSETTS.

THE landscape here is of changeful character, and often strikingly beautiful, embracing not a few of the most famous scenes in the Union. In the south-eastern part of the State the surface is flat and sandy, though the sea coast is, in many places, very bold, and charmingly varied with fine pictures of rocky bluff and cliff. It abounds in admirable summer-houses, where the lovers of sea-breezes and bathing may find every means and appliance for comfort and pleasure.

In the eastern and central regions, the physical aspect of the country, though agreeably diversified, is eclipsed in attraction by the lavish art-adornments of crowding city and village, and happy homesteads, nowhere so abundant and so interesting as here.

The Green Mountains traverse the western portion of Massachusetts in two ridges, lying some 25 miles apart, with picturesque valley lands between. Here are the favorite summer resorts of Berkshire, and other parts of the Housatonic region. Saddle Mountain, 3,505 feet, is a spur of the most western of the two ridges we have mentioned, known as the Taconic or the Taughkannic hills. Mount Washington, another fine peak of this line, has an altitude of 2,624 feet. It rises in the extreme southern corner of the State, while Saddle Mountain stands as an outpost in the north-west angle. The more eastern of the two hill-ranges here is called the Hoosic Ridge. Noble isolated mountain peaks overlook the winding waters and valleys of the Connecticut—some of them though not of remarkable altitude, commanding scenes of wondrous interest, as Mount Holyoke and Mount Tom, near Northampton. North of the middle of the State is the Wachusett Mountain, with an elevation of 2,018 feet.

On Hudson's Brook, in Adams, there is found a remarkable natural bridge, 50 feet high, spanning a limestone ravine 500 feet in length. In New Marlborough, the tourist will see a singular rock poised with such marvellous art that a finger can move it; and on Farmington River, in Sandisfield, he will delight himself with the precipices, 300 feet high, known as the Hanging Mountain.

Massachusetts has some valuable mineral springs, though none of them are places of general resort. In Hopkinton, mineral waters impregnated with carbonic acid and carbonates of iron and lime; in Winchendon, a chalybeate spring, and one in Shutesbury, containing muriate of lime. But we need not make further mention of these points of interest here, as we shall have occasion to visit them all, under the head of one or other of the group of New England States, as we follow the net-work of routes by which they may be reached.

MASSACHUSETTS.

While the most thoughtless traveller will thus find, in the physical aspect of Massachusetts, ample sources of pleasure, the more earnest will not fail to draw yet higher delight from the strongly-marked *morale* of the country. Though small in area, compared with some other States of the confederacy, it is yet, in all the qualities which make national fame, one of the greatest of them all. Nowhere are there records of historical incident of higher sequence; nowhere a more advanced social position, or a greater intellectual attainment; nowhere a nobler spirit of commercial enterprise; nowhere a more inventive genius, a more indomitable industry.

In Massachusetts, more than in any other section of the Union, the dullest perception will be impressed with the evidences of all the highest and best characteristics of the American mind and heart, those wise and persistent qualities, which the Pilgrim Fathers first planted upon the shore at Plymouth, where the history of the State began, with the landing of the May Flower, on the memorable 22d of December, 1620: the same righteous and unyielding nature which commenced the struggle for the national independence, in 1775, at Lexington and Bunker Hill.

RAILWAYS.—In a State so crowded with active and prosperous cities and homes as Massachusetts, and among a people of such wonderful will and energy and ambition, there is, of course, no lack of railway communication—that great modern test of national enterprise; and so Massachusetts, while excelling all her sister States in every phase of industrial and mechanical achievement, has built (her area and population considered) more miles of railway than either of them. The iron tracks cover all the land, uniting all parts of the State to all others, and to every section of the Republic. We forbear to catalogue these routes at this point, as we shall follow them all, closely, in our visits to the several sections of the State.

ROUTES TO BOSTON FROM NEW YORK.

ROUTE 1. *Railway*—From the depot in Fourth av. cor. 27th st., via New Haven, Hartford, Springfield, and Worcester, 236 miles, twice a day; or by the new *Shore Line*, via New Haven, New London, Stonington, and Providence. A pleasant and very speedy route to the latter city.

ROUTE 2. *Stonington*—By steamer, daily, from pier No. 18 North River, to Groton, Ct.; thence by railway, via Providence, R. I.

ROUTE 3. *Fall River*—Steamer, daily, at 5 P. M., from pier No. 3. North River, via Newport, R. I.; thence by railway.

ROUTE 4. *Norwich Line*—Steamer, daily, from pier No. 39, North River, to New London, Conn.; thence by railway, *via* Norwich, Ct., and Worcester, Mass.

The most expeditious way between New York and Boston is that which we have marked No. 1, Railroad Routes—generally called the New Haven Line and the Shore Line. The time on these lines is between 8 and 9 hours, leaving one city in the morning and reaching the other in the afternoon, or leaving in the afternoon and arriving before midnight. All the other routes, by steamboat and railway, occupy the night, starting about 5 P. M., and arriving by dawn next day.

The New Haven route (No. 1), is upon the N. Y. and N. H. road for 76 miles, to New Haven all the distance along the south line of the State of Connecticut, near the shore of the Long Island Sound. To Williams Bridge, 13 miles from New York, the track is the same as that of the Harlem R. R. to Albany. Leaving William's Bridge, we pass the pretty suburban villages of New Rochelle, Mamaroneck, Rye, and Port Chester, and reach Stamford, 36 miles from New York. (The *Shore Line* leaves the other route at New Haven and extends through New London, Stonington, and Providence.)

Norwalk.—HOTELS :—*Allis House.*

Norwalk (New Haven Line—44 m.) is a pleasant village upon Norwalk river. The Norwalk and Danbury railroad, 24 miles, comes in at this point. The quiet, rural beauties of Norwalk make it one of the most available of the summer homes of Connecticut; particularly as it is scarcely beyond suburban reach of New York.

Bridgeport, Conn.—HOTELS :—*The Stanley House.*

Bridgeport, 58 miles from New York, is the southern terminus of the Housatonic R. R., which traverses the valley of the Housatonic 110 miles to Pittsfield, Mass. This route is through the most picturesque portions of Connecticut and Massachusetts—the western or mountain regions. (See *Valley of the Housatonic.*) The Naugatuck R. R. extends hence via Waterbury to Winsted. Steamers ply between New York and Bridgeport. Bridgeport is upon an arm of the Long Island Sound, at the mouth of the Pequannock River. A terrace height of 50

MASSACHUSETTS.

feet, occupied by beautiful private mansions and cottages, commands a charming view of the town and the Sound.

New Haven.—HOTELS:—*The Tontine.* An excellent house.

New Haven, 76 miles from New York, is one of the most beautiful and most interesting places in New England. It is known as the city of Elms in a land of Elms, from the extraordinary number of beautiful trees of this species by which the streets are so gratefully shaded and so charmingly embellished. New Haven is a semi-capital of Connecticut. It is famous as the seat of Yale College, which has sent out more graduates than any other Institution in America. The buildings of the college, which occupy nearly a square, are among the chief attractions of the city; especially the apartments devoted to the Fine Arts, and occupied by the large collection of the works of the eminent painter Trumbull. The *American Journal of Science and Art*, edited by Professors Silliman and Dana, and other literary periodicals, are published here. Steamboats connect the city with New York. The New Haven and Northampton or Canal R. R. extends 76 miles to Northampton, and the New Haven and New London R. R. 50 miles to New London.

Hartford.—HOTELS:—*The United States,* etc.

Hartford. Leaving New Haven our route turns northward from the Sound, over the New Haven, Hartford, and Springfield R. R. Hartford, a semi-capital of Connecticut, is 36 miles from New Haven and 112 from New York, and 124 from Boston. It is upon the right bank of the Connecticut River, navigable to this point by sloops and small steamboats, 50 miles up from Long Island Sound. Among the literary and educational institutions of Hartford are Trinity College, the Wadsworth Athenæum, the Connecticut Historical Society. Among its chief benevolent establishments, for which it is famous, are the American Asylum for the Deaf and Dumb, and the Retreat for the Insane. That old historic relic, the Charter Oak, held in so much reverence, stood in Hartford until 1856, when it was prostrated by a violent storm. Mrs. L. H. Sigourney, the poetess, resides here. The population of Hartford is 29,000.

Springfield, Mass.—HOTELS:—*Massasoit House.* A popular establishment.

Springfield is upon the Connecticut River, 26 miles north of Hartford, 98 miles from Boston, and 138 from New York. The U. S. Arsenal, located here, is the largest in the Union. It is charmingly perched upon Arsenal Hill, looking down upon the beautiful town, the river, and the fruitful valleys. This noble panorama is seen with still better effect from the cupola which crowns one of the arsenal buildings. This establishment employs nearly 300 hands, and 175,000 stands of arms are kept constantly on hand. This is a famous gathering point of railroads. The Connecticut valley routes start hence, and furnish one of the pleasantest ways from New York to the White Mountains, through Northampton, Brattleboro', Bellows Falls, to Wells River and Littleton, N. H. (See "Valley of the Connecticut" and White Mountain routes, No. 10.) The Western railway from Albany to Boston passes through Springfield also, and continues our present route to Worcester. Population about 20,000.

Worcester.—HOTELS:—*The Lincoln House; the Bay State House.*

Worcester is a flourishing city of 25,000 people, 45 miles from Boston, in the centre of one of the most productive agricultural regions of Massachusetts. It is noted for its schools and manufactures. Quite a net-work of railways connects the city with all parts of the country. The Western road, direct from Boston to Albany; the Worcester and Nashua, communicating through other routes with the St. Lawrence River; the Worcester and Providence; the Norwich and Worcester, and the Boston and Worcester, which we now follow to the end of our present journey.

The Stonington Route, (No. 2)—This route, as well as Nos. 3 and 4, by Fall River and by Norwich, takes us from New York by steamboat around the Battery and Castle Garden, along the whole eastern line of the city, and by the cities of Brooklyn and Williamsburg, up the beautiful East River by the suburban villages on the Long Island shore, by Blackwell's, Ward's, and Randall's Islands (covered by the public asylums and prisons), through the famous passage of Hell Gate, and up Long Island Sound—a gallery of admirable pictures, seen as they are from the Boston boats, in the declining evening light.

Stonington.—HOTELS:—*The Wadawannuck House.*

Stonington, Conn., is on the Stonington and Providence route from New York to Boston, and was the terminus of its water travel on the Sound from New York, until the late transfer of the steamers to the more convenient port of Groton, opposite New London, on the Thames.

The Fall River Route. (No. 3.)—By steamer

on Long Island Sound, round Point Judith, and up Narragansett Bay to Newport, R. I. (see Newport), and thence by rail to Fall River. From Fall River 51 miles to Boston, by Old Colony and Fall River R. R.

Fall River.—HOTELS :—*Richardson House.*

Fall River is a thriving town of nearly 12,000 inhabitants and very extensive manufactures. It is at the entrance of Taunton River into Mount Hope Bay, an arm of the Narragansett. The historic eminence of Mount Hope, the home of King Philip, is admirably seen across the bay. Steamboats connect Newport with Providence indirectly by this route via Narragansett Bay.

The Norwich Route, (No. 4.)—This line is also by steamboat from New York, via Long Island Sound to the mouth of the Thames River, which it ascends to New London, and passengers there take cars and follow the course of the Thames through Connecticut, directly north to Worcester; thence with other lines to Boston.

BOSTON AND VICINITY.

Hotels. The most fashionable are the Tremont House, on Tremont street; the Revere House, on Bowdoin square; the Winthrop House, Tremont street; the American House, Hanover street; the Adams; the United States, &c.

Boston is one of the most interesting of the great American cities, not only from its position as second in commercial rank to New York alone, but from its thrilling traditionary and historical associations, from the earliest days of discovery and colonization on the western continent; and through all the trials and triumphs of the childhood, youth, and manhood of the Republic—from its dauntless public enterprise, and from its high social culture and morals; from its great educational and literary facilities; from its numerous and admirable benevolent establishments; from its elegant public and private architecture, and from the surpassing natural beauty of all its suburban landscape.

Boston is divided into three sections, Boston Proper, East, and South Boston. The old city is built upon a peninsula of some 700 acres, very uneven in surface, and rising at three different points into eminences, one of which is 138 feet above the sea. The Indian name of this peninsula was Shawmut, meaning "Living Fountains." It was called by the earlier inhabitants Tremont or Trimountain, its sobriquet at the present day. The name of Boston was bestowed on it in honor of the Rev. John Cotton, who came hither from Boston in England. The first white inhabitant of this peninsula, now covered by Boston Proper, was the Rev. John Blackstone. Here he lived all alone until John Winthrop—afterwards the first Governor of Massachusetts—came across the river from Charlestown, where he had dwelt with some fellow-emigrants for a short time. About 1635 Mr. Blackstone sold his claim to the now populous peninsula for £30, and removed to Rhode Island. The first church was built in 1632; the first wharf in 1673. Four years later a postmaster was appointed, and in 1704 (April 17), the first newspaper, called the "Boston News-Letter," was published.

A narrow isthmus, which is now called the Neck, joins the peninsula of Old Boston to the main land on the south, where is now the suburb of Roxbury.

Many bridges, most of them free, link Cambridge, Charlestown, Chelsea, and South Boston with the Peninsula. These structures are among the peculiarities of the place, in their fashion, their number, and their length. The first one which was built was that over Charles River to Charlestown, 1,503 feet long. The Old Cambridge Bridge, across the Charles River to Cambridge, is 2,758 feet in length, with a causeway of 3,432 feet. The South Boston Bridge, which leads from the Neck to South Boston, is 1,550 feet long. The Canal Bridge between Boston and East Cambridge, is 2,796 feet, and from E. Cambridge another bridge extends 1,820 feet, to Prison Point, Charlestown. Boston Free Bridge to South Boston is 500 feet; and Warren Bridge to Charlestown is 1,390 feet. Besides these bridges, a causeway of a mile and a half extends from the foot of Beacon street to Sewell's Point, in Brookline. This causeway is built across the bay upon a substantial dam. Other roads lead into Boston over especial bridges, connecting the city with the main as closely as if it were a part thereof. Thus the topography of Boston is quite anomalous as a *mountain* city in the sea!

South Boston extends some two miles along the south side of the harbor, from Old Boston to Fort Independence. Near the centre, and some two miles from the State House, are *Dorchester Heights,* the memorable battle-ground where, in the Revolution, the enemy were driven from Boston. A fine view of the city, of the vicinity, and the sea, may be obtained from these Heights. Here, too, is a large reservoir of the Boston water works.

East Boston (the "Island Ward") is in the western part of Noddle's Island. It was the homestead of Samuel Maverick, while John Blackstone was sole monarch of the peninsula, 1630. Here is the wharf, 1,000 feet long, of the Cunard steamers. E. Boston is connected by two ferries with the city proper. It is the terminus

MASSACHUSETTS.

of the Grand Junction R. R. Chelsea is near by.

The streets of Boston, which grew up according to circumstances, are many of them very intricate, and troublesome to unravel, a difficulty which is being gradually obviated in a degree. The fashionable promenades and shopping avenues are Washington, Tremont, Summer, and Winter streets.

Boston Common is a large and charming public park in the old city, and is, very justly, the pride of the people and the admiration of strangers. It contains nearly 50 acres, of every variety of surface, up-hill and down, and around, all covered with inviting walks, grassy lawns, and grand old trees. A delicious pond and fountain occupy a central point in the grounds, and around them are many of the old mansions of the place—led, on the upper hill, by the massive, dome-surmounted walls of the State Capitol. The Common drops from Beacon street, the southern declivity of Beacon Hill, by a gentle descent to Charles River.

Faneuil Hall. This famous edifice, called the "Cradle of Liberty," is in Dock square. It is about 109 years old, and is an object of deep interest to Americans. Here the fathers of the Revolution met to harangue the people on the events of that stirring period; and often since that time the great men of the State and nation have made its walls resound with their eloquence. It was presented to the city by Peter Faneuil, a distinguished merchant, who, on the 4th of July, 1740, made an offer, in a town-meeting, to build a market-house. There being, at that time, none in the town, it was, as a matter of course, accepted. The building was begun the following year, and finished in 1742. The donor so far exceeded his promise, as to erect a spacious and beautiful Town Hall over it, and several other convenient rooms. In commemoration of his generosity, the town, by a special vote, conferred his name upon the Hall; and, as a further testimony of respect, it was voted that Mr. Faneuil's full-length portrait be drawn at the expense of the town, and placed in the Hall. This, with other pictures, can be seen by visitors.

The State House is on the summit of Beacon Hill, and fronting the "Common." Its foundation is 110 feet above the level of the sea. Length 173 feet, breadth 61. The edifice was completed in 1798, at a cost of $133,330, about three years having been occupied in its construction. On the entrance floor is to be seen Chantrey's statue of Washington. Near by is the staircase leading to the dome, where visitors are required to register their names, and from the top of which is obtained a fine view of the city, the bay, with its islands, and the country around.

The Exchange, on State street, was completed in the fall of 1842. It is 70 feet high and 250 feet deep, covering about 13,000 feet of ground. The front is built of Quincy granite, with four pilasters, each 45 feet high, and weighing 55 tons each. The roof is of wrought iron, and covered with galvanized sheet iron; and all the principal staircases are fire-proof, being constructed of stone and iron. The centre of the basement story is occupied by the Post Office. The great central hall, a magnificent room, is 58 by 80 feet, having 18 very beautiful columns in imitation of Sienna marble, with Corinthian capitals, and a sky-light of colored glass, finished in the most ornamental manner. This room is used for the merchants' exchange and subscribers' reading-room.

The Custom House is located at the foot of State street, between the heads of Long and Central wharves. It is in the form of a cross; the extreme length being 140 feet, breadth 95 feet. The longest arms of the cross are 75 feet wide, and the shortest 67 feet, the opposite fronts and ends being alike. The entire height to the top of the dome is 90 feet.

The Court House, a fine building in Court square, fronting on Court street, is built of Quincy granite.

The City Hall is near the Court House, and fronting on School street, with an open yard in front. Here, in September, 1856, a colossal bronze statue of Benjamin Franklin, who was a native of Boston, was erected, with great public parade and rejoicing. This fine work was modelled by R. B. Greenough, Esq., a brother of the distinguished sculptor, Horatio Greenough.

The Massachusetts Hospital covers an area of four acres on Charles River, between Allen and Bridge streets. Near by, at the foot of Bridge street, is the Mass. Medical College. The Boston Music Hall fronts on Winter street and Bumstead place.

The Boston Athenæum occupies an imposing edifice on Beacon near Tremont street. It was incorporated in 1807, and is one of the best endowed literary establishments in the world. There are in the library 50,000 volumes, and an extensive collection of manuscripts. The Athenæum possesses a fine gallery of paintings, in connection with which the annual displays of the Boston artists are made.

The Massachusetts Historical Society, organized in 1794, possesses 12,000 volumes, and many valuable manuscripts, coins, charts, maps, &c. The Boston Public Library is on Boylston street. It possesses, at this time, about 50,000 volumes.

The American Academy of Arts and Sciences, one of the oldest societies of the kind in the country, has 15,000 volumes. It occupies an apartment in the Athenæum. Besides these libraries, Boston has many others, as, the State Library, the Bowditch, the Social Law Library, &c.

The Lowell Institute provides for regular courses of free lectures upon natural and revealed religion, and many scientific and art topics. We may mention, also, among literary, scientific, and art societies of the city, the Lyceum, the Natural History, the American Oriental, the American Statistical, the Musical Educational, and the Handel and Haydn Societies, and the Boston Academy of Music.

Harvard University. This venerable seat of learning is at Cambridge, three miles from the city of Boston. It was founded in 1638, by the Rev. John Harvard. The University embraces, besides its collegiate department, law, medical, and theological schools. The buildings are 15 in number, all located in Cambridge, except that of the medical school in North Grove street, in Boston.

The *Old Washington Headquarters*, at Cambridge, known as the Cragie House, where the poet Henry W. Longfellow has resided for many years, is near the Harvard University. It was at Cambridge that the painter, Washington Allston, lived and died.

Boston, always so much distinguished for its literary character, as to have won the name of the Athens of America, has, besides its innumerable libraries and institutions of learning, more than a hundred periodical publications and newspapers, dealing with all themes of study and all shades of opinion and inquiry.

The churches of the city are numerous, as might be expected of the home of the Puritans. They are more than 100 in number—the Unitarians having the largest share. Many of the churches are very costly and imposing edifices. The oldest is Christ Church, built in 1723, and the next is the famous "Old South," erected in 1730. This is a building of great historical interest.

Theatres. The Boston Academy of Music, 361 Washington street; the Howard Athenæum, Howard street; the Museum, Tremont street. The *New Public Garden* is a very beautiful resort.

Bunker Hill Monument, commemorative of the eventful battle fought on the spot, is in Charlestown. The top of this structure commands a magnificent view, embracing a wide extent of land and water scenery. The journey up is somewhat tedious, traversing nearly 300 steps—yet this is forgotten in the charming scene and delightful air. On the hill is a stone marking the spot where Warren fell. Horse cars run from the head of Tremont st. to the monument. Near at hand is the United States Navy Yard, containing, among other things, a ropewalk—the longest in the country.

Mount Auburn Cemetery, about a mile from Harvard University, and about four from Boston, by the road from Old Cambridge to Watertown. It is the most beautiful of American rural burying-places, embellished by landscape, and horticultural art and taste, and by a most picturesque chapel, and many elegant and costly monuments. Its walks and lanes, and lawns make it the most delightful of all the resorts in the vicinage of the city. Cars run from the station in Bowdoin square every 15 minutes, during the day, and until half past eleven o'clock at night. *Mr. Cushing's Garden*, a place of great beauty, is a short distance beyond Mount Auburn, in Watertown. Tickets may be obtained gratis on application at the Horticultural Store in School street. *Fresh Pond*, another charming place of resort, is about four miles from Boston, and about half a mile from Mount Auburn. The other fine sheets of water in the vicinity of Boston, well worthy the attention of visitors, are *Horn*, *Spot*, *Spy*, and *Mystic* Ponds.

WATERING AND OTHER PLACES IN THE VICINITY OF BOSTON.

Nahant.—HOTELS:—*The Nahant House*, an elegant and spacious edifice.

Nahant, a delightful watering-place, is situated about 12 miles from Boston, by water, and 14 by land. During the summer season, a steamboat plies daily. *Fare, 25 cents.* This is a most agreeable excursion, affording an opportunity, in passing through the harbor, for seeing some of the many beautiful islands with which it is studded. Nahant may also be reached by taking the Eastern Railway cars as far as Lynn, and thence walking or riding a distance of three miles, along the hard sandy beach, in full view of the open sea; or by omnibuses, which intersect the railroad cars, at Lynn, several times daily.

The peninsula is divided into Great and Little Nahant, and Bass Neck. The former is the largest division, containing 300 acres—a part of which is under cultivation—many handsome dwellings, and a spacious hotel, with a piazza on each floor. From this place the visitor has a boundless sea-coast view.

On the south side of Great Nahant is the dark cave or grotto, called the *Swallow's Cave*, 10 feet wide, 5 feet high, and 70 long, increasing in a short distance, to 14 feet in breadth, and 18 or 20 in height. On the north shore of the peninsula is a chasm 20 or 30 feet in depth, called the *Spouting Horn*, into which, at about half tide, the water rushes with great violence and noise, forcing a jet of water through an aperture in the rock to a considerable height in the air.

Philip's Beach, a short distance north-east of Nahant, is another beautiful beach, and a noted resort for persons in search of pleasure or health.

Nantasket Beach, 12 miles from Boston, is situated on the east side of the peninsula of Nantasket, which forms the south-east side of Boston Harbor. The beach, which is remarkable for its great beauty, is four miles in length, and celebrated for its fine shell-fish, sea-fowl, and good bathing.

Chelsea Beach, about three miles in length, is situated in the town of Chelsea, and is another fine place of resort. A ride along this beach on a warm day is delightful. It is about five miles from Boston, and may be reached through Charlestown over Chelsea Bridge.

Lynn.—HOTELS :—*Sagamore House.*

Lynn is nine miles from Boston, on the Eastern Railroad. It is charmingly situated on the north-east shore of Massachusetts Bay, in the vicinity of Nahant, and is a famous place for the manufacture of ladies' shoes. This business, here, employs 150 establishments and 10,000 hands, half of whom are females. It is estimated that 4,500,000 pairs of ladies' and misses' shoes are made here every year, amounting, in value, to $3,500,000. Besides the product of the city, another half a million pairs are made in the neighborhood.

Salem is a beautiful city, 16 miles from Boston, by the Eastern Railroad. It extends about two miles along and three quarters of a mile across the peninsula formed by the north and the south rivers. It is distinguished for its literary institutions, and for its commercial enterprise. Next to Plymouth, it is the oldest town in New England. Salem was the chief scene of the "witchcraft" madness in 1692. Upon Gallows Hill—a fine eminence overlooking the city—19 persons of the town and the neighborhood were executed for this supposed crime.

Salem is also distinguished for its services in the war of the Revolution. *Marblehead* is 5 miles from Salem, by a branch road.

Beverly is upon an arm of Ann Harbor, two miles from Salem, with which it is connected by a bridge of 1,500 feet, and from Boston 18 miles, via Eastern Railroad.

Wenham, 22 miles from Boston, via Eastern Railway. Wenham Pond, a beautiful sheet of water, about a mile square, affords abundance of excellent fish, and is much visited by persons fond of angling. It is also noted for the quality and quantity of its ice, a large amount of which is yearly exported.

Newburyport, Mass.—HOTELS :— *The Merrimac House.*

Newburyport, Mass., 36 miles from Boston, via Eastern Railroad, lies on a gentle acclivity, on the south bank of the Merrimac River, near its union with the Atlantic. It is considered one of the most beautiful towns in New England. In consequence of a sand-bar at the mouth of the harbor, its foreign commerce has greatly declined. The celebrated George Whitefield died in this town in September, 1770.

Salisbury Beach, celebrated for its beauty and salubrity, is much visited during the warm season. It is from four to five miles distant from Newburyport.

Hampton, 43 miles from Boston, via Eastern Railroad, is pleasantly situated near the Atlantic coast. From elevations in the vicinity there are fine views of the ocean, the Isle of Shoals, and of the sea-coast from Cape Ann to Portsmouth. Hampton Beach has become a favorite place of resort for parties of pleasure, invalids, and those seeking an invigorating air. Great Boar's Head, in this town, is an abrupt eminence extending into the sea, and dividing the beaches on either side. There is here a hotel for the accommodation of visitors. The fishing a short distance from the shore is very good.

The Isle of Shoals is distant about nine miles from Hampton and from Portsmouth. These shoals are seven in number. Hog Island, the largest, contains 350 acres, mostly rocky and barren. Its greatest elevation is 59 feet above high-water mark. Upon this island is a hotel, recently erected. These rocky isles are a pleasant resort for water parties; and the bracing air, while refreshing to the sedentary, cannot be otherwise than salutary to invalids.—*Rye Beach* is another noted watering-place on this coast, much frequented by persons from the neighboring towns.

Portsmouth, N. H.—HOTELS :— *Rockingham House.*

Portsmouth, N. H., 56 miles from Boston, and 51 from Portland, Maine, by the Eastern Railroad, the principal town of the State, and the only seaport, is built on the south side of the Piscataqua River. Its situation is a fine one, being on a peninsula near the mouth of the river. It is connected by bridges with Kittery in Maine, and Newcastle on Grand Island, at the mouth of the river. The harbor is safe and deep, and is never frozen, its strong tides preventing the formation of ice. There is here a United States Navy Yard, one of the safest and most convenient on the coast. The North America, the first line-of-battle ship launched in this hemisphere, was built here during the Revolution.

Andover, the seat of Phillips' Academy and of the Andover Theological Seminary, is 23 miles from Boston, on the Boston and Maine Railway.

Lowell.—HOTELS :— *Washington* and *Merrimac.*

Lowell. This famous manufacturing city, the first in the Union, is upon the Merrimac, 26 miles from Boston, by the Boston and Lowell Railroad. Lowell was incorporated as a town in 1826, and in 1854 its population was about 37,000. There are over 50 mills in operation in Lowell, employing a capital of $13,900,000, and nearly 13,000 hands, of whom about 9,000 are females.

Concord is situated on the river of the same name, 20 miles from Boston, by the Fitchburg Railroad. It is celebrated as the place where the first effectual resistance was made, and the first British blood shed, in the Revolutionary war. On the 19th of April, 1775, a party of British troops was ordered by General Gage to proceed to this place to destroy some military stores, which had been deposited here by the province. The troops were met at the north bridge by the people of Concord and the neighboring towns, and forcibly repulsed. A handsome granite monument, erected in 1836, commemorates the heroic and patriotic achievement.

Lexington.—HOTELS :—

Lexington, the scene of the memorable battle of Lexington, at the commencement of the Revolution, April 19, 1775. From Boston by Fitchburg and branch railways through West Cambridge—a fine ride of eleven miles; fare, 35 cents.

Brighton is five miles from Boston, on the Boston and Worcester Railroad. This is a beautiful suburban town, on the south side of Charles River. It is also a noted cattle market.

Quincy is eight miles from Boston, by the Old Colony and Fall River Road. Famous for its granite quarries, and as the birth-place of John Hancock, Josiah Quincy, Jr.; Presidents John and John Quincy Adams, and other eminent men. The fine estate of the Quincy family is here.

Plymouth.—HOTELS :—*The Samoset House.*

Plymouth is 37 miles from Boston, by the Old Colony and Fall River and branch roads. It is a spot of especial interest, as the landing-place of the Pilgrim Fathers, and as the oldest town in New England. The immortal Plymouth Rock lies at the head of Hedge's Wharf. It is now much reduced from its ancient proportions, being only 6½ feet across its greatest breadth, and but 4 feet thick. The surface only is visible above the ground. The landing of the Pilgrims from the Mayflower occurred on the 22d December, 1620.

Marshfield, interesting as the home of Webster, is 28 miles south-east of Boston, by the Old Colony and the South Shore Railways.

Hingham is 17 miles from Boston, by the Old Colony and the South Shore Railways; or may be reached by a pleasant sail down the bay.

Cohassett, three miles from Hingham (South Shore Road), is a popular sea-side resort.

Taunton, Mass. — HOTELS : — *Taunton House.*

Taunton, Mass., is a beautiful town of some 15,000 inhabitants, situated at the head of navigation, on the Taunton River. It may be reached from Boston, 35 miles; and from Providence, 30 miles, by the Boston and Providence Railroad, taking the New Bedford and Taunton Road at Mansfield, about midway between Boston and Providence.

New Bedford.—HOTELS :—

New Bedford is a charming maritime city, of some 22,000 people; situated on an estuary of Buzzard's Bay. It is famous for its whale fisheries, in which enterprise it employs between 300 and 400 ships. New Bedford is the terminus of the New Bedford and Taunton Railway, by which route, via Mansfield, on the Boston and Providence Road, it may be reached from those two cities. It is accessible also from New York and Boston, by the Fall River route. Distance from Boston, 55 miles.

Martha's Vineyard, and Nantucket. These famous sea-islands lie off New Bedford, with which port they are in daily steamboat communication. Nantucket may be still more easily reached via Cape Cod Railway to Hyannis.

Middleborough.—HOTELS :—

Middleborough is a prosperous town on the Fall River route, between New York and Boston, at its point of junction with the Cape Cod Railroad. It is pleasantly situated upon the Taunton River, 40 miles from Boston. It is the seat of a very popular Scholastic Institution, under the direction of the Rev. Mr. Jenks.

Cape Cod and the Sea Islands. Those who delight in the sea breezes, in salt water bathing, and fishing, and in the physical beauties and wonders of the ocean changes, will find ample gratification everywhere upon the Atlantic borders, and especially upon the bold islanded coast of New England.

Besides the well-known haunts of the Long Island and the Jersey shores, of Newport, and of the numerous suburban resorts of Boston, to which we have elsewhere alluded, the Isle of Shoals, off Portsmouth, Martha's Vineyard, and Nantucket, off New Bedford, &c., we commend the summer wanderer to a tour through the towns and villages, and along the coasts of that very secluded portion of Massachusetts— Cape Cod. Let him journey from "Plymouth Rock," the inner point, to Provincetown, the outer verge, and he will find novelties in both physical nature and social life, which will be most agreeable. The Cape Cod Railway from Boston will take him far along the Cape, 74 miles to Barnstable; 76 to Yarmouth; and 80 to Hyannis; or he may go thence by steamboat, and afterwards continue from point to point by stage.

THE CONNECTICUT VALLEY, RIVER, AND RAILWAYS.

The beautiful valleys watered by the Connecticut, are among the most inviting portions of the New England landscapes; whether for rapid transit, or for protracted stay. The whole region is speedily and pleasantly accessible from every point, and may be traversed *en route* to most of the principal summer resorts of New England, since many important and very attractive towns and villages lie within its area and since it is crossed, and recrossed, everywhere, by the intricate railway system which unites Boston so intimately, not only with all the Eastern States, but by connection infinite, with the whole country.

The Connecticut, the Queen of New England rivers—the chiefest and most beautiful— rises in the hills of New Hampshire and Vermont, near the Canada borders; and flowing nearly southward, for 400 miles, separates the two States of its mountain birth; traverse the entire breadth of Massachusetts and Connecticut, to the Long Island Sound. Its waters are swelled by the tribute of the Passumpsic, the White, the Deerfield, the Westfield, the Ammanoosuc, and other rivers. It is navigable for sloops 50 miles up to Hartford, and with the help of numerous canals, very much farther. The Connecticut Valley is perhaps 300 miles long in a straight line, with a mean width of 40 miles. The soil is as fertile as the landscape is beautiful.

Railroads from New York, Boston, Albany, and other places, meet at Springfield, the southern threshold of the most picturesque part of the Connecticut; so we will commence our tour here—referring the traveller to the route from New York to Boston, for mention of Springfield itself.

Northampton.—HOTELS:—

The Mansion House, an elegant establishment, upon the upper edge of the village; Warner's Hotel, in the business street; and the Nonotuck House at the Railway station.

Northampton, Mass., is 17 miles above Springfield, on the line of railway which follows the Connecticut, up to the diverging lines for the White Hills of New Hampshire, and for Vermont and Canada. It is in every way one of the most charming villages in New England, and none other is more sought for summer residences. It lies about a mile west of the Connecticut, surrounded by rich alluvial meadows, sweeping out in broad expanse, from the base of grand mountain ridges. The village is not too large for country pleasures, the population of the township falling within 6,000; yet, its natural advantages are so great, and so many pleasant people have established themselve here in such pleasant and beautiful places, and the hotels are so admirable, that the tourist, the most *difficile*, will not miss either the social or the physical enjoyments of his city home. Even the little business part of Northampton has a cosy, rural air, and all around are charming villas, nestled on green lawns, and among fragrant flowers. Among the specialties of Northampton, are several distinguished water-cure establishments, the chief of which is that known as Round Hill, a large and beautiful place, upon the fine eminence after which it is named, just west of the village. The schools here have always been in very high repute. Its chief academy is the Collegiate Institute. The vicinage of Northampton is, perhaps, the most beautiful portion of the Connecticut Valley, the most fertile in its interval land, and the most striking in its mountain scenes; for it looks out directly upon the crags and crests of those famous hills— Mount Holyoke and Mount Tom.

Northampton is united to New Haven, by the New Haven and Northampton Railroad, 76 miles long, as well as via Springfield.

Mount Holyoke is directly across the river from Northampton; a good carriage road winds to the summit, 1,120 feet above the sea, where there is a little inn and an observatory. There are not of its kind many scenes in the world more beautiful than that which the visitor to Mount Holyoke looks down upon: the varied features of the picture—fruitful valleys, smiling villages and farms, winding waters; and, afar off, on every side, blue mountain peaks innumerable, will hold him long in happy contemplation.

"Mount Holyoke," says Mr. Eden's Handbook to the region, "is a part of a ridge of greenstone, commencing with West Rock near New Haven, and proceeding northerly across the whole of Connecticut; but its elevation is small until it reaches Easthampton, when it suddenly mounts up to the height of nearly 1,000 feet, and forms Mount Tom. The ridge crosses the Connecticut, in a north-east direction, and curving still more to the east, terminates 10 miles from the river, in the north-west part of Belchertown. All that part of the ridge east of the river is called Holyoke, though the Prospect House is erected near its south-western extremity, opposite Northampton, and near the Connecticut. This is by far the most commanding spot on the mountain, though several distinct summits, that have as yet received no uniform name, afford delightful prospects.

Mount Tom, upon the opposite side of the river, is not yet so much visited as are its neighboring cliffs of Holyoke, though it is considerably higher, and the panorama from its crest is no less broad and beautiful.

The village of Easthampton is situated on the west side of Mount Tom, four miles from Northampton. It contains a very extensive button manufactory, well deserving of a visit from those who can appreciate mechanical ingenuity. The principal feature of the place, however, is its noble seminary for the youth of both sexes, which was founded and liberally endowed by the Hon. Samuel Williston, at an expense of $55,000, and has been in successful operation 15 years, having now an average attendance of about 200 pupils.

On the east side of Mount Tom and on the river is the village of South Hadley, famous as the seat of the Mount Holyoke Female Seminary, founded and for many years conducted by Miss Mary Lyon. This institution has sent out hundreds of graduates, as teachers, into all parts of the land. South Hadley has many spots which afford most agreeable prospects. Standing on the elevated bank of the river and facing the north-west, you look directly up the Connecticut, where it passes between Holyoke and Tom; those mountains rising with precipitous boldness, on either side of the valley through the opening, the river is seen for two or three miles, enlivened by one or two lovely islands, while over the rich meadows, that constitute the banks, are scattered trees, through which, half hidden, appears in the distance the village of Northampton, its more conspicuous edifices being only visible.

The village of Hadley is connected with Northampton by a bridge over the Connecticut. The river immediately above the town, leaving its general course, turns north-west; then, after winding to the south again, turns directly east; and thus having wandered five miles, encloses, except on the east, a beautiful interval, containing between two and three thousand acres. On the isthmus of this peninsula lies the principal street —the handsomest, by nature, in New England. It is a mile in length, running directly north and south; is sixteen rods in breadth; is nearly a perfect level; is covered during the fine season with a rich verdure; abuts at both ends on the river, and yields everywhere a delightful prospect.

In this town resided for fifteen or sixteen years Whalley and Goff, two of those who composed the court for the trial of King Charles the First, and who signed the warrant for his execution. They came to Hadley in 1664. When the house which they occupied was pulled down, the bones of Whalley were found buried just without the cellar wall, in a kind of tomb formed of mason work, and covered with flags of hewn stone. After Whalley's death, Goff left Hadley and went, it was thought, to New York, and finally to Rhode Island, where he spent the rest of his life with a son of his deceased *confrere*.

Amherst, the seat of the famous College, is built upon an eminence, four miles east of Hadley. The College Observatory and especially its rich cabinet, should receive due consideration from the visitor here.

The Sugar-Loaf Mountain comes now into view, as we journey on up the valley. This conical peak of red sandstone rises almost perpendicularly five hundred feet above the plain, on the bank of the Connecticut, in the south part of Deerfield. As the traveller approaches this hill from the south, it seems as if its summit were inaccessible. But it can be attained without difficulty on foot, and affords a delightful view on almost every side. The Connecticut and the peaceful village of Sunderland on its bank, appear so near, that one imagines he might almost reach them by a single leap. This mountain overlooks a spot which was the scene of the most sanguinary conflicts that occurred during the early settlement of this region. A little south of the mountain the Indians were defeated in 1675 by Captains Lathrop and Beers; and one mile north-west, where the village of Bloody Brook now stands (which derived its name from the circumstance), in the same year, Captain Lathrop was drawn into an ambuscade, with a company of "eighty young men, the very flower of Essex County," who were nearly all destroyed.

The spot where Captain Lathrop and about thirty of his men were interred, is marked by a stone slab; and a marble monument, about twenty feet high and six feet square, is erected near by.

Deerfield Mountain rises some 700 feet above the plain on which the village stands. From the western verge of this summit the view is exceedingly interesting.

The alluvial plain on which Deerfield stands is sunk nearly 100 feet below the general level of the Connecticut valley; and at the south-west part of this basin, Deerfield River is seen emerging from the mountains, and winding in the most graceful curves along its whole western border. Still more beneath the eye is the village, remarkable for regularity, and for the number and size of the trees along the principal street. The meadows, a little beyond, are one of the most verdant and fertile spots in New England. Upon the whole, this view is one of the most perfect rural peace and happiness that can be imagined.

A few miles north of Deerfield, and in the same valley, but on higher ground, can be seen the lovely village of Greenfield.

Mount Toby lies in the north part of Sunderland, and west part of Leverett, and is separated from Sugar-Loaf and Deerfield Mountains by the Connecticut river. On various parts of the mountain, interesting views may be obtained, but at the southern extremity of the highest ridge, there is a finer view of the valley of the Connecticut than from any other eminence. Elevated above the river nearly 1,000 feet, and but a little distance from it, its windings lie directly before you; and the villages that line its banks—Sunderland, Hadley, Hatfield, Northampton and Amherst, appear like so many sparkling gems in its crown.

Mount Warner is a hill of less altitude than any before named, being only 200 or 300 feet in height, but a rich view can be had from its top of that portion of the valley of the Connecticut just described. It lies in the north part of Hadley, not more than half a mile from the river, and it can be easily reached by a carriage. A visit to it can, therefore, be performed by the invalid, and will form no mean substitute for an excursion to Holyoke or Toby.

Greenfield.—HOTELS :—*The American—The Mansion House.*

Greenfield, in its business quarter, is a lively little place. The wonted New England quiet, however, is all around it, in elm-shaded streets and garden-surrounded villas. The high hills in the neighborhood open fine pictures of the valleys and windings of the great river.

Greenfield is the terminus of a railway from Boston via Fitchburg, 100 miles from the former, and 56 miles from the latter place. Other routes now unite it with the railway systems of the west and of the north-west. The Green River, which flows near the village, is a pretty stream, and hard by are the Deerfield and Greenfield rivers. Among the manufactures of Greenfield there is a tool shop, in which are made 382 different shapes of carpenter's planes. In an extensive cutlery establishment upon Green River, 300 operatives are employed.

Vernon.—At Middle Vernon there is a charming view up the river, as seen from the railway track. Mount Chesterfield, in New Hampshire, opposite Brattleborough, rising up stoutly in the back-ground.

Brattleborough.—HOTELS :—*The Revere House.*

Brattleborough brings us fairly out of the rich alluvial lands into the upper and more rugged portions of the Connecticut. The intervals now grow narrower, and the hills more stern. This beautiful village is in a very picturesque district upon the west side of the river. It is, deservedly, one of the most esteemed of the summer resorts of the Connecticut, so pure and health-restoring are its airs, and so pleasant all its belongings, within and without. There are here several large and admirable water-cure establishments. The village cemetery on a lofty terrace overlooking the river, above and below, is a beautiful rural spot. West River, above the town, is an exceedingly picturesque stream. The buildings and grounds, in this vicinity, of the Asylum for the insane, have a fine manorial appearance.

Our next stage is 24 miles, from Brattleborough to Bellows' Falls, over the Vermont valley road.

Bellows' Falls. — HOTELS :— *The Island House.*—A sumptuous establishment.

Bellows' Falls is a famous congregating and stopping place of railways. With the exception of some bold passages of natural scenery, and a most sumptuous summer hotel, called the Island House, there is not much here, comparatively, to allure the traveller. Railways come in from Boston on the east, from the valley of the Connecticut on the south, from Vermont and Canada on the north, and from Albany and Troy, via Rutland, on the west.

The Falls are a series of rapids in the Connecticut, extending about a mile along the base of a high and precipitous hill, known as "Fall Mountain," which skirts the river on the New Hampshire side. At the bridge which crosses the river at this place, the visitor can stand directly over the boiling flood; viewed from whence, the whole scene is effective in the extreme. The Connecticut is here compressed into so narrow a compass that it seems as if one could almost leap across it. The water, which is one dense mass of foam, rushes through the chasm with such velocity, that in striking on the rocks below, it is forced back upon itself for a considerable distance. In no place is the fall perpendicular to any considerable extent, but in the distance of half a mile the waters descend about 50 feet. A canal three-fourths of a mile long, with locks, was constructed round the Falls, many years since, at an expense of $50,000.

Keene is one of the prettiest towns of New Hampshire in this vicinity. It is situated on a flat, east of the Ashuelot river, and is upon the route of the Cheshire railway, by which it is connected with Boston, and with the Connecticut river roads. It is particularly entitled to notice for the extent, width, and uniform level of its streets. The main street, extending one mile in a straight line, is almost a perfect level, and is well ornamented with trees. It is a place of considerable business, there being several manufacturing establishments here.

From Bellows' Falls we pass on to Windsor, 26 miles, by the Sullivan railway.

South Charleston and **Charleston** are quiet little aside villages on the east bank of the Connecticut, in Sullivan County, New Hampshire, 50 miles west of Concord. A bridge crosses the river to Springfield, Vermont. Charleston was the extreme northern outpost in the early days of the New England colonies. There was then a rude military work here called Fort No. 4.

Claremont is also on the east bank of the Connecticut, and in Sullivan County, N. H. It is a pleasant little manufacturing village. The scenery in this neighborhood is extremely fine. The banks of the Sugar River are very picturesque, and the changing aspects of Mount Ascutney, which we now approach, are of the highest interest. It is upon this side this noble hill, standing solitary and alone, a brave outpost of the coming Green Mountains on the one hand, and of the White Mountains on the other, is seen in its greatest grandeur. Its rugged precipitous summits and its dark ravines have here a very vigorous and massive character. Ascutney is sometimes called the Three Brothers from its trio of lofty peaks, all visible from the southern approach. From the eastward and northward, at Windsor and from the west, its appearance is totally different, but always fine. It may be very comfortably ascended from Windsor, in a good day's tramp; and the view from the summit is scarcely inferior in extent, variety, and magnificence, to that from any other peak of the Vermont chain. Its height is 1,732 feet above the river.

Windsor is one of the pleasantest rural retreats of all this charming region, with its vicinage to Mount Ascutney, and other attractive scenes of land and water. It is the centre of a fine agricultural and wool-growing neighborhood. There is an excellent, quiet, summer hotel here. Windsor is the seat of the Vermont State Prison, and the terminus of the Vermont Central Railway, from Burlington through the valley of the Winooski river.

At Windsor, the Sullivan road ends, and we continue our journey along the Connecticut, 14 miles, to White River Junction, by the Vermont central route.

White River Junction. From this point the Vermont central road continues, via Northfield and Montpelier, to Burlington; and we leave it for the Connecticut and Passumpsic, upon which we continue, 40 miles, to Wells River.

Hanover, is in this neighborhood, half-a mile east of the Connecticut, and 55 miles north west of Concord. It occupies a broad terrace 180 feet above the water. Here is the venerable Dartmouth College, founded in 1769, and name in honor of William, Earl of Dartmouth. Daniel Webster was one of the alumni of this esteemed institution.

The College buildings are grouped around a square of 12 acres, in the centre of the plain upon which the village stands.

Wells River.—Hotels :— *Coosac House.*

At this point, the railway to Littleton, 20 miles, and thence by stage to the White Mountains, diverges. Here, too, comes in the Boston, Concord, and Montreal route, sending its passengers, via Littleton, to the White Hills or onward, by the Connecticut and Passumpsic road, via St. Johnsbury, to Canada. The Connecticut now assumes the appearance of a mountain stream, the railways follow its banks no farther, and we leave our traveller to proceed on either hand, as we have indicated, to New Hampshire, or to Canada.

CONNECTICUT.

The scenery of Connecticut is delightfully varied by the passage of the Connecticut, the Housatonic, and other picturesque rivers; and of several low hill ranges. Spurs of the Green Mountains rise, here and there, in isolated groups or points through the western portions of the State. The Talcot, or Greenwood's Range, extends from the northern boundary almost to New Haven. Between this chain and that in the extreme west, lies another ridge, with yet two others on the eastward, the Middletown Mountains, and the line across the Connecticut, which is a continuation, most probably, of the White Hills of New Hampshire. Lying between these mountain ranges, are valleys of great luxuriance and beauty. The lakes among the mountains of the northwestern corner of the State are extremely attractive.

Excepting a trading house built by the Dutch at Hartford, in 1631, the first colony planted in Connecticut was the settlement of some of the Massachusetts emigrants at Windsor. Soon afterwards Hartford fell into the possession of the English colonists. Wethersfield was next occupied, in 1636, and New Haven, in 1638. The State had its share of Indian troubles in its earlier history, and of endurance, later, in the days of the Revolution.

THE HOUSATONIC VALLEY, RIVER, AND RAILROAD.

The valley of the Housatonic, traversed by the Housatonic river and railroad, extends for about 100 miles northward from Long Island Sound, through the extreme west of Connecticut and Massachusetts, including the famous county of Berkshire in the latter State. The whole region is replete with picturesque and social attractions, and has long been resorted to for summer travel and residence. It is a county of bold hills, pleasant valleys, and beautiful streams—more particularly that portion lying in Berkshire. Saddle Mountain, in the north part of this county, is the highest land in Massachusetts. The natural beauties of Monument Mountain, also in Berkshire, have been heightened by traditionary story, and by the verse of Bryant. Stockbridge and Great Barrington—very popular summer homes—are here. Lenox, honored by the residence of the authoress, Miss C. M. Sedgwick; and Pittsfield, the home of Melville and Holmes. North and South Adams, too, and Williamstown, the seat of Williams College—but we will follow the line of the valley, and glance, briefly, at its points of interest in due order.

From New York, take the New Haven Railroad, 58 miles, to Bridgeport, on the Sound, thence up the valley, on the Housatonic road; or take the Hudson river, or the River Railroad route, 116 miles, to the city of Hudson, and thence by Hudson and Boston Railroad, 34 miles, to West Stockbridge; or the Harlem Railroad, to its intersection with the Hudson and Boston, at Chatham Four Corners.

From Albany, by the Albany and Boston road, 38 miles, to State Line (Housatonic road); or onward to Pittsfield. From Boston by western (Albany) road, 151 miles, to Pittsfield.

Falls Village, 67 m. above Bridgeport. The Falls here, which are the largest in Connecticut, are very bold and picturesque. The waters traverse a ledge of limestone, and make a descent of 60 feet.

The Salisbury Lakes. The country west of Canaan, as all this part of the State, is beautifully embellished with hill and lake scenery. The Twin Lakes, in Salisbury, are very charming waters.

Sheffield is a prosperous village, famous for its manufactures, and for its varied attractions in hills and cascades, and other forms of natural beauty.

Great Barrington.—HOTELS:—*The Berkshire House.*

Great Barrington, with excellent hotels for summer travel, is a place of favorite resort. Mount Peter, on the southern edge, overlooks the village pleasantly; and it is most agreeably seen approaching, on the river road from the north.

The Taughkanic Mountains, a range extending from the Green Hills of Vermont, lie between the Housatonic Valley and the Hudson River. Mount Washington, Mount Riga, and other peaks, are interesting places of pilgrimage and exploration. The Falls of Bashpish are in this hill range. Following the Housatonic, and passing Monument Mountain, we reach

Old Stockbridge.—HOTELS:—*Stockbridge House.*

Old Stockbridge is one of the quietest and most winsome retreats in the world, lying in the lap of a fertile, hill-sheltered valley. The houses, which are all far apart, and buried in dense verdure, stand back in gardens, upon either side of a broad street or road, thickly lined with noble specimens of the over-attractive New England elm. There is a pleasant, well-ordered hotel here. Miss Sedgwick has, in her stories, woven much romantic interest about many spots in this vicinity, and about her own home of Lenox near by.

Lebanon Springs (N. Y.), and the Shaker village are hereabouts. (See New York.)

Pittsfield.—Hotels :— *Berkshire House.*

Pittsfield, Berkshire county, Mass., is a large manufacturing and agricultural town, elevated 1,100 feet above the level of the sea. It is 151 miles west from Boston, and 40 east from Albany. The village is beautifully situated, and contains many elegant public edifices and private dwellings. In this village there is still standing one of the original forest trees—a large elm, 120 feet high, and 90 feet to the lowest limb—an interesting relic of the primitive woods, and justly esteemed a curiosity by persons visiting this place. The town received its present name in 1761, in honor of William Pitt, (Earl of Chatham).

Upon a fine spacious square in the heart of the town, are the principal hotels, the Berkshire Medical School, a popular institution founded in 1823, and the First Congregational Church, a Gothic structure of stone, erected in 1853. There is, too, a prosperous Young Ladies' Institute here, which occupies several admirable buildings, surrounded by well-embellished grounds. Pittsfield is a large depot of manufactures, being extensively engaged in the production of cotton and woollen goods, machinery, fire-arms and railroad cars. The population of the township is nearly 9,000. It is upon the Western Railway, from Boston to Albany, at the northern terminus of the Housatonic valley route, and at the southern terminus of the Pittsfield and North Adams Railway.

The scenery of this region, traversed by the western road through Berkshire, from Boston to Albany, is often of very impressive aspect.

After leaving the wide meadows of the Connecticut, basking in their rich inheritance of alluvial soil, and unimpeded sunshine, you wind through the narrow valleys of the Westfield river, with masses of mountains before you, and woodland heights crowding in upon you, so that, at every puff of the engine, the passage visibly contracts. The Alpine character of the river strikes you. The huge stones in its wide channel, which have been torn up and rolled down by the sweeping torrents of spring and autumn, lie bared and whitening in the summer sun. You cross and recross it, as in its deviations it leaves space, on one side or the other, for a practicable road.

At "Chester Factories" you begin your ascent of 80 feet in a mile for 13 miles. The stream between you and the precipitous hill-side, cramped into its rocky bed, is the Pontoosne, one of the tributaries of the Westfield river. As you trace this stream to its mountain home, it dashes along beside you with the recklessness of childhood. It leaps down precipices, runs forth laughing in the dimpling sunshine, and then, shy as the mountain nymph, it dodges behind a knotty copse of evergreens. In approaching the "summit level," you travel bridges built 100 feet above other mountain streams, tearing along their deep-worn beds; and at the "deep cut" your passage is hewn through solid rocks, whose mighty walls frown over you.

The Pittsfield and North Adams Railroad Route. This road extends 20 miles, via Packard's, Berkshire, Cheshire, Cheshire Harbor, Maple Grove, and South Adams to North Adams.

Adams. The villages of North and South Adams are in the immediate neighborhood of Saddle Mountain. This noble peak has an elevation of 3,500 feet, and is the highest land in Massachusetts. There is a notable natural bridge upon Hudson's brook near North Adams.

Williamstown, near North Adams, is the seat of Williams College, founded in 1793. This institution is well endowed, and holds high rank among the best educational establishments of the country. The village is in one of the most picturesque portions of picturesque Berkshire.

New Haven, Hartford, &c. For mention of these and other cities and scenes of Connecticut, see Index and Routes to Boston from New York, under the head of Massachusetts.

RHODE ISLAND.

RHODE ISLAND is the smallest of the many States of the great American confederacy, her entire area not exceeding 1306 miles, with an extreme length and breadth, respectively, of 47 and 37 miles.

The country is most pleasantly varied with hill and dale, though there are no mountains of any great pretensions. Ample compensation for this lack in the natural scenery, is made by the numberless small lakes which abound everywhere, and especially by the beautiful waters and islands and shores of the Narragansett Bay, which occupy a great portion of the little area of the State. Its capitals, Providence and Newport, are among the most ancient and most interesting places in the land, and the latter has of late years become the most fashionable of all the numerous American watering-places.

Rhode Island was first settled at Providence, in 1636, by Roger Williams. To the enlightened and liberal mind of Williams in Rhode Island, and to the like true wisdom of Penn in Pennsylvania, and of Lord Baltimore in Maryland, America owes its present happy condition of entire freedom of conscience; perfect religious toleration having been made a cardinal point in the policy of these colonies.

The people of Rhode Island were early and active participants in the war of the Revolution, and many spots within her borders tell thrilling tales of the stirring incidents of those memorable days.

PROVIDENCE AND VICINITY.

Hotels. The Aldrich House (new), near the R. R. depot, and the City Hotel (old), Broad street.

Providence, one of the most beautiful cities of New England, and surpassed only by Boston in wealth and population, is a semi-capital of Rhode Island, on the northern arm of the Narragansett Bay, called Providence River. It is an ancient town, dating back as far as 1635—when its founder, Roger Williams, driven from the domains of Massachusetts, sought here that religious liberty which was denied to him elsewhere. It bears its venerable age, however, bravely, and looks to-day as youthful and vigorous as the Aladdin cities of yesterday—yet with the accumulated refinements and amenities, in its social character, of very many cultivated generations. This city makes a charming picture seen from the approach by the beautiful waters of the Narragansett, which it encircles on the north by its business quarter, rising beyond and rather abruptly to a lofty terrace, where the quiet and gratefully shaded streets are filled with dainty cottages and grand manorial homes. Providence was once a very important commercial depot, its rich ships crossing all seas—and at the present day the city is equally distinguished for its manufacturing wealth and enterprise. In this department of human achievement it took the lead, which it still keeps,—the first cotton mill which was built in America, being still in use, in its suburban village of Pawtucket, and some of the heaviest mills and print-works of the Union being now in operation within its borders. The value of the annual product of the cotton mills and print-works of Providence, is estimated at nearly four millions of dollars; that of the manufacturers of jewelry of various kinds—its establishments in this labor being no less than from sixty to seventy in number—at two and a half millions. It has also extensive manufactories of steam machinery, and of tools and implements of all sorts, and it furnishes the major part of all the screws used in the United States. The workshops of the Eagle Screw Company, where these little implements are made, are among the best-appointed in the world. The total capital invested here, in manufactures, is six millions of dollars.

Providence is the seat of BROWN UNIVERSITY, one of the best educational establishments in America, founded in Warren, R. I., 1764, and removed to Providence in 1770. Its library is very large, valuable, and is remarkably rich in rare and costly works.

The ATHENÆUM has a fine reading-room, and a collection of 12,000 books. The BUTLER HOSPITAL for the Insane, upon the banks of the Seekonk River, is an admirable institution, occupying large and imposing edifices. In the same part of the city, and lying also upon the Seekonk River, is the Swan Point Cemetery, a spot of great rural beauty. There are about fifty public schools in Providence, in which instruction is given to between six and seven thousand pupils, at an annual expense of over $45,000, one-fourth only of which is contributed by the State. The Dexter Asylum for the Poor is upon an elevated range of land east of the river. In the same vicinage is the yearly meeting boarding-school, belonging to the Society of Friends. The Reform School occupies the large mansion in the south-east part of the city, formerly known as the Tockwotton House. The Exchange, the Railroad Depot, some of the banks, and many of the churches of Providence, are imposing structures.

The topography and the natural scenery of Providence and its vicinity are great temptations to tourists, and to those seeking pleasant summer abodes. Situated upon the shore of the Narragansett Bay, and connected with it at all points by railway and steamboat, it unites all the pleasures of city and country life.

Upon the immediate edge of the city, on the shore of a charming bay in the Seekonk River, is the famous WHAT CHEER ROCK, where the founder of the city, Roger Williams, landed from the Massachusetts side, to make the first settlement here. From this rock he was greeted by the Indians, with the salutation which gives name to the spot.

At HUNT'S MILL, three or four miles distant, is a beautiful brook with a delicious little cascade, a drive to which is among the morning or evening pleasures of the Providence people and their guests.

VUE DE L'EAU is the name of a picturesque and spacious summer hotel, perched (four or five miles below the city) upon a high terrace overlooking the Bay and its beauties for many miles around.

Gaspee Point, below, upon the opposite shore of the Narragansett, tells a stirring story of the initial days of the Revolution, when some citizens of Providence, after adroitly beguiling an obnoxious British revenue craft upon the treacherous bar, stole down by boats in the night and settled her business by burning her to the water's edge.

Rocky Point is a wonderful summer retreat, among shady groves and rocky glens, upon the west shore of the Bay. In summer-time half a dozen boats ply, each twice a day, on excursion trips from Providence to various rural points down the Bay, charging 25 cents only for the round voyage.

Rocky Point is the most favored of all these rural recesses. Hundreds come here daily, and feast upon delicious clams, just drawn from the water, and roasted on the shore in heated seaweed, upon true and orthodox "clam-bake" principles. Let no visitor to Providence fail to eat clams and chowder at Rocky Point, even if he should never eat again.

The charming towns of Warren and Bristol are across the Bay, each worthy of a long visit. They may both be reached several times a day from Providence, by the Providence, Warren and Bristol Railroad.

Mount Hope, the famous home of the renowned King Philip, the last of the Wampanoags, is just below Bristol, upon Mount Hope Bay, an arm of the Narragansett on the east. The bare crown of this picturesque height presents a glorious panorama of the beautiful Rhode Island waters. Upon the shore of Mount Hope Bay, opposite, is the busy village of Fall River, which we have already visited, on our route to Boston from New York. Off on our right, as we still descend towards the sea, is Greenwich, and near by it, the birth-place and home of General Nathaniel Greene—the revolutionary hero—and just below is the township and (lying inland) the village of Kingston. In this neighborhood there once stood the old snuff-mill in which Gilbert Stuart, the famous American painter, was born.

Prescott's Head-quarters is a spot of Revolutionary interest, on the western shore of the large island filling the lower part of the bay, the island after which the State is named.

At the southern extremity of the island is the venerable town of Newport, at this day the most fashionable of all summer watering-places. Leaving Newport for a chapter by itself, let us, now that we have run rapidly down the 35 miles of the Narragansett waters, return for another moment to Providence.

We may get here any day from New York, by the Fall River route for Boston, round Point Judith from the sea, up the Narragansett (calling at Newport) to Fall River, in Mount Hope Bay, and thence by a Providence steamer. Or we may come by the Stonington route from New York, to Groton, Ct., by steamer, thence by rail; or more speedily, morning or afternoon, by the charming route along the marge of Long Island Sound—the new "Shore Line," via New Haven, New London, etc.; or we may come from Boston, any hour, almost, by the Boston and Providence road.

Distance from New York to Providence, about 175 miles—usual fare, $3 to $4. Distance from Boston, 43 miles—fare, $1 50. Population of Providence, about 51,000.

Newport.—HOTELS:—*The Ocean House*, Touro street, the most fashionable and most delightfully situated; the *Atlantic*, at the head of Pelham street, the *Bellevue*, the *Fillmore*, and the *Aquidneck*.

ROUTE. From New York (pier No. 3, North River), at 5 P. M., daily, in the superb steamers of the Fall River line. Also, by the steamers of the Stonington and Providence line, from Pier 18, every evening, to Groton, Ct. Remain on board till morning, breakfast, and take the cars for Greenwich at 7.45 A. M., connecting with boat on Narragansett Bay. Reach Newport by this route at 11. A. M. From Boston, by railway to Fall River, thence by steamer; or by rail to Providence, and thence by boat down Narragansett Bay.

If Newport were not, as it is, the most elegant and fashionable of all American watering-places, its topographical beauties, its ancient commercial importance, and its many interesting historical associations, would yet claim for it distinguished mention in these pages. Coming in from the sea round Point Judith, a few miles bring the traveller into the waters of the Narragansett Bay, where he passes between fort Wolcott, on Goat Island, and the stronghold of Fort Adams, upon Brenton Point on the right, and enters the harbor of the ancient town, once among the commercial capitals of the Union.

In the Revolution, the British long held possession of Newport, during which time, and at their departure, it became almost desolate. Before leaving, they destroyed 480 buildings, burned the lighthouse, cut down all the ornamental and fruit trees, broke up the wharves, used the churches for riding-schools, and the State House for a hospital, and carried off the church bells and the town records of New York; disasters which reduced the population from 12,000 to 4,000. But the incidents of this period have left some pleasant memories for the present day, and remembrances of the fame of Commodore Perry, the gallant commander on Lake Erie, who was born in Narragansett, R. I., across the bay, and whose remains lie now in Newport; of the residence of Rochambeau, and other brave officers of the French fleet; and of the visits of General Washington, and the *fetes* given in his honor; the venerable buildings associated with all these incidents being still to be seen.

The old town lies near the water, but of late years, since the place has grown so great as a summer residence, a new city of charming villas and sumptuous mansions has sprung up, extending far along upon the terraces which overlook the sea. Of the old buildings, and of those which belong to Newport *per se*, instead of in its character of a watering-place, are, the ancient State House (for Newport is a semi-capital of Rhode Island), the Redwood Library and Athenæum, the Old Stone Mill, an interesting relic of a period past remembrance, and almost of tradition; Tammany Hall Institute, Trinity Church, the Vernon family mansion, the Perry monument, Com. Perry's house, the City Hall, the fortifications in the harbor, Fort Adams, Fort Wolcott, Fort Brown, and the Dumplings.

The chief picturesque attractions of the town and its immediate vicinity, are the fine ocean-shores, known as the First, the Second, and the Third Beach. It is the First Beach which is chiefly used as a bathing-ground by the Newport guests. At the Second Beach are the famous rocks called Purgatory, and the Hanging Rocks, within whose shadow it is said that Bishop Berkeley wrote his "Minute Philosopher."

The Glen, and the Spouting Cave are charming places to ride to, when the weather invites.

Newport was the birthplace of the gifted miniature painter Malbone, and Gilbert Stuart's place of nativity may be seen in Narragansett, across the bay. Stuart made two copies of his great Washington picture for Rhode Island, one of which may be seen in the State House at Newport, and the other in that at Providence.

A steamboat passes every day up and down the bay between Newport and the city of Providence, enabling the traveller to see at his leisure the many attractions of the neighborhood.

NEW HAMPSHIRE.

NEW HAMPSHIRE contains some of the grandest hill and valley and lake scenery in America. The White Mountains here are popularly supposed to be the highest land east of the Mississippi River, as, indeed, they are, with the single exception of Black Mountain in North Carolina. These noble hills occupy, with their many outposts, a very considerable portion of the State, and form the speciality in its physical character. The reader will find a detailed mention of all these features, and of the beautiful intermediate lake-region, in subsequent pages.

On his route from Boston to the mountain regions, the tourist will find much to interest him, if his interest lies that way, in the enterprising manufacturing towns of the lower part of the

NEW HAMPSHIRE.

State. In its historical records, New Hampshire has no very striking passages—no important reminiscences, either of the Revolutionary War or of the later conflict with Great Britain in 1812.

The railway lines of New Hampshire are numerous enough to give ready access to all sections of her territory, and to the neighboring States. Occasion will occur for ample mention of the facilities which they afford for travel, as we follow them, severally, hither and thither.

ROUTES TO THE WHITE MOUNTAINS, N. H.

FROM NEW YORK, BOSTON, PORTLAND, ETC.

Via Boston.

ROUTE 1. From Boston, by Lake Winnipiscogee and Conway Valley. (See routes from New York to Boston.) From Haymarket Square, Boston, at 7 30 A. M. and 12 M., 26 miles to Lawrence by Boston and Maine road; 27 miles to Manchester, upon Manchester and Lawrence R. R.; 18 miles to Concord, upon Concord R. R., 33 miles to Weirs, on Lake Winnipiscogee, by Boston and Concord and Montreal R. R.; 10 miles by steamer Lady of the Lake, on Lake Winnipiscogee to Centre Harbor (dine at Centre Harbor); 30 miles by stage to Conway; arrive at Conway in the evening, remain there all night, and proceed, 24 miles, to Crawford House, White Mountain Notch, next day. Total distance from Boston to the Crawford House, 168 miles; time, 2 days and 1 night; fare, $7 45. Distance from New York, 431 miles; time, 2 days and 2 nights; fare, about $12 45. Passengers by the Boston morning train only reach Conway the same evening. Those taking No. 2, or noon train, will pass the night at Centre Harbor, on Lake Winnipiscogee, and the next night at Conway, reaching the mountains on the third day.

ROUTE 2. From Boston. (See routes from New York to Boston.) Leave Haymarket Square (as in route 1) at 7 30 A. M. and 12 M.; 68 miles to Dover, N. H., upon Boston and Maine R. R.; thence to Alton Bay, 28 miles, upon Cocheco R. R.; thence, 30 miles, by steamer Dover (dine on board) to Wolfboro' and Centre Harbor, on Lake Winnipiscogee; thence by stage, via Conway, to the mountains, as in Route No. 1.

Passengers by morning train only, from Boston, reach Conway same night. Those by second, or noon train, will pass the night at Wolfboro' or Centre Harbor. From Boston to Crawford House, by this route, 96 miles by railroad, 30 by steamboat, and 54 by stage; total, 180. Time, 2 days and 1 night, from Boston; fare, $7 45.

ROUTE 3. From Boston. At 7 30 A. M. and 12 M. (station, Causeway street), 26 miles to Lowell, by Boston and Lowell R. R.; 15 miles to Nashua, upon Nashua and Lowell R. R.; 35 miles to Concord, upon Concord R. R.; afterwards as in Route No. 1. Distance, time, and fare the same.

ROUTE 4. From Boston, same as in Routes Nos. 1 and 3, as far as Weirs on Lake Winnipiscogee; thence, continuing upon the railroad, 18 miles from Weirs, to Plymouth, N. H.; dine at Plymouth, and proceed by stage, 24 miles, through West Campton, etc., to the Flume House, Franconia Notch, the western end of the mountains. Passengers by the morning train from Boston will reach the Flume House, Franconia Notch, same evening. Those taking the second train will stay over until next day at Plymouth. Distance from Boston to Flume House, 148 miles, being 124 by railway and 24 by stage. Time, from Boston by morning line, 10 hours; fare, $5 15. Stages daily from Flume House, 5 miles, to the Profile House, 22 miles to White Mountain House; thence, 5 miles, to Crawford House, terminus of Routes 1, 2, and 3, on the east side. Distance from Flume House to Crawford House, 32 miles. Fare, $3.

ROUTE 5. From Boston, same as in Routes 1 and 3, to Weirs; thence (as in Route 4) to Plymouth (dine), continuing upon the railroad, 42 miles, from Plymouth to Wells River; thence upon White Mountain R. R., 20 miles, to Littleton; thence by stage, 11 miles, to Profile House, and 5 miles further to Flume House, or 23 miles to Crawford House. Passengers by the early train only reach the mountains the same night. Those taking second train stay till next day at Plymouth. From Boston to Profile House, 193 miles; to Flume House, 198 miles; to Crawford House, 205 miles. 182 miles by railroad, rest by stage. Fare from Boston to Profile House, $6 15; to Crawford, $6 90. Time, 12 hours.

From New York or Boston, via Portland, Maine.

ROUTE 6. (See routes from New York to Boston.) Leave Boston for Portland, 111 miles by railway, morning and evening, from Haymarket street, via Reading, Lawrence, Haverhill, Exeter, etc. Through baggage for the White Mountains to be marked, "*Portland East.*" Passengers by first train will dine in Portland, and take Grand Trunk Railway through Cumberland, Yarmouth, etc., 91 miles, to Gorham, N. H. Second train passengers will pass the night at Portland, and proceed to Gorham next day.

NEW HAMPSHIRE.

From Gorham, 8 miles, by stage to Glen House, foot of Mount Washington. Stages leave Glen House every morning for Crawford House, 34 miles distant, via Pinkham Notch, also via Cherry Mountain. From Boston to Gorham, 202 miles; from Boston, via Portland, Gorham, Glen House, and Pinkham Notch, to Crawford House, 244 miles.

ROUTE 7. From Boston, via Portland. Leave Causeway street, morning and evening, by railway, through Lynn, Salem, Beverly, Newburyport, Portsmouth, etc., to Portland, and thence as in Route No. 6.

ROUTE 8. From Boston by steamer to Portland, every night, from end of Central Wharf; thence, as in Route No. 6. Fare by this line, $3, from Boston to Gorham.

ROUTE 9. From Boston to Portland, by railway or steamer, as in Routes 6, 7, and 8, and thence by Sabago Lake and Pleasant Mountain to Conway; thence to Crawford House, etc., as in Route 1.

From New York, NOT via Boston.

ROUTE 10. From New York by railway, via New Haven, Hartford, and Springfield; thence by railway up the Valley of the Connecticut to Wells River, and from thence to Littleton, N. H.; from Littleton by stage, as in route 5.

ROUTE 11. By steamboat from Pier 18, North River, N. Y., every evening to New London; thence by railway to Worcester, Nashua, and Concord; and from Concord on the east side by Conway to Crawford House, route 1; or on the west side by Campton to the Franconia Notch, route 5. A very charming route, full of ever-changing and always increasingly attractive points.

ROUTE 12. From New York by Hudson River, or Hudson River Railway, to Albany and Troy; thence to Whitehall, and down Lake Champlain to Burlington, Vermont; thence by Vermont Central Railroad through the Winooski Valley and Green Mountains (via Montpelier), to connections with the Connecticut Valley roads to Littleton, N. H.

ROUTE 13. From New York by Hudson River to Albany; thence to Whitehall, foot of Lake Champlain, or other routes to Rutland, Vermont; thence to Bellows' Falls, on the line of the Connecticut Valley road, to Littleton, N. H.

We might much extend our list, but as all roads lead to Rome, so the ways to the favorite summer haunts in the White Mountains are infinite.

DESCRIPTION OF ROUTES TO WHITE MOUNTAINS.

ROUTE 1. By Lake Winnipiseogee and Conway Valley. From Boston, 26 miles, to Lawrence, via Boston and Maine Railroad (Boston and Portland), passing numerous suburban villages, for which see "Boston and Vicinity." Lawrence is a large manufacturing city (population 18,000), upon the Merrimac River. It is connected with Lowell (13 miles distant), with Salem, 21 miles, and with all surrounding points by railway. It has grown up suddenly within a few years, having been incorporated only in 1845.

From Lawrence by Manchester and Lawrence Railroad, 26 miles to Manchester, N. H., still following the Merrimac River, upon which Manchester, like Lawrence, is situated. At this point we are 53 miles above Boston, and 18 miles below Concord. This place has, like Lawrence and others, suddenly grown under the development of manufacturing enterprise—from an inconsiderable village, into a large and populous city. Its charter was granted in 1846, and in 1860 it had 20,000 inhabitants.

Concord.—HOTELS :—*Eagle House.*

Concord, the capital of the State, is upon the banks of the Merrimac, 18 miles above Manchester, by the Concord Railroad. The State Capitol, the Lunatic Asylum, the State Prison, are public edifices of interest. A Methodist General Biblical Institute was founded here in 1847. We might suggest to the tourist a brief halt at Concord, were he not now so near yet more attractive scenes. Concord is 71 miles from Boston, via Lawrence; 47 miles from Portsmouth, by Concord and Portsmouth Railroad; 25 miles from Bradford, by the Concord and Claremont Railroad; 35 miles from Nashua, by Concord Railroad; 93 miles from Wells River, by Boston, Concord, and Montreal Railroad. Concord is an eligible place for summer abode. It is the home of ex-President Pierce.

From Concord, our route follows the Boston, Concord, and Montreal Railroad, 33 miles to Wiers, on Lake Winnipiseogee, where we take the steamer Lady of the Lake, 10 miles to Centre Harbor. Our White Mountain route, No. 5, continues on this road past Weirs, to Wells River, and Littleton, N. H.

Lake Winnipiseogee. The little voyage on this beautiful lake, is among the most agreeable passages in our present journey to the White Mountains, and well deserves a pilgrimage to itself alone. Winnipiseogee is an enchanting reach of pure, translucent waters, very irregular in form; some 23 miles long, and from one to ten miles wide. It is crowded with exquisite island groups, indented with surprising bays; and bold mountain peaks cast their shadows everywhere into its still, deep floods.

Red Mountain, about 1,600 feet high, a remarkably beautiful eminence, is situated on the N. W. of the above lake. The ascent to the summit, although steep and arduous, can be effected for a portion of the distance in carriages, and all the way on horseback. From the S.E. there is a fine panoramic view of the lake and the adjacent country. On the south ascends Mount Major, a ridge of a bolder aspect and loftier height. On the N.E. the great Ossipee raises its chain of elevations, with a bold sublimity, and looking down in conscious pride upon the regions below.

Squam Lake lying W. from Red Mountain; and two miles N.W. from Winnipiseogee Lake, is another splendid sheet of water. It is about six miles in length, and in its widest part not less than three miles in breadth, and, like its neighbor (Winnipiseogee), is studded with a succession of romantic islands. This lake abounds in trout of the finest kind.

Centre Harbor.—HOTELS :—*Senter House.*

Centre Harbor, with its excellent summer hotel upon the margin of Winnipiseogee, is the halting-place for the explorer of the many beauties of this region. White Mountain tourists dine here *in transitu*, and proceed for the rest of the way by stage-coach, first for thirty miles through a country of picturesque delights to Conway Valley, where they might well linger till their summer days all went by.

Conway Valley is a wide stretch of delicious interval lands upon the Saco River, hemmed in upon all sides by bold, mountain summits, chief among which are the stern cliffs of Mount Washington itself. It is a delightful place for artistic study, and for summer residences ; and within a few years past, it has been a favorite resort of the American Landscapists, and has grown to be a veritable " watering-place," in the great numbers of tourists who not only pass, but linger within its borders. Pleasant hotels and boarding-houses are springing up, and country villas even are beginning to dot its knolls, and to lurk in its verdant glens. The picturesque portion of this valley, *par excellence*, is North Conway, where the Kearsarge House (Thompson's), or the Washington House, offers all desirable hotel appliances. Beside the distant views of the White Mountain ranges, proper, which are of surpassing interest here, Conway is full of local and neighboring attractions of the greatest beauty, as are the broad meadows, and the wooded, winding banks of the Saco ; the nooks and turns of the Artists' Brook, and other elfish waters ; the Paquawket Mountain, those grand perpendicular cliffs, 650 and 950 feet in height, called the Ledges ; the magnificent peaks of Kearsarge and Chicorua ; the Echo Lake, the Crystal Falls, and Diana's Bath.

Conway village and Conway corners are a few miles below North Conway. They are most agreeable places, *en route*, amply supplied with hotel accommodations. Leaving Conway, as the tourist does, the morning following that of his departure from Boston, he continues on through valley and over hill, 24 miles to the Crawford House, where we shall meet him when we have followed over other routes to the threshold of the mountains. We will, however, accompany him yet on his journey from Conway, through Bartlett and Jackson, by the Old Crawford House ; and by the famous Willey House, the scene of the awful avalanches of 1826, when the entire Willey family were destroyed. (*See further mention later.*)

ROUTE 2. From Boston, 68 miles via Lawrence to Dover, N. H., on the Boston and Maine Railroad. Dover is a pleasant town of some 8,000 people ; upon the banks and at the falls of the Cocheco River, a tributary of the Piscataqua. Our route leads hence by the Cocheco Railroad to Alton Bay ; southern extremity of Lake Winnipiseogee. Here, we take the steamer Dover for Centre Harbor, traversing the entire length of the lake, and proceed thence via Conway, as in Route 1.

ROUTE 3. From Boston, 26 miles, to the famous manufacturing city of Lowell. (See Boston and Vicinity.) From Lowell, 15 miles, to Nashua—an important manufacturing town, at the confluence of the Nashua with the Merrimac river ; thence, 35 miles, to Concord, N. H., and from Concord to Weirs and Centre Harbor, on Lake Winnipiseogee, and on, via Conway, as in Routes 1 and 2.

ROUTE 4. From Boston, as in Route 1, or 3, to Weirs, on Lake Winnipiseogee, thence on, without stopping, to Plymouth, N. H., where passengers dine and take stage for the rest of the way ; or where they remain all night, if they leave Boston by the noon, instead of the morning train. Plymouth is in the midst of a noble mountain landscape, being the extreme southern threshold of the Franconia range of the White Hills. It is upon the banks of the beautiful Pemigewasset river, near its confluence with Baker's river. The Pemigewasset House here, at the railway station, is an inviting place for summer tarry. The Wells River and Littleton route from Boston to the mountains by the west passes Plymouth.

Leaving Plymouth in the stage, after dinner, we reach the Flume House, at the Franconia Notch, 24 miles distant, the same evening, unless we stop by the way, as would be very reasonable

—for the whole journey is through most inviting spots and places. The villages on the route are but little affairs; and there is not a fashionable hotel in all the distance, until we reach the Flume; but there are numerous small inns, where artists and their families are well content to pass the summer. There is such an one at

West Campton, a little hamlet on Pemigewasset river, seven miles above Plymouth. West Campton is becoming a greater resort of the landscape painters than North Conway, on the south-east slope of the mountains, has been for several years past. Other tourists will follow, and hotels and boarding-houses will grow up with the demand. The views here, of the Franconia Hills, are especially fine, and the river and brook landscape, with its wealth and variety of vegetation, is of extraordinary interest. The Pemigewasset river, which rises in the little lakes of the Franconia Mountains, winds through all the wonderful valley which we traverse between Plymouth and the Flume House. We shall rejoin our tourist, by and by, at the Flume.

ROUTE 5. To Boston, same as in Route 1, to Weirs on Lake Winnipiseogee, thence on, without halt (as in route 4), to Plymouth, N. H. Thence, after dinner (*morning* train from Boston), still upon the railroad, 42 miles, to Wells River, Vermont.

Wells River is at the junction of the Connecticut and Wells river—a famous meeting-point of railway lines. Our present route meets here with the Connecticut valley road to the White Mountains. The Vermont Central Road, from Burlington, on Lake Champlain, comes in at White River Junction, 40 miles below. From Wells River our route proceeds by White Mountains Railroad, 20 miles, to Littleton, and for the rest of the way, by stage, either to the Franconia Notch, 12 miles (Profile House), or to the Eastern or White Mountain Notch (Crawford House), 22 miles.

ROUTE 6. Via Portland, and through Maine, on the east side of the mountains. This route, as Routes 7, 8, and 9, are all agreeable approaches to the White Hills, but more circuitous from New York or Boston, than either of the Routes 1 to 5. The Boston and Maine, one of the two railways from Boston, to Portland, runs (111 miles) east of north, and always near the Atlantic coast, through portions of Massachusetts and through New Hampshire. (See "Boston to Portland.") From Portland our present route is by the Grand Trunk Railway, 91 miles, to Gorham, N. H. The Rev. Mr. Willey, in his "Incidents in White Mountain History," says of Gorham that "it is a rough, unproductive township, lying on the northerly base of the mountains. The opening of the Atlantic and St. Lawrence Railway (the Grand Trunk) brought the little town out from the greatest obscurity, and it has become one of the favorite resorts for the travelling community. Its peculiarly favorable situation for viewing the mountains was never known until travellers, posting through its borders, for other destinations, were compelled to admire its beauties.

"Immediately on the completion of the railroad to this point, the Alpine House was erected, and the announcement made that the cars set passengers down at the very base of the White Mountains. People, for a moment, were dumb with astonishment. It had never been supposed that there was any north or south, or east or west, to these old heights; but that every one who visited them must make up his mind for a long stage-coach ride through Conway or Littleton, and ultimately be set down at the Crawford or Fabyan's. That the cars should actually carry visitors to the base of the mountains was something which every one had supposed would take place in the far-off future, but not until they themselves had ceased to travel; but it was certainly so; and the Alpine House and Gorham had become familiar words to travellers.

"The Alpine House is a large hotel, owned by the railroad company. It is some distance from the base of the mountains, which are seldom ascended from this point; but for quiet and comfort, and beautiful drives, it is surpassed by no house in the White Hills. A beautiful little village has sprung up around it, consisting mostly of buildings owned by the company. The Post Office is kept here, and the telegraph affords an excellent opportunity to business men to visit the mountains and attend to their business at the same time. Mount Moriah, Randolph Hill, Berlin Falls, and Lary's, should all be visited before the traveller takes his departure."

The Glen House, our next point, (seven miles from Gorham), is, says Mr. Willey, whom we have just quoted, "in the valley of the Peabody river, immediately under Mount Washington, and in the midst of the loftiest summits in the whole mountain district. The house is situated in Bellows' clearing, which contains about 100 acres. For a base view of the mountains, no spot could be selected so good. Several huge mountains show themselves proudly to view, in front of the piazza, nothing intervening to obscure their giant forms. You see them before you in all their noble, calm, and silent grandeur, severally seeming the repose of power and strength. On the left is the *mountain* bearing the *worthiest name* our country ever gave us. Toward the right of its rock-crowned summit rise, in full view, the celebrated peaks of Adams and Jefferson—the one pointed, the other rounded. On both wings of these towering summits are the tops of lesser elevations. In an opposite direction, fronting the 'patriot group' of gigantic forms, is the long, irregular rise of Carter Mountains."

It is from the Glen House that a famous carriage way was to lead to the summit of Mount Washington. This road was to be eight miles long, and was to be made to the very crown of the lofty mountain; 15 feet wide, clear of all obstructions, and macadamized throughout. The average grade was to be an ascent of one foot to eight and a half feet, with frequent stations at eligible points of view. The estimated cost of this road was 100,000 dollars. A magnificent hotel was to be built upon the mountain top. This bold project has, since our earlier editions, been in a great measure realized; the carriage-way having been completed and successfully used.

From the Glen House we must now reach the Crawford House, on the mountain, 34 miles distant, via the Pinckham Notch, or by Cherry Mountain.

ROUTE 7, is from Boston to Portland, by the Eastern (the lower) Railway, through Lynn, Newburyport, and other towns in Massachusetts, and Portsmouth, in New Hampshire. From Portland we proceed by Grand Trunk road to Gorham, as in Route 6.

ROUTE 8. From Boston to Portland, by steamer, and thence as in Route 6.

THE WHITE MOUNTAINS—SCENES AND INCIDENTS.

These mountains cover an area of about 40 miles square, in Northern New Hampshire; though the name of White Mountains is, in the neighborhood, given to the central group only—the half-dozen lofty peaks, of which Mount Washington is the royal head and front. The western cluster is contra-distinguished as the Franconia range. We will suppose our tourist to have made his approach on the south-east, to the Central or White Mountain group, via Route 4, Lake Winnipiseogee and Conway valley, and thus meet him at the Crawford House, near the

Great Notch. The mountains, which have gradually gathered about us, in our steep ascent here have all closed in. The magnificent pass—the Gateway of the Notch, is a chasm between two perpendicular masses of rock, approaching each other to within 22 feet. Dark overhanging cliffs stand as sentinels over this solemn pass, and it has been a work of toil to cut a pathway through the frowning barrier. This gorge is some three miles long, descending the valley of the Saco, towards "the Willey House." Upon the north, the bold cliffs of Mount Willard rise to the height of 2,000 feet above quiet vales below. The rugged flanks of the devoted Mount Willey, bearing yet the fatal tempest scars of 1826, stop the view on the left, while Mount Webster—dark, and massive, and grand, as was he whose name it bears—fills the landscape on the right.

The White Mountains (specifically so called) extend from the Notch, in a north-easterly direction, some 14 miles, increasing from each end of the line gradually in height towards Mount Washington, in the centre. These respective elevations are, in the order in which they stand, beginning at the Notch—Mount Webster, 4,000 feet above the level of the sea; Jackson, 4,100; Clinton, 4,200; Pleasant, 4,800; Franklin, 4,900; Monroe, 5,300; Washington, 6,500; Clay, 5,400; Adams, 5,700; Jefferson, 5,800; and Madison. 5,400.

Passing westward from the Notch, we reach the valley of the Ammanoosuc, after a distance of four miles, through dense woods, and enter abruptly into a spacious clearing, from which the whole mountain group bursts upon our wondering sight. Here, upon the "Giant's Grave," an eminence of some 60 feet, the panorama is marvellous. In the centre of the amphitheatre of hills, Mount Washington, barren, and seamed and whitened by the winter tempests of centuries, looks down, upon the right and upon the left, on the hoary heads of Webster and Madison—each, on its side, the outpost of the mountain army.

The Hotels. The Crawford House—a most excellent establishment—bears the name of the earliest hosts of these mountain gorges. The story of the adventures and the endurance of the early settlers here, is extremely interesting. How Captain Eleazar Rosebrook, of Massachusetts, built a house on the site of the Giant's Grave, four miles from the Notch, afterward occupied by Fabyan's Mount Washington Hotel *—how his nearest neighbors were 20 miles away, excepting the Crawford family, 12 miles down in the Notch valley—the present old Crawford House, at the base of the mountains, coming from Conway, on the south-east. How the Rosebrook children were often sent, for family supplies, over the long and dangerous path to the Crawfords', returning, not unfrequently, late at night—how Ethan Allen Crawford was heir to the Rosebrook estate, and how he became known as the "Giant of the Hills"—how he and his family made the first mountain paths,† and were for long years the only guides over them of the rare visitors, which the brief summers brought —and how they have since seen their home

---

* Destroyed by fire and never rebuilt.
† The first bridle-path was cut by Ethan Crawford, in 1821.

thronged, for weeks together, like a city saloon, with beauty and fashion. The Crawfords are a large, athletic race. Abel, the father, called the "Patriarch of the mountains," would walk five mountain miles, to his son's, before breakfast, at the age of 80. At 70, he made the first ascent ever made on horseback, to the top of Mount Washington. His sons were all over six feet tall; and one of them was six and a half feet, and another, Ethan Allen, was seven feet in height.

Ascent of Mount Washington. The chief exploit of the visitor, at this group of the White Hills, is to ascend Mount Washington; and a toilsome, and even dangerous feat it is to this day, despite the improved facilities of travel. The journey from the Crawford House is nine miles, made on the backs of Canadian ponies, over the old Crawford bridle-paths, though a grand carriage-way is now completed, from the Glen House on the opposite side of the mountain. (See Glen House.) The excursion occupies a long day, with the utmost industry. We made it, on one occasion, in midsummer, with a party of thirty ladies and gentlemen, besides our guides, and it was a gay scene—the getting *en route*, and a singular cavalcade; miles onward as we wound, in Indian file, cautiously along the rugged, narrow path, trusting to our trusty ponies to walk with us upon their backs, over logs, and rocks, and chasms, which we would not have dared to leap ourselves; and surprising was the picture, as we at length bivouacked, and ate our grateful lunch, upon the all-seeing crest of the grand old mountain. At another time, we ascended, in the middle of October, when we could muster no larger group than our friend, ourself, and our guide. For two miles from the summit, the way was blocked with snow; so we left our ponies to take care of themselves, and completed the tramp on foot. The day, though so bitterly cold as to remind us of Webster's salutation upon a like occasion—"Mount Washington! I have come a very long distance, have toiled hard to arrive at your summit, and now you give me but a cold reception"—was happily a brilliant one; the atmosphere was exceedingly clear: and we had the delight of seeing all the delicious panorama, which has been thus catalogued:—

View from the Summit. "In the west, through the blue haze, are seen, in the distance, the ranges of the Green Mountains; the remarkable outlines of the summits of Camel's Hump and Mansfield Mountain being easily distinguished when the atmosphere is clear. To the north-west, under your feet, are the clearings and settlement of Jefferson, and the waters of Cherry Pond; and, further distant, the village of Lancaster, with the waters of Israel's river. The Connecticut is barely visible; and often its appearance for miles is counterfeited by the fog rising from its surface. To the north and north-east, only a few miles distant, rise up boldly the great north-eastern peaks of the White Mountain range—Jefferson, Adams, and Madison—with their ragged tops of loose dark rocks. A little further to the east are seen the numerous and distant summits of the mountains of Maine. On the south-east, close at hand, are the dark and crowded ridges of the mountains of Jackson; and beyond, the conical summit of Kearsarge, standing by itself, on the outskirts of the mountains; and, further over the low country of Maine, Sebago Pond, near Portland. Still further, it is said, the ocean itself has sometimes been distinctly visible.

"The White Mountains are often seen from the sea, even at 30 miles distance from the shore; and nothing can prevent the sea from being seen from the mountains, but the difficulty of distinguishing its appearance from that of the sky near the horizon.

"Further to the south are the intervals of the Saco, and the settlements of Bartlett and Conway, the sister ponds of Lovell, in Fryburg; and, still further, the remarkable four-toothed summit of the Chocorua, the peak to the right being much largest, and sharply pyramidal. Almost exactly south are the shining waters of the beautiful Winnipiseogee, seen with the greatest distinctness in a favorable day. To the south-west, near at hand, are the peaks of the south-western range of the White Mountains; Monroe, with its two little alpine ponds sleeping under its rocky and pointed summit; the flat surface of Franklin, and the rounded top of Pleasant, with their ridges and spurs. Beyond these, the Willey Mountain, with its high, ridged summit; and, beyond that, several parallel ranges of high-wooded mountains. Further west, and over all, is seen the high, bare summit of Mount Lafayette, in Franconia."

Tuckerman's Ravine is a marvellous place, seen in the ascent of the mountains, by the Davis' Road leading from the Crawford House. It lies upon the right in passing over the high spur directly south-east of Mount Washington. Turning aside, the edge of the precipice is reached, and may be descended by a rugged pathway. It is a long, deep glen, with frowning walls, often quite inaccessible. It is filled, hundreds of feet deep, by the winter snows, through which a brook steals, as summer suns draw near, gradually widening its channel, until it flows through a grand snow cave, which was found, by measurement, to be, one season, 84 feet wide on the *inside*, 40 feet high, and 180 feet long. The snow forming the arch was 20 feet thick! The engineers of the projected carriage road went through this arch one July, in the bed of the stream, to the foot of the cataract, which flows for 1,000 feet, down the wild mountain side.

Oakes' Gulf is another fathomless cavern, seen, far down on the right, in winding around the summit of Mount Monroe. Near the summit of Mount Washington, a few rods northward, is yet another black abyss, which is called the Great Gulf. Its descent here is 2,000 feet, rugged and precipitous.

The Summit House. The building of the rude inn, which now stands upon the summit of Mount Washington where the great hotel is to be when the grand projected road is completed, was a daring and painful enterprise. It is said the bold scheme was suggested by Jos. S. Hall, for many years a guide from the Notch House. "The house was located," says Mr. Willey, "under the lee of the highest rocks on Mount Washington, and was laid out forty feet long, and twenty-two feet wide. The walls were four feet thick, laid in cement, and every stone had to be raised to its place by muscular strength alone.

While these were laying the walls, the materials for finishing and furnishing were being dragged up from the Glen House, a distance of six miles. Lime, boards, nails, shingles, timbers, furniture, crockery, bedding and stoves, all had to be brought up by piecemeal on the men's or horses' backs. No one ever went up without taking something—a chair; or door, or piece of crockery. Four boards (about sixty feet) could be carried up at once on a horse's back, and but one trip could be made daily. Mr. Rosebrook, a *young giant*, carried up at one time a door of the usual length, three feet wide, three and one-half inches thick, ten pounds of pork, and one gallon of molasses.

The walls were raised eight feet high, and to these the roof was fastened by strong iron bolts; while over the whole structure were passed strong cables, fastened to the solid mountain itself. The inside was thrown, primitive fashion, into one room, in which the beds were arranged, berth-like, for the most part on one side of the room, in two tiers, with curtains in front. A table, capable of seating thirty or forty persons, ran lengthwise of the room. At one end of the room a cooking-stove and the other furniture of a kitchen were placed, with a curtain between it and the table. At the other end was a small stove, in which was burned mountain moss. The walls are perfectly rough, outside and in; a little plaster upon the inside merely fills up the chinks. The Tip Top House was the second structure erected here.

Much as we have necessarily left unseen, on the mountains, we must now descend, and with a hasty peep at some yet unmentioned scenes, in the vicinage of the Notch, pass on, thirty-six miles, to the Franconia range, in the west.

The Silver Cascade is a favorite scene, about half a mile south of the entrance to the Notch. It is one of the most charming waterfalls imaginable, seen at a distance of two miles, bubbling down the mountain side, eight hundred feet above the neighboring valley.

The Flume is another cascade yet further down the Notch. It descends two hundred and fifty feet, in two rills, over two precipices, and there are three streams over a land ledge, reuniting in a small rocky basin below.

The Devil's Den is a mysterious cavern, near the top of Mount Willard, opposite the Silver and the Flume cascades.

The Crystal Falls, of eighty feet, and the *Glen Ellis Falls*, of seventy feet, are on the Ellis River, the one on the left and the other on the right of the road from Jackson to the Glen House.

The Ammanoosuc River, rising in this group of the White Mountains, and followed in the journey towards the Franconia Hills, is a stream of wonderful beauty. It falls 6,000 feet from its source on the mountain, to the Connecticut River, and is said to be the wildest and most impetuous river in New Hampshire. It abounds in rapids and cascades.

The Willey House is passed some miles below, at the commencement of the ascent to the Notch. The spot will be for ever of memorable interest, from its tragic story of the fearful avalanches of 1826, when the entire family which then occupied the house—Mr. Willey, his wife, five children, and two hired men—were all buried beneath the mighty *debris* of the mountain slides.

The ride through the hills and by the waterfalls, 23 cheery miles from the Crawford House to the Profile House, in the Franconia Pass, might detain us pleasantly enough at many points, but we bear our traveller on, at once, to the last chapter of our White Mountain story.

The Franconia Hills, though in popular estimation inferior in interest to the eastern cluster, are really not so; except it be in the wonders of the mountain ascents; and even in this, the panorama, from the summit of Lafayette, is scarcely less extensive or less imposing than the scene from the crown of Mount Washington, while the exquisite little lakes, and the singular natural eccentricities in the Franconia group, have no counterpart in the other. In this, as in other ranges of the White Hills, the mountains are densely wooded at their base, while their rock-ribbed summits are barren, and scarred by time and tempest. The hills approach, at one point, to within half a mile of each other, and form the wild procrustean portal, called the Notch.

Mount Lafayette, or the Great Haystack, is the monarch of the Franconia kingdom, towering up, skyward, to the height of 5,200 feet. Its lofty pyramidal peaks are the chief objects, in all views, for many miles around.

Eagle Cliff is a magnificently bold and rocky promontory, near Mount Lafayette. It casts its dark shadows down many hundred feet into the glen, traversed by the road beneath.

Cannon Mountain, 2,000 feet above the road and 4,000 above the sea, is nearly opposite Lafayette, and forms the western side of the Notch. Away up upon its crown is a group of mighty rocks, which, as seen from the Profile House below, bear an exact resemblance to a mounted cannon. It is upon this mountain, also, that we find that marvellous freak of nature,

The Profile Rock, or The Old Man of the Mountain.—This wonderful eccentricity, so admirably counterfeiting a human face, is 80 feet long, from the chin to the top of the forehead, and is 1,200 feet above the level of the road, being yet far below the summit of the mountain. This strange apparition is formed of three distinct masses of rock, one making the forehead, another the nose and upper lip, and a third the chin. The rocks are brought into the proper relation to form the Profile, at one point only, upon the road, through the Notch, a quarter of a mile south of the Profile House. The face is boldly and clearly relieved against the sky, and, except in a little sentiment of weakness about the mouth, has the air of a stern, strong character, well able to bear, as he has done unflinchingly, for centuries, the scorching suns of summer and the tempest blasts of winter. Passing down the road a little way, the "Old Man" is transformed into a "toothless old woman in a mob cap;" and, soon after, melts into thin air, and is seen no more. Hawthorne has found in this scene the theme of one of the pleasantest of his "Twice Told Tales," that called the "The Old Stone Face."

The Profile Lake is a beautiful little pond, lying at the base of the mountain, and immediately under the over-watchful eye of the stern "Old Man." This lakelet is sometimes called the "Old Man's Wash-bowl." It is a quarter of a mile long and about half as wide.

Echo Lake, one of the greatest charms of this part of the White Mt. region, is a diminutive but very deep and beautiful pond, north of the Cannon Mountain. It is entirely enclosed by high mountains. From the centre of this fairy water, a voice, in ordinary tone, will be echoed distinctly several times, and the report of a gun breaks upon the rocks like the roar of artillery. The Indian superstition was, that these echoes were the voice of the Great Spirit, speaking in gentleness or in anger.

The Basin, another remarkable scene of this neighborhood, is five miles south of the Notch. It lies near the road side, where the Pemigewasset has worn deep and curious cavities in the solid rocks. The basin is 45 feet in diameter, and 28 feet from the edge to the bottom of the water. It is nearly circular, and has been gradually made by the whirling of rocks round and round in the strong current. The water, as it comes from the Basin, falls into most charming cascades. At the outlet, the lower edge of the rocks has been worn into a very remarkable likeness of the human leg and foot.

The Flume, the last and most famous, perhaps, of all the Franconia wonders, is quickly and easily reached from the Flume House. Leaving the road, just below the Basin, we turn to the left among the hills, and after a tramp of a mile, reach a bare granite ledge a hundred feet high and about thirty feet wide, over which a small stream makes its varied way. Near the top of this ledge we approach the ravine known as the Flume. The rocky walls here are fifty feet in height, and not more than twenty feet apart. Through this grand fissure comes the little brook which we have just seen. Except in seasons of freshets, the bed of the stream is narrow enough to give the visitor dry passage up the curious glen, which extends several hundred feet, the walls approaching, near the upper extremity, to within ten or eleven feet of each other.

About midway, a tremendous boulder, several tons in weight, hangs suspended between the cliffs, where it has been caught in its descent from the mountain above. A dangerous bridge for a timid step has been sprung across the ravine, near the top, by the falling of a forest tree.

The Pool, a supplemental or tail piece to the great picture of the Flume, is a deep natural well in the solid rock. The diameter of the Pool is about sixty feet; the depth to the surface of the water is 150 feet, and the water itself extends 40 feet yet below. Some years ago, a poor fellow was unlucky enough to fall into this plutonian *cul de sac*, but he clung to a crag just above the water until ropes were lowered, and he was, wonderful to relate, fished up, though bruised and not a little scared, yet alive!

We have now peeped hastily at the leading points of interest in the grand Granite Hills; but the enterprising tourist of an enquiring turn of mind, may, very easily, discover for himself a thousand other marvels and delights; or, if he cares not to explore new scenes for himself, he may bend his way northward, via Littleton, and between Colebrook and Errol, penetrate the well-known, but as yet very little visited slate-stone gorge, called the Dixville Notch.

VERMONT.

The thousand points of interest in the Green Hills of Vermont have not yet received their due meed of favor from tourists, but their claims to especial homage are now being fully admitted. The mountain chain extends from near New Haven, in Connecticut, northward through Massachusetts and Vermont, into Canada; though, properly speaking, it lies in Vermont alone, where are the chief summits of Mansfield, Camel's Hump, Connoll's Peak, Shrewsbury Mountain, South Peak, Killington Peak, Ascutney (on the Connecticut), and others. After the White Mountains of New Hampshire, the Green Hills rank with the noblest groups west of the Rocky Mountains—with the Blue Ridge in North Carolina, Georgia, and Virginia, the Alleghanies in Pennsylvania, and the Catskill and the Adirondacks, New York.

The Vermont Central Railway from Burlington, on Lake Champlain, eastward via Montpelier, the capital of Vermont, to the shores of the Connecticut River, traverses the valley of the Winooski, by the banks of the Winooski River, and gives easy access to Mount Camel's Hump and Mount Mansfield.

The Valley and River of Winooski.—The Winooski traverses almost the entire breadth of northern Vermont. Rising in Caledonia, its course is generally westward to Lake Champlain, 40 miles from which it passes through Montpelier, the capital of Vermont. The Vermont Central Railway from Burlington to the Connecticut runs through the valley, and very closely follows the banks of the river as far eastward as Montpelier. Some of its valley passages are scenes of great pastoral beauty, strongly contrasted with high mountain surroundings, the singularly-formed peak of Camel's Hump continually reoccurring, sometimes barely peeping over intervening ranges, and again—as near the middle of the valley stretch—coming into full display. In places, the Winooski is a wild turbulent water, dashing over stern precipices and through rugged defiles. It is found in this rough mood just above the village of Winooski, a few miles from Burlington, where the waters rush in rapid and cascade through a ravine a hundred feet. This picture is well seen from the railway. Passing on into the open valley lands, which succeed, Mount Camel's Hump comes finely into view, as the central and crowning point of one of the sweetest pictures of all this region. Continuing always through scenes of great picturesque interest, the tourist comes to the village of Richmond, 13 miles from Burlington, and three miles, yet beyond, to Jonesville, a little scattered village of fine farms, lying upon both sides of the river. The inn here was a famous half-way house in the journey between Burlington and Montpelier, before the railway was built, and it is still a pic-nic and excursion resort for all the region. It is an excellent and very inexpensive place for a little quiet tarry. Mount Camel's Hump is accessible from this vicinity, and a stage runs from the hotel, some nine or ten miles, to Underhill Centre, at the foot of Mount Mansfield. On the south side of the Winooski, at Jonesville, the Huntington River comes in, after an exceedingly wild journey for the last two or three miles of its course, through fine rocky ravines, which it traverses with many bold foaming leaps. About three miles yet east of Jonesville, near Bolton, there is the most striking picture in all the Winooski gallery. It is beautifully seen from the cars on the left, but yet very inadequately. From the bottom of the glen through which the maddened waters here make their way, the huge cliffs on either hand, the torrent foaming at their feet, and the gentle bit of verdant interval, with the tall crest of Camel's Hump, seen beyond, within the frame of the opposing precipices, make altogether a scene well worth the looking for. The pictorial interest of this valley transit is admirably sustained at all points, as far as we yet follow it onward, through Waterbury and Middlesex, to Montpelier, where some of the best scenes the traveller will find, when he looks abroad from the hill-tops there, have been kept to the last.

Mount Camel's Hump, the most salient feature in the Winooski landscape, is the highest of all the Green Mountain peaks, having an elevation of 4,188 feet. It may be ascended, without much difficulty, from any side, though the usual point of access is at Duxbury, from whence carriages can pass to within three miles of the summit. The mountain is crowned by jagged, barren rocks, and the imposing scene which the lofty heights overlook is in no way obstructed by the forest veil, which often disappoints the hopeful climber of forbidding mountain tops.

Mansfield Mountain, the second in dignity of the Green Hills, is very accessible from the village of Underhill Centre on the north, or yet more easily from Stow on the south, both of which points may be reached from the Vermont Central road—Underhill, from Jonesville station, and Stow from Waterbury. The views of the mountain itself, its cliffs and peaks, are very grand from many points in the path upwards, and the panorama unfolded upon the summit is, if possible, finer than that from the Camel's Hump. Lake Champlain and the Adirondack peaks lie to the westward, while the White Mountains of New Hampshire make yet new pictures on the east; and, again, the many crests of the Green Hills, with their intervening vales and lakes and villages, stretch out towards the south.

Montpelier.—HOTELS :— *The Pavilion.*

Montpelier on the east, and Burlington on the west, are the two extremities of the Winooski section of the Green Mountain scenery. Burlington is upon the Lake Champlain shore, and is the largest town in Vermont. We have already mentioned it in our tour from New York to Canada. Montpelier is the State capital. It has a population of between two and three thousand; is a very pretty town, and with the charm of most beautiful natural surroundings, and the comforts of an excellent hotel, it is perfectly eligible as a summer residence. The Winooski river passes here. A little branch railway, of a mile or so, connects Montpelier with the Central road. The State House, which was destroyed by fire January 6th, 1857, was an imposing granite edifice, in the form of a cross. It was 150 feet in length, and 100 deep. It had a fine portico of six columns, each six feet in diameter at the base, and 36 feet high. The apex of the dome which crowned the structure was 100 feet from the ground. Among the objects of interest in this edifice were two pieces of cannon taken from the Hessians at the battle of Bennington.

Rutland to Burlington.

Rutland.—HOTELS :— *Bardwell House,* and *Central House.*

Rutland is near the western borders of Vermont, south of the centre of the State, and nearly east of Whitehall, at the lower extremity of Lake Champlain. It is a centre of railway lines for all points of the compass. The Troy, Whitehall, and Castleton R. R., 95 miles, unites Rutland with Troy and Albany, via Whitehall and Saratoga Springs. It is also connected, again, with Troy and Albany, by the Albany, Vermont, and Canada line (formerly Albany Northern) to Eagles Bridge, and thence by the Rutland and Washington road, 95 miles, as via Whitehall and Saratoga; and yet again, via North Bennington, by the Troy and Boston and Western Vermont roads, 84 miles. The famous Hoosick tunnel is near the line of this route. Northward, it is connected with Burlington, and all the routes which intersect at that point, by the Rutland and Burlington R. R., 50 miles and eastward with Bellows' Falls, on the Connecticut, by another division of the same line. Rutland is a pleasant town, with a population of about 4,000, situated in the midst of some of the finest of the Vermont hill and valley scenery, at the foot of the western slope of the mountains. Otter Creek, a most picturesque stream in all its course, passes by the village, and Killington Peak is admirably seen as the leading feature in the landscape around.

The Clarendon Springs, a favorite place of resort, is a few miles south of Rutland, on the Rutland and Washington Railroad. The medical virtues of these waters, the varied and beautiful scenery, the pleasant drives around, and the excellent hotel accommodations, make this watering-place a very desirable summer halt.

The Otter Creek Falls, at Vergennes, are upon the Otter Creek, about seven miles from Lake Champlain. The brook is 500 feet in width, divided by a fine island, on either side of which the Fall leaps bravely some 30 or 40 feet. There are many other beautiful cascades in the Otter Creek. Some at Middlebury, above Rutland; and a few miles below Middlebury, still others of yet greater interest.

The Elgin Spring is in the neighborhood of the Otter Creek cascades.

Killington Peak, rising grandly on the east of Rutland, is the third in rank of the mountains of Vermont. A visit to this peak makes a pleasant excursion from the neighborhood. To the foot of the mountain the distance is seven miles, and two miles more to the summit. On the north side is a perpendicular ledge of 200 feet, called Capitol Rock. Mount Ira, too, is hereabouts, and beyond Killington Peak, as seen from Rutland, and northward are Mount Pico and Castleton Ridge, shutting out the view of Lake Champlain.

Lake Dunmore is a delicious water, 30

miles above Rutland. It is on the railway to Burlington, a few miles (by stage) from Middlebury. Dunmore is a wonderfully picturesque lake, surrounded at most points by bold hills, seen here in verdant slopes, and there in rocky bluff and cliff. The lake is about five miles in length and three in breadth. A good summer hotel is upon its banks.

Lake Castleton, in this neighborhood, is also a most interesting scene.

Eastward from Rutland, the route lies amidst the beauties of the Queechy Valley, replete with delightful pictures of running and falling waters, and of grassy meadows and wooded hills.

Middlebury is upon the railway to Burlington, 30 miles from Rutland. It is upon the Otter Creek, near some fine falls on that stream, and is also a few miles only from Lake Dunmore. It has a population (the township) of some 4,000, and, like nearly all the villages in Vermont, is a very beautiful place, surrounded at all points by most attractive mountain scenery. It is distinguished as one of the first manufacturing towns in the State, and also as the seat of Middlebury College. This institution was founded in 1800. Its chief edifice is 100 feet long and four stories high, built of stone.

Brandon, on the route of the Rutland railroad, is a flourishing town, finely watered by Otter Creek, Mill River, and Spring Pond, on which streams are good mill-seats. Minerals of fine quality are found in this town. There are here two curious caverns formed of limestone, the largest containing two apartments, each from 16 to 20 feet square. It is entered by descending from the surface about 20 feet.

Bennington is at the meeting of the Troy and Boston and the Western Vermont railways, in the extreme south-west corner of the State. It is famous as the scene of the battle of Bennington (August 16, 1777), in which a detachment of the enemy's forces, under Gen. Burgoyne, was terribly beaten by the Green Mountain Boys, led by the intrepid Major Stark. It was upon the occasion of this memorable engagement that Stark is reported to have made the famous address to his troops: "See there, men ! there are the red coats ! Before to-night they are ours, or Molly Stark will be a widow !" Two pieces of artillery, taken in the battle of Bennington, were, until recently, preserved in the Capitol at Montpelier. The manufactories of the United States Pottery Company at Bennington are well deserving of a visit. Fine porcelain and Parian ware are made here, the vicinage yielding the necessary materials in abundant and excellent supply. The landscape about Bennington is not of especial attraction.

Willoughby Lake is a popular resort in Orleans County, Vermont, lying upon the Canada line. This lake is a beautiful water, nearly five miles long. It lies upon the great railway route from Boston, via St. Johnsbury, to Canada, leaving the Connecticut Valley route at Wells River.

Lake Memphremagog is 30 miles long, and from one to four in breadth. About eight miles only of its waters are in Vermont, the rest lying in Canada. It unites its waters, by Magog outlet, with those of the St. Francis river, in Canada.

For Mount Ascutney, Windsor, Bellows' Falls, Brattleboro', and other scenes and places in Vermont, on and near the Connecticut River, see route through that region.

NEW YORK.

NEW YORK is very appositely called the Empire State; being the first in the confederacy in population, first in wealth, and in commercial importance; exceeded by none in soil and climate, unsurpassed in the variety and beauty of her natural scenery, and in her historical associations.

The earliest settlements here were made by the Dutch, at Fort Orange, now called Albany, and at New Amsterdam, now New York City. This was in 1614, seven years after the voyage of Hendrick Hudson up the waters of that river which now bears his name.

In 1664, the colony fell into the possession of the English—was recaptured by the Dutch in 1673, and finally came again under British rule in 1674—and so continued until the period of the Revolution. Many stirring events transpired within this territory during the wars between France and England, in 1690, 1702, and 1744, and through all the years of the War of Independence. Of these events the traveller will find some chronicle as we reach the various locations where they transpired, hereafter, in the course of our proposed travels.

Every variety of surface and every character of physical aspect is found within the great area of New York; vast fertile plains and grand mountain ranges; meadows of richest verdure, and wild forest tracts; lakes innumerable and of infinite variety in size and beauty; waterfalls unequalled in the world in extent and grandeur; and rivers matchless in picturesque charms. We need not now catalogue these wonders, as our rambles will afford us, by and by, abundant opportunity to see them all in turn and time—the peaks and gorges of the Adirondacks and the Catskills—the floods of Niagara and the ravines of Trenton; the pure waters of Lake George, the mountain shores of Champlain, the deer-filled wildernesses and the Highland passes of the Hudson, and all the intricate reticulation of cities, towns, villages, villas, and watering-places.

RAILWAYS IN NEW YORK.

See index for description of the routes.

The New York and Erie, 459 miles through the State, from the city of New York to Dunkirk, or 422 to Buffalo (Branch) on Lake Erie. A route to the far west.

The Hudson River Railway, 146 miles to Albany, or 152 to Troy, along the banks of the Hudson River, from New York City.

The Harlem Railway, 154 miles from New York to Albany.

New York Central, from Albany to Buffalo, 398 miles, or to Niagara Falls, 327 miles—unites eastward with the Western Railway from Boston, and with the Hudson River and Harlem roads from New York and at the western extremity, with routes for the Mississippi regions.

Rensselaer and Saratoga, and Saratoga and Whitehall. From Troy to Saratoga Springs, 32 miles; to Whitehall, 72 miles.

Troy and Boston, and Albany and Rutland railways.

Montreal and New York, and Plattsburg and Montreal railways, 62 miles from Plattsburg, on Lake Champlain, to Montreal, Canada.

Northern (Ogdensburg) Railway, across the northern part of the State, 118 miles, from Rouse's Point, on Lake Champlain, to Ogdensburg on the St. Lawrence.

Black River and Utica Railway, 35 miles from Utica, on the New York Central Road to Boonville.

Watertown and Rome, 97 miles from Rome on the New York Central, to Cape Vincent, on Lake Ontario.

Potsdam and Watertown, from Watertown junction (Watertown and Potsdam Road) to Potsdam on the Northern (Ogdensburg) Railway.

Newburgh Branch of New York and Erie Railway; from Newburgh on the Hudson, to Chester, New York and Erie Railway.

Oswego and Syracuse; 35 miles from Syracuse, New York Central Road, to Oswego, Lake Ontario.

Syracuse and Southern; 80 miles from Binghamton (Erie Railway) to Syracuse (New York Central Railway).

Oswego (New York and Erie Railway), 35 miles to Ithaca, on Cayuga Lake.

Elmira, Canandaigua and Niagara Falls; 168 miles from Elmira (Erie Railway), to Suspension Bridge, Niagara.

Buffalo, Corning and New York; from Corning (Erie Railway), 100 miles to Batavia, or 94 miles to Rochester (New York Central Road).

Williamsport and Elmira; 78 miles from Elmira (Erie Railway), south to Williamsport, Pa.

Corning and Blossburg and Tioga; 41 miles from Corning (Erie Railway) to Blossburg, Pa.

Lake Shore Railway, from Buffalo, via Dunkirk, by the shore of Lake Erie, to Cleveland, Ohio, and westward.

Hudson and Boston; from Hudson, on the Hudson River, eastward, to West Stockbridge, 34 miles.

Western Railway; from Albany, 200 miles, to Boston.

New York and New Haven; 75 miles, from New York to New Haven, Ct., thence to Boston, etc.

Long Island Railway; 95 miles from New York (Brooklyn Ferry), through the entire length of Long Island, to Greenport.

THE CITY OF NEW YORK.

Hotels. The Astor House, in Broadway, opposite the Park, is the only leading hotel left in the lower part of the city. The St. Nicholas, one of the most splendid of all the hotels constructed in New York, is in Broadway, between Broome and Spring streets. The Metropolitan Hotel, corner of Broadway and Prince—in the rear of this hotel is Niblo's Theatre. The Fifth Avenue Hotel is a palatial marble edifice, of the most imposing extent and of the highest fashion. It covers the ground between 23d and 24th streets on Fifth Avenue, at its intersection with Broadway—opposite Madison square. Rapidly as the large hotels have been creeping "up town" of late, the "Fifth Avenue" outstrips all other strides, and stands now higher up on the island than any of its competitors. The Prescott House is on Broadway and Spring street. The Lafargo House is a large hotel of the highest class. It has an imposing façade of white marble upon Broadway, between Bond and Amity streets. The "Winter Garden" Theatre is in the rear, with entrance on Broadway, through the hotel edifice. The New York Hotel is one of the largest and most fashionable in the city: It is "up town," on Broadway, extending from Washington to Waverley Place. The St. Dennis, corner of Broadway and Eleventh street, of unique architecture, is one of the best appointed and most fashionable houses in the city. Union Place Hotel is an elegant establishment upon the corner of Broadway and Union Square. In front, northward, is the bronze equestrian statue of Washington, erected July 4th, 1856, the Union Park and fountain. The Everett House, erected within the past few years, is an imposing, ornamented brick edifice fronting on Fourth Avenue and the north line of Union Square. It is directly across, opposite the Union Place Hotel. It has both *table d'hôte* and restaurant. The Clarendon is yet higher up, on Fourth avenue, corner of Eighteenth street. It is a *recherché* house of high fashion, in especial favor with English travellers. The Brevoort House, Fifth Avenue and Clinton Place, and the St. Germain, Fifth Avenue and Broadway, and Twenty-second street, are new up-town hotels of the highest rank. They are, like the Everett House, conducted on the European plan, with both *table d'hôte* and café, rooms and board, or either alone. The Julian, in Washington Place, near Broadway, is a fashionable hotel and boarding-house.

There are, besides, very many most excellent hotels, and hundreds of the second and third class—but we have mentioned enough for all the uses of the traveller.

Supposing our traveller to be at home in New York, or a stranger already comfortably lodged at his hotel, we will (leaving it to a later moment to see whether or not he has got into the right place) gossip, for a brief while, touching the past of the scenes, of which we propose to show him the present.

The rapid growth of this great city—so little time gone by a wild, forest settlement, and now magnificent in its million people—is evidence enough of the mind and will of the race, which is now everywhere making the once wilderness of the west to blossom as the rose. Though settled by the Dutch as early as 1612, the metropolitan character of New York scarcely dates back to the beginning of the present century: for it is within the past 50 years, or less, that all its present municipal glories and fame have grown up. Not so long, indeed, for the city which now covers nearly the whole Island of Manhattan, and is running over every day into other cities, villages, and suburbs, wherever it can find vent, was, within a shorter period, composed in the small triangular area, of which the Battery is the apex, and Canal street the base. The City Hall was then built with less care, on the upper or north side, because that, at the time, overlooked, and was seen only from lanes and fields. To go *above*, or even to Canal street, (then, literally, a canal), was a rural excursion; while, to go *below* it, at this day, is to exceed the general down-town travel, on any but business errands. Of the palatial private houses, the public structures, the magnificent churches, the parks, and even the streets, in all the middle and upper parts of the city, no mention would have been made in an edition of this work twenty-five years ago; which leads us to ask, what story it may be necessary to tell, in the revised edition of a quarter of a century hence! But, thinking no longer of past or future, let us come at once to the present, and see New York as it is.

Panorama of the City. The visitor will do well to accompany us to the lofty outlook from the top of the tower of Trinity Church, in the lower part of Broadway. Here he can pick up some general idea of the topography and extent of the City. He may go there at any time when the building is not in use for sacred service, paying the porter a fee for his guidance. At the landing, on a level with the ceiling of the church, a fine view is had of the beautiful interior. At the head of another flight of stairs, the belfry, with its pleasant chimes, is reached. Here, too, is a balcony affording a fine view of the City; but it is still higher up that the scene is spread forth in all its glory—a boundless array of charms, in city, and town, and village, river, and bay, and island, all teeming with bright and busy life and action.

With this superb picture, or rather galleries of pictures, before him, the observer gets a better idea than he may, perhaps, have had before of one of the natural advantages which has made New York the great metropolis of this wide country; its noble position at the meeting of great waters, leading inland, and its unrivalled harbors upon the sea.

Yonder stretches the beautiful bay—one of the safest and easiest of access in the world—eight miles out to those great portals famous as the "Narrows," which open its way to the ocean. The circumference of the harbor is 25 miles, within which the combined navies of the world might lie in comfort and security. The scenery here is of infinite attraction, in all the protean shapes and effects of mingled land and water. Great ships and little crafts innumerable seem to jostle each other, and cities, and villages, and villas crowd the shore, from the water's edge to the bold hill-tops. The outer harbor, or the bay proper, extends from the Narrows to Sandy Hook Light, 18 miles from the city. Within the harbor are the picturesque fortifications on Governor's, on Bedlow's, and on Ellis's Islands. Fort Columbus occupies the centre of Governor's Island, and at its north-east point is Castle William, a round tower some 600 feet in circumference, and 60 feet high, with three tiers of guns; while at the north-west is a battery, commanding the entrance to Buttermilk Channel, by which the island is separated from the City of Brooklyn. The defences on the Long Island shore, at the Narrows, are Forts Hamilton, and Lafayette—formerly Fort Diamond. This neighborhood is a popular summer resort and residence of the people of New York. Opposite these fortifications, on the Staten Island shore, separated here by the passage of the Narrows, about two-thirds of a mile in width, are Forts Tompkins and Richmond.

Staten Island, a favorite suburban home of New York, and to which the Bay is indebted for so much of its beauty, is about six miles below the city, with which it has frequent daily connection. The island is 14 miles long, and from four to eight wide. It constitutes the county of Richmond, and forms the southern extremity of the State of New York. It is separated from New Jersey, on the west, by Staten Island Sound. Richmond Hill, at the north end of the island, commands all the grand scenes which might be expected in this vicinity, at an elevation of 307 feet above the sea. Elegant residences cluster about these heights, and, from the summit, a marine telegraph overlooks them and the sea. Upon a bluff, on the east side of the island, is Prince's Bay Light House.

City Parks and Squares. The Central Park is a new public domain of such grand extent that it will, before many years, rival the most famous places of the kind in the world. It is situated in the upper part of the city, between the Fifth and the Eighth Avenues, East and West, and 59th and 110th streets, South and North—a noble area of 843 acres, extending 2½ miles in length by ½ a mile in breadth. Millions were expended in the purchase of the ground, and millions more will soon have been generously laid out in embellishing it. The work of improvement—with the help of an army of 3,000 laborers—went on so fast and so magically, that the avenues, and drives, and walks, hills and dales, and lawns and lakes, already make it a very popular resort. In the winter time its frozen ponds are covered with myriads of rollicking skaters, many of whom are of the *beau sex*. Access—(direct) by the city railways—the 3d, 6th, 8th, Broadway, or 7th avenue roads—fare, 6 cents.

The Battery, which contains about 11 acres, is situated at the extreme south end of the City, at the commencement of Broadway, and is planted with trees and laid out in gravel walks. From this place is a delightful view of the harbor and its islands, of the numerous vessels arriving and departing, of the adjacent shores of New Jersey, and of Staten and Long Islands. Castle Garden, on the Battery, was at one time a popular public hall. Here Jenny Lind first sang in America. Here, too, the fairs of the American Institute were once held. It is given over now to the Emigrant Office for a receptacle of the debarking foreign populations.

The Bowling Green, situated near the Battery, and at the commencement of Broadway, is of an oval form, and surrounded by an iron railing. Within its enclosure is a fountain, the water of which falls in pleasant whispers, to the dusty streets, of the freshness and beauty of forest-wilds.

The Park is a triangular enclosure in the lower part of the city; it has an area of 11 acres, containing the City Hall and other buildings.

St. John's Park. Small but beautiful grounds in Hudson street, belonging to the vestry of Trinity Church. St. John's Church, a Chapel of Trinity, is on the east side of the square.

Washington Square is a pleasant uptown park, a little west of Broadway, with the elegant private residences of Waverley Place and Fourth street on the north and south sides, and upon the east the grand marble edifice of the New York University, and Dr. Hutton's beautiful Gothic church. A superb fountain occupies the centre of these grounds.

Union Park, a most charming bit of wood and lawn, is in Union Square, at the bend in the upper part of Broadway, extending from Fourteenth to Seventeenth streets. On the south-east corner of Union Square is the Union Place Hotel and the fine bronze equestrian statue of Washington, by Henry K. Brown. On the upper side is the Everett House, and, near by, the Clarendon Hotel. Upon the west is Dr. Cheever's "Church of the Puritans."

Gramercy Park is a little to the north-east of Union Square, a charming ground, belonging to the owners of the elegant private homes around it.

Stuyvesant Park is divided in the centre by the passage of the Second Avenue. It extends from Fifteenth to Seventeenth streets. The Saint George's Church (Rev. Dr. Tyng) is upon the west side of this park.

Tompkins Square, one of the largest parks of New York, is between Avenues A and B, and Seventh and Tenth streets.

Madison Square is up town, just above the intersection of Broadway and Fifth Avenue at Twenty-third street.

Hamilton Square, newer ground, still above.

PUBLIC BUILDINGS—MUNICIPAL.

The City Hall is an imposing edifice; the south front is built of marble, and the rear, or north side, of Nyack freestone. It was constructed between the years 1803 and 1810. It occupies the centre of the Park, in the lower part of the city, and is surrounded by other city offices. It is at present proposed to enlarge it very greatly. In this building are twenty-eight offices, and other public apartments, the principal of which is the Governor's room, appropriated to the use of that functionary on his visiting the city, and occasionally to that of other distinguished individuals. The walls of this room are embellished with a fine collection of portraits of men celebrated in the civil, military, or naval history of the country. In the Common Council room is the identical chair occupied by Washington when President of the first American Congress, which assembled in this city.

The *Custom House* (formerly the Exchange) is on Wall street. It is built of Quincy granite, and is fire-proof, no wood having been used in its construction, except for the doors and window frames. It is erected on the site occupied by the Exchange building destroyed by the great fire of 1835. The present one, however, covers the entire block, and is 200 feet long by 171 to 144 wide, and 124 to the top of the dome. Its entire cost, including the ground, was over $1,800,000.

The *U. S. Treasury* (once the Custom House) is on Wall and Nassau streets. It is built of white marble, in the Doric order, similar in model to the Parthenon at Athens. It is 200 feet long, 90 wide, and 80 high. The great hall for the transaction of business is a circular room, 60 feet in diameter, surmounted by a dome, supported by 16 Corinthian columns, 30 feet high, and having a skylight, through which the hall is lighted.

The *Post Office* is in Nassau street, between Cedar and Liberty streets. The building is in no way remarkable for any architectural beauty, but merely as being one of the remnants of the past, having been formerly used as a church by one of the old Dutch congregations.

The *Hall of Justice*, or "Tombs," is located in Centre street, between Leonard and Franklin streets. It is a substantial-looking building, in the Egyptian style of architecture, 253 feet long and 200 wide, constructed of a light-colored granite.

Literary Institutions and Libraries.

Columbia College has been recently removed from the foot of Park Place, near Broadway, far up town, having resigned the old grounds which it has occupied for so many years. The extension of Park Place has already destroyed the ancient green lawns, and its venerable buildings. Columbia College was chartered by George II. in 1754, under the title of King's College. Students, 150. Library, 10,000 vols.

The New York University occupies a grand Gothic edifice of white marble, upon the east side of Washington Park. This structure is a fine example of pointed architecture, not unlike that of King's College, Cambridge, England. The chapel—in the central building—is, with its noble window, 50 feet high and 24 feet wide, one of the most beautiful rooms in the country. The whole edifice is 180 feet long. Founded in 1831.

The Free Academy, Lexington Avenue and 23d street, up town, reached by Harlem cars or Fourth Avenue stages. This is a public collegiate academy of the highest rank. Its students are chosen from the pupils of the public schools only. The building is a fine structure, in the style of the town halls of the Netherlands. It will accommodate 1,000 pupils.

The Cooper Union occupies a magnificent brown stone edifice opposite the Bible House on Astor Place, at the point where the union of the Third and Fourth avenues forms the Bowery. This establishment is familiarly known as the Cooper Institute. It was founded by the generous munificence of Peter Cooper, Esq., an eminent merchant of New York. The building erected for its uses cost about 600,000 dollars. It is devoted to the free education of the *people* in the practical arts and sciences. It was publicly opened in November, 1859, with over 2,000 students. It contains a noble free reading-room. One of its departments is a School of Design for women.

The General Theological Seminary of the Protestant Episcopal Church, is charmingly situated on West 20th street, between Ninth and Tenth Avenues.

St. Francis Xavier, 39 West 15th street.—Union Theological Seminary, 9 University Place, just above the New York University.

The New York Historical Society occupies a dainty edifice on Second Avenue, corner of 11th street. Its library is large and valuable; besides which it possesses a fine collection of works of art.

American Geographical Society has rooms in Clinton Hall, Astor Place.

Lyceum of Natural History, in the building of the New York University Medical School,

14th street near Fourth Avenue. Incorporated in 1808, for scientific advancement. The Society possesses a large library, and a fine cabinet of mineralogical specimens.

New York Law Institute. City Hall.

The Astor Library is a public collection of high order, founded by the munificence of the late John Jacob Astor. It numbers at present about 110,000 volumes. The building, on Lafayette Place, is one of the chief architectural attractions of the city.

The Mercantile Library, Clinton Hall (late Astor Place Opera House), Eighth street, near Broadway. This old and popular institution has at present some 54,000 volumes, in every department of letters. Its members number between 4,000 and 5,000. The winter courses of lectures before the Mercantile Library Association are among the greatest pleasures of the season.

New York Society Library now occupies a new and beautiful building in University Place, near 12th street. It possesses about 36,000 books.

Apprentices' Library (14,000 vols.), is in the Mechanics' Hall, Broadway, near Grand street.

The American Institute is at present in apartments on the first floor of the Cooper Institute, in Astor Place. The Annual Exhibitions of mechanic art and industry, of this Society, make a feature in the autumn pleasures of the metropolis.

The Mechanics' Institute has a library of about 3,000 volumes; 20 Fourth Avenue.

ART SOCIETIES AND GALLERIES.

The National Academy of Design—the chief Art institution of America—was founded in 1826, since which time it has steadily advanced in influence and usefulness. It numbers among its academicians and associates nearly all of the eminent artists of the city and vicinity. It supports free schools for the study of the antique and the living model; possesses an extensive and valuable Art library; makes Annual Exhibitions of original works by American and foreign painters and sculptors, &c. The Exhibitions of the National Academy are the great event of the spring season in New York. A noble marble edifice has just been built for the Academy on 23d street and 4th avenue.

Studios. In Tenth street, near the Sixth av., there is a spacious quadrangular edifice, called the Studio Building, occupied entirely by artists. A fine gallery, for the uses of the fraternity, fills the court. Dodworth's, 212 Fifth Avenue, Madison Square, is another famous lair of the knights of the easel, and so too is the University in Washington Park. The artist brotherhood of New York is large and potent in character, both socially and professionally. The stranger or the citizen may while away pleasant days and weeks in exploring their lofty abodes.

The *Artists' Fund Society*, founded in 1859, makes an Annual Exhibition and sale of works of art, in the months of November and December. The Society has yet no fixed abode.

Free Galleries for the exhibition and sale of works of Art: Schaus' Gallery, 749 Broadway; Goupil's, Broadway and Ninth street; Williams', 353 Broadway; Snedecor's, 768 Broadway.

The New Bible House is one of the largest structures in the city. It covers the entire area between Third and Fourth avenues on the west and east, and Eighth and Ninth street on the south and north. The printing rooms and other offices of the American Bible Society are here.

The New York Hospital (founded in 1771) stands back on a lawn upon Broadway, opposite Pearl street.

Medical Schools. New York University Medical Department, 107 East Fourteenth street. College of Physicians and Surgeons, East Twenty-third street and Fourth Avenue. New York Academy of Medicine, meets the first Wednesday of each month at the University.

Institution for the Blind, occupies a large and imposing Gothic edifice of granite, on Ninth avenue, in the north-west part of the city. Reached by the Ninth and Eighth avenue stages. Visitors received on Tuesdays, from 1 to 5 P. M. The institution has about one hundred pupils.

Deaf and Dumb Asylum, Fiftieth street and Fourth avenue, via Harlem railroad. Visitors admitted from 1 to 4 P. M. The large and commodious building of this Institution accommodates about two hundred and fifty pupils.

The Bloomingdale Asylum for the Insane and the *New York Orphan Asylum*, are in the upper part of the Island, 7 miles from the City Hall, on the line of one of the pleasantest drives about New York.

Blackwell's Island, the City Penitentiary, the Lunatic Asylum, the Alms House, Hospital and Work House, on Blackwell's Island, in the East River, are worth the especial attention of the stranger. This Island, as also Ward's and Randall's Islands, may be reached by steamboat, from foot of Grand street, East River, at 12 M. daily; or by the Harlem stages to Sixty-first street. Stages leave 25 Chatham street every fifteen minutes.

Ward's and Randall's Islands, near by, are occupied by the public charitable Institutions. The elegant and massive structures which cover this famous group of islands make a striking feature in the landscape, as we sail up the East River to the suburban villages on Long Island, or en route for Long Island Sound.

Churches. New York has nearly 300 churches, many of which are very costly and imposing edifices. Among those most worthy the notice of the stranger are *Trinity Church* (Episcopal), in the lower part of Broadway; *St. Paul's* (Episcopal), not far off, in Broadway; *St. John's* (Episcopal), in St. John's Park; *St. Thomas's* (Episcopal), Broadway and Houston street; *Grace Church* (Episcopal), Broadway and Tenth street; *Church of the Puritans,* Union sq.; *St. Paul's* (Methodist), Fourth Avenue; *Dutch Reformed* (Dr. Hutton), Washington Square; *St. Mark's* (Episcopal, Dr. Anthon), Stuyvesant street; *St. George's* (Dr. Tyng, Episcopal), East Sixteenth street, Stuyvesant Square; *First Baptist*, corner of Broome and Elizabeth streets; *Amity Street*, Dr. Williams (Baptist), 31 Amity; *Madison Av.* (Baptist), Dr. Hague; *16th Baptist*, near Eighth av., Rev. W. S. Mikels; *St. Patrick's Cathedral* (R. C.), corner of Prince and Mott streets; *Dr. Potts'* (Presbyterian), in University Place, corner Tenth street; *Church of the Divine Unity* (Universalist), Dr. Chapin, 548 Broadway; *Church of the Messiah* (Unitarian), Dr. Osgood, 728 Broadway; *Church of All Souls* (Unitarian), Dr. Bellows, Fourth Avenue, corner of Twentieth street; *Church of the Holy Communion* (Episcopal), Dr. Muhlenburg, Sixth Avenue and Twentieth street; *Fifth Avenue Presbyterian Church* (Dr. Alexander), corner of Nineteenth street; *French Church* (Protestant Episcopal), Dr. Verren, corner of Church and Franklin streets; *Trinity Chapel* (Episcopal), Twenty-sixth street, near Broadway; *Church of the Annunciation* (Episcopal), Dr. Seabury, Fourteenth street, between Sixth and Seventh Avenues; *Church of the Ascension* (Episcopal), J. C. Smith, Fifth Avenue, corner of Tenth street; *Shaarai Tephila* (Gates of Prayer), Hebrew, 112 Wooster street.

Theatres, etc. The *Academy of Music,* or Italian Opera House, is at the corner of Fourteenth street and Irving Place. Seats 4,600 persons. This is one of the grandest edifices of the kind in the world.

Niblo's Garden and Saloon, rear of Metropolitan Hotel, Broadway.

Winter Garden, one of the leading theatres, is in the rear of the Lafarge House, 641 Broadway.

Wallack's Theatre, another of the most popular establishments, is corner of Broadway and 13th st.

The *Olympic Theatre* is in Broadway, between Bleecker and Houston streets.

Broadway Theatre (formerly Wallack's) is at 485 Broadway, near corner of Broome street.

Bowery Theatre, Bowery.

Barnum's Museum, cor. of Broadway and Ann.

The Croton Aqueduct, the greatest public work of the city, brings abundant supplies of water from the Croton Lake, 40 miles distant. The original cost of this magnificent labor was over thirteen millions of dollars. The Receiving Reservoir in the Central Park, and the great Distributing Reservoir in Fifth Avenue bet. 40th and 42d sts., are well worth seeing.

High Bridge is a noble work, constructed for the passage of the Croton Aqueduct over the Harlem River, from Westchester County to the Island of New York. The High Bridge may be pleasantly reached by the Third Avenue cars or the Harlem Railway (Fourth Avenue) to Harlem, and thence up the Harlem River a mile or two in excursion steamboats. Fare, 6 cents on the boats—same on the cars.

The **New Arsenal,** which takes the place of the old edifice now within the Central Park grounds, is on the Seventh Avenue. It may be reached by the Sixth Avenue or the Broadway railroads.

FIRST-CLASS BUSINESS HOUSES.

Banking House. Duncan, Sherman & Co., cor. Pine and Nassau. Travellers desiring letters of credit will find this house a good one.

Newspapers. The magnificent printing establishments of the New York Herald, the Times and the Tribune are well worth seeing.

Publishing Houses. The palatial edifice, occupied by the Appletons, 443 and 445 Broadway, and the wonderful establishment of the Harpers in Franklin Sq., will, each, well reward a visit.

Jewellers. See the world of rich treasures of Tiffany & Co., 550 Broadway, and of Ball, Black & Co., cor. of Broadway and Prince st.

Dry Goods. Explore the great dry goods palaces of A. T. Stewart & Co., Broadway, and of Lord & Taylor, 461 Broadway and Grand st.

Piano-fortes. Steinway & Sons, 71 East 14th st.

Rubber fabrics. The N. Y. Belting and Packing Co., 37 and 38 Park Row.

Life Insurance. The Guardian Life Ins. Co., 7 Nassau st., is worth visiting, and (with its high character) worth *using*, amidst the risks of travel.

Fire Insurance. The Home Ins. Co., 112 & 114 Broadway, New York, is a beautiful building, its interior arrangements most complete, and the company one of the best in the country.

Clothing. Supply your wardrobes, to suit your most fickle fancy, at the great dépôt of Derby & Co., 57 Walker st.

Sewing Machines. Grover & Baker, 495 Broadway, and Wheeler & Wilson, 625 Broadway.

Furniture and Housekeeping Articles. J. & C. Berrian, 601 Broadway. A curious and interesting place.

Art Materials and Picture Gallery. Goupil & Co., Broadway and 9th st., Schaus, 749 Broadway.

Billiard Tables. Phelan & Collender.

Stationers. Francis & Loutrell, Maiden Lane.

The private palatial abodes on the *Fifth Avenue* and its vicinity, should be seen, if one would get any fair idea of the architectural beauty and splendor of the metropolis.

Harlem, a part of the city, at the north end of the Island, is upon the Harlem River. Cars from City Hall Harlem R. R. depots, or Third Avenue cars, City Hall, seven miles.

Bloomingdale and **Manhattanville** are at the north end of the Island of New York.

THE ENVIRONS OF NEW YORK.

Places of interest in the vicinage of the city.

Hoboken and **Weehawken,** charming rural resorts—in summer-time—across the Hudson River, on the New Jersey shore. Here are delightful walks, for miles, along the margin of the river, on high ground, overlooking the Bay and city; and all the country round—in shady woods, and upon verdant lawns, and among wild forest glens. Ferry, every few minutes (fare three cents), from Barclay, Canal, Christopher, and West Nineteenth streets.

Astoria, a suburban village on Long Island, six miles up the East River, near the famous whirlpool of Hell Gate, a place of beautiful villas. Steamboat, foot of Fulton street, East River, or by stage every hour, from 23 Chatham street, to foot of Eighty-sixth street—cross by Hell Gate Ferry.

Staten Island. *New Brighton, Port Richmond, and Sailors' Snug Harbor.* Ferry every hour and a half, from 8¼ A. M. to 6¼ P. M., from foot of Whitehall street. To Quarantine, Stapleton, and Vanderbilt's Landing, ferry every hour, foot of Whitehall street. Nothing can be more enjoyable than a sail down the Bay to any of the villages and landings of Staten Island; and nothing more agreeable than the sight of its many suburban villas, or of the superb views over land and sea which its high grounds command. Brighton is a particularly beautiful little village, with good hotels and boarding-houses. Near it is the Sailors' Snug Harbor, an ancient foundation for dilapidated mariners. Two miles east of Brighton is the Marine hospital and the village of Tompkinsville, and its 3,000 people. The voyage to Staten Island occupies about half an hour.

Fort Hamilton, 8 miles down the Bay, commands, in connection with Forts Lafayette and Tompkins, opposite, the passage seaward of the Narrows. A summer residence and resort for sea bathing. Via boats to Coney Island.

Coney Island, belonging to the town of Gravesend, is five miles long, and one broad, and and is situated about 12 miles from New York. It has a fine *beach* fronting the ocean, and is much frequented, but not by ladies. On the north side of the island is an hotel. Steamboats ply regularly between the city and Coney Island during the summer season. *Fare,* 12½ cents each way.

Rockaway Beach, a celebrated and fashionable watering-place on the Atlantic sea-coast, is in a south-east direction from New York. The *Marine Pavilion,* a splendid establishment, erected in 1834, upon the beach, a short distance from the ocean, is furnished in a style befitting its object, as a place of resort for gay and fashionable company. There is another hotel here which is well kept; also several private boarding-houses, where the visitor, seeking pleasure or health, may enjoy the invigorating ocean breeze, with less parade, and at a more reasonable cost than at the hotels. The best route to Rockaway is by the Long Island Railroad to Jamaica, twelve miles, 25 cents; thence by stage eight miles, over an excellent road, to the beach, 50 cents.

Long Branch.—HOTELS:—*The Metropolitan; the Ocean House.*

ROUTES. Steamboats Rip Van Winkle and Alice Price leave foot of Robinson street, New York, daily (except Sunday), at 8 A. M. and 4 P. M., and (extra) on Saturdays at 5¼ P. M.: connect at Port Monmouth with cars on the Raritan and Delaware Bay R. R.

There is admirable sport in this vicinity for the angler. The Shrewsbury river on the one side, and the ocean on the other, swarm with all the delicate varieties of fish with which our markets abound.

Shrewsbury, Red Bank, and **Tinton Falls,** in the vicinity of the above, are also places of great resort.

Flushing, on Long Island, 10 miles from the metropolis, is upon an arm of the Sound called Flushing Bay. The Linnæan Botanic Garden is here. Boat at Fulton street.

Flatbush, about five miles from Brooklyn, **Flatlands, Gravesend,** ten miles, are small but handsome places. Shores of the latter place abound with clams, oysters, and fowl, and are much resorted to.

Jamaica, another suburban town on Long Island, is 12 miles distant by the Long Island Railroad.

Greenwood Cemetery is in the south part of Brooklyn, at Gowanus, about three miles from New York and Brooklyn ferries. One of the numerous railways which so thickly and so conveniently link all parts of Brooklyn, extends to the Cemetery; the cars leaving the Fulton Ferry every five minutes, and speedily transporting the traveller for the small fare of five

4*

cents. On Sunday, only the owners of lots are admitted within the grounds.

This Cemetery was incorporated in 1838, and contains 242 acres of ground, about one half of which is covered with wood of a natural growth. It originally contained 172 acres, but recently 70 more have been added by purchase, and brought within the enclosure. Free entrance is allowed to persons on foot during week-days, but on the Sabbath none but the proprietors of lots and their families, and persons with them, are admitted; others than proprietors can obtain a permit for carriages on week-days. These grounds have a varied surface of hills, valleys, and plains. The elevations afford extensive views; that from Ocean Hill, near the western line, presents a wide range of the ocean, with a portion of Long Island. Battle Hill, in the northwest, commands an extensive view of the cities of Brooklyn and New York, the Hudson River, the noble bay, and of New Jersey and Staten Island. From the other elevated grounds in the Cemetery there are fine prospects. Greenwood is traversed by winding avenues and paths, which afford visitors an opportunity of seeing this extensive Cemetery, if sufficient time is taken for the purpose. Several of the monuments, original in their design, are very beautiful, and cannot fail to attract the notice of strangers. Those to the memory of Miss Canda, of the Indian Princess, Dohumme, and the "mad poet," McDonald Clark, near the Sylvan Water, are admirable; so also are the memorials to the Pilots and to the Firemen.

Visitors, by keeping the main avenue, called *The Tour*, as indicated by *guide-boards*, will obtain the best general view of the Cemetery, and will be able again to reach the entrance without difficulty. Unless this caution be observed, they may find themselves at a loss to discover their way out. By paying a little attention, however, to the grounds and guide-boards, they will soon be able to take other avenues, many of which pass through grounds of peculiar interest and beauty.

The New York Bay Cemetery is reached in a pleasant sail down the harbor. It is one of the most beautiful rural spots in all the beautiful vicinage of New York.

The U. S. Navy Yard is across the East River at Brooklyn. The *United States Naval Lyceum*, in the Navy Yard, is a literary institution, formed in 1833 by officers of the navy connected with the port. It contains a splendid collection of curiosities, and mineralogical and geological cabinets, with numerous other valuable and curious things worthy the inspection of the visitor. A *Dry Dock* has been constructed here at a cost of about $1,000,000. On the opposite side of the Wallabout, half a mile east of the Navy Yard, is the *Marine Hospital*, a fine building, erected on a commanding situation, and surrounded by upwards of 30 acres of well-cultivated ground.

At the Wallabout were stationed the Jersey and other prison-ships of the English during the Revolutionary war, in which it is said 11,500 American prisoners perished, from bad air, close confinement, and ill-treatment. In 1808, the bones of the sufferers, which had been washed out from the bank where they had been slightly buried, were collected, and deposited in 13 coffins, inscribed with the names of the 13 original States, and placed in a vault beneath a wooden building erected for the purpose, in Hudson Avenue, opposite to Front street, near the Navy Yard.

The Atlantic Dock, about a mile below the South Ferry, Brooklyn, is a very extensive work, and worthy the attention of strangers. The Hamilton Avenue Ferry, near the Battery, lands its passengers close by. The company was incorporated in May, 1840, with a capital of $1,000,000. The basin within the piers contains 42¼ acres, with sufficient depth of water for the largest ships. The piers are furnished with many spacious stone warehouses.

Jersey City.—HOTELS:—*American*, 9 and 11 Montgomery street.

Jersey City, N. J., is on the Hudson, opposite the City of New York, with which it is connected by continual ferry, from foot of Cortlandt street (fare three cents). The present population is about 30,000. Jersey City is the New York terminus of the Philadelphia and New York and Erie Railroad routes, and of the Morris Canal. It is also the berth of the Cunard line of Atlantic steamers.

For *Newark* and other places near New York, upon the Philadelphia routes, see index. For suburban villages on the Hudson, see route from New York to Albany.

CITY OF BROOKLYN.

Hotels. The Pierrepont House is an elegant establishment on the Heights; the Mansion House is in the same eligible part of the city; the Globe Hotel is at 244 Fulton street.

We have already spoken of many of the objects of interest in Brooklyn, in the preceding article upon New York; as the Navy Yard, Greenwood Cemetery, the Atlantic Dock, the neighboring Long Island villages of Astoria, Jamaica, Flushing, Rockaway Beach, etc. Besides these points, there is much else of interest across the river—many fine churches, and other public buildings.

NEW YORK.

Brooklyn possesses more than eighty church edifices, of various denominations. Among the most costly and imposing are—

The Church of the Holy Trinity, Clinton street, Episcopal.

The Church of the Pilgrims, Congregational (Rev. R. S. Storrs).

The Church of the Saviour, Pierrepont st., cor. of Monroe, First Unitarian Congregational (Rev. F. A. Farley).

Grace Church, Brooklyn Heights, Epis. (Rev. Jared B. Flagg).

Christ Church, Clinton st. (Dr. Canfield).

Plymouth Church, Orange street (Rev. Henry Ward Beecher).

First Reformed Dutch Church, Joralemon st. (Rev. Mr. Willets).

First Presbyterian Church, Henry, near Clark street.

Dutch Reformed Church, Pierrepont st. (Rev. Mr. Eells).

St. Ann's Church, Washington, near Sands street (Dr. Cutler).

Second Presbyterian Church, Fulton street, cor. of Clinton.

The *City Hall* (Court and Fulton streets), is one mile distant from the ferry. It is a handsome building of white marble, from the Westchester quarries. Its length is 162 feet, and its height to the top of the cupola is 153 feet. Cost, $2,000,000.

The *Post Office* is opposite the City Hall.

The *Brooklyn Athenæum,* corner of Atlantic and Clinton streets, in South Brooklyn, is a fine edifice of brick, with brown stone facings. It has an admirable library, reading-rooms, and a spacious lecture or concert hall.

The *Lyceum,* containing the city Library and a good lecture-room, is at the corner of Washington and Concord streets.

The new *Academy of Music*—a superb structure, is in Montague street.

Brooklyn, which now comprehends also the city of Williamsburg, is in population (which is no less than 270,000) the second city in the State of New York, though its close vicinage to the metropolis absorbs it, and destroys its distinctive importance. A great portion of its residents do business in New York, and live in Brooklyn only for the convenience and comfort of purer air, more quiet, and less cost.

The city is in many parts elegantly built, and the bold position on the Heights, directly looking down upon the river and bay, is a charming site for a summer abode. Some of the avenues of Brooklyn are wide, and delightfully lined with cottage residences.

The numerous ferries across the East river afford pleasant and perpetual access to Brooklyn.

Fulton Ferry—From Fulton st., N. Y., to Fulton st., Brooklyn, every five minutes in the day time, at a fare of *two cents.*

South Ferry—From Whitehall street, N. Y., to Atlantic st., Brooklyn.

Hamilton Ferry—Whitehall st., N. Y., to Hamilton avenue and Atlantic Docks, Brooklyn.

Catharine Ferry—Catharine st., N. Y., to Main st., Brooklyn.

Jackson Ferry—From Jackson st., N. Y., to Bridge st., Brooklyn.

Wall Street Ferry—Wall st., N. Y., to Montague st., Brooklyn.

To Brooklyn, E. D., or Williamsburg.

Peck Slip, Roosevelt st., Grand street, and Houston street, N. Y., every ten minutes.

NEW YORK TO ALBANY AND TROY.

It is fortunate for the gratification and the cultivation of the public taste for the sublime and beautiful in natural scenery, when our great highways of travel chance to lead through such wondrous landscape, as does our present journey up the Hudson River, from New York to Albany. Even to the wearied or the hurried traveller this voyage is ever one of pleasure, in its unique and constantly varying attractions, its thousand associations, legendary, historical, poetical, and social.

The Hudson received its name in honor of Hendrick Hudson, a Dutch navigator, who discovered it and ascended its waters for the first time, in 1607, in his good barque, the Half-Moon. It is also known as the North River, which name was given to it by the original Dutch colonists, to distinguish it from the South (Zuyd), as they called the neighboring floods of the Delaware. Its source is in the mountain region of the Adirondack, in the upper portions of New York, whence it flows in two small streams, the one from Hamilton, and the other from Essex County. These waters, after a journey of 40 miles, unite in Warren County. The course of the Hudson varies from south by east to east for some distance, but at length drops into a straight line, and continues thus, nearly southward, until it falls into the Bay of New York. Its entire extent is about 300 miles; its navigable length, from the sea to Albany, is half that distance. Its breadth, near the head of steamboat navigation, varies from 300 to 900 yards; and, at the Tappan Bay, 20 miles above the city of New York, it widens to the extent of from four to five miles. Ships of the first class may ascend to Hudson, a distance of 117 miles, and small sailing craft may reach the head of tide water (166 miles), at Troy. The number of steamboats and other vessels upon the river may be counted by thousands.

To the Hudson belongs the honor, not only of possessing the finest river steamers in the world, but of *having borne upon its waters the first steamboat which ever floated*, when Robert Fulton ascended the river in the Clermont, in 1807, exactly two centuries after the first voyage of Hendrick Hudson in the Half-Moon.

Every possible facility is now at command for the passage of the Hudson, either by steamer or by railway, morning, noon, and night. The commercial traveller, thinking more of his destination than of the pleasures by the way, will take the railroad route, while the pleasure-seeking tourist, in quest of the picturesque, and with time to enjoy it, will assuredly go by water.

RAILWAY ROUTE.

The journey by the Hudson River Railway, 144 miles, to Albany, is a poem in prose. The road lies on the eastern bank of the river, kissing its waters continually, and ever and anon crossing wide bays and the mouths of tributary streams. Incredible difficulties have been surmounted in its mountain, rock, and water passage, and all so successfully and so thoroughly, that it is one of the securest railway routes in the world. With all its immense business, its history is happily free from any considerable record of collision or accident whatever. This is owing as much to the vigilant management, and the admirable police, as to the substantial nature of the road. Flag-men are so stationed along the entire line, at intervals of a mile, and at curves and acclivities, as to secure unbroken signal communication from one end to the other.

Trains leave Chambers street and College Place almost hourly. Fare, usually, $3. Time, about five hours.

STATIONS.

(For Descriptions of places and scenes, see Steamboat Route following.)

Chambers street, New York; Thirty-first street, New York; Manhattan, 8 miles from New York; Yonkers, 17; Dobb's Ferry, 22 (Ferry to Piermont, Erie Railway); Tarrytown, 27; Sing Sing, 32; PEEKSKILL, 43; Garrison's 51 (Steam Ferry to West Point and Cozzens' Hotel); Cold Spring, 54; Fishkill, 60 (will be the junction of Providence, Hartford, and Fishkill Railroad, Steam Ferry to Newburgh, terminus of Newburgh branch of Erie Railway); New Hamburgh, 66; POUGHKEEPSIE, 75 (Half-way and refreshment station); Hyde Park, 81; Staatsburg, 85; Rhinebeck, 91; Barrytown, Tivoli, 100; Germantown, 105; Oakhill, 110 (Ferry to Catskill village, route to Catskill Mountains); HUDSON, 116 (Railway route to Boston, via Hudson and Berkshire road); Stockport, 120; Coxsackie, 125; Stuyvesant, 126; Schodack, 133; Castleton, 136; East Albany, 144 (Ferry to Albany); Troy, 152 miles.

STEAMBOAT ROUTE.

If the traveller accompany us up the Hudson, he will take passage in one of the noble steamers (very fittingly called floating palaces), which leave New York every morning and night.

The size and beauty of the boats, and the conveniences, comfort, and luxury of all their appointments, will be matter for pleasant wonder and thought, even to those most accustomed to them, whenever a moment can be stolen from the endless attractions on the way.

We start as the morning sun is falling upon the thousand sail which fill the grand Bay of New York; but scarcely have our eyes taken in half the beauties of this superb panorama—the roofs, and spires, and domes of the great metropolis on one side, Jersey City upon the opposite shore, the fortresses of Governor's Island, of Bedloe's, of Ellis's Islands, and of Fort Hamilton; the shores of Old Long Island, and the villa banks of Staten Island beyond, with the far-off perspective of the hill-bound "Narrows"—before we must turn our backs upon it all, to gaze upon the yet more charming scenes which are presented to us as our steamer turns its prow northward.

Along we sail, past the streets and wharves of the city, which seem interminable in succession, but our eyes fall upon the wooded shores at last, upon the elegant country villas peeping out from among the trees on the one hand, and the tall cliffs of the far-reaching Palisades on the other. The wilderness of brick and stone is behind us and forgotten, in the presence of green fields and rustling woods. Even the suburban charms of Hoboken, and the precipices of Weehawken, with its grave and memories of the unfortunate Hamilton, give place, in our esteem, to the more rural landscape upon which we now enter. Let us peep as closely as our rapid flight may permit at each passing village, city, and scene. First come

The Palisades. These grand precipices, rising to the height of 500 feet, follow, in unbroken line, as far as that great bay of the river called the Tappan Sea, a distance of 20 miles. The rock is trap, columnar in formation, something after the fashion of the famous Giant's

Causeway and of Fingal's Cave in Ireland. They lend great beauty to.the picture as we start upon our journey, and to all the pictures of the river, into which they come.

Bull's Ferry, six miles from the city, now lies upon our left. It is a favorite summer resort and residence of the people of New York. In the hot months, the ferry boats, continually plying thither, at a fare of only 12½ cents, are ever well-freighted with merry passengers.

Bloomingdale, a suburban village five miles from the City Hall is now upon our right. The Orphan Asylum here, with its emerald lawns, sloping down to the quiet waters, is a pleasant picture for both eye and heart.

Fort Lee, ten miles up the river, and opposite 160th street, New York, now calls us back again to the western shore. It crowns the lofty brows of the Palisades. Some interesting memories of the days of the American Revolution are awakened here. The anxious thoughts of Washington and his generals turned to this point in that eventful period. A fortification here stood upon the heights, which was called Mount Constitution, and here it was attempted, by the express command of Congress, to obstruct the navigation of the river by every art, and at whatever expense, "as well to prevent the egress of the enemy's frigates lately gone up as to hinder them from receiving succors." A large force of Americans, in retreating from Fort Lee, were overpowered, and either slain or taken prisoners by a greatly superior body of Hessian troops.

Fort Washington, another spot of deep historical interest, lies nearly opposite to Fort Lee, and, like that locality, reminds us of the most trying hours of the trying times in American story. It fell into the hands of the enemy, November 16th, 1776, and the garrison of 3,000 men became prisoners of war. Two days after, November 18th, Lord Cornwallis, with 6,000 men, crossed the river, at Dobb's Ferry, and attacked Fort Lee. The garrison there, then commanded by General Greene, made a hasty retreat to the encampment of the main army, under Washington, five miles back, at Hackensack. All the baggage and stores fell into the hands of the enemy. Had the English general followed up his successes at this period, with proper celerity and energy, he would most likely have effectually crippled the American army. Fort Washington is situated upon the highest part of Manhattan Island, between 181st and 186th streets, New York. It is between 10 and 11 miles from the City Hall. The fort was a strong earthwork of irregular form, covering several acres. Some 20 heavy cannons, besides smaller arms, bristled upon its walls, though its strength lay chiefly in its position. The very spot where the old fort once stood, as well as all the region round, is now covered by the peaceful and fragrant lawns and gardens of elegant villa residences. Just below the high ground once occupied by Fort Washington, and close by the river, is the promontory of Jeffrey's Hook: a redoubt was constructed here as a covering to the *chevaux-de-frise* in the channel. The banks of this work are still plainly to be seen. Above Fort Washington, and still upon the eastern side of the river, was Fort Tryon. The site now lies between 195th and 198th streets, New York. Not far beyond, is the northern boundary of Manhattan Island—the little waters, famous in history and story, as Spuyten Duyvel Creek (Spite the Devil). Hard by (217th street), was a redoubt of two guns called Cock Hill Fort; and upon Tetard's Hill, across the creek, was Fort Independence, a square redoubt with bastions.

There was still another military work here, strengthened by the British in 1781, and named Fort Prince. The upper end of the island of New York, where we have lingered so long, is rich in scenes and memories of interest; and the beautiful landscape is yet embellished by abundant traces of all its ancient history.

Yonkers.—HOTELS:—*Getty House.*

Yonkers, 17 miles up the river, is an ancient settlement at the mouth of the Neperan, or Saw Mill River. Since the opening of the railway, it has become a fashionable suburban town of New York, as the short distance thence permits pleasant, and speedy, and cheap transport by land or water.

Yonkers was the home of the once famous family of the Phillippses, of which was Mary Phillippse, the first-love of General Washington. The family exercised manorial rule in the neighborhood, and their ancient mansion is still to be seen. East of the manor-house of the Phillippses, is Locust Hill, where the American troops were encamped, in 1781. Near the village is the spot where Colonel Gist was attacked (1778) by a combined force under Tarleton and others. In 1777, a naval action occurred in front of Yonkers, between the American gun-boats and the British frigates, Ross and Phœnix.

Mr. Frederic Cozzens, the writer, resides at Yonkers, and some pleasant reminiscences of his home may be found in his genial "Sparrowgrass" papers.

Font Hill—Academy of Mount St. Vincent. The "Castle" of Mr. Edwin Forrest, known as Fonthill, is just below Yonkers. It is now, together with a larger and more imposing edifice, owned and occupied by the R. C. School of Mt. St. Vincent.

Hastings, three miles north of Yonkers, is a thriving little village, and its fortunes are daily improving with the favors of the citizens of New York, who eagerly seek homes amidst its pleasant and healthful places. Some of the country seats in the neighborhood—and they are numerous—are very elegant and luxurious establishments.

Dobb's Ferry, two miles yet beyond, and still upon the eastern bank of the river, is an ancient settlement, with a new leaven of metropolitan life, like all the places within an hour or two's journey from New York. The village has a pleasant air, lying along the river slope, at the mouth of the Wisquaqua Creek. Its name is that of an old family which once possessed the region and established a ferry. We are led back again here, to the times of the Revolution, and especially to that dramatic episode—some of the scenes of which transpired here and hereabouts—the story of Arnold and André. Remains of military work still exist at Dobb's Ferry.

Irvington and "Sunnyside." Irvington, to which we now come, still on the right or eastern bank, was once called Dearman, and it was expected to grow into a large town, as an outlet of the Great Erie Railway, which touches the river opposite, at Piermont; but the Erie travel was afterwards led to the metropolis through another terminus at Jersey City, and so Irvington is little more than a railway station to this day.

Dearman was rechristened Irvington in honor of the late beloved author, Washington Irving, whose unique little cottage of SUNNYSIDE is close by upon the margin of the river, hidden from the eye of the traveller only by the dense growth of the surrounding trees and shrubbery.

Piermont is on the left or western bank of the widest part of the Hudson, called the Tappan Bay or Sea, in the heart of which we are now sailing. It was born of, and has grown up from, the business of the Erie Railway, of which it is the terminus, and was once the only eastern terminus, the route of the road having originally been entirely continued, as it is now in part, thence down the Hudson to New York. The river here is three miles in width, and the shores, particularly upon the west, are so varied and bold, as to present most striking and attractive pictures. Piermont, rising from the water's edge to the villa-crowned summits of lofty hills, and with its grand railway pier reaching out a mile or more into the river, is not one of the least pleasing features of the scenery of the Tappan Bay.

Mr. Lewis Gaylord Clark, editor of the "Knickerbocker Magazine," lives upon the eminence here, in a little house, which he calls "CEDAR HILL COTTAGE."

Two or three miles back of Piermont is the old town of Tappan, interesting as having been one of the chief of Washington's head-quarters during the Revolution; and as the spot also where Major André was imprisoned and executed. The home of the commander-in-chief and the jail of the ill-fated officer are still in good preservation, though the latter house has been somewhat modified in its interior arrangements of late years, to suit its present occupancy as a tavern, under the style and title of the "'76 Stone House." The old Dutch church, in which André was tried, stood near by, but it was torn down in 1836, and a new structure reared upon its site. The spot where the execution took place (October 2d, 1780) is within a little walk of the old Stone House, in which the prisoner was confined.

Nyack is the next village above Piermont, and upon the same side of the river, while Tarrytown lies directly opposite it, and is connected therewith, and with New York, by a steam-ferry. Beds of red sandstone were once industriously quarried at Nyack.

Tarrytown.—HOTELS :—*The Franklin House.*

Tarrytown is a very active, prosperous little place on the eastern bank of the Hudson. It has many attractions, historical, pictorial, and social. Elegant villas, chiefly occupied by New York gentlemen, having gathered thickly around it, as about all this part of the river's marge, within the past few years; among them is the Irving homestead of Sunnyside, at Irvington, two miles below, and a mile or more distant, in the opposite direction, is the quiet little valley of Sleepy Hollow, which he has wreathed with such a garland of poetic remembrances and fancy, through his charming legends and tales. The visitor at Tarrytown will neglect many things before he denies himself the pleasure of a stroll to Sleepy Hollow, where Diedrich Knickerbocker roamed and meditated in days gone by; and of a walk by the Pocantico, and across the bridge, over which Ichabod Crane was pursued by the Headless Horseman. The scenes are all there still; and so the old Dutch church, to which the luckless pedagogue fled for sanctuary.

During the Revolution, Tarrytown witnessed many stormy fights between those lawless marauding bands of both British and Americans, known as Skinners and Cow-boys. The ground suited their wants, as it lay between the encampments of the two armies, and was in possession of neither. It was upon a spot, now in the heart of Tarrytown, that Major André was arrested, while returning to the British lines, after a visit to General Arnold. A simple monument—an obelisk of granite—now occupies the ground.

Sing Sing.—HOTELS:—*American House.*

Sing Sing, on the right as we ascend, is 33 miles from the city. In its acclivitous topography, upon a hill-slope of 200 feet, it makes a fine appearance from the water. The greatest breadth of the Hudson, nearly four miles, is at this point. Many fine country seats crown the heights of this pleasant village. It is distinguished for its educational establishments; for its vicinage to the mouth of the Croton river, from whence the city of New York derives its abundant supply of water; and for being the seat of the chief prison of the State.

The Croton enters the Hudson two miles above the village, where its artificial passage to the metropolis is begun. The great aqueduct at this point is especially interesting, being carried over the Sing Sing Kill by an arch of stone masonry 88 feet between the abutments, and 100 feet above the water. The State Prison, which no visitor will fail to see, is located on the bank of the Hudson, nearly three-quarters of a mile south of the village. The buildings are large structures, erected by the convicts themselves, with material from the marble and limestone quarries which abound here, and which many of them are continually employed in working. The prisons form three sides of a square. The main edifice is 484 feet long, 44 feet wide, and 5 stories high, with cells for 1,000 occupants, 869 of which were filled in 1852. The female prisoners are lodged in a fine edifice, some 30 or 40 rods east of the male departments. The prisoners are guarded by sentinels, instead of being enclosed by walls. The whole area covered by the establishment is 130 acres of ground. The railway passes through and beneath the prisons, but from the river they are all seen to advantage. The convicts not employed in working the marble quarries are engaged in the pursuit of various mechanical arts and trades. Sing Sing is a bustling business town. Its population over 3,000. Though the river communication with New York is not so great since as before the building of the railroad, way-steamboats from the city yet touch here daily.

Verdritege's Hook, opposite Sing Sing, is a commanding height, with such a deceptive appearance from the water above and below of a grand headland, that it has been christened Point-no-Point. Coming near it, its promontory look entirely disappears, and it proves to be only, as we once called it elsewhere, a topographical will-o'-the-wisp.

Upon this mountain summit there is a charming pellucid water called **Rockland Lake.** It is about four miles in circumference, and forms the source of the Hackensack River. Though not more than a mile from the Hudson, it is yet 250 feet above it. It is from this crystal lake that New York gets its best supplies of ice, which is cut into large square blocks. These blocks are then slid down to the level of the river, and when the winter passes they are transported to the city. Every voyager will bestow a pleasant thought upon the Rockland Lake, as he passes, in gratitude for the cooling beverages it gives him in the hot summer months, be that beverage julep, cobbler, cocktail, or Croton.

Haverstraw is also on the west side of the river, 36 miles up. It is a pleasant and prosperous place, of much picturesque vicinage. Some charming brooks, upon which artists delight to study, come into the Hudson here. We touch now again upon sacred ground, as we reënter amidst the scenes of our Revolutionary history; for directly opposite is Verplanck's Point, and in the immediate vicinity is the famous battleground of Stony Point.

Verplanck's Point, on the east side of the Hudson, is the spot at which Hendrick Hudson's ship, the Half-Moon, first came to anchor, after leaving the mouth of the river. Great was the marvel and terror of the astonished natives at that extraordinary event. "Filled with wonder," says Lossing, "they came flocking to the ship in boats, but their curiosity ended in a tragedy. One of them, overcome by acquisitiveness, crawled up the rudder, entered the cabin window, and stole a pillow and a few articles of wearing apparel. The mate saw the thief pulling his bark for land, and shot at and killed him. The ship's boat was sent for the stolen articles, and when one of the natives, who had leaped into the water, caught hold of the side of the shallop, his hand was cut off by a sword, and he was drowned. This was the first blood shed by these voyagers. Intelligence of it spread over the country, and the Indians hated the white man ever after." The creek which winds through the marsh, south of Verplanck's Point, as, afterwards, the peninsula itself, was called Meahagh by the Indians. Stephen Van Cortlandt purchased it of them in 1683, and it passed from his possession into that of his son, whose only daughter and heiress married Philip Verplanck, from whom its present name. Topographically, Verplanck's Point may be described as a peninsula, gradually rising from a gentle surface, until it terminates in the river in a bold bluff of from 40 to 50 feet elevation. The railway recedes here from the river-shore, and takes a seemingly inland route across the neck of the peninsula. "Here," says Mr. Lossing, in his Field-Book, from which we have already quoted, "during the memorable season of land and town speculation, when the water-lot mania emulated that of the tulip and the South Sea games, a large village was mapped out, and one or two fine mansions were erected. The bubble burst, and many fertile acres there, where corn and potatoes once yielded a profit to the cultivator, are scarred and made barren by intersecting streets, not de-populated, but un-populated, save by the beetle and the grasshopper."

The narrowness of the river between this bluff and the opposite promontory of Stony Point, makes it the lower gateway of the Hudson Highlands, and renders it easily defensible against any possible hostile force. A small fortification, called Fort Fayette, once existed at the western extremity of Verplanck's Point, many remains of which are yet distinctly visible. This fort, and that of Stony Point opposite, were taken by the English, under Sir Henry Clinton, June 1, 1779. The garrisons at the time consisted, respectively, of only 70 and 40 men. Sir Henry Clinton immediately proceeded to strengthen his new possessions, while Washington was meditating their recapture, as the passage which they controlled was important to the free communication between the northern and southern portions of his army. We must now look across to

Stony Point.—The old lighthouse here calls this scene loudly to the notice of all passers. The beacon is placed amidst the remains of the ancient fort, and exactly upon the former site of the magazine. As we have said, the fort here, together with that upon Verplanck's Point opposite, fell into the hands of the enemy on the 1st of June, 1779. Despite its natural defences, and the additional strength which the enemy industriously gave it, the Americans determined to regain their lost possession. General Wayne, who was to command the proposed assault, is reported to have said to Washington, with daring emphasis, apropos of the dangers before him in this perilous venture: "General, I'll storm hell, if *you* will only plan it!" He did storm Stony Point on the night of July 15th, 1779, and next day he wrote to the commander-in-chief that the fort and garrison were his! It was a gallant exploit, and we wish we had the opportunity to review the whole story; but there are many miles yet between us and Albany, and we must move on to

Peekskill.—HOTELS:— *The Eagle*

We now enter Haverstraw Bay, the second of the great extensions of the Hudson, and the commencement of the magnificent scenery of the Highlands. On our left is the rugged front of the Dunderburg Mountain, at whose base the little hamlet and landing of Caldwell are nestled; on the right, the village of Peekskill ascends from the shore to the lofty hill summit, and before us is the narrow passage of the river, around the point of the Dunderburg, the grand base of Anthony's Nose, and other mountain cliffs and precipices. Let us look a moment, before we pass on, at Peekskill and its memories. The village was named after John Peek, one of the early Dutch navigators, who mistook the creek which comes into the river just above for the continuation of the Hudson itself (not an unreasonable mistake, so uncertain seems its direction at this highland pass), and thus thinking himself at the end of his journey, ran his craft ashore, and commenced his settlement. The present village was first settled in 1764. Its position is exceedingly picturesque. A romantic brook comes down a deep glen in the centre of the town, as it descends from the elevated plateau to the river, disfigured not a little at this day by the houses and foundries near it.

Noble views may be found everywhere here, and in every direction, of the river and the surrounding country. From Gallows Hill northward (so called in remembrance of the execution there of a spy in the days of the Revolution), a grand panorama is exposed. Here, to the west, overlooking the village, the river, and its mountain shores; there, southward, hill and valley, as far as the high grounds of Tarrytown below; and above, the Canopus valley, in the shadow of the Highland precipices. The division of the American army under Putnam, in 1777, was encamped upon Gallows Hill. Beneath this lofty ground, and upon the banks of Canopus Creek, is Continental Village, which was destroyed by General Tryon (Oct. 9, 1777), together with the barracks, capable of accommodating 2,000 men, and also much public store and many cattle.

The Van Cortlandt House, in this vicinage, is an object of interest, as the ancient seat of an ancient family, and as the temporary residence of Washington. Near by is a venerable church, erected in 1767, within whose grave-yard there is a monument to the memory of John Paulding, one of the captors of Major André.

At the date of the first edition of this work (1857) there stood, in the streets of Peekskill, another of those venerable roofs, sacred as having at one time sheltered the head of the American chieftain. But, alas! it has since passed away, as are fast passing all its fellow-shrines. Thus fewer and fewer is the number of such spots to which we shall be able to direct the traveller as the years speed on.

A pleasant ride from Peekskill is to Lake Mahopac, a fashionable summer resort for the pleasure-seekers of New York. See Index.

The population of Peekskill in 1854 was 2,500. It is 43 miles from the city of New York, by rail.

Caldwell's Landing, at the foot of Dunderburg Mountain, was a calling-place for the river steamers, when the chief travel was by water, instead of by rail, as at the present day. The passengers for Peekskill, opposite, were then always landed at Caldwell. This spot is memorable for the search so seriously and actively made for the pirate treasure which the famous Captain Kidd was supposed to have secreted at the bottom of the river here. Remains of the apparatus used for this purpose are still seen, in bold, black relief, at the Dunderburg point, as the boat rounds it, towards the Horse Race. This Quixotic exploration has at least proved to a certainty that much valuable treasure *now* lies buried here, however uncertain the matter was before!

The Highlands. This grand mountain group, through which the Hudson now makes its way, extends from north-east to south-west, over an area of about 16 by 25 miles. The landscape which these noble heights and their picturesque and changeful forms present, is of unrivalled magnificence and beauty, whether seen from their rugged summits, or from the river gorges.

Thus says Theodore Fay of these scenes—

"By wooded bluff we steal, by leaning lawn,
By palace, village, cot, a sweet surprise
At every turn the vision breaks upon,
Till to our wondering and uplifted eyes
The Highland rocks and hills in solemn grandeur rise."

"Nor clouds in heaven, nor billows in the deep
More graceful shapes did ever heave or roll;
Nor came such pictures to a painter's sleep,
Nor beamed such vision on a poet's soul!
The pent-up flood, impatient of control,
In ages past here broke its granite bound,
Then to the sea in broad meanders stole,
While ponderous ruin strew'd the broken ground,
And these gigantic hills for ever closed around."

This powerful river, says another writer, writhes through the Highlands in abrupt curves, reminding one, when the tide runs strongly down, of Laocoon in the enlacing folds of the serpent. The different spurs of mountain ranges which meet here abut upon the river in bold precipices, from five to fifteen hundred feet from the water's edge; the foliage hangs to them from base to summit, with the tenacity and bright verdure of moss; and the stream below, deprived of the slant lights which brighten its depths elsewhere, flows on with a sombre and dark-green shadow in its bosom, as if frowning at the narrow gorge into which its broad-breasted waters are driven.

Passing round the point of Dunderburg (or Donderbarrack, the *Thunder Chamber*) we enter the swift channel called the Horse Race. On our right, in this wild and narrow gorge of the giant hills, are the rugged flanks of **Anthony's Nose**—bold, rocky acclivities, which rise to the height of 1,128 feet above the water. Two miles above is the **Sugar Loaf** mountain, with an elevation of 806 feet. Near by, and reaching far out into the river, is a sandy bluff, on which Fort Independence once stood. Further on is Beveridge Island, and in the extreme distance, Bear Mountain. Forts Clinton and Montgomery, taken by the British troops, after traversing the Dunderburg mountain, are in this vicinity; and so, too, a little lake called Skinnipink, or Bloody Pond, where a disastrous skirmish occurred on the eve of the capture of the forts and the consequent opening to the enemy of the passage of the Highlands. On this (the west) side of the river, the **Buttermilk Falls** are seen descending over inclined ledges, a distance of 100 feet. They form a pleasant passage in the river landscape, though in themselves they are not especially picturesque.

In the heart of the Highland Pass and just below West Point, on the west bank, is **Cozzens'**—a spacious and elegant summer hotel, which comes most charmingly into the pictures of the vicinage. It is accessible, as is West Point at the same time, from the railway on the opposite side of the river, by a steam ferry from **Garrison's Station**, between Peekskill below and Cold Spring above. The concourse of sail sometimes wind-locked in the angles of this mountain pass, is a wonderful sight. "This channel," says Mr. Willis, "is narrow and serpentine, the breeze baffling, and small room to beat; but the little craft will work merrily and well; and dodging about, as if to escape some invisible imp in the air, they gain point after point, till at last they get the Donderbarrack behind them, and fall once more into the regular current of the wind."

Constitution Island, with the rocky plateau of West Point, now bars our view of the upper portion of the Highland passage. Rounding it, we come into that wonderful reach of the river, flanked on the west by the royal cliffs of Cronest and Butter Hill, or Storm King, and upon the east by the jagged acclivities of Breakneck and Bull Hill, with the pretty village of Cold Spring beneath. From the heights of West Point delicious views of this new chapter of the river beauties may be obtained. Constitution

Island was called, prior to the Revolution, Martelear's Rock. It was fortified together with West Point, hard by, in 1775–6, when **Fort Constitution** was built, the remains of which still exist. Those of the magazine especially are well preserved, on the highest point, near the western extremity of the bluff. From this island to West Point a chain was thrown across the river as an obstruction to the enemy's ships. Some links of this defence are yet to be seen.

West Point.—HOTELS :—*The West Point*, on the terrace, and *Cozzens'* below.

West Point, both from the unrivalled charms of its scenery and from its position as the seat of our most famous military school, is one of the most attractive spots upon the Hudson. It is replete with interest, too, as the centre of the important interests and incidents which in the days of the Revolution wove such a web of story and romance about all this portion of the beautiful river.

The edifices of the United States Military Academy, in full view as we approach, occupy a noble plateau, about a mile in circuit, and 188 feet above the water; and grand hills, which were fortified in the war time, leaving at this day romantic ruins to embellish the landscape, rise hundreds of feet yet above. It was the same bold and varied physical aspect of this spot which now delights the lover of Nature's wonders, that in other days gave it its grand value, and its memorable fame as a site of military operations and achievements. The visitor will delight his eye at all points, whether he gaze upon the superb panorama of the river as he sits upon the piazza of the hotel upon the plateau, or whether he looks upon the scene from the yet loftier eminence above, crowned by the ruins of ancient fortresses; or whether he stroll amidst the interlacing walks, with new vistas of beauty and fresh memories of a gallant gone-by at every turn and step. When the remains of the old forts Putnam, Clinton, Webb and Wylly's have been seen, together with the little glen below the Parade Ground, called "Kosciusko's Garden," and embellished with an obelisk erected to the honor of the gallant Pole, the visitor will be ready to explore the edifices of the Academy establishment, and the many objects of interest which they contain; among them, Revolutionary relics and cannon captured in the war with Mexico. If his visit be in the month of July or August the pleasures of the place will be agreeably increased by the picturesque scene of the annual encampment, on the broad terrace, of the Cadets, and by the daily practice of the military band. If he can gain the entrée of the studio of the distinguished painter Weir, who resides here, he will be fortunate.

The United States Military Academy at West Point was established by Congress in 1802, and it is entirely controlled and supported by the Government. The education of the Cadets is gratuitous, but each one is required to spend eight years in the public service, unless he be sooner excused. The course of study lasts five years, and embraces every theme required for a thorough mastery of the military art. The graduates number more than 3,000.

West Point, in the Revolution, was the great key of the river, which Arnold, then in command of the post, would have betrayed into the possession of the enemy, but for the providential arrest of his co-plotter, André, at Tarrytown below.

The Robinson or **Beverly House**, occupied by Arnold at the time of his meditated treason, at which he received intelligence of the arrest of Major André, and from whence he made his escape to the British vessel, the Vulture, lying near by in the river, is on the opposite (east) bank, a pleasant drive of four or five miles south from Cold Spring. It is situated upon a fertile meadow, at the foot of the Sugar Loaf mountain—the lofty elevation on the east, which proves so Protean in form—now a bold cone, and now a ridgy line, as seen from below or from above. This homestead is now occupied by Lieutenant Thomas Arden, and is called "Ardenia." It has been kept in thorough repair, and its old aspect has been always religiously preserved.

Cronest casts its broad evening shadow upon us as we continue our voyage up from West Point. This is one of the grandest mountains found in the Highland group. Its height is 1,428 feet. From the summit, which may be readily reached, wonderful pictures of far and near are exposed to view.

The poet, George P. Morris, has happily sung the beauties of these bold cliffs—

"Where Hudson's waves o'er silvery sands
Wind through the hills afar,
And Cronest, like a monarch stands,
Crown'd with a single star!"

The tourist, as he passes this romantic ground, will not fail to recall the scenes and incidents of Drake's charming story of the Culprit Fay, with its classic whispers of the dainty Fairy doings here.

Butter Hill, or **Storm King,** as Mr. N. P. Willis has re-named it, is the next mountain crest, and the last of the Highland range upon the west. The jealous people on the opposite shore say that Butter Hill is only a corruption of But-a-hill! It would, though, be irreverent to believe in this derivation, for the Storm King, with its 1,500 and more feet of bold cliff and crag, is not an object to be spoken or thought lightly of.

Between Cronest and Storm King, (if we may adopt Mr. Willis's nomenclature,) and in the laps of both, is a lovely valley, replete with forest and brook beauties. It is called Tempe, and will one day be a Mecca to the nature-loving tourists.

"*Idlewild*," the residence of the poet N. P. Willis, is hidden from view now, only by the front of Butter Hill; and were it not for the forest of verdure around it we might descry "Undercliff," the home of George P. Morris, near the village of Cold Spring, across the river on the east.

Cold Spring and "Undercliff." Cold Spring is one of the most picturesque of the villages of the Hudson, whether seen from the water or from the hills behind, or in detail amidst its little streets and villa homes. It is built upon a steep ascent, and behind it is the massive granite crown of Bull Hill. This noble mountain overshadows the beautiful terrace upon which the poet Morris has lived in the rural seclusion of "**Undercliff**" for many years. It is scarcely possible to find a spot of sweeter natural attractions than the site of Undercliff, looking over the pretty village to the castellated hills of West Point, across the blue Hudson to old Cronest, or northward beyond the Newburgh Bay, to the far away ranges of the Catskills.

The West Point Iron Foundry, which is located here, supports much of the population and business of the village.

Two miles below are the Indian Falls, a romantic cascade, on the Indian Brook, a wild rocky stream which enters the river hereabouts.

The Beverly House, memorable for its associations with the history of the treason of Arnold, is a few miles below. See previous pages for further mention of this locality. The population of Cold Spring is 1,200. Its distance from New York, 54 miles.

Beyond Cold Spring, and still on the east bank of the river, the Highland range is continued in the jagged precipices of the Break neck, and Beacon Hills, in height, respectively, 1,187 and 1,685 feet. These mountains are among the most commanding features of the river scenery. As we leave them to the south we approach Pollopel's Island, and enter the wide Newburgh Bay, with the villages of Cornwall, New Windsor and Newburgh upon our left, and Fishkill on our right, all imposingly displayed from the water.

Cornwall Landing, on the west bank, comes first to our reach. It is a rugged and picturesque little place. On the lofty Highland Terrace back, is Canterbury, a quiet village, much in favor as a summer residence by the seekers of repose and rural pleasure, rather than of fashionable display and distraction.

"**Idlewild**," Mr. Willis's romantic home, on a lofty plateau above and north of the village, is the chief object of interest. A wonderful ravine, full of the most delightful cascades, with its neighborhood of hill side, rock and forest, occupies one part of the domain, and a fertile terrace sweep, upon which his cottage stands, fills the rest. In its multiplicity of charms, it is a retreat which any poet might be content to enjoy.

There is an extensive paper manufactory, under the conduct of Mr. Carson, just back of Idlewild, in the out-of-the-way little village of Moodna. The Moodna Creek, a romantic stream, comes into the river at the northern point of Idlewild.

New Windsor, between Idlewild and Newburgh, and once the rival of the latter, is a straggling hamlet, of no special present attraction; though it has some old historical memories of interest. The chief camp ground of the Revolutionary army, during the operations on the Hudson, lies back of it, with memories and scenes yet remaining, of the residence of Greene and Knox, and other distinguished generals of the period; of the site of the memorable old building which was known as the Temple, and was erected at the command of Washington for a chapel for the army; a hall for the free mason fraternity, which existed among the officers, and for general public assemblies. This structure was baptized the "Temple of Virtue," at the time of its erection, a name which it lost even in the orgies of the dedicatory festival!

On the shore of Plum Point, the elegant promontoried estate of Philip A. Verplanck, Esq., at the mouth of the Moodna Creek and the river, are preserved some curious débris of old military defences, and of buildings long before the days of the Revolution.

Washington established his headquarters at New Windsor, first on June 23d, 1779, and again in 1780. His residence, a plain Dutch house, has long since passed away.

"**Cedar Lawn.**" Joel T. Headley, the distinguished author, possesses a charming river estate, which is called "Cedar Lawn," between the villages of New Windsor and Newburgh.

Asher B. Durand, the eminent landscape painter, at one time possessed and occupied an elegant country seat in the same neighborhood.

Newburgh.—Hotels:—

The Powelton is an elegant summer house, picturesquely located in the upper and more rural part of the village. In the business centre is the Orange Hotel, a large and well-ordered establishment of old fame. Near the river landings is the United States.

Newburgh, with its population of 12,000, and its social and topographical attractions, is one of the largest and most delightful towns on the Hudson. Rising, as it does, rather precipitously from the water to an elevation of 300 feet, it presents a very imposing front to the voyager. The higher grounds are occupied by beautiful residences, and the luxurious villas of gentlemen retired from metropolitan life. There are a dozen churches here, three banks, five newspapers, one a daily. Newburgh is the eastern terminus of a branch of the Erie Railway, connecting daily with that great thoroughfare at Chester, N. Y. It is united by steam ferry to Fishkill, on the opposite shore, and here is its station on the Hudson River, and Hartford, Providence, and Fishkill Railroads. It has large manufactories of various kinds, and an extensive trade in farm and dairy products. The home of the lamented landscape gardener and horticultural writer, A. J. Downing, was here. The village, too, is honored by the residence of H. K. Brown, the eminent sculptor.

Newburgh was the theatre of many interesting events in the war of the Revolution. It was the site of one of Washington's chief headquarters, and the house in which he lived is now the principal boast of the town. It occupies a bold position, overlooking the great pass of the Highlands. It was here that the Revolutionary army was finally disbanded at the close of the war, June 23d, 1783.

Fishkill Landing, on the eastern shore, opposite Newburgh, is, like that village and all the region round, opulent in natural beauties, and prolific in elegant residences of retired city gentlemen. It is a small place, with a population, in 1854, of 1,600. It lies in the lap of a lovely fertile plain, which reaches far back to the base of a bold mountain range. It is, like all the neighborhood, replete with memories of Revolutionary and Ante-revolutionary interest. A portion of the Continental army was encamped here. The building occupied as barracks was the property of a Mr. Wharton, and has thence been since known as the Wharton House. It is, like most of the buildings of the period, a plain, Dutch, wooden construction. It may be found about half a mile south of the village.

Fishkill is the scene of many of the incidents in Cooper's novel of The Spy: a Tale of the Neutral Ground. Enoch Crosby, who was supposed to be the actual character represented in Mr. Cooper's tale as Harvey Birch, was subjected to a mock trial before the Committee of Safety in the Wharton House, mentioned above.

Two miles north-east of Fishkill landing is the Verplanck House, interesting as having once been the head quarters of the Baron Steuben, and the place in which the famous *Society of the Cincinnati* was organized, 1783.

Low Point, three miles above Fishkill landing, is a small river hamlet.

New Hamburg comes next, near the mouth of Wappinger's Creek, and a little north is the village of *Marlborough*, with *Barnegat*, famous for its lime-kilns, two miles yet beyond.

Poughkeepsie.—HOTELS :—*Gregory House.*

Poughkeepsie is 75 miles from New York, and thus the half way station on the river railroad. It is a pleasant city, and the largest place between New York and Albany. Its population is some 15,000. It contains about sixteen churches, four banks, and three or four newspapers. It has a variety of manufactories; and the rich agricultural region behind it makes it the depot of a busy trade.

College Hill, the site of the Collegiate Institute, half a mile north-east, is a commanding elevation, overlooking the river and the region around.

Poughkeepsie was founded by the Dutch more than 150 years ago. It is symmetrically built, upon an elevated plain, half a mile east of the river. It has no historic associations of especial interest. Professor Morse, the inventor of the electric telegraph, and Benson J. Lossing, the historian, reside here.

New Paltz Landing is on the opposite side of the river, west.

Hyde Park, and "Placentia."—Hyde Park, 80 miles above New York, is a quiet little village on the east side, in the midst of a country of great fertility, and thronged with wealthy homesteads and sumptuous villas. Near the village, on the north is " Placentia," famous as the home of the late veteran author, Jas. K. Paulding. Here that distinguished pioneer in American letters reached a kindly age, his time divided between his books and his fields. Placentia commands a magnificent view of the river windings, far above, even to the peaks of the distant Catskills.

Staatsburg is upon the railway, a few miles above.

Rondout and Kingston lie on the western side, the former on the Rondout Creek, one mile from the Hudson, and the latter on an elevated plain, three miles distant from the river. At Rondout is the terminus of the Delaware and Hudson Canal, through which large supplies of coal are brought to market. The Rondout Creek is a singularly picturesque stream, in all its course from the mountains, westward.

Kingston is a thriving and pleasant place. Its population in 1855, was nearly 5,000, and that of the township, 13,000. It was settled by the Dutch (1663) about the time of the settlement of Albany and New York. In the times of the Revolution it was burned by the British (1777). The first constitution of New York was framed and adopted in a house still standing here.

Kingston was the birth-place of Vanderlyn, the eminent painter. He died here in 1853.

Rhinebeck is on the railway, opposite Kingston, and is connected with that village by a ferry.

In our voyage up the Hudson, we have now, as we have had for some miles back, new and magnificent features in the landscape. Far away on the west, lie the bold ranges of the Shawangunk and the Catskill Mountains, forming fresh and charming pictures at every step of our progress.

Saugerties and Tivoli, the one on the west, and the other on the east bank of the river, now attract our attention. Saugerties is a picturesque and prosperous village, at the debouchure of the beautiful waters of Esopus Creek.

Passing *Malden* on the left, and *Germantown* on the right, we come to *Oakhill*, the station on the railway for the opposite town of

Catskill.—Hotels:—

Catskill, at the mouth of the Catskill Creek, on the west bank of the Hudson. In its pictorial attractions, this is one of the most interesting points of our present route. The village, which is a pleasant and thriving one, rises from the margin of the creek, to an elevated site on the north, where it is dissipated in many beautiful country villas, overlooking the river on the east, and the valley and mountains on the west.

Among these homes is that of the family of the painter, Thomas Cole. This great artist was buried in the village cemetery here. His studio, seen from the water, is still preserved in all its arrangements, as it was when he last occupied it.

Catskill is chiefly interesting to the tourist as the point of detour towards the wonders of the mountain ranges, which lie over the intervening valley, 10 miles westward. See Tour to the Catskill Mountains.

Hudson. — Hotels:—*The Worth House*, Main street.

In the voyage above and below Hudson, there are displayed some of the finest passages of the river scenery. With a varied shore on the east, and the Catskill peaks and ridges on the west, the tourist will scarcely regret that he has left even the Highlands behind him. Passing Mount Merino, about four miles above Catskill, the city of Hudson, lying upon the water and upon a high terrace, spreading away to higher lands on the east, comes imposingly into view. It is one of the most important river towns commercially, and one of the most attractive topographically and pictorially. The main street, which lies through the heart of the city, from east to west, terminates at the river extremity in a pleasant little park called Promenade Hill, on a bold promontory, rising abruptly 60 feet above the water; while the other terminus climbs to the foot of Prospect Hill, an elevation of 200 feet. From these lofty heights the views of the Catskills, of the far-spreading river, and of the beautiful city itself are incomparable. There are nearly a dozen churches, some of them elegant structures, in Hudson; a fine court-house of marble, and other public edifices, among them a famous Lunatic Asylum. It has various educational establishments, and newspapers and other publications to the number of half a dozen. Hudson is a depot of large business, and at one time it had an extensive Indian and whaling trade. It is at the head of ship navigation on the river. There are also large manufacturing interests here, maintaining upwards of seventy establishments of various kinds. It is the chief terminus of the Hudson and Boston Railway, extending eastward 34 miles to West Stockbridge, Mass., and uniting with the trains from Albany to Boston, and with other routes.

Passengers for the Shaker Village at New Lebanon, 36 miles from Hudson, take the Hudson and Boston cars to within seven miles of the Springs, which are much sought in summer time.

Columbia Springs, five miles distant, is a summer resort of great value to invalids, and of interest to all. The Claverack Falls, some eight miles off, should not be overlooked by the visitor.

Athens is a little village, with a population of 1,400, directly opposite Hudson, and connected with it by a steam ferry.

Stockport, Coxsackie and *Stuyvesant* come now in succession along the east side of the river. These are bustling and thriving little places.

Kinderhook Landing, and "**Lindenwold.**" The village of Kinderhook, about five miles east of the landing, on the east side of the river, is the birth-place of Martin Van Buren, Ex-President of the United States. His present residence is upon his estate of "Lindenwold," two miles south of the village.

New Baltimore and Coeymans are now passed on the left, and Schodack and Castleton on the right, after which we yet journey some eight miles, and then reach East Albany, where we may continue on to Troy, or cross the river by ferry to the end of our present route at the city of

Albany.—HOTELS:—*The Delavan House.*

We are now at the capital of the Empire State, after our voyage of 145 miles (by railway, 144), from the city of New York. For the continuation elsewhere of our travels from this point to Boston, Canada, Saratoga Springs, Niagara Falls, and the Great West--for railways in all directions meet here--the tourist is referred to our Index of routes and places.

Albany was founded by the Dutch, first as a trading post on Castle Island, directly below the site of the present city, in 1614. Fort Orange was built where the town now stands, in 1623; and, next to Jamestown in Virginia, was the earliest European settlement in the original thirteen States. It was known as Beaver Wyck, and as Williamstadt, before it received its present name in honor of James, Duke of York and Albany, afterwards James the Second, at the period when it fell into British possession, 1664. The population in 1860, was about 68,000.

It has a large commerce, from its position at the head of the sloop navigation and tide water upon the Hudson, as the *entrepôt* of the great Erie Canal from the west, and the Champlain Canal from the north, and as the centre to which many routes and lines of travel converge. The boats of the canal are received in a grand basin constructed in the river, with the help of a pier 80 feet wide, and 4,300 long.

Albany, seen from some points upon the river, makes a very effective appearance, the ground rising westward from the low flats on the shore, to an elevation of some 220 feet, in the range of a mile westward. State street ascends in a steep grade from the water to the height crowned by the State Capitol.

Among the public buildings are the Capitol, the State Hall, the City Hall, the Hospital, the Penitentiary (a model prison), the Alms-House, and more than 40 church edifices. Of the latter the new Cathedral is a noble structure.

The Dudley Observatory, founded by the munificence of Mrs. Blandina Dudley, was erected at a cost of $25,000, and has been further endowed to the amount of $100,000. The University of Albany was incorporated in 1852. The Law Department is now one of the best in the Union. The Medical College, which was founded 1839, is a prosperous establishment. The State Normal School was organized successfully in 1844, "for the education and practice of teachers of common schools, in the science of education, and the art of teaching." The Albany Institute, for scientific advancement, has a library of 5,000 volumes. The Young Men's Association has a collection of 8,000 volumes; the Apprentices' Library, 3,000; and the State Library (accessible to public use) has 46,000 volumes.

The edifice on State street, where are deposited the public collections in natural history, and in geology, and in agriculture, is most interesting. The Orphan Asylum, and other benevolent establishments of the city, are well worth the consideration of the tourist.

The distinguished sculptor E. D. Palmer resides here. His studio is a place of especial attraction.

Troy.—HOTELS:—*American Hotel, Mansion House, Troy House, Temperance House, Northern Hotel, Washington Hall, Union Hall, and the St. Charles.*

Troy is a large and beautiful city of over 45,000 inhabitants. It is upon both banks of the Hudson, at the mouth of the Poestenkill Creek. It is built upon an alluvial plain, overlooked on the east side by the classic heights of Mount Ida, and on the north by the barren cliffs of Mount Olympus, 200 feet high. These elevated points command superb views of the city and its charming vicinage, and of the great waters of the Hudson. Troy lies along the river for the length of three miles, and drops back a mile from east to west. Troy is a busy city, with its manufacturing industry, and as a great *entrepôt* of railway travel from and to all points. It boasts many fine churches and public buildings, and many admirable private mansions and cottages. Here is the well-known Female Seminary, established by Mrs. Emma Willard, in 1821. It is the seat, too, of the Troy Polytechnic Institute.

The cars leave Troy and Greenbush every hour during the day and evening. Steamboats and stages also run between Albany and Troy. Railway trains extend to all points. See Index.

West Troy, a suburb of Troy, on the opposite side of the river, is a rapidly growing place. The inhabitants are employed principally in manufactures. A fine Macadamized road leads from West Troy to Albany, a distance of six miles.

At **Gibbonsville** is a United States Arsenal, where is kept a large and constant supply of small arms, and the various munitions of war. This is one of the most important of the national depots, and is worthy the attention of the traveller.

NEW YORK TO LAKE ERIE.

By the New York and Erie Railroad.

This great route claims especial admiration for the grandeur of the enterprise which conceived and executed it, for the vast contribution it has made to the facilities of travel, and for the multiplied and varied landscape beauties which it has made so readily and pleasantly accessible. Its entire length, from New York to Dunkirk, on Lake Erie, is 460 miles (including the Piermont and the Newburgh branch, it is 497 miles), in which it traverses the southern portion of the Empire State in its entire extent from east to west, passing through countless towns and villages, over many rivers, through rugged mountain passes now, and anon amidst broad and fertile valleys and plains. In addition, it has many branches, connecting its stations with other routes in all directions, and opening yet new stores of pictorial pleasures.

The road was first commenced in 1836. The first portion (46 miles, from Piermont to Goshen) was put in operation September 23d, 1841; and, on the 15th of May, 1851, the entire line to Lake Erie was opened amid great rejoicings and festivals, in which the President of the United States and other distinguished guests of the company assisted.

Some idea of the extent of this noble route may be gathered from the fact, that, in 1854, it employed about 200 locomotives, nearly 3,000 cars, 4,000 employés (682 of which are engaged in repairing engines and cars). The cost of the road and equipments, up to 1854 (including the Newburgh branch), was nearly $34,500,000. The earnings for the year 1856 were $6,349,050 15, and the expenses for the same period were $5,002,754 48.

An interesting feature of this road, and one of great convenience to the Company and security to the traveller, is its own telegraph, which runs by the side of the road through its whole extent, and has its operator in nearly every station-house. This telegraph has a double wire the entire length of the road; enabling the Company to transact the public as well as their own private business. Daily trains leave for the West on this route, from the foot of Duane street, morning, noon, and night.

STATIONS.

New York, Jersey City, Bergen, 2 miles; Boiling Spring, 9; Passaic Bridge, 11; Huyler's, 12; Paterson, 16; Godwinville, 21; Hohokus, 23; Allendale, 25; Ramsey's, 27; Suffern's, 31; Ramapo, 33; Sloatsburg, 35; Southfields, 41; Greenwood, 44; Turner's, 47; Monroe, 49; Oxford, 52; Chester, 55; Goshen, 59; Hampton, 63; Middletown, 66; Howell's, 70; Otisville, 75; Port Jervis, 88; Shohola, 106; Lackawaxen, 110; Mast Hope, 118; Narrowsburg, 122; Cochecton, 130; Callicoon, 135; Hankins, 142; Basket, 146; Lordville, 153; Stockport, 159; Hancock, 163; Hale's Eddy, 171; Deposit, 176; Susquehanna, 192; Great Bend, 200; Kirkwood, 206; Conklin, 210; Binghamton, 214; Hooper, 220; Union, 223; Campville, 228; Owego, 236; Tioga, 242; Smithboro, 246; Barton, 248; Waverley, 255; Chemung, 260; Wellsburg, 266; Elmira, 273; Junction, 277; Big Flats, 283; Corning, 291; Painted Post, 292; Addison, 301; Rathboneville, 306; Cameron, 314; Adrian, 322; Canisteo, 327; Hornellsville, 331; Almond, 336; Alfred, 340; Andover, 349; Genesee, 358; Scio, 361; Phillipsville, 365; Belvidere, 369; Friendship, 373; Cuba, 382; Hinsdale, 389; Olean, 395; Alleghany, 398; Great Valley, 411; Little Valley, 420; Cattaraugus, 428; Dayton, 437; Perrysburg, 440; Smith's Mills, 447; Forestville, 451; Sheriden, 455; Dunkirk, 460.

To Suffern's (31 miles) *via New Jersey, and via Piermont.* The first 31 miles of the Erie route, that through the State of New Jersey, from Jersey City, opposite New York, to "Suffern's," consists of parts of three different railways, though used of late years for all the general passenger travel of the Erie road, and with its own broad track and cars. The original line of the road is from Suffern's eastward, 18 miles, to Piermont, and thence 24 miles down the Hudson River. This route is now employed only for freight, and for local travel. It leads through a rude but not uninteresting country, with here and there a fine landscape or an agreeable village.

We pass now, without halt, through the New Jersey towns—Paterson, with its "Falls of the Passaic" among them, and begin our mention of places and scenes of interest on the Erie route, at Suffern's station, where the original Piermont and the present Jersey City lines meet. The Ramapo Valley commences at this point, and, in its wild mountain passes, we find the first scenes of especial remark in our journey. Fine hill farms surround us here, and on all our way through the region of the Ramapo for 18 miles, by *Sloatsburg, Southfields, Greenwood,* and *Turner's* to *Monroe.* The chief attraction of the Ramapo Gap is the Torn Mountain, variedly seen, on the right, near the entrance of the valley and about the Ramapo station. This is historical ground, sacred with memories of the movements of the Revolutionary army, when it was driven back into New York from the Hudson. Washington often ascended to the summit of the Torn to overlook the movements of the British. On one such occasion, anecdote says that he lost his watch in a crevice of a rock, of which credulity afterwards heard the ticking in the percolations of unseen waters. Very near the railway at Suffern's the débris of old intrenchments are still visible; and marks of the camp fires of our French allies of the period may be traced in the woods opposite. Near by is an old farm house, once occupied by the commander-in-chief. The Ramapo is a great iron ore and iron manufacture region; and it was here that the great chain which was stretched across the Hudson to check the advance of the English ships, was forged, at the spot once called the Augusta Iron Works, and now a poetical ruin by a charming cascade with overhanging bluff, seen close by the road, on the right, after passing Sloatsburg. The Ramapo Brook winds attractively through the valley, and beautiful lakelets are found upon the hill tops. There are two such elevated ponds near Sloatsburg. At Sloatsburg passengers for the summer resort of Greenwood Lake, 12 miles off, take stage tri-weekly. See Greenwood Lake.

From *Monroe* onward through *Oxford, Chester, Goshen, Hampton, Middletown, Howell's*, and *Otisville* to *Port Jervis* (or Delaware), we are in the great dairy region of Orange County, New York, which sends a train of cars laden only with milk daily to the New York market. A very charming view is seen south from the station at Oxford, led by the cone of the Sugar Loaf, the chief hill feature of the vicinage. At Chester, the branch road from Newburgh, on the Hudson river, 19 miles long, comes in. From this point, as well as from Sloatsburg, passengers for Greenwood Lake (eight miles) take stage. At Howell's, 70 miles from New York, the country gives promise of the picturesque displays to be seen through all the way onward to Port Jervis. Approaching Otisville, the eye is won by the bold flanks of the Shawangunk Mountain, the passage of which great barrier (once deemed almost insurmountable) is a miracle of engineering skill. A mile beyond Otisville, after traversing an ascending grade of 40 feet to the mile, the road runs through a rock cutting, 50 feet deep and 2,500 feet long. This passed, the summit of the ascent is reached, and thence we go down the mountain side many sloping miles to the valley beneath. The scenery along the mountain slope is grand and picturesque, and the effect is not lessened by the bold features of the landscape all around—the rugged front of the Shawangunk, stepping, like a colossal ghost, into the scene for one instant, and the eye anon resting upon a vast reach of untamed wilderness. In the descent of the mountain the embankment is securely supported by a wall 30 feet in height and 1,000 feet long. The way onward grows momently in interest, until it opens upon a glimpse, away over the valley of the mountain spur, called the Cuddleback; and, at its base, the glittering water seen now for the first time, of the Delaware and Hudson Canal, whose *débouchure* we have looked upon at Kingston, in our voyage up the Hudson River. Eight miles beyond Otisville we are imprisoned in a deep earthy cut for nearly a mile, admirably preparing us for the brilliant surprise which awaits us. The dark passage made, and yet another bold dash through rocky cliffs, and there lies suddenly spread before us, upon our right, the rich and lovely valley and waters of the Neversink. Beyond sweeps a chain of blue hills, and at their feet, terraced high, there gleam the roofs and spires of the village of Port Jervis; while onward, to the south, our eye first beholds the floods of the Delaware, which is to be so great a source of delight in all our journey hence, for nearly 90 long miles, to Deposit.

Port Jervis, or Delaware, as the station was called, is the terminus of the eastern division, one of four great sub-sections into which the road is measured. It is the point at which the tourist who can spend several days in viewing the route, should make his first night's halt. The vicinage is replete with pictorial delights, and with ways and means for rural sports and pleasures. Charms of climate and of scenery, with the additional considerations of a pretty village and a most excellent hotel (the Delaware House at the station), have made Port Jervis a place of great and continuous summer resort and tarry.

There is a stage route hence, 6 miles, to the neighboring "Falls of the Sawkill." This stream, after flowing sluggishly for some miles through level table-land, is here precipitated over two perpendicular ledges of slate-rock—the first of about 20 feet, and the second about 60 feet—into a wild gorge. The brook still continues, dashing and foaming on for a quarter of a mile, over smaller precipices, and through chasms scarcely wide enough for the visitor to pass. The beetling cliffs that form the sides of the gorge are surmounted and shaded by cedars and hemlocks, that lend a peculiarly sombre air to the scenery. The sojourner here must not omit a tramp to the top of Point Peter, overlooking the village, and all the wonders for miles around.

We now continue the transit of the second grand division of the road, which carries us onward, 104 miles further, to *Susquehanna*, and from New York, all told, 192 miles. The canal keeps us company, nearer or more remote, for some miles, and by and by we cross the Delaware on a fine bridge of 800 feet, built at a cost of $75,000. The river, from this point, is seen, both above and below, to great advantage. Here we leave Orange County and New York for a little incursion into the Keystone State, for which privilege the company pays Pennsylvania ten thousand dollars per year.

The canal, and its pictures and incidents, are still the most agreeable features of our way, though at Point Eddy we open into one of the wide basins so striking in the scenery of the Delaware.

Near *Shohola* (106 miles from New York), we are among some of the greatest engineering successes of the Erie route, and some of its chief pictorial charms. Here the road lies on the mountain side, several feet above the river, along a mighty gallery, supported by grand natural abutments of jagged rock. It is a pleasant scene to watch the flight of the train upon the crest of this rocky and secure precipice; and the impressiveness of the sight is deepened by its contrast with the peaceful repose of the smiling meadow slopes, on the opposite side of the river below. Upon three miles alone of this Shohola section of the road, no less than three hundred thousand dollars were expended.

At Lackawaxen (111 miles from New York), there is a charming picture of the village, and of the Delaware bridged by the railway, and by the grand aqueduct for the passage of the canal, supported by an iron wire suspension bridge.

We pass on now by Mast Hope to Narrowsburg.

Narrowsburg (122 miles from New York and 337 from Dunkirk), is a pleasant place for quiet summer rest and rural pastime, with its inviting hotel comforts, and its piscatory and field recreations and sports.

Beyond Narrowsburg, for some miles, the traveller may turn to his newspaper or book for occupation a while, so little of interest does the scene, without, present, with the exception, now and then, of a pleasing bit of pastoral region. Some compensation may be found in recalling the stirring incidents of Cooper's novel of the "Last of the Mohicans," of which this ground was the theatre.

At Callicoon, a brook full of wild and beautiful passages and of bright trout, comes down to the Delaware.

As we approach Hancock, once called Chehocton, we come near the charming picture of the meeting of the two branches of the Delaware, seen on our left.

Hancock is one of the most important places on this division of our route, and in every way a pleasant spot for sojourn.

At *Deposit*, 13 miles beyond Hancock, and 176 from New York, we bid good-bye to the Delaware, which we have followed so long; refresh ourselves at the excellent café, and prepare for the ascent of a heavy grade over the high mountain ridge which separates it from the lovely waters of the Susquehanna. We go up 58 feet per mile to an elevation 865 feet above Deposit. The way is wild and desolate, covered with the jagged *debris* left in the strong battle with the mountain fastnesses. The grand pass of the summit reached, we descend again by a grade of 60 feet, into that most beautiful region of the Erie road, the Valley of the Susquehanna.

For a little while, as we go down, there seems no promise of the wonders which are awaiting us, but they come suddenly, and before we are aware, we are traversing the famous

Cascade Bridge, a solitary arch, 250 feet wide, sprung over a dark ravine of 184 feet in depth.

No adequate idea of the bold spirit and beauty of the scene can be had from the cars, and especially in the rapid transit often passed before the traveller is aware of its approach. It should be viewed leisurely from the bottom of the deep glen, and from all sides, to be realized aright. To see it thus, a half a day's halt should be made at the next station, to which we shall soon come.

The Cascade Bridge crossed, the view opens almost immediately, at the right—deep down upon the winding Susquehanna, reaching afar off amidst a valley and hill picture of delicious quality, a fitting prelude to the sweet river scenes we are henceforth to delight in. This first grateful glimpse of the brave Susquehanna is justly esteemed as one of the finest points on the varied scenery of the Erie Railway route. It may be looked at more leisurely and more lovingly by him who tarries to explore the Cascade Bridge hard by, and the valley of the Starrucca, with its grand viaduct, which we are now rapidly approaching.

The Starrucca Viaduct (190 miles from New York and 269 from Dunkirk), is one of the chief art-glories of our present route—perhaps the chiefest. The giant structure is made of stone from the ravine, two miles above, crossed by the fairy Cascade Bridge. It is 1,200 feet in length and 110 feet high, and has 18 grand arches, each 50 feet span. The cost was $320,000. The landscape is of exceeding beauty, whether seen from the viaduct or from any one of many points, near or afar off, below. From the vicinity of Susquehanna, the next station, the viaduct itself makes a most effective feature in the valley views.

A little way beyond, and just before we reach the Susquehanna station, we cross a fine trestle bridge, 450 feet long, over the Cannewacta Creek, at Lanesborough. We are now fairly upon the Susquehanna, not in the distance but near its very marge, and, anon, we reach the end of the second grand division of our route, and enter the busy depot of Susquehanna, from New York 192 miles, and from Dunkirk 267.

At **Susquehanna** we are passing beyond the wild scenery on our route, and in a few miles further we shall fall in with and follow, for many miles, through broad valley tracts, coursed by the great winding river—a country which we shall find replete with interest, and very often of marked natural beauty, however unlike the scenes upon which we have looked in our transit of the wild hills and forest region of the Delaware.

The Susquehanna station is one of the busiest points on our route, being the place where divisions meet—where the great massive engines, or *pushers*, which are used to push the heavy trains hence to the top of the grand hill "Summit," are housed, and where the workshops for the repairs of disabled locomotives and cars are located. 200 hands are employed here by the company. Indeed, the place is all railroad, from which it was born and from which it has grown. If the hotels at this station are too noisy for the tarrying stranger, we may go a mile backwards to Lanesborough, and from thence review the scenes of the Starrucca and of the Cascade Bridges, with many other points of pictorial attraction.

Just beyond the Susquehanna depot we cross to the right bank of the river, and, after two more miles ride, yet amidst mountain ridges, we reach

Great Bend, 200 miles from New York, and 259 from Dunkirk. The village of this name lies close by, in the State of Pennsylvania, at the base of a bold cone-shaped hill.

At Great Bend there comes in to the Erie Road the Delaware, Lackawanna, and Western Railway, leading nearly south into Pensylvania, through the coal regions of Scranton, the neighborhood of the valley of Wyoming, the Water Gap of the Delaware, and ending upon that river, five miles yet below; here it is connected by other railway routes with New York, Philadelphia, etc. See Index.

Leaving Great Bend we enter upon the more cultivated landscape of which we lately spoke, and approach villages and towns of greater extent and elegance.

Near Kirkwood, the next station, six miles from Great Bend, there stands an old wooden tenement, which may attract the traveller's notice as the birth-place of the Mormon prophet, Joe Smith.

Binghamton.—HOTELS :— *The American House, The Lewis House.*

Binghamton, 210 miles from New York, is, with its population of ten or eleven thousand people, one of the most important places on the Erie route, and indeed in southern New York. It is a beautiful town, situated upon a wide plain, in an angle made by the meeting of the Susquehanna and the Chenango rivers.

Binghamton was settled in 1804, by Mr. Bingham, an English gentleman, whose daughters married the brothers Henry and Alexander Baring, the famous London bankers. One of those gentlemen was afterwards created Lord Ashburton.

The Chenango Canal, extending along the Chenango river, connects Binghamton with Utica, N. Y., 95 miles distant; and it is also the southern terminus of the Syracuse and Binghamton Railroad, 80 miles long. See Index. Passing on by the stations of Hooper, Union, and Campville, we come to

Owego.—HOTELS :—*Ah-wa-ga House.*

Owego, another large and handsome town, almost rivalling Binghamton in beauty and importance. Owego is surrounded by a landscape, not of bold nor of very beautiful features. Many noble panoramas are to be seen from the hilltops around, overlooking the village and the great valley. The Owego Creek, which enters the Susquehanna here, is a charming stream. Just before its meeting with the greater waters, it passes through the meadow, and at the base of the hill slopes of "Glenmary," once the home of N. P. Willis, and now one of the Meccas of the vicinage, to which all visitors are won by the charms and spells the genius of the poet has cast about it. It was here that Mr. Willis wrote his famous "Letters from under a Bridge."

The Cayuga and Susquehanna Railroad diverges here, some 30 miles, to Ithaca, on Cayuga Lake. See Index.

The Owago House, in the heart of the town and on the banks of the Susquehanna, is a large and elegant summer hotel.

Elmira.—HOTELS :—*Brainard House.*

Passing the half dozen intermediate stations, we jump now 37 miles, to Elmira, 273 miles from New York. This beautiful town is a peer of Binghamton and Owego, with the same charming valley nest and the same environing hill-ridges.

The Newton Creek and the Chemung River, near the junction of which waters Elmira is built, lend a world of picturesque beauty to the vicinage.

The Elmira, Canandaigua, and Niagara Falls Railway diverges here, and connects the village with the Canada lines. This road is one of the pleasantest from New York to the Falls of Niagara.

The Williamsport and Elmira Railroad conducts hence into Pennsylvania, and unites with other lines for Philadelphia. The Chemung Canal also connects Elmira with Seneca Lake, 20 miles distant. It is a delightful excursion from the village to Geneva and other places on the Seneca Lake, by the railway travel. See Index for routes from Elmira.

Five miles beyond Elmira, our route is over the Chemung River, bringing us to "Junction," the starting point of the Chemung Railroad for Jefferson and Niagara.

Corning, 290 miles from New York, is an important point on the Chemung River. The feeder of the Chemung Canal extends hither from Elmira. It is the depôt of the Corning and Blossburg Railroad, 40 miles distant from the coal beds of Pennsylvania. At Corning there terminates also the Buffalo, Corning, and New York Railroad, 94 miles, via Avon (Springs), and Batavia to Rochester, on the great routes west from Albany. See Index.

Hornellsville. Passing half a dozen stations, we now reach Hornellsville, where passengers for Buffalo, Niagara, &c., follow the Branch Road north, for 91 miles. (See Buffalo division.) At this point we enter upon the fourth and last division of the Erie route, being now 331 miles from New York, and having 128 miles yet to travel to Dunkirk. The country through the rest of our way is comparatively new, and no important towns have yet grown up within it. Pictorially this division is the least attractive of the whole route, though beautiful scenes occur still, at intervals, all along. Beyond Hornellsville, we enter the valley of the Caniacadea, a fine mountain passage, filled with the merry waters of the Caniacadea Creek. *Almond* and *Alfred* lie upon the banks of this charming stream.

Reaching Tip Top Summit (the highest grade of the Erie Road, being 1,700 feet above tide water), we commence the descent into the valley of the Genesee. The country has but few marks of human habitation to cheer its lonely and wild aspect, and for many miles onward, our way continues through a desolate forest tract, alternated only by the stations and little villages of the road. Beyond Cuba Summit, there are many brooks and glens of rugged beauty.

Passing Olean, on the Alleghany River, we come into the lands of the Indian Reservation, where we follow the wild banks of the Alleghany, between lofty hills, as wild and desolate as itself.

At Cattaraugus, 426 miles from New York, and 31 from Dunkirk, we traverse a deep valley, where the eye is relieved for a little while, with scenes of gentler aspect than the unbroken forest we have long traversed, and are to traverse still.

Three miles beyond Perrysburg we catch our first peep at the great Erie waters, towards which we are now rapidly speeding. Yet a few miles, and we are out of the dreary woods, coursing again through the more habitable lands which lie upon the lakes.

Dunkirk.—HOTELS:— *The Eastern.*

Reaching Dunkirk at last, we may pursue our journey westward by any one of many routes by land and by water—on the blue waves, or still upon the rapid rail. We shall follow all these routes in other pages—the steamers to Cleveland or Detroit, or the lake shore road to Cleveland. Thence by railway to Columbus, and Cincinnati, southward, or to Toledo, westward. From Toledo onward by the Michigan and Northern Indiana Railway to Chicago, and thence again by the Rock Island Road to the Mississippi, or by the Illinois Central route to St. Louis, and by other ways still onward, to the far west.

NEW YORK TO BUFFALO AND NIAGARA FALLS.

VIA BUFFALO BRANCH OF NEW YORK AND ERIE RAILROAD.

Follow the main trunk of the New York and Erie Road, from Jersey City, 331 miles to Hornellsville, N. Y.

Trains continue on immediately to Buffalo, on their arrival at Hornellsville, by the Branch Route, formerly the New York City and Buffalo Railroad.

STATIONS.

HORNELLSVILLE; Burns, 9; Whitney's, 13; Swainville, 17; Nunda, 24; Hunt's Hollow, 26; PORTAGE, 30; Castile, 34; Gainesville, 37; Warsaw, 44; Middlebury, 49; Linden, 53; ATTICA, 60; Darien, 64; Alden, 71; Town Line, 76; Lancaster, 81; BUFFALO, 91.

The road follows by the side of the Dunkirk track through the village, and then bends northward. For nearly 30 miles, along very elevated ground, there is but little to interest the tourist, until he comes in sight of the village of Portage, lying in a deep valley to the north-west.

Portage is deservedly a Mecca to the lover of the picturesque, abounding, as it does, in the wildest wonders of mountain gorge and cataract. The Genesee River steals and tumbles through the lawns and ravines of this region in a very wonderful way. At Portage, it enters a grand rocky defile, and in passing, falls in many a superb cascade. Near the station, this gorge is crossed by the railroad, upon a bridge of great magnitude and remarkable construction. From below, it rises upon the view like story upon story of solid and symmetrical scaffolding, to a height of 234 feet; its length is 800 feet. Beneath its huge masses of timber, foams the river, and by its angry side are the placid waters of the Genesee Valley Canal.

To see the wonders of Portage aright, one must tarry for days in the village, or better yet, at the hotel near the station house.

The Genesee makes a bold descent of 40 feet (seen from the cars), as it rushes beneath the great bridge, onward to yet deeper beds. A quarter of a mile northward is the second cataract of 80 feet; huge high cliffs soar yet far above it. To see the scene properly, the visitor will cross the bridge over the Genesee above the mill, and place himself immediately in front of the fall.

Some distance beyond, a staircase conducts to the bottom of the ravine, whence you may pass in a boat, or pick your way along beneath the spray of the tumbling floods. The walls of this gorge are of slate stone; they rise to a height of more than 300 feet, and in the many and sudden turnings of the way, offer a grateful succession of noble pictures.

A mile and a-half still down the glen, and we reach the third, and, perhaps, the grandest of the cascades; placed as it is in an exceedingly deep and narrow passage of the ravine. This leap is 60 feet.

The canal far up above the descending bed of the Genesee in this vicinage, is a most telling feature in the landscape—a strain of gentleness in the wild anthem of the rugged ravine.

We leave the traveller here to pursue the rest of his way, 61 miles to Buffalo, and to go thence to Niagara or elsewhere, as he may find directions in other parts of our Hand Book.

NEW YORK TO THE CATSKILL MOUNTAINS.

We can commend to the traveller no pleasanter or more profitable summer excursion for a day, or a month, or even a season, than a visit to the Catskills—one of the grandest and most picturesque of the mountain ranges of the United States.

To reach the Catskills from New York we will follow our previous routes up the Hudson to the village of Catskill (111 miles), or the river railway to Oakhill station opposite, crossing thence to Catskill by ferry.

At Catskill good stages are always in waiting to convey passengers to the Mountain House, on the crest of the hills, 12 miles westward. This ride will occupy about four hours, at a cost of one to one and a quarter dollars.

The Catskills are a part of the great Appalachian chain which extends through all the eastern portion of the Union, from Canada to the Gulf of Mexico. Their chief ranges follow the course of the Hudson River, from some 20 to 30 miles, lying west of it, and separated by a valley stretch of 10 to 12 miles. These peaks lend to all the landscape of that part of the Hudson from which they are visible, its greatest charm.

The Mountain House is reached by a pleasant stage-coach ride through ever-charming scenes of valley and hill. The last three miles of the journey is up the side of the mountain, made easy by a good winding way. Within a mile of his destination, the tourist halts upon the spot universally conceded to be the site of the famous 15 years' nap of Mr. Irving's myth, Rip Van Winkle.

The Mountain House stands near the brink of some bold rocky ledges, upon the summit of one of the eastern ranges, commanding all the landscape round for miles and miles away. Lifting its grand front thus, it is a curious and wonderful object, no less within its own shadow than at every point from which it may be seen. It is a massive and elegant structure of wood, with a grand facade of columns reaching the entire height of the eaves. It was originally built by the people of Catskill, at an expenditure of 20,000 dollars: but it has since been from time to time enlarged and improved, until now it possesses every reasonable, if not every possible, hotel convenience and comfort—capacious and well-furnished parlors, halls, and chambers—a luxurious table, and attentive hosts and waiters—and bathing, billiard, and bowling appointments. In the summer the house is a post-office, with daily mails.

The superb panorama of the river and valley of the Hudson, and of the New England hill ranges to the eastward, which the bold site of the Mountain House commands, will first fix the attention and admiration of the guest. Of this unrivalled sight he will never weary, so varied is it in the changing hours and atmospheres, and so imposing under every aspect. It is thought, at the dawn of the day and at the rising of the sun, when his magic beams are lifting the mystical vapor and cloud-curtain, which the night has invisibly spread over the scene, that the enchantment will reach its highest point. Luckily for the tourist who is not an enthusiast, but is contented with the simple, solid fact of a subject, like Mr. Gradgrind, these marvellous exhibitions of sun-rise effects may be comfortably seen from his warm, secure chamber-window, when the morning air is, as it often chances to be, at this mountain altitude, rather too chill and damp for comfort.

A visit to the locality, called the "North Mountain," will be a remunerative morning or afternoon's walk. It is only a mile or two through the forest, on the lofty ridge; but a guide will be desirable, for the path is more easily lost than found. At the end of the stroll he will look back upon his wilderness home over a brace of dainty little lakelets, smilingly sleeping on the mountain top; and beyond, towards the south and east, his eye will follow the windings of the Hudson far down in the sunny valley. Some stories may be told him of the fondness of the bear for this particular locality, but he need not be alarmed, for it is rarely indeed, except it be when the winter snows envelope the earth, that these gentry are about.

Another agreeable excursion will be in the opposite direction, from the house to the spot known as "South Mountain," where, upon the brink of huge cliffs, may be seen the river and valley, and the wonderful pass of the Kauterskill, through the mountain chain westward.

The Two Lakes, which we have just overlooked from the North Mountain, make one of the leading items in the Catskill programme. They lie side by side, in gentle beauty, in the heart of the lofty plateau, upon the eastern brink of which the Mountain House is perched. They may be reached in a pleasant little walk back of the hotel. Onwards, and on the way to the Great Falls of the Kauterskill, a few minutes' stroll, indeed, is sufficient to bring us to the nearest of these twin waters, the Upper or Sylvan Lake. This is a spot for repeated and habitual visits, with its pleasures by the forest shore; in the skiff, upon the quiet and lonely flood; or, with angle in hand and trout in prospect.

The High Falls lie two miles back of the Mountain House, overleaping the western brink of the great plateau. A wagon road leads thither; and there is, besides, a footpath in the forest, by which the way is shortened one-half. A good team is sent down with passengers (fare 25 cents) at least once a day from the hotel. At the very brink of the cascades there is another small but pleasanter summer inn, called the Laurel House, kept by Mr. Scutt, the proprietor of the Falls. It is a wonderful sight to overlook the ravine below, and the giant crests of Round Top and High Peak—the proudest of all these hills—from the windows or piazzas of the Laurel House, or from the platform in front, which overhangs the glen. This view enjoyed, with refreshments if you please, we commence the descent to the base of the cataracts, by many straggling flights of wooden steps. Coming to the base of the first Fall, we may steal along a narrow ledge behind the descending torrent, as one gets to Termination Rock, at Niagara. On the opposite bank parties often pic nic, the means and appliances, if duly ordered before at the Laurel House, being lowered down, upon a signal, in a basket, over the edge of the projecting platform above. The descent of the first cascade is 175 feet, and of the second, 75 feet, with many a tumble of the vexed waters afterwards in their way for a mile down the ravine into the main branch of the Kauterskill or Catskill Creek, which dashes down the great clove, of which the Mountain House stream is only an arm held at a right angle.

Fenimore Cooper, in his story of the "Pioneer," thus describes these cascades—"The water comes croaking and winding among the rocks, first, so slow that a trout might swim in it, then starting and running like any creature that wanted to make a fair spring, till it gets to where the mountain divides like the cleft foot of a deer, leaving a deep hollow for the brook to tumble into. The first pitch is nigh 200 feet, and the water looks like flakes of snow before it touches the bottom, and then gathers itself together again for a new start; and maybe flutters over 50 feet of flat rock before it falls for another 100 feet, when it jumps from shelf to shelf, first running this way and that way, striving to get out of the hollow, till it finally gets to the plain."

This branch of the Kauterskill comes from the waters of the two lakes on the plateau above; and, as the supply has to be economized in order that the cascades may look their best when they have company, the stream is dammed, and the flood is let on at proper times only. For this service, and for the use of the steps, perchance of guide also, every visitor pays a toll of 25 cents. This is a reasonable although a disagreeable bit of prose in the poem of the Catskill Falls.

We have now peeped at all the usual "sights" of the region; but there are other chapters of beauty, perhaps, yet more inviting. Let the tourist, if he be adventurous and is a true lover of nature, follow the brook down from the base of the cataracts we have just described, into the principal clove; then let him ascend the main stream for a mile over huge boulders, through rank woods, and many by cascades, which, if smaller, are still more picturesque than those "nominated in the bond;" or, let him descend the creek, two miles, sometimes by the edge of the bed of the waters, and, when that is impracticable, by the turnpike road, which traverses the great clove or pass. At every turn and step there will be a new picture—sometimes a unique rapid or fall, sometimes a soaring mountain cliff, sometimes a rude bridge across the foaming torrent, sometimes a little hut or cottage, and, at last, as he comes out towards the valley on the east, the humble village of Palenville. This portion of the Catskills is that most preferred by artists for study, and the inns at Palenville are often occupied by them, though they offer no inviting accommodation to the case and comfort-loving tourist.

At one time (when the hemlock was abundant on the mountain sides) this clove was a den of tanneries, and a few establishments of the kind yet linger here.

Stony Clove. Another nice excursion from the Mountain House, is a ride along the ridge, five or six miles, to the entrance of the Stony Clove, and thence on foot, or still in your vehicle (though the wagon road is execrable), through the wilderness of this fine pass.

High Peak, the most elevated of the Catskill summits, towering 4,000 feet towards heaven, should certainly be climbed, in order to see the region fairly. It is a long and toilsome journey, especially for ladies—six miles thither *on foot*—but we have accompanied the fairest of women through the difficulties and dangers of the way. Once we "assisted" at a night camp on the very crown of High Peak, of a party which included a dozen damsels. If they had not been brave, as they all were, they would not have deserved the glorious sunrise effects, which they saw never to be forgot, from their ambitious bivouac. Even the Mountain House, on its grand perch, looked from High Peak like a pigmy in the vale.

Plauterkill Clove is another grand pass on the hills, five miles below the Kauterskill passage. A mountain torrent, full of beauties in glen, and rock, and cascade, winds through it. A post-road also traverses the pass. High Peak rises on the north of the Plauterkill, and the South Mountain, on which is a lovely lake, ascends on the opposite side. It is not yet a scene of much resort, being out of the very convenient reach of the Mountain House, and having no hotel attractions in its neighborhood.

The tourist here will recall with pleasure Bryant's dainty poem of the Katterskills, from which we borrow a few lines to end our own intimations—

"Midst greens and shades the Catterskill leaps
From cliffs where the wood-flower clings;
All summer he moistens his verdant steeps,
With the light spray of the mountain springs;
And he shakes the woods on the mountain side,
When they drip with the rains of the autumn tide.

"But when, in the forest bare and old,
The blast of December calls,
He builds in the star-light, clear and cold,
A palace of ice where his torrent falls,

With turret, and arch, and fretwork fair,
And pillars clear as the summer air."

The Cataracts of the Catskills in winter, when the spray is frozen into a myriad fantastic forms, all glowing like the prism, as the clear cold sunlight reveals these mystical wonders, is a sight so grand and novel as to well repay the exposure and fatigue of a visit thither through bleak January's snows and ice.

The Mountain House is then closed, but Mr. Scutt inhabits his Laurel Inn all the year, we believe. This is a hint to the enthusiast in the search for the strange and beautiful in Nature. Most tourists will care to see the Catskills only in July or August.

Charges at the Mountain House are, as in most of the fashionable summer resorts in the United States, $2.50 per day. At the Laurel Inn, by the High Falls, about half that price, we believe.

Stages will take you back to Catskill village, as they have brought you thence, in season for steamboats and railways, for elsewhere.

To visit the Catskills comfortably, three days will suffice for the journey thence by rail from New York, for the stay and the return to the city. Not less than four, however, ought to be thus invested, if one would make sure of a satisfactory dividend; and if a week is at command, so much the happier he who commands it.

NEW YORK TO ALBANY.

Via Harlem Railroad.

This route extends from the heart of the city of New York to the State capital, skirting in its course the eastern portions of all those counties lying upon the Hudson and traversed by the river railway. The distance between the termini is 154 miles, a few miles longer than that of the Hudson River. Time, about the same. The stations and towns upon the Harlem Road are, for the most part, inconsiderable places, many of them having grown up with the road. The country passed through is varied and picturesque in surface, and much of it is rich agricultural land. It does not compare with the river route in scenic attractions.

STATIONS.

New York—corner of White and Centre streets—Yorkville, 6 miles; Harlem, 7; Mott Haven, 8; Melrose, 9; Morrisania, 10; Tremont, 11; Fordham, 12; Williams Bridge, 14; (Junction of the N. Y. and N. Haven Road) Hunt's

Bridge, 16; Bronxville, 18; Tuckahoe, 19; Scarsdale, 22; Hart's Corners, 24; WHITE PLAINS, 26; Kensico, 29; Unionville, 31; Pleasantville, 34; Chapequa, 36; Mount Kisco, 40; Bedford, 42; Whitlockville, 45; Golden Bridge, 47; Purdy's, 49; Croton Falls, 51; Brewster's, 55; Dykman's, 58; Towner's, 61; Patterson's, 63; Pawling's, 67; South Dover, 73; Dover Furnace, 76; Dover Plains, 80; Wassaic, 84; Amenia, 88; Sharon Station, 91; Millerton, 96; Mount Riga, 99; Boston Corners, 103; Copake, 108; Hillsdale, 112; Bains, 115; Martindale, 118; Philmont, 122; Ghent, 128; CHATHAM FOUR CORNERS, 130; (Junction with railway route from Albany and from Hudson, for Boston) EAST ALBANY, 153 miles.

All the stations from New York to White Plains (26 miles) are suburban, being escape-valves of the overgrown population of the city, where the business of the principal part of their population lies, and to which they go daily by the railway. Many of the villages are picturesque, pleasant, and prosperous.

On leaving the city streets, the road passes under a considerable extent of tunnelling and continued bridging across thoroughfares overhead, making merry diversion for the passengers. At the extremity of the Island and city of New York at Harlem, the road crosses the Harlem River into Westchester County.

White Plains (Westchester County), is interesting as the scene of important events in the Revolution. An eventful battle was fought here, October 28, 1776. A residence of Washington) in which are some attractive relics) is yet standing in the vicinage.

Croton Falls, upon the river which supplies the great Croton Aqueduct to the city of New York.

Lake Mahopac. Passengers for Lake Mahopac take stage thence (distance two hours) at the Croton Falls Station. See "Lake Mahopac."

Dover Plains, 20 miles east of Poughkeepsie, is surrounded by much pleasing landscape.

For Albany, and routes thence by Hudson River and the river railway, see Index.

Saratoga Springs.—HOTELS:—The most desirable hotels at Saratoga are the *United States*, the *Union Hall*, *Congress Hall*, and the *Clarendon*. Besides this famous quartette of houses, there are many of less fashion and price, besides numerous private boarding-houses, where one may live quietly at a moderate cost. The hotels which we have named have each accommodations for six or seven hundred guests, for all of which the demand is more than ample. Fine bands of music discourse on the broad, shady piazzas, and in the ball-rooms, at the dinner and evening hours. The *Clarendon* is a new spacious house, just erected in the heart of the village.

From Boston by the Western Railway, 200 miles to Albany; or, from New York, by the Hudson River line or steamboats, 144 miles to Albany, or 150 miles to Troy. From either place, by the Rensselaer and Saratoga Railway, through Ballston Springs. Journeying from New York by water, the traveller may pass pleasantly up the river, during the night, breakfast in Albany or Troy, and reach the Springs in good season next morning.

The little ride from Troy to the Springs is a most agreeable one, as the route crosses and follows the Hudson and the Mohawk Rivers, as it passes Waterford at the meeting of these waters, four miles above Troy, and near the Cohoes Falls, a much-admired and frequented resort upon the Mohawk, as it thence continues upon the west bank of the Hudson, eight miles further to Mechanicsville. It afterwards crosses the canal, passes Round Lake, and enters Ballston Springs.

During the summer, a car on the Hudson River express trains from New York passes *through* to the Springs without change. Passengers via Albany for the Springs change cars at Albany.

Ballston Spa is upon the Kayaderosseros Creek, a small stream which flows through the village, 25 miles from Troy. Its mineral waters, which were discovered in 1769, are celebrated for their medicinal qualities, although not so popular as they were formerly, those of Saratoga being now generally preferred. The *Sans Souci Hotel* is a pleasant house near the centre of the village.

Five miles distant is Long Lake, a resort of the angler. Saratoga Lake is six miles from Ballston.

Saratoga has been for many years, and still is, and probably always will be, the most famous place of summer resort in the United States, frequented by Americans from all sections, and by foreign tourists from all climates. During the height of the fashionable season no less than two or three thousand arrivals occur within a week. There is nothing remarkable about the topography or scenery of Saratoga; on the contrary, the spot would be uninteresting enough but for the virtues of its waters and the pleasures of its brilliant society. The village streets, however, are gratefully shaded by fine trees, and a little "let up" in the gay whirl may be got on the walks and lawns of the pretty rural cemetery close by.

The health-giving Springs of which the fame of

Saratoga has been born, however much Fashion may have since nursed it, are all in or very near the village. There are many different waters in present use, but the most sought after of all are those of the Congress Spring, of which Dr. Chilton gives us an analysis thus:—One gallon, of 231 cubic inches, chloride of sodium, 363.829 grains; carbonate of soda, 7.200; carbonate of lime, 86.143; carbonate of magnesia, 78.021; carbonate of iron, .841; sulphate of soda, .651; iodine of sodium and bromide of potassium, 5.920; silica, .472; alumina, .321; total, 543.998 grains. Carbonic acid, 284.65; atmospheric air, 5.41; making 290.06 inches of gaseous contents.

This Spring was discovered in 1792, though it was long before known to and esteemed by the Indians.

After the Congress waters, which are bottled and sent all over the world, as everybody knows, the Springs most in favor and use at Saratoga are the Empire, the Columbian, the High Rock, the Iodine, the Pavilion and Putnam's. The Empire Spring, the most northerly one in the village, has grown greatly in repute of late years. So far its landscape surroundings have received but poor attention. The High Rock Spring, not far from the Empire, is much esteemed both for its medical virtues and for the curious character of the rock from which it issues, and after which it is named. This singular rock has been formed by the accumulated deposits of the mineral substances (magnesia, lime and iron) held in solution by the carbonic acid gas of the Springs. The circumference of the rock, at the surface of the ground, is 24 feet 4 inches, its height 2½ feet, with an aperture of nearly one foot diameter.

The Alpha and the Omega of the daily Saratoga programme, is to drink and to dance—the one in the earliest possible morning, and the other at the latest conceivable night. Among the outside diversions is a jaunt to Saratoga Lake, a pleasant water six miles away. Here they have nice boating fun, and sometimes "make believe" to fish. This lake is nine miles in length and very near three in width. The marshes around it prevent access, except here and there. *Snake Hill* steps into the water, and lifts up its head 40 feet or so, upon the eastern side of the lake.

A visit to Lake George, 28 miles distant, by rail and plank road, is a delightful episode and variation in Saratoga life. See Lake George.

Lake George.—HOTELS:—The *Fort William Henry Hotel* and the *Lake House*, at the south end of the lake (Caldwell's).

The route from New York, Boston, and the West, to Lake George, is through Saratoga, and thus far is the same as to that point; thence to Moreau station, 15 miles, by the Troy and Whitehall line, and from there to Caldwell, at the south end or head of the lake, by plank road.

Glen's Falls, in the upper Hudson, is on the way, nine miles from the lake. The wild and rugged landscape is in striking contrast with the general air of the country below—there, quiet pastoral lands; here, rugged rock and rushing cataract. This is a spot trebly interesting, from its natural, its poetical, and its historical character. The passage of the river is through a rude ravine, in a mad descent of 75 feet over a rocky precipice 900 feet in length. Within the roar of these cataracts were laid some of the scenes in Cooper's story of the "Last of the Mohicans." They are gently associated with our romantic memories of Uncas and Hawk's Eye, David, Duncan Haywood and his sweet wards, Alice and Cora Monroe.

When within four miles of the lake, we pass a dark glen, in which lie hidden the storied waters of *Bloody Pond*, and close by is the historic old boulder, remembered as *Williams' Rock*. Near this last-named spot, Colonel Williams was killed in an engagement with the French and Indians, Sept. 8, 1755. The slain in this unfortunate battle were cast into the waters near by, since called Bloody Pond. It is now quiet enough, under its surface of slime and dank lilies.

The first broad view of the beautiful lake, seen suddenly as our way brings us to the brink of the high lands, above which we have thus far travelled, is of surpassing beauty, scarcely exceeded by the thousand-and-one marvels of delight which we afterwards enjoy in all the long traverse of the famous waters.

Our road now descends to the shore, the gleaming floods and the blue cliffs of Horicon still, ever and anon, filling our charmed eye. We halt at the Lake House at Caldwell village, or at the Fort William Henry Hotel, a new and elegant establishment near by, at the ruins of the Old Fort, on the right.

About a mile south-east from the site of Fort William Henry are the ruins of Fort George. These localities are seen from the piazza of the Lake House, which commands also a fine view of the French Mountain and Rattle-Snake Hill, and of the islands and hills down the lake.

The passage of Lake George, 36 miles, to the landing near the village of Ticonderoga, and four miles from the venerable ruins of Fort Ticonderoga, on Lake Champlain, is made by steamboat,*

---

* The "John Jay," which has plied the waters of Lake George for some years past, was destroyed by fire in the summer of 1856, opposite Garfield's, near Sabbath Day Point.

the trip down to the Fort and back occupying the day very delightfully.

Leaving Caldwell after breakfast, we proceed on our voyage down the lake. The first spot of especial interest which we pass is Diamond Island, in front of Dunham Bay. Here, in 1777, was a military depot of Burgoyne's army, and a skirmish between the garrison and a detachment of American troops.

North of Diamond Isle, lies Long Island, in front of Long Point, which extends into the lake from the east. Harris Bay lies between the north side of this Point and the mountains. In this bay Montcalm moored his boats and landed, in 1757.

Dome Island is passed, in the centre of the lake, some 12 miles north of Caldwell. Putnam's men took shelter here while he went to apprise General Webb of the movements of the enemy, at the mouth of the North-West bay. This bay lies in one of the most beautiful parts of Lake George, just beyond Bolton Landing, where there is an inviting place of sojourn called the "Mohican House." The bay extends up on the west of the Tongue Mountain some five miles. On the east side of the bay, the Tongue Mountain comes in literally like a tongue of the lake, into the centre of which it seems to protrude, with the bay on one side and the main passage of the waters on the other. On the right or east shore, in the neighborhood, and just as we reach the Tongue and enter the "Narrows," is the bold semicircular palisades called Shelving Rock. Passing this picturesque feature of the landscape, and, afterwards, of the point of the Tongue Mountain, we enter the Narrows at the base of the boldest and loftiest shores of Horicon. The chief peak of the hills here is that of Black Mountain, with an altitude of 2,200 feet. The islands are numerous, though many of them are merely peeps out of the water. The best fishing-grounds of Lake George are in that part of the waters which we have already passed, in the vicinity of Bolton Landing, Shelving Rock, and thence to Caldwell, though fine trout and bass are freely caught from one end of the lake to the other.

Sabbath Day Point. Emerging from the Narrows, on the north, we approach a long projecting strip of fertile land, called Sabbath Day Point—so named, by General Abercrombie, from his having embarked his army on the spot on Sunday morning, after a halt for the preceding night. The spot is remembered, also, as the scene of a fight, in 1756, between the colonists and a party of French and Indians. The former, sorely pressed, and unable to escape across the lake, made a bold defence and defeated the enemy, killing very many of their men. Yet, again,

5*

in 1776, Sabbath Day Point was the scene of a battle between some American militia and a party of Indians and Tories, when the latter were repulsed, and some 40 of their number were killed and wounded. This part of Horicon is even more charming in its pictures, both up and down the lake, than it is in its numerous historical reminiscences. On a calm sunny day the romantic passage of the Narrows, as seen to the southward, is wonderfully fine; while, in the opposite direction, the broad bay, entered as the boat passes Sabbath Day Point, and the summer landing and hotel at "Garfield's," are soon to be abruptly closed on the north by the huge precipices of Anthony's Nose on the right, and Rogers' Slide on the left. This pass is not unlike that of the Highlands of the Hudson as approached from the south.

Rogers' Slide is a rugged promontory, about 400 feet high, with a steep face of bare rock, down which the Indians, to their great bewilderment, supposed the bold ranger, Major Rogers, to have passed, when they pursued him to the brink of the precipice.

Two miles beyond is *Prisoner's Island*, where, during the French war, those taken captive by the English were confined; and directly west is *Lord Howe's Point*, where the English army, under Lord Howe, consisting of 16,000 men, landed previous to the attack on Ticonderoga. We now approach the termination of our excursion on this beautiful lake, and in a mile reach the steamboat landing near the village of *Ticonderoga*, whence stages run a distance of three miles, over a rough and romantic road, to *Fort Ticonderoga*—following the wild course of the passage by which Horicon reaches the waters of Lake Champlain—a passage full of bold rapids and striking cascades.

After exploring the picturesque ruins of the ancient fort, and dining satisfactorily at the excellent hotel, which stands upon the marge of a beautiful lawn, sloping to the Champlain shore, our stage will take us back to the landing we have left on Lake George, and our steamboat thence to Caldwell again, in time for tea and a moonlight row among the countless green isles; or we may take the Champlain boat to Whitehall; or from Whitehall *en route* for Canada.

Fort Ticonderoga, of which the ruins only are visible, was erected by the French in 1756, and called by them "Carrillon." It was originally a place of much strength; its natural advantages were very great, being surrounded on three sides by water, and having half its fourth covered by a swamp, and the only point by which it could be approached, by a breastwork. It was afterwards, however, easily reduced, by an expedient adopted by General Burgoyne—that of placing a piece of artillery on the pinnacle of *Mount Defiance*, on the south side of the Lake George outlet, and 750 feet above the lake, and entirely commanding the fort, from which shot was thrown into the midst of the enemy's works. Fort Ticonderoga was one of the first strongholds taken from the English in 1775, at the commencement of the Revolutionary war. Colonel Ethan Allen, of Vermont, at the head of the Green Mountain Boys, surprised the unsuspecting garrison, penetrated to the very bedside of the commandant, and waking him, demanded the surrender of the fort. "In whose name, and to whom?" exclaimed the surprised officer. "In the name of the great Jehovah, and the Continental Congress!" thundered the intrepid Allen, and the fort was immediately surrendered.

NEW YORK TO MONTREAL AND NIAGARA, via LAKE CHAMPLAIN.

One of the most delightful of American summer tours is from New York, via the Hudson River, Lake Champlain, and the St. Lawrence River, to the Falls of Niagara, returning by the lower routes—the Central or the Erie Railways. From Boston and Portland, lines of railroad connect conveniently with the St. Lawrence routes. A thousand places and objects of interest fall within the direct line of this journey; besides which, it has many alluring asides, which may be readily reached.

From Portland, Maine, take the Grand Trunk Route, to Montreal (or to Quebec); from Boston, take the Boston, Concord, and Montreal Routes.

From New York, take the Hudson River Route, which we have already travelled, to Albany or Troy; thence, by rail, via Saratoga Springs, to Whitehall, at the southern extremity of Lake Champlain. We resume the programme at Moreau Station, on this line, to which point we have already followed it in our visit to Lake George. At Ticonderoga, above, on Lake Champlain, we shall meet those who prefer, as many do, to pursue the journey to that point by the way of Lake George, instead of via Whitehall and the lower end of Champlain.

To Whitehall the country is exceedingly attractive, much of the way, in its quiet, sunny valley beauty, watered by pleasant streams, and environed, in the distance, by picturesque hills. The Champlain Canal is a continual object of interest by the way; and there are, also, as in all the long journey before us, everywhere spots of deep historic charm, if we could tarry to read their stories—of the memorable incidents which they witnessed, both in the French and Indian, and afterwards in the Revolutionary war. In the valley regions of the Hudson, which lie between Albany and Lake Champlain, are many scenes famous for the struggles between the Colonists and Great Britain—the battle-grounds of Bemis Heights and Stillwater (villages of the Upper Hudson), and of Saratoga, which ended in the defeat of Burgoyne and his army. Then there is the tale of the melancholy fate of Jane M'Crea, so cruelly murdered by the Indians at Fort Edward; and many histories, which it is pleasant to recall, ever so vaguely, as we pass along.

Whitehall.—Hotels :—

Whitehall was a point of much consideration during the French and Indian war, and through the Revolution. In former times it was called Skenesborough. It is at the south end or the head of Lake Champlain, within a rude rocky ravine, at the foot of Skene's Mountain. Its position, as a meeting-place of great highways of travel, has made it quite a bustling and prosperous village. There is nothing in the vicinage, however, to delay the traveller. From Whitehall our journey lies down Lake Champlain, 156 miles, to St. John's, though we might, instead, go by railway through Vermont, via Castleton, Rutland, Burlington, &c., to Rouse's Point, and thence, still by railway, to Montreal.

The narrowness of the lower part of Lake Champlain gives it much more of a river than lake air. For 20 miles the average breadth does not exceed half a mile; and, at one point, it is not more than 40 rods across. However it grows wide enough as we pass Ticonderoga, where passengers by the Lake George *détour* are picked up, and in the vicinity of Burlington there are too many broad miles between the shores for picturesque uses. Whether it is broad though or narrow, the voyage, in large and admirable boats, over its mountain-environed waters, is always a pleasure to be greatly enjoyed and happily remembered. On the east rise the bare peaks of the Green Hills of Vermont, the bold Camel's Hump leading all the long line; and on the west are the still more varied summits and ridges of the Adirondack Mountains in New York.

Mount Independence lies in Vermont, opposite Ticonderoga, about a mile distant. The remains of military works are still visible here.—*Mount Hope*, an elevation about a mile north from Ticonderoga, was occupied by General Burgoyne previous to the recapture of Ticonderoga, which took place in 1777, nearly two years after its surrender to the gallant Allen. St. Clair, the American commander, being forced to evacuate, it again fell into the possession of the British, and was held during the war.

Not far above, and upon the opposite shore, is the village of Crown Point; and, just beyond, the picturesque and well-preserved ruins of the fortifications of the same name. Opposite is Chimney Point; and, just above, on the left, at the mouth of Bulwaggy Bay, is Port Henry.

Burlington.—HOTELS:—The *American;* the *Lake House.*

Burlington, the largest town on the lake, is upon the eastern or Vermont shore, about midway between Whitehall and St. Johns. Rising gradually to an elevation of several hundred feet, it is imposingly seen from the water. It is the seat of the University of Vermont, and is a place of much commercial importance, connected by railways with all parts of the country. Across the lake is

Port Kent, from which vicinity, whether on land or on water, the landscape in every direction is exceedingly striking and beautiful.

The Walled Banks of the Ausable. The remarkable Walled Banks of the Ausable are a mile or two west of Port Kent, on the way to the manufacturing village of Keeseville.

It is at the Ausable House, an excellent summer hotel in the picturesque village of Keeseville, that the traveller will establish himself, if he would visit this wonderful ravine, with its grand walls and its rushing waters. The Falls of the Ausable, though they are but little known as yet, will one day be esteemed among the chief natural wonders of the country.

Plattsburg.—HOTELS:—"*Fouquet's*"

Above and opposite Burlington is the pleasant village of Plattsburg, where the Saranac river comes in from its lake-dotted home, at the edge of the great wilderness of northern New York, 30 miles westward.

Battle of Lake Champlain. Plattsburg was the scene of the victory of M'Donough and Macomb over the British naval and land forces, under Commodore Downie and Sir George Provost. Here the American commodore awaited at anchor the arrival of the British fleet, which passed Cumberland Head about eight in the morning of the 11th September, 1814. The first gun from the fleet was the signal for commencing the attack on land. Sir George Provost, with about 14,000 men, furiously assaulted the defences of the town, whilst the battle raged between the fleets, in full view of the armies. General Macomb, with about 3,000 men, mostly undisciplined, foiled the repeated assaults of the enemy; until the capture of the British fleet, after an action of two hours, obliged him to retire, with the loss of 2,500 men and a large portion of his baggage and ammunition. Here we might land and take the Plattsburg and Montreal Railway, 62 miles direct to Montreal.

Rouse's Point, on the west side of the lake, is the last landing-place before we enter Canada. Railways from the Eastern States, through Vermont, come in here, and are prolonged by the Champlain and St. Lawrence road to Montreal. If the traveller towards Canada continues his journey, neither via Plattsburg nor Rouse's Point, he may go on by steamboat to the head of navigation on these waters to St. Johns, and thence by Lachine to Montreal.

See Canada for the tour of the St. Lawrence and Lake Ontario from Montreal to Niagara.

NEW YORK TO TRENTON FALLS.

Via Hudson River to Albany, thence by the New York Central Railroad as far as Utica, and thence, over a plank road, or by railroad, 15 miles.

Trenton Falls, says Mr. Willis, "is the most *enjoyably beautiful* spot among the resorts of romantic scenery in our country. The remembrance of its loveliness becomes a bright point, to which dream and reverie oftenest return. It seems to be curiously adapted to enjoy, being somehow, not only the *kind* but the *size* of a place which the (after all) *measurable* arms of a mortal heart can hold in its embrace. Niagara is too much as a roasted ox is, a thing to go and look at, though one retires to dine on something smaller."

Trenton Falls is the place, above all others, where it is a luxury to *stay*—which one oftenest revisits, which one most commends to strangers to be sure to see.

"In the long corridor of travel between New York and Niagara, Trenton," Mr. Willis says again, "is a sort of alcove aside—a side-scene out of earshot of the crowd—a recess in a window, whither you draw a friend by the button for the sake of chit-chat at ease."

Trenton Falls is rather a misnomer, for the wonder of nature which bears the name is a tremendous torrent, whose bed, for several miles, is sunk fathoms deep into the earth—a roaring and dashing stream, so far below the surface of the forest, in which it is lost, that you would think, as you come suddenly upon the edge of its long precipice, that it was a river in some inner world (coiled within ours, as we in the outer circle of the firmament), and laid open by some Titanic throe that had cracked clear asunder the crust of this "shallow earth." The idea is rather assisted if you happen to see below you, on its abysmal shore, a party of adventurous travellers; for at that vast depth, and in contrast with the gigantic trees and rocks, the same number of well-shaped pismires, dressed in the last fashion, and philandering upon your parlor floor, would be about of their apparent size and distinctness.

Trenton Falls are upon the West Canada Creek, a branch of the Mohawk. The descent of the stream, 312 feet in a distance of two miles, is by a series of half a dozen cataracts, of wonderful variety and beauty. Every facility of path and stairway, and guide, for the tour of the Trenton ravine has been provided by Mr. Moore, who has for many years resided on the spot, and been always its Prospero, and its favorite host.

A walk of a few rods through the woods brings the visitor to the brink of the precipice, descended by secure stairways for some hundred feet.

The landing is a broad pavement, level with the water's edge, often, in times of freshet, the bed of foaming floods. Here is commanded a fine view of the outlet of the chasm, 45 rods below, and also of the first cascade, 37 rods up the stream.

The parapet of the First Fall, visible from the foot of the stairs, is, in dry times, a naked perpendicular rock, 33 feet high, apparently extending quite across the chasm, the water retiring to the left, and being hid from the eye by intervening prominences. But in freshets, or after rain, it foams over, from the one side of the gorge to the other, in a broad amber sheet. A pathway to this fall has been blasted at a considerable cost, under an overhanging rock and around an extensive projection, directly beneath which rages and roars a most violent rapid. The passage, though at first of dangerous aspect, is made secure by chains well riveted in the rocky wall. In the midst of this projection, five tons were thrown over by a fortunate blast, affording a perfectly level and broad space, where 15 or 20 persons may find ample footing, and command a noble view of the entire scene. A little to the left, the rapid commences its wild career. Directly underneath, it rages and foams with great fury, forcing a tortuous passage into the expanded stream on the right. In front is a projection from the other side, curved to a concavity of a semicircle by the impetuous waters. The top of this projection has been swept away, and is entirely flat, exhibiting from its surface downwards, the separate strata as regular and distinct, and as horizontal as mason-work in the lock of the grand canal. Here, in the old time, was a lofty fall, now reduced to the rapid we have described.

Beyond, massive rocks, thrown over in flood times, lie piled up in the middle of the river. Passing to the left, yet a few rods above, we come into the presence of Sherman's Fall, so named in memory of the Rev. Mr. Sherman, whose account of the spot we are now closely following. He was one of the earliest pioneers of the Trenton beauties, and it was by him that the first house, called the "Rural Resort," for the accommodation of visitors, was built. It has formed an immense excavation, having thrown out thousands of tons from the parapet rock, visible at the stairs, and is annually forcing off slabs at the west corner, against which it incessantly forces a section of its powerful sheet. A naked mass of rock, extending up 150 feet, juts frowningly forward, which is ascended by natural steps to a point from which the visitor looks securely down upon the rushing waters.

Leaving this rocky shelf, and passing a wild rapid, we come suddenly in sight of the High Falls, 40 rods beyond. This cascade has a perpendicular descent of 100 feet, while the cliffs on either side, rise some 80 feet yet higher. The whole body of water makes its way at this point —divided by intervening ledges into separate cataracts, which fall first about 40 feet, then reuniting on a flat below, and veering suddenly around an inclination of rocky steps, they plunge into the dark caldron beneath.

Passing up at the side, we mount a grand level, where in dry times the stream retires to the right and opens a wide pavement for a large party to walk abreast. Here a flight of stairs leads to a refreshment house, called the Rural Retreat, 20 feet above the summit of the High Falls.

The opening of the chasm now becomes considerably enlarged, and a new variety of scene occurs. Mill Dam Fall, 14 feet high, lies some distance beyond, reaching across the whole breadth of the chasm.

Ascending this Fall, the visitor comes to a still larger platform of level rock, 15 rods wide at low water, and 90 in length, lined on each side by cedars. At the extremity of this locality, which is known as the Alhambra, a bare rock 50 feet in height reaches gradually forward from the mid-distance; and, from its shelving top, there descends a perpetual rill, which forms a natural shower-bath. A wild cataract fills the picture on the left.

Here the wide opening suddenly contracts, and a narrow aperture only remains, with vistas of winding mountain, cliff and crag. Near by is a dark basin, where the waters rest from the turmoil of the wild cascade above. In this vicinage is an amphitheatre of seemingly impossible access, replete with even new surprises and delights. Yet beyond is the Rocky Heart, the point at which the traverse of the ravine usually ends, though despite the difficulties and dangers of the way, even ladies frequently penetrate beyond as far as the falls at Boon's Bridge, the terminus of the gorge.

The scene at Trenton varies much, according as drought or freshet dries or fills the stream, and passages are easy enough at one time, which are utterly impracticable at others. It is difficult to say when the glen is the most beautiful, whether with much or with little water.

Trout once inhabited these waters, but they are gone now. Game, too, is scarce in the vicinage, though partridges, wild ducks, snipes, black and gray squirrels, woodcock, and the rabbit may yet be taken. Trenton is a spot for a long sojourn, though it may be run over pleasantly in a day.

NEW YORK TO BUFFALO.

To Albany by the Hudson River, 144 miles, and thence by the New York Central—a chain of railways 298 miles.

This great route traverses, from east to west, the entire length of the Empire State. It has two termini at the eastern end, one at Albany and the other at Troy, which meet, after 17 miles, at Schenectady. It then continues, in one line, to Syracuse, 148 miles from Albany; when it is again a double route for the remainder of the way; the lower line being looped up to the other about midway, between Syracuse and Buffalo, at Rochester. The upper route is the more direct, and the one which we shall now follow. The great Erie Canal traverses the State of New York from Albany to Buffalo, nearly on the same line with the Central Railroad.

Trains leave Albany and Troy for Buffalo and all points west to the Mississippi and beyond, on the arrival there of the cars from the south, east, and north—New York, Boston, and Canada.

Schenectady.—HOTELS:—*Given's Hotel.*

At Schenectady the railways from Albany and Troy meet, and the Saratoga route diverges. Schenectady is upon the bank of the Mohawk. It is one of the oldest towns in the State, and is distinguished as the seat of *Union College.* The council-grounds of the Mohawks were once on this spot. In the winter of 1690, a party of two hundred Frenchmen and Canadians, and fifty Indians, fell at midnight upon Schenectady, killed and made captive its people, and burned the village to ashes. Sixty-nine persons were then massacred and twenty-seven were made prisoners. The church and sixty-three houses were destroyed. It was afterwards taken in the French war of 1748, when about seventy people were put to death.

Leaving Schenectady, the road crosses the Mohawk River and the Erie Canal, upon a bridge nearly one thousand feet in length.

At **Palatine Bridge**, 55 miles from Albany, passengers for Sharon Springs leave the road and proceed by stage. See Sharon Springs.

At **Fort Plain**, 68 miles from Albany, passengers for Otsego Lake, Cooperstown, and Cherry Valley, proceed by stage.

Little Falls is remarkable for a bold passage of the Mohawk River and the Erie Canal through a wild and most picturesque defile. The scenery, with the river rapids and cascades, the locks and windings of the canal, the bridges, and the glimpses, far away, of the valley of the Mohawk, is especially beautiful.

Utica.—HOTELS:— *Bagg's*, connected with the Railway Depot.

At Utica, 95 miles from Albany, a railway and canal come in from Binghamton, on the line of the Erie Road. Here passengers leave for Trenton Falls (see Trenton Falls), 15 miles distant. Utica is a large and thriving place, with many fine public and private buildings. It is built upon the site of old Fort Schuyler, and has now a population of over 22,000.

Syracuse.—HOTELS:—*The Globe, the Syracuse, and the Onondaga.*

At Syracuse, 148 miles from Albany, the Central Road connects by rail with Binghamton on the Erie route, and with Oswego, northward. The most extensive salt manufactories in the United States are found here. It is famous, too, as the meeting-place of State political and other conventions. Syracuse is a large and elegant city, with a population of over 28,000.

Auburn.—HOTELS:— *The American.*

This important city is near Owasco Lake, a beautiful water, 12 miles long. It is the seat of the Auburn State Prison.

Skaneateles is five miles distant, by a branch railway, at the foot of Skaneateles Lake, a charming water, 16 miles long, with picturesque shores, and good supplies of trout and other fish.

Cayuga is a pleasant village upon the eastern shore of Cayuga Lake. *Ithaca* is 38 miles off, at the other extremity of the Lake. These fine waters are traversed daily by steamboat,

NEW YORK.

connecting Cayuga with Ithaca, and by railway with Oswego, on the New York and Erie route.

Geneva is upon Seneca Lake, one of the largest and most beautiful of the many lakes of western New York. It is 40 miles long, and from two to four wide. Steamboats connect its towns and villages with the great routes of travel. The Hobart Free College, under the direction of the Episcopalians, is here; also the Medical Institute of Geneva College and the Geneva Union School.

Canandaigua is a beautiful village, at the north end of Canandaigua Lake. The railroad from Elmira, on the New York and Erie route to Niagara Falls, passes through Canandaigua. The lake is about 15 miles in length, and is well stocked with fish.

Rochester.—HOTELS:—The *Osborne* (new), the *Eagle*, and the *Congress Hotels*, are among the many excellent houses here.

Rochester is the largest and most important city upon our present route, between Albany and Buffalo, its population being nearly 45,000. It is the seat of the Rochester University, founded by the Baptists in 1850. There is also here a Baptist Theological Seminary, founded in 1850. Among its picturesque attractions, are the Falls of the Genesee, upon both sides of which river the city is built. The Mount Hope Cemetery, in the vicinity, is also a spot of much natural beauty.

Rochester is connected by railway with the New York and Erie route at Corning, and with Niagara Falls direct; by the Rochester, Lockport, and Niagara Falls division of the New York Central Road, and by steamboats, with all ports on Lake Ontario.

The **Genesee Falls** are seen to the best advantage from the east side of the stream. The railroad cars pass about one hundred rods south of the most southerly fall on the Genesee River, so that passengers in crossing lose the view. These falls have three perpendicular pitches and two rapids; the first great cataract is 80 rods below the aqueduct, the stream plunging perpendicularly 96 feet. The ledge here recedes up the river from the centre to the sides, breaking the water into three distinct sheets.

From *Table Rock*, in the centre of these falls, Sam Patch made his last and fatal leap. The river below the first cataract is broad and deep, with occasional rapids to the second fall, where it again descends perpendicularly 20 feet. Thence the river pursues its course, which is noisy, swift, and rapid, to the third and last fall, over which it pours its flood down a perpendicular descent of 105 feet. Below this fall are numerous rapids, which continue to Carthage, the end of navigation on the Genesee River from Lake Ontario.

At Rochester the two routes of the Central Road unite, and again diverge to reunite at Buffalo. By the upper route the traveller will pass through Lockport direct to Niagara, leaving Buffalo to the south-west. The lower route, direct to Buffalo, is intersected at Batavia by the Buffalo and Corning Road, from Corning on the Erie Railway, via Rochester to Niagara.

Buffalo.—HOTELS:—*The American.*

We have now reached the shores of Lake Erie, and are at the end of our route, whence we may proceed at our pleasure, by steamboat or railway, to any place northward or southward, in the Far West; for Buffalo is the point where routes of travel most do meet.

This important commercial and manufacturing city has grown so great and so fast, that although it was laid out as late as 1801, and in 1813 had only 200 houses, its population now numbers nearly 80,000. It is an earnest of the wonderful progress which we shall see by and by, when we continue our travels hence, towards the further West. Stop at the American.

Niagara Falls. — HOTELS: — Upon the American side of the river, the Cataract House and the International Hotel are most excellent homes for the tourist. On the Canada side stop at the Clifton.

ROUTES.—From *New York*, via Hudson River and Hudson River Railroad, to Albany, 146 miles; from Albany to Buffalo, via N. Y. Central R. R., 298 miles; from Buffalo, by Buffalo, Niagara Falls, and Lewiston R. R. (to Niagara), 22 miles. Total, 466 miles.

From *New York*, via New York and Erie R. R., to Buffalo, 422 miles; Buffalo (as above), by Buffalo, Niagara Falls, and Lewiston R. R. (to Niagara), 22 miles. Total, 444 miles.

From *New York*, by New York and Erie R. R. to Elmira, 273 miles; from Elmira to Niagara, by Elmira, Canandaigua, and Niagara Falls R. R., 166 miles. Total, 439 miles.

Passengers can leave the main N. Y. Central Railway (from Albany to Buffalo) at Rochester, and take the Rochester, Lockport, and Niagara division, 76 miles, thence to Niagara.

From *New York* to *Albany*, by Hudson River, 146 miles; thence to Troy, 6 miles. Railway from Troy to Whitehall, 65 miles; from Whitehall by steamer on Lake Champlain, to St. Johns, 150 miles; St. Johns to La Prairie Railroad, 15 miles; La Prairie, steamboat on the St. Lawrence to Montreal, 9 miles; from Montreal (Grand Trunk Railroad and other lines to Niagara), of railroad and steamboat, 436 miles. Total, 727 miles.

This great Mecca of the world's worshippers of landscape beauty, the mighty wonder of Niagara, is on its namesake river, a strait connecting the flood of Lakes Erie and Ontario, and dividing a portion of the State of New York on the west from the Provinces of Canada. The cataracts thus lie within the territory both of Great Britain and of the United States. They are some 20 miles below the entrance of the river, at the north-east extremity of Lake Erie, and about 14 miles above its junction with Lake Ontario.

The waters for which the Niagara is the outlet, cover an area of 150,000 square miles—floods so grand and inexhaustible as to be utterly unconscious of the loss of the *ninety millions of tons* which they pour every hour, through succeeding centuries, over these stupendous precipices.

Fortunately, the most usual approach to Niagara—that by the American shore—is the best, all points considered. "The descent of about 200 feet, by the staircase, brings the traveller directly under the shoulder and edge of the American Fall, the most imposing scene, for a single object, that he will ever have witnessed. The long column of sparkling water seems, as he stands near it, to descend to an immeasurable depth, and the bright sea-green curve above has the appearance of being set into the sky. The tremendous power of the Fall, as well as the height, realizes his utmost expectations. He descends to the water's edge and embarks in a ferry-boat, which tosses like an egg-shell on the heaving and convulsed water, and in a minute or two he finds himself in the face of the vast line of the Falls, and sees with surprise that he has expended his fullest admiration and astonishment upon a mere thread of Niagara—the thousandth part of its wondrous volume and grandeur. From the point where he crosses to Table Rock, the line of the Falls is measurable at three-quarters of a mile; and it is this immense extent which, more than any other feature, takes the traveller by surprise. The tide at the ferry sets very strongly down, and the athletic men who are employed here keep the boat up against it with difficulty. Arrived near the opposite landing, however, there is a slight counter current, and the large rocks near the shore serve as a breakwater, behind which the boat runs smoothly to her moorings." *

It is from the American side of the river that access is had to the hundred points of interest and surprise in the famous Goat Island vicinage, with its connecting bridges, its views of the Rapids, of the Cave of the Winds, of the scene of Sam Patch's great leap, and of its bold over-topping tower; and in other neighborhoods of the Whirlpool, of the Chasm Tower, and the Devil's Hole.

A totally different and not less wonderful gallery of natural master-pieces is opened upon the Canada shore. The terrible marvels of the Table Rock above, and of Termination Rock behind the mighty Horse-Shoe Fall; the noble panorama from the piazzas of the Clifton House, the Burning Spring, the historical village of Chippewa, and the battle field of Lundy's Lane; Bender's Cave, etc.

Goat Island. (American side.)—Leaving the Cataract House, take the first left-hand street, two minutes' walk to the bridge, which leads to the toll-gate on Bath Island. This bridge is itself an object of curious wonder, in its apparently rash and dangerous position. It is, however, perfectly safe, and is crossed hourly by heavily-laden carriages.

The Rapids are seen in grand and impressive aspect on the way to Goat Island. The river descends 51 feet in a distance of three-quarters of a mile by this inextricable turmoil of waters. It is one of the most striking incidents in the Niagara scenery. Standing on the bridge, and gazing thence up the angry torrent, the leaping crests seem like "a battle-charge of tempestuous waves animated and infuriated against the sky. The rocks, whose soaring points show above the surface, seem tormented with some supernatural agony, and fling off the wild and hurried waters, as if with the force of a giant's arm. Nearer the plunge of the Fall, the Rapids become still more agitated, and it is impossible for the spectator to rid himself of the idea that they are conscious of the abyss which they are hurrying, and struggle back in the very extremity of horror. This propensity to invest Niagara with a soul and human feelings is a common effect upon the minds of visitors, in every part of its wonderful phenomena. The torture of the Rapids, the clinging curves with which they embrace the small rocky islands that live amid the surge; the sudden calmness at the brow of the cataract, and the infernal writhe and whiteness with which they reappear, powerless, from the depths of the abyss—all seem, to the excited imagination of the gazer, like the natural effects of im-

---

* This passage is from "American Scenery," and since it was written a fairy little steamer has been employed to traverse the vexed river, and the timid cross readily upon the grand Suspension Bridge.

pending ruin—desperate resolution and fearful agony on the minds and frames of mortals."

Chapin's Island is upon the right of the bridge, within a short distance of the American Fall. It is named in memory of a workman whose life was imperilled by falling into the stream, as he was laboring upon the bridge. Mr. Robinson went gallantly and successfully to his relief in a skiff.

The Toll Gate is upon Bath Island, where baths, warm and otherwise, are accessible at all times to visitors. A fee of 25 cents, paid here, gives you the freedom of Goat Island, during all your stay, be it for the year or less. Near this point are Ship and Big Islands. There is here a very extensive paper-mill.

Another small bridge, and we are upon Iris, or Goat Island. The only place of habitation here is a house at which the traveller can supply himself with refreshments of all inviting kinds, and store his trunks with every variety of samples of Indian ingenuity and labor. The place is called the *Indian Emporium*. Three routes over the island diverge at this point. The principal path followed by most visitors is that to the right, which keeps the best of the sights, as Wisdom always does, until the last; affording less striking views of the Falls than do the other routes at first, but far surpassing them both in its grand revealments at the end. This way conducts to the foot of the island, while the left-hand path seeks the head, and the middle winds across. Taking the right-hand path, then, from the Toll Gate, we come, first, to the centre Fall, called **The Cave of the Winds**, mid-distance nearly between the American and the Horse-Shoe Falls. This wonderful scene is best and most securely enjoyed from the spacious flat rock beneath. The cave is 100 feet high, and of the same extent in width. You can pass safely into the recess behind the water, to a platform beyond. Magical rainbow-pictures are formed at this spot; sometimes bows of entire circles and two or three at once, delight the vision.

At the foot of Goat Island the *Three Profiles* is an object of curious interest. These profiles, seemingly some two feet long, are to be seen, one directly above the other, as you look across the first sheet of water, directly under the lowest point of rock.

Luna Island is reached by a foot bridge, from the right of Goat Island. It has an area of some three-quarters of an acre. The effective rainbow forms, seen at this point, have given it the name it bears. The venturesome visitor may get some startling peeps far down into the great caldron of waters. A child of eight years once fell into the torrent at this point, and was lost, together with a gallant lad who jumped in to rescue her.

Sam Patch's Leap.—It was upon the west side of Goat Island, near Biddle's Stairs, which we shall next look at, that the immortal jumper, Sam Patch, made two successful leaps into the waters below, saying, as he went off, to the throng of spectators, that "one thing might be done as well as another!" The fellow made his jump too much, within the same year (1829), over the Genesee Falls, at Rochester.

Biddle's Stairs, on the west side of the island, was named after Nicholas Biddle, of United States Bank fame, by whose order they were built. "Make us something," he is reported to have said to the workmen, "by which we may descend and see what is below." At the base of these spiral stairs, which are secured to the rocks by strong iron fastenings, there are two diverging paths. The *up* river way, towards the Horse-Shoe Fall, is difficult, and much obstructed by fallen rocks; but down the current a noble view is gained of the centre Fall or Cave of the Winds. Reascending the Biddle Stairs, we come, after a few rods' travel, to a resting-place at a little house, and thence we go down the bank, and, crossing a bridge, reach

Prospect Tower.—The precarious looking edifice, which seems to have "rushed in, as fools do, where angels fear to tread," is very near the edge of the precipice, above which it rises some 45 feet in air. From the top, which is surrounded by an iron railing, a magnificent scene is presented—a panorama of the Niagara wonders—the like of which can be seen from no other point.

The Horse-Shoe Falls, which leads the host of astonishments in this astonishing place, is the connecting link between the scenes of the American and of the Canadian sides of the river, always marvellous from whatever position it is viewed. This mighty cataract is 144 rods across, and, it is said by Prof. Lyell, that fifteen hundred millions of cubic feet of water pass over its ledges every hour. One of the condemned lake ships (the Detroit) was sent over this Fall in 1829, and, though she drew 18 feet of water, she did not touch the rocks in passing over the brink of the precipice, showing a solid body of water, at least some 20 feet deep, to be *above* the ledge. We shall return to the Horse-Shoe Fall from the Canada side.

Gull Island, just above, is an unapproachable spot, upon which it is not likely or possible that man has ever yet stood. There are three other small isles seen from here, called the *Three Sisters*. Near the Three Sisters, on Goat Island, is the spot remembered as the resort of an eccentric, and which is called, after him, the

Bathing Place of Francis Abbott the Hermit. At the head of Goat Island is Navy Island, near the Canada shore. It was the scene of incidents in the Canadian rebellion of 1837-8, known as the McKenzie War. Chippewa, which held at that period some 5,000 British troops, is upon the Canadian shore below. It was near Fort Schlosser, hard by, that, about this period, the American steamboat Caroline, was set on fire, and sent over the Falls, by the order of Col. McNabb, a British officer. Some fragments of the wreck lodged on Gull Island, where they remained until the following spring.

Grand Island, which has an area of 17,240 acres, was the spot on which Major M. M. Noah hoped to assemble all the Hebrew populations of the world.

Near the Ferry (American side still) there was once an observatory or Pagoda, 100 feet high, from which a grand view of the region was gained. This spot is called *Point View.*

The Whirlpool.—Three miles below the Falls (American side) is the Whirlpool, resembling in its appearance the celebrated Maelstrom on the coast of Norway. It is occasioned by the river making nearly a right angle, while it is here narrower than at any other place, not being more than 30 rods wide, and the current running with such velocity as to rise up in the middle 10 feet above the sides. This has been ascertained by measurement. There is a path leading down the bank to the Whirlpool on both sides, and, though somewhat difficult to descend and ascend, it is accomplished almost every day.

The Devil's Hole is a mile below the Whirlpool. It embraces about two acres, cut out laterally and perpendicularly in the rock by the side of the river, and is 150 feet deep. An angle of this hole or gulf comes within a few feet of the stage-road, affording travellers an opportunity, without alighting, of looking into the yawning abyss. But they should alight, and pass to the further side of the flat projecting rock, where they will feel themselves richly repaid for their trouble.

Chasm Tower, 3¼ miles below the Falls, is 75 feet high, and commands fine views (seen, if you please, of all hues, through a specular medium) of all the country round. A fee of 12½ cents is required.

The Maid of the Mist.—The landing of that singular feature of these wild scenes, the steamboat Maid of the Mist, is two miles below the Falls, whose troublous brink she touches in her frequent trips across the river.

The Great Suspension Bridge spans the chasm at this point. Its total length, from centre to centre of the towers, is 800 feet; its height above the water, 258 feet. The first bridge, which was built by Mr. Charles Ellett, was a very light and fairy-like affair, in comparison with the present substantial structure. The bridge, as it now stands, was constructed under the directions of Mr. John A. Roebling, at a cost of $500,000. It was first crossed by the locomotive March 8, 1855. Twenty-eight feet below the floor of the railway tracks a carriage and footway is suspended. This bridge is used at present by the New York Central, the Erie, and the Great Western, Canada roads.

We will now cross the river on the Suspension Bridge, and explore the wonders of the opposite shore.

Taking a carriage at our hotel, on the American side, we may "do" the Canadian shore very comfortably between breakfast and dinner, if we have no more time to spare. The regular price of carriage hire at the livery stables is one dollar per hour. On the plank road, going and returning, five cents; at the bridge, for each foot passenger, going and returning the same day, 25 cents, or 12½ each way. If the passenger does not return, the bridge toll is still 25 cents. For each carriage (two horses), going and returning, 50 cents for each passenger, and 50 cents besides for the carriage. A plank road leads from the opposite terminus of the bridge to the Clifton House.

At the bridge is shown a basket in which Mr. Elliott, his wife, and other ladies and gentlemen, crossed over the river on a single wire, about one inch in diameter. A perilous journey across such a gorge and at an elevation in the air of 280 feet! Two or three persons thus crossed at a time, the basket being let down on an inclined plane to the centre of the towers (this was during the building of the first Suspension Bridge), and then drawn up by the help of a windlass to the opposite side. The usual time in crossing was from three to four minutes. By the means of this basket the lives of four men were once saved, when the planks of the Foot Bridge were blown off in a violent storm, and they were suspended over the river by only two strands of wire, which oscillated with immense rapidity, 60 or 70 feet. The basket was sent to their relief, at a moment when the hurricane grew less fearful, and they descended into it by means of a ladder, one at a trip only, until all were released from their terrible position.

Bender's Cave is midway between the Suspension Bridge and the Clifton House. It is a recess six feet high and twenty in length, made by a decomposition of the limestone.

If the tourist prefer it, he may cross the river by the ferry, the only route of other days. From the ferry-house the cars descend to the water's edge on an inclined plane of 31 degrees. They are worked by water-power. The time required to make this descent and to cross to the Canada shore is about ten minutes. During the forty years it is said that this ferry has been in operation, not one life has been lost, nor has any serious accident occurred. We have described the passage of the river in the opening of our article. Upon landing, plenty of carriages will be always found in readiness, as at all other starting and stopping places about the Falls. It will be well to ascertain the fares before employing any of them.

The Clifton House is an old and very favorite resort here, for its home luxuries and for its noble position, overlooking the river and Falls. It was the residence of Mdlle. Jenny Lind during her visit to Niagara. "The Clifton House," writes Mr. Willis, from whose descriptions of these scenes we have already quoted, "stands nearly opposite the centre of the irregular crescent formed by the Falls ; but it is so far back from the line of the arc, that the height and grandeur of the two cataracts, to an eye unacquainted with the scene, are respectively diminished. After once making the tour of the points of view, however, the distance and elevation of the hotel are allowed for by the eye, and the situation seems most advantageous. This is the only house at Niagara where a traveller, on his second visit, would be content to live."

"The *ennui* attendant upon public-houses can never be felt at the Clifton House. The most common mind finds the spectacle, from its balconies, a sufficient and untiring occupation. The loneliness of uninhabited parlors, the discord of baby-thrummed pianos, the dreariness of great staircases, long entries, and bar-rooms filled with strangers, are pains and penalties of travel never felt at Niagara. If there is a vacant half-hour to dinner, or if indisposition to sleep create that sickening yearning for society which sometimes comes upon a stranger in a strange land, like the calenture of a fever—the eternal marvel going on without is more engrossing than friend or conversation—more beguiling from sad thoughts than the Corso in carnival time. To lean over the balustrade, and watch the flying of the ferry-boat below, with its terrified freight of adventurers, one moment gliding swiftly down the stream in the round of an eddy, the next lifted up by a boiling wave, as if it were tossed from the scoop of a giant's hand beneath the water ; to gaze, hour after hour, into the face of the cataract ; to trace the rainbows, delight like a child in the shooting spray-clouds, and calculate fruitlessly and endlessly, by the force, weight, speed, and change of the tremendous waters—is amusement and occupation enough to draw the mind from any thing—to cure madness or create it."

Table Rock. The grand overhanging platform called Table Rock, and the fearful abysmal scene at the very base of the mighty Horse Shoe Fall, which it presents, is one of the cardinal wonders of Niagara. If one would listen to the terrible noise of the great cataract, let him come here where the sound of its hoarse utterance drowns all lesser sounds, and his own speech is inaudible to himself.

Termination Rock is a recess behind the centre of the Horse Shoe Fall, reached by the descent of a spiral stairway from Table Rock, the traverse for a short distance of the rude marge of the river, and then of a narrow path over a frightful ledge and through the drowning spray, behind the mighty Fall.

Before descending the visitors make a complete change of toilette, for a rough costume more suitable for the stormy and rather damp journey before them. When fully equipped, their ludicrous appearance excites for a while, a mirthful feeling, in singular contrast with the solemn sentiment of all the scene around them. This strange expedition, often made even by ladies, has been thus described : "The guide went before, and we followed close under the cliff. A cold, clammy wind blew strong in our faces from the moment we left the shelter of the staircase, and a few steps brought us into a pelting fine rain, that penetrated every opening of our dresses and made our foothold very slippery and difficult. We were not yet near the sheet of water we were to walk through ; but one or two of the party gave out and returned, declaring it was impossible to breathe ; and the rest, imitating the guide, bent nearly double to keep the heating spray from their nostrils, and pushed on, with enough to do to keep sight of his heels. We arrived near the difficult point of our progress ; and in the midst of a confusion of blinding gusts, half deafened, and more than half drowned, the guide stopped to give us a hold of his skirts and a little counsel. All that could be heard amid the thunder of the cataract beside us was an injunction to push on when it got to the worst, as it was shorter to get beyond the sheet than to go back ; and with this pleasant statement of our dilemma, we faced about with the longest breath we could draw, and encountered the enemy. It may be supposed that every person who has been dragged through the column of water which obstructs the entrance to the cavern behind this cataract, has a very tolerable idea of the pains of drowning. What is wanting in the density of the element is more than made up by the force of the contending winds, which rush into the mouth, eyes, and nostrils, as if flying from a water-fiend. The "courage of worse behind" alone persuades the gasping sufferer to take one desperate step more.

It is difficult enough to breathe within; but with a little self-control and management, the nostrils may be guarded from the watery particles in the atmosphere, and then an impression is made upon the mind by the extraordinary pavilion above and around, which never loses its vividness. The natural bend of the falling cataract, and the backward shelve of the precipice, form an immense area like the interior of a tent, but so pervaded by discharges of mist and spray, that it is impossible to see far inward. Outward the light struggles, brokenly, through the crystal of the cataract; and when the sun shines directly on its face, it is a scene of unimaginable glory. The footing is rather unsteadfast, a small shelf composed of loose and slippery stones; and the abyss below boils like—it is difficult to find a comparison. On the whole, this undertaking is rather pleasanter to remember than to achieve.

The *Museum*, near Table Rock, contains more than 10,000 specimens of minerals, birds, fish, and animals, many of which were collected in the neighborhood of the Falls. Admittance, 25 cents. The Burning Spring is near the water, two miles above the Falls. The carbonated sulphuretted hydrogen gas here, gives out a brilliant flame when lighted. Charge 12¼ cents.

The height of the Falls is 165 feet. The roar of the waters has been heard at Toronto, 44 miles away, and yet in some states of wind and atmosphere, it is scarcely perceptible in the immediate neighborhood. Niagara presents a new and most unique aspect in winter, when huge icicles hang from the precipices, and immense frozen piles of a thousand fantastic shapes glitter in the bright sun light. Father Hennepin a Jesuit missionary, was the first European who ever saw Niagara. His visit was in 1678. Niagara is an Indian word of the Iroquois tongue, from Ongakarra, meaning mighty or thundering water.

In the vicinity of Niagara is Lewistown, seven miles distant, at the head of navigation on Lake Ontario—and directly opposite Lewistown is Queenstown, under Queenstown Heights—a famous battle-ground. Brock's Monument, a column of 126 feet, crowns the Heights.

THE ADIRONDACK MOUNTAINS—THE SARANAC LAKES, ETC.

The Upper part of the State of New York, lying west and south of Lake Champlain and the St. Lawrence River, respectively, is still a wild primitive forest region, of the highest interest to the tourist, for its wonderful natural beauties and for the ample facilities it offers for the pleasures of the rod and the rifle. Fine mountain peaks stud the whole region, and charming lakes and lakelets are so abundant that travel here is made by water instead of by land—traversing the ponds in row-boats which are carried by easy portage from one lovely brook or lake to another. Deer fill the woods, and trout are unsuspecting in the transparent floods everywhere. This wilderness land is visited at various points under distinctive names, as the hunting-grounds of the Saranacs, of the Chateaugay woods, of the Adirondacks, and of Lake Pleasant, etc. We shall speak of these several divisions, briefly, in order.

The Saranac Lakes. These wonderful links of the great chain of mountain waters in upper New York, are about a dozen in number, large and small. They lie principally in Franklin County, and may be most readily reached by stage from Westport or from Keeseville, about midway on the western shore of Lake Champlain—taking stage or private conveyance thence (30 miles) to the banks of the Lower Saranac—which is the outer edge of civilization in this direction. There is a little village and an inn or two at this point, and here guides and boats, with all proper camp equipage for forest life, may be procured. For this route the tourist must engage a boatman, who, for a compensation of two or three dollars per day—the price will be no more if he should have extra passengers—will provide a boat, with tent and kitchen apparatus, dogs, rifles, etc. The tourist will supply, before starting, such stores as coffee, tea, biscuit, etc., and the sport by the way, conducted by himself or by his guide, will keep him furnished with trout and venison. If camp life should not please him, he may, with some little inconvenience, so measure and direct his movements as to sleep in some one or other of the shanties of the hunters or of the lumber-men found here and there on the way. The tent in the forest, however, is much preferable.

Leaving the Lower Saranac, we will pass pleasantly along some half a dozen miles—then make a short portage, the guide carrying the huge boat by a yoke on the back, to the Middle Saranac—there he may go on to the Upper Lake of the same name, and thence by a long portage of three miles to Lake St. Regis. These are all large and beautiful waters, full of delicious islands and hemmed in upon all sides by fine mountain ranges. Trout may be taken readily at the inlets of all the brooks, and deer may be found in the forests almost at will.

Returning from St. Regis, and back via the Upper to the Middle Saranac, we continue our journey, by portage, to the Stony Creek ponds—

thence three miles by Stony Creek to the Rackett River—a rapid stream with wonderful forest vegetation upon its banks. This water followed for some 20 miles, brings us to Tupper's Lake—the finest part of the Saranac region. Tupper's Lake is the largest of this chain, being seven miles long and from one to two miles broad. The shores and headlands and islands are especially bold and picturesque, and at this point the deer is much more easily found than elsewhere in the neighborhood. Below Tupper's Lake—the waters commingling—is Longhneah, another charming pond. The chain continues on yet for miles, but the Saranac trip, proper, ends here. This mountain voyage and the return to Lake Champlain might be made in a week, but two or three, or even more, should be given to it. It is seldom that ladies make the excursion, but they might do so with great delight. The boatmen and hunters of the region are fine, hearty, intelligent and obliging fellows. That wonderful ravine, the "Walled Banks of the Ausable," (see Index,) should be seen by the Saranac tourist, on his way from Lake Champlain to Keeseville.

The Adirondack Mountains. The Adirondack region may be reached by private conveyance (only) over a rude mountain road from Schroon Lake, above Lake George, or more conveniently from Crown Point village, just beyond the ruins of Fort Ticonderoga, on Lake Champlain. The distance thence is some 30 miles, and requires a day to travel. The tourist in this region will move about by land more than by water, as among the Saranacs; for, although the lakes are numerous enough, it is among and upon the hills that the chief attractions are to be found. The accommodations are rude enough—the only inn being the boarding-house at the village of the Adirondack Iron Works. Stopping at this point, as head-quarters, he may make a pleasant journey down Lake Sandford near by, on one side, and upon Lake Henderson on the other hand. In one water he ought to troll for pickerel, and in the other, cast his fly for trout; and upon both enjoy the noble glimpses of the famous mountain peaks of the Adirondack group, the cliffs of the Great Indian Pass, of Mounts Colden, M'Intyre, Echo Mountain, and other bold scenes. It will be a day's jaunt for him afterwards to explore the wild gorge of the Indian Pass, five miles distant; another day's work to visit the dark and weird waters of Avalanche Lake; and yet another to reach the Preston Ponds, five miles in a different direction. He will find indeed, occupation enough for many days, in exploring these and many other points, which we may not tarry to catalogue; and, in any case, he must have 48 hours to do the tramp,

par excellence, of the Adirondack—the visit to the summit of the brave Tehawus, or Mount Marcy, the monarch of the region. Tehawus is 12 miles away, and the ascent is extremely toilsome.

The Adirondacks (named after the Indian nation which once inhabited these fastnesses) lie chiefly in the county of Essex, though they extend into all the jurisdiction around. Mount Marcy, or Tehawus, "the Cloud Splitter," is 5,467 feet High. Mount M'Intyre has an elevation almost as great. The Dial Mountain, M'Marten, and Colden are also very lofty peaks, impressively seen from the distance, and inexhaustible in the attractions which their ravines, and waterfalls present. White Face and other grand hill peaks belong to the neighboring range called the Keene Mountains. The Hudson River rises in this wilderness.

Lake Pleasant. To reach Lake Pleasant and the adjoining waters of Round, Piseco, and Louis Lakes—a favorite and enchanting summer resort and sporting-ground—take the Central Railway from Albany as far as Amsterdam, and thence, by stage or carriage, to Holmes' Hotel, on Lake Pleasant. The ride from Amsterdam is about 30 miles. The stage stops over night at a village, *en route.* Mr. Holmes' house is an excellent place, with no absurd luxuries, but with every comfort for which the true sportsman can wish. It is a delightful summer home for the student, and may be visited very satisfactorily by ladies. The wild lands and waters here are a part of the lake region of northern New York, of which we have already seen something on the Saranacs, and among the Adirondacks. The Saranac region is connected with Lake Pleasant by intermediate waters and portages. The deer, and other game, is abundant here in the forests, and fine trout may be taken in all the brooks and lakes. Lake Pleasant and its picturesque *confrères,* lie in Hamilton County.

All this northern part of New York is quite similar in its attractions to the wilderness in the upper part of the State of Maine.

TO LAKE MAHOPAC.

ROUTE:—From New York, via Harlem Railroad, the depot up town, at the corner of Fourth Avenue and Twenty-sixth street. New York, 51 miles, to Croton Falls Station. (See Harlem Route from New York to Albany.)

Stages leave Croton Falls for Lake Mahopac, five miles, on the arrival of the cars; stage fare, 25 cents.

Lake Mahopac. — HOTELS : — *Gregory's, Baldwin's,* and *Thompson's.*

NEW YORK.

Lake Mahopac, a favorite summer resort, in the immediate vicinity of New York, and much frequented by its citizens, both for a day's excursion and as a continued home, lies in the western part of the town of Carmel, Putnam County, New York, 13 miles east from Peekskill, on the Hudson, and five miles from Croton Falls Station, on the Harlem Railroad. The lake is nine miles in circumference, and is about 1,800 feet above the sea. It is one of the principal sources of supply to the Croton. Though the landscape has no very bold features, but little to detain the *artist*, yet its quiet waters, its pretty wooded islands, the romantic resorts in its vicinage, the throngs of pleasure-seeking strangers, the boating, and fishing, and other rural sports, make it a most agreeable spot for either a brief visit or long residence. There are many attractive localities of hill and water scenery around Mahopac. The pleasant hotels are well filled during the season by boarders or by passing guests. It is a nice retreat to those whose business in the great city below forbids their wandering far away.

Lebanon Springs and Shaker Village.
—HOTELS :—*Columbia Hall.*

ROUTE. From New York, by the Hudson River Railway or Boats, to the city of Hudson, and thence, by the Hudson and Boston R. R., to Canaan; from Canaan, by stage, seven miles to the Springs. From Boston, 167 miles to Canaan by the Western Railway. From Albany, by same route, to Canaan, 33 miles. Travellers taking the steamboat in the evening from New York will breakfast in Hudson and proceed comfortably by rail.

There are ample accommodations for the traveller at this favorite watering-place, in a well-appointed hotel, a water-cure establishment, &c., pleasantly perched on a hill slope, overlooking a beautiful valley. There are pleasant drives all around, over good roads, to happy villages, smiling lakelets, and inviting spots of many characters. Trout, too, may be taken in the neighborhood. The water of the Springs flows from a cavity 10 feet in diameter, and in sufficient volume to work a mill. Its temperature is 72°. It is soft and pleasantly suited for bathing uses, is quite tasteless and inodorous. For cutaneous affections, rheumatism, nervous debility, liver complaint, &c., it is an admirable remedial agent.

The village of New Lebanon, or the celebrated Shaker settlement, is two miles from the Springs, and is a point of great interest to the visitors there, especially on Sunday, when their singular forms of worship may be witnessed.

Sharon Springs.—HOTELS :—The *Pavilion* is a large and well appointed establishment. The *Eldridge*, also, is a good and less expensive house.

ROUTE :—From Albany, New York, by the Central Railroad for Buffalo, as far as Palatine Bridge, 55 miles; thence by stage, 10 miles, over a plank road.

The waters are pure and clear, and although they flow for one-fourth of a mile from their source with other currents, they yet preserve their own distinct character. The fall here is of sufficient force and volume to turn a mill. It tumbles over a ledge of perpendicular rocks, with a descent of some 65 feet. The magnesia and the sulphur springs much resemble the White Sulphur of Virginia.

Cherry Valley is in the vicinage of Sharon Springs, accessible also from Palatine Bridge, on the Albany and Buffalo road, and from Canajoharie, on the Erie Canal, from which it lies about 26 miles in a south-west direction.

Otsego Lake and Cooperstown, famous as the home of the late Fenimore Cooper, the novelist, are near by.

Columbia Springs.

From New York, by Hudson River Railway, or steamboats to Hudson; thence, by carriage or stage, four miles.

The Columbia Springs have of late years grown into great popular favor. They are easily accessible, lying only four miles from the City of Hudson. They are within the town of Stockport, Columbia County, New York. The site and grounds are highly varied and picturesque, jumping delightfully from hill to dale, from forest glen to grassy lawn.

There is, too, a merry brooklet, which winds coquettishly through the landscape, affording now a quiet slope for some "melancholy Jacques," now a dashing cascade for him of brighter mood. In the immediate neighborhood, moreover, there flows a larger water, offering all the country charms of boating and fishing. The hotel here is large and well-appointed, and Mr. Charles B. Nash, the enterprising proprietor, is every year swelling its conveniences and comforts, and adding to the seductions of the occupations and enjoyments, and to the beauty of the scenery out of doors.

Avon Springs.—HOTELS :—

The Avon Springs may be reached by the Central Railway from Albany to Buffalo, via Rochester, from which city they are distant 20 miles. The village of Avon is upon the Genesee

River, which it overlooks from a charming terrace 100 feet above. On this lofty position the picturesque landscape of the neighborhood is seen to great advantage. The Springs are near at hand, a little to the south-west. With ample hotel conveniences and enjoyments, the Avon Springs meet the popular favor they so well deserve.

Richfield Springs. — HOTELS: — *Spring House.*

Richfield Springs are reached from Herkimer, 81 miles from Albany, on the Central Road to Buffalo. They are in the town of Richfield, Otsego County, south-east of Utica, near the head of Canaderaga, one of the numerous lakes of all this part of New York. Otsego Lake is six miles distant; and another six miles will take the traveller to Cooperstown. Cherry Valley, Springfield, and other villages are near by.

LONG ISLAND.

Long Island, part of the State, is 115 miles in length, and, at some points, 20 in breadth; with the Atlantic on the south, and the Long Island Sound on the north. The upper part of the island is agreeably diversified with hills, though the surface is for the most part strikingly level. The coast is charmingly indented with bays; and delicious fresh-water ponds, fed by springs, are everywhere found on terraces of varying elevation. These little lakes, and the varied coast-views give Long Island picturesque features, which, if not grand, are certainly of most attractive and winning character, yet heightened by the rural beauty of the numerous quiet little towns and charming summer villas.

The places in the immediate vicinity of New York, we have already mentioned among the suburban resorts of the city, and we might almost have included the whole island in that classification, so easily is every part reached either by the steamboats or by the railway which traverses the length of the island, from Brooklyn 95 miles to Greenport.

The lower shore of the island, which is a net-work of shallow, land-locked waters, extending 70 miles, is the resort of innumerable flocks of aquatic fowl, and thither go the New York sportsmen or gunners for pastime and glory for themselves, and for delights for the tables of their city friends. In no other part of the Union is there a greater variety and abundance of wild birds than on this coast, and nowhere else are they more systematically sought. To answer the wants of the sportsmen, excellent accommodations have been everywhere provided, in the way of comfortable hotels and boat equipage.

Cedarmere, the home of the Poet Bryant, is near the pretty village of Roslyn, at the head of Hempstead Bay, about two hours' journey from New York by steamboat to Glen Cove, and thence by stage; or by the Long Island Railway 20 miles to Hempstead Branch, and by connecting stages. Cedarmere is a spot of great, though quiet picturesque beauty, overlooking Hempstead Bay, and the Connecticut shore across the Sound. Many of the charming, terraced, spring-water lakes of which we have spoken already, as among the pleasant and unique features of the Long Island landscape, are found within the domain of Cedarmere, in the village of Roslyn, and, indeed, through all the vicinage for miles around. Within a pleasant stroll of Mr. Bryant's residence is Hempstead Hill, the highest land on Long Island. This fine eminence overlooks the Sound and its inlets on the one hand, and the ocean beach on the other; at its base the village of Roslyn is nestled among green trees and placid lakelets. The house at Cedarmere makes no architectural pretensions, though it falls most agreeably into all the charming pictures, which every changing step over the hills, or along the margin of the ponds presents to view.

Battle of Long Island (August, 1776). The thoughts of the tourist on the quiet pastoral plains of Long Island, will revert with interest to that eventful night when the British troops under Sir Henry Clinton, Lord Cornwallis, and General Howe, made their silent, unsuspected march from Flatlands, through the swamps and passes to Bedford Hills, stealing upon the rear, and almost surrounding the patriot lines; "that able and fatal scheme which cost the Americans the deadly battle of Long Island, with the loss of nearly 2,000 out of the 5,000 men engaged."

The surprise of the attack, the obstinacy of the conflict, the bold retreat, and the loss of the city of New York, to which it led, makes this battle one of the most romantic episodes in the history of the Revolution.

"Never," says Mr. Irving, "did retreat require greater secrecy and circumspection. Nine thousand men, with all the munitions of war, were to be withdrawn from before a victorious army, encamped so near that every stroke of the spade and pick-axe from their trenches could be heard.

"The retreating troops, moreover, were to be embarked and conveyed across a strait, three-quarters of a mile wide, swept by rapid tides. What with the greatness of the stake, the darkness of the night, the uncertainty of the design, and the extreme hazard of the issue, it would be difficult to conceive a more deeply solemn and interesting scene.

"Washington wrung his hands in agony at the sight of this fatal battle. 'Good God!' cried he, as his troops were swept down, 'what brave fellows I must lose to-day!'"

NEW JERSEY.

SETTLEMENTS were made in this State at Bergen, by the Dutch, soon after their arrival in New York. In 1627 a Swedish colony was founded near the shores of the Delaware, in the south-western part of the State. A droll account of the quarrels of these Swedish folk with the Dutchmen of New Amsterdam may be found in Diedrick Knickerbocker's solemn history of the Amsterdam colonists. New Jersey is one of the old Thirteen States. She did her part nobly in the long war of Independence. The famous battles of Trenton, and of Princeton, and of Monmouth, at all of which Washington was present and victorious, occurred within her limits. Morristown was the winter camp of the American army in 1776-'77.

New Jersey has not a very wide territory, yet she presents many natural attractions to the traveller. Her sea-coast abounds in favorite bathing and sporting resorts, much visited by the citizens of New York and Philadelphia. Among these Summer haunts are Cape May, Long Branch, Sandy Hook, Absecum Beach, Deal, Squam Beach, and Tuckerton.

In the southern and central portions of this State, the country is flat and sandy; in the north are some ranges of picturesque hills, interspersed with charming lakes and ponds. Some of the Alleghany ridges traverse New Jersey, forming the spurs known as Schooley's Mountain, Trowbridge, the Ramapo, and Second Mountains. In the north-western part of the State are the Blue Mountains. The Nevisink Hills, rising nearly 400 feet on the Atlantic side, are usually the first and last land seen by ocean voyagers as they approach and leave New York. The celebrated Palisade Rocks of the Hudson River are in this State.

NEW YORK TO PHILADELPHIA.

There are two great routes between the cities of New York and Philadelphia, one known as the New Jersey railway line, and the other as the Camden and Amboy route. The former is the most expeditious; the latter, being partly by water, is the most agreeable in summer time.

The New Jersey Railway Route.

This route passes over the New Jersey, and the Philadelphia and Trenton Railroads. Leaves New York at foot of Cortlandt street (by ferry across the Hudson to Jersey City), several times each day. Distance, 87 miles; time (express trains), four hours.

STATIONS.

NEW YORK—Jersey City, 1 mile; Newark, 9; Elizabeth, 15; Rahway, 19; Uniontown, 23; Metuchin, 27; NEW BRUNSWICK, 31; Dean's Pond, 39; Kingston, 43; PRINCETON, 48; Trenton, 57; Bordentown, 62; Burlington, 68; CAMDEN, 89; Bristol, 70; Cornwell's, 74; Tacony, 79; Kensington, 85; PHILADELPHIA, 90.

This route, lying as it does, between the two greatest cities on the continent of America, is an immense thoroughfare, over which floods of travel pour unceasingly by day and by night. The region is populous and opulent, and necessarily thronged with towns and villages, and villas; for 20 or 25 miles from each terminus, over which the two cities spread their suburbs, the crowded trains are passing and repassing continually.

Leaving Jersey City (see New York and vicinity), the track over which we pass for two miles is that used also by the great Erie Railway, and which is traversed by all the thousands daily voyaging from every part of the Canadas, the New England States, and New York, for any and all regions of the wide South. Perhaps no other two miles of railway in the world bears such prodigious freights of men and merchandise as this.

Newark.—HOTELS :—

NEW JERSEY.

Newark, 9 miles from New York, and 78 from Philadelphia; settled in 1666; population (in 1860) 72,000; is upon the right bank of the Passaic River, 4 miles from its entrance into Newark Bay. It is built on an elevated plain, regularly laid out in wide streets, crossing at right angles. Many portions of the city are very elegant, and in its most recherché quarter are two charming parks, filled with noble elms. Among its most imposing public edifices are the Court House, the Post Office, the Custom House, and several of the Banks.

Of the literary institutions, the most noteworthy are the Library Association, the State Historical Society, and the Newark Academy.

The city contains over forty churches, some of which are very interesting structures, as the Catholic, on Washington street; the Presbyterian, near the Lower Park, and in High street; the Methodist, on Market, and on Broad street; Grace (Epis.), and the Baptist, on Academy street.

The city is divided into twelve wards, and possesses some forty public schools, which are attended by more than 9,000 pupils.

Newark is distinguished for its manufactures, which are large and prosperous. Steamboats, as well as railways, connect it with New York. It is the eastern terminus of the Morris and Essex R. R., and the Morris Canal passes through it on its way to Jersey City.

Elizabeth.—Hotels :—*American Hotel.*

Population at this time, about 12,000; 15 miles from New York; is situated upon the Elizabethtown Creek, two miles from its entrance into Staten Island Sound. It was once the capital and chief town of the State. Here diverges the N. J. Central R. R., 61 miles hence, into the interior, at Easton, Penn., on the Delaware River.

Rahway, N. J.—Hotels :—

Population (in 1860), about 7,000; 19 miles from New York; lies upon both sides of the Rahway River. Rahway is noted for its manufactures of carriages, stoves, hats, earthenware, etc. Some 3,000 vehicles are annually sent hence to the Southern markets.

New Brunswick.—Hotels : — *Williams Hotel.*

Population (in 1860), 12,000, is at the head of steamboat navigation on the Raritan River. This is the seat of Rutgers College and School, and also of a Theological Seminary of the Dutch Reformed Church. The streets on the river are narrow and crooked, and the ground low; but those on the upper bank are wide, and many of the dwellings are very neat and elegant, surrounded by fine gardens. From the site of Rutgers College on the hill there is a wide prospect, terminated by mountains on the north, and by Raritan Bay on the east. The Delaware and Raritan Canal extends from New Brunswick to Bordentown, on the Delaware River, 42 miles. This canal is 75 feet wide and 7 feet deep, and is navigated by sloops and steamboats of 150 tons. This fine work cost $2,500,000. The Railway here crosses the Raritan River.

Princeton.—Hotels :—

Built on an elevated ridge, is a pleasant little town, of literary and historical interest. It is the seat of the Princeton College, one of the oldest and most famous educational establishments in the country, founded by the Presbyterians at Elizabethtown, 1746, and removed to Princeton in 1757. Here, also, is the Theological Seminary of the Presbyterian church, founded 1812. In this vicinity was fought the memorable battle of January 3, 1777, between the American forces under General Washington, and those of the British, under Lieutenant-Colonel Mawhood, in which the latter were vanquished.

Trenton.—Hotels :— *Trenton House.*

Trenton, the capital of New Jersey; population (in 1860), 17,000; is on the left bank of the Delaware, 30 miles from Philadelphia, and 57 from New York. The city is regularly laid out, and has many fine stores and handsome dwellings. The State House, which is 100 feet long and 60 wide, is built of stone, and stuccoed, so as to resemble granite. Its situation, on the Delaware, is very beautiful, commanding a fine view of the river and vicinity. Here is the State Lunatic Asylum, founded in 1848, and also the State Penitentiary. Trenton has two daily and two other newspapers, 17 churches, and a State Library. The city is lighted with gas. The Delaware is here crossed by a handsome covered bridge, 1,100 feet long, resting on five arches, supported on stone piers, and which is considered a fine specimen of its kind. It has two carriage-ways, one of which is used by the railroad. The Delaware and Raritan Canal, forming an inland navigation from New Brunswick, passes through Trenton to the Delaware at Bordentown. It is supplied by a navigable feeder, taken from the Delaware 23 miles north of Trenton. It was completed in 1834, at a cost of $2,500,000. The Delaware and Raritan Canal passes through the city, and connects it with New York and Philadelphia. At this point the New Jersey Railroad, which we have thus far travelled, 57 miles from New York, ends, and the Philadelphia and Trenton, upon which we make the rest of our journey, begins. A branch road, six miles long, connects with the Camden and Amboy Railway at Bordentown. The Belvidere, Delaware, and Flemington Railroad leads hence, 63 miles, to Belvidere, in the interior, along the Delaware River.

Here was fought the famous

Battle of Trenton.—On Christmas night, in 1776, and during the most gloomy period of the Revolutionary war, General Washington crossed the Delaware with 2,500 men, and early on the morning of the 26th commenced an attack upon Trenton, then in possession of the British. So sudden and unexpected was the assault, that of the 1,500 German troops encamped there, 906 were made prisoners. This successful enterprise revived the spirit of the nation, as it was the first victory gained over the German mercenaries. General Mercer, a brave American officer, was mortally wounded in the attack.

It was here, upon Trenton Bridge, that occurred the memorable and beautiful reception of Washington, while on his way from New York to Mount Vernon, twelve years after the glorious victory.

Trenton was settled about 1680, and was named in 1720, in honor of Wm. Trenton, one of the early Provincial judges.

[Here we take the Branch road, six miles to Bordentown, and then by Camden and Amboy line, or continue, as we now shall, by Philadelphia and Trenton route.]

Bristol is a beautiful village, on the west bank of the Delaware, nearly opposite Burlington. The Delaware division of the Pennsylvania Canal, which communicates with the Lehigh at Easton, terminates here in a spacious basin on the Delaware. Pop., 2,570.

Tacony and *Kensington* are within the corporate limits of Philadelphia, at which city we have now arrived. See description of Philadelphia for hotels. We will now follow the line of the second, or Camden and Amboy route.

CAMDEN & AMBOY (OR STEAMBOAT) ROUTE.

FROM NEW YORK TO PHILADELPHIA.

Steamboat for Philadelphia, via Camden and Amboy route, leave Pier No. 1 North River, New York, daily (Sundays excepted), at 6 A. M. and 2 P. M., for South Amboy, 27 miles, and thence by rail. Fare by morning line is $2 25; by the afternoon (*Express*) line, $3.

Camden and Amboy Railroad from South Amboy.—STATIONS : *New York*, South Amboy, 27 miles; South River, —; Spotswood, 38; Jamesburg, 42; Prospect Plains, —; Cranberry Station, 45; Hightstown, 49; Centreville, 53; Newtown, 56; Sandhills, 58; *Bordentown*, 63; Hammel's Turn, 68; Burlington, 71; Beverly, 77; Rancocas, 78; Palmyra, 83; Fish House, 85; *Camden*, 89; *Philadelphia*, 90.

In the summer season, no more delightful journey can be made than the first twenty-seven miles of our present route across the lovely Bay and Harbor of New York, to South Amboy, past the villaed and villaged shores of Staten Island, and the Raritan River. The scenery of this region is described in our chapter upon New York City and its vicinity.

South Amboy is the landing place, and also the terminus of the Camden and Amboy Railroad. Upon our arrival there, we are transported, in a short space of time, from the steamboat to the railroad cars; and after a slight detention, we proceed on our journey up the steep ascent from the river, and soon enter a line of deep cutting through the sandhills. The road is then continued through a barren and uninteresting region of country, towards the Delaware at

Bordentown, 35 miles from Amboy. Here are the extensive grounds and mansion formerly occupied by the late Joseph Bonaparte, ex-King of Spain, which are among the most conspicuous objects of the place.

Bordentown is situated on a steep sandbank, on the east side of the Delaware. Although in a commanding situation, the view is greatly obstructed from the river. This is a favorite resort of the Philadelphians during the summer season. The Delaware and Raritan Canal here connects with Delaware River. A branch road running along the canal and river, unites this town with Trenton. Population, 3,000.

Burlington.—HOTELS :—*City Hotel.*

Burlington, settled in 1670, and with a present population of about 5,000, is a port of entry on the Delaware, 19 miles from Philadelphia. Burlington College, founded by the Episcopalians in 1846, is located here, and there are besides, upon the banks of the river, two large boarding-schools, one for each sex. Burlington is connected with Philadelphia by steamboat, and is a place of great summer resort thence.

Beverly, built on the banks of the Delaware since 1848, has now a population of from 1,000 to 1,500. It is a suburb of Philadelphia, distant thirteen miles.

Camden is at the terminus of our route, upon the banks of the Delaware River, immediately opposite the city of Philadelphia, to which we now cross by ferry. For further mention of Camden, and for hotels, etc., see description of Philadelphia.

The Falls of the Passaic occur in the town of Paterson, 16 miles from New York, on the route of the Erie Railway. This bold passage on the Passaic, though it has of late years lost much of its ancient beauty, is still a scene of great attraction, particularly when the stream chances to be generously swollen after heavy rains. Paterson itself is an agreeable town, of very considerable importance. It has a population of some 20,000.

Cape May.—HOTELS :—*Congress Hall, United States*, the *Columbia*, and many others.

ROUTE.—From New York every evening, by steamboat line to Philadelphia ; from Philadelphia by same line. Fare from New York, $2 ; from Philadelphia, $1.

Cape May is at the extreme southern point of New Jersey, where the floods of the Delaware are lost in the greater floods of the Atlantic. The beach is excellent for bathing or riding. The little village of the Cape (Cape Island) is thronged in the summer season by thousands of gratified pleasure seekers. They come chiefly from Philadelphia and Baltimore.

Atlantic City.—HOTELS :—

Atlantic City, 61 miles from Philadelphia, may be reached thence twice each day by the Camden and Atlantic Railroad. It is one of the best bathing-places on the coast.

Schooley's Mountain.—HOTELS :—*Heath House.*

ROUTE. From New York, by the N. J. Central and Morris and Essex Railway, 53 miles to Hackettstown, and thence, 2¼ miles by stage. Southerners proceed via Philadelphia and New Brunswick, connecting with the New Jersey Central Railway at Bound Brook, and from this line as above.

The height of the mountain is about 1,100 feet above the sea. The spring is near the summit. It contains muriate of soda, of lime, and of magnesia, sulphate of lime, carbonate of magnesia, and silex, and carbonated oxide of iron.

Budd's Lake.—HOTELS :—*Forest House.*

ROUTE. From New York by the N. J. Central and the Morris and Essex Railway, 43 miles to Stanhope, and thence, 2¼ miles by stage.

Budd's Lake is a beautiful mountain water, deep, pure, and well supplied with fish.

Greenwood Lake.—HOTELS :—

ROUTE. From New York by Erie Railway, 35 miles to Sloatsburg, and thence by stage.

To Greenwood Lake, sometimes called Long Pond, is a very agreeable jaunt from the metropolis, whether for the pure air of the hills, the pleasant aspects of nature, or for the sports of the angle and the gun. Greenwood lies half in New York and half in New Jersey, in the midst of a very picturesque mountain region. It is a beautiful water of seven miles in extent, and all about it, in every direction, are lesser, but scarcely less charming, lakes and lakelets, some of which, as you ride or ramble over the country, delight your surprised eyes where least dreamed of. Such an unexpected vision is Lake Macopin, and the larger waters of Wawayandah. This last-mentioned lake is situated on the Wawayandah Mountains, about 3½ miles from the New York and New Jersey boundary line. The word Wawayandah signifies winding stream, and is very characteristic of the serpentine course of the outlet of this lake towards the Wallkill. Wawayandah is almost divided by an island into two ponds, and thus gets its *home* name of "Double Pond." It is very deep, and abounds in fine trout. This varied hill and lake neighborhood presents in its general air an admirable blending of the wild ruggedness of the great mountain ranges and the pastoral sweetness of the fertile valley lands ; for it possesses the features of both, though of neither in the highest degree.

For other places in New Jersey, near New York, see New York City and Vicinage.

DELAWARE.

DELAWARE is, after Rhode Island, the smallest State in the Union—her greatest length and breadth being, respectively, only 96 and 37 miles. The first settlements here were made by the Swedes and Finns, about the year 1627. In 1655, the country fell into the possession of the Dutch, and in 1664 passed under British rule. It was originally a portion of Pennsylvania, and was governed by the rulers of that Colony, until the time of the Revolution.

The landscape of the northern portion of Delaware is agreeably varied with modest hills and pleasant vales. In the central and southern portions of the State the country is level, ending in marsh and swamp lands. The only considerable waters are the Delaware River and Bay, on the eastern boundary. The Brandywine is a romantic stream, famous for the Revolutionary battle fought upon its banks near the limits of this State, September, 1777. Lords Cornwallis and Howe, Generals Washington, Lafayette, Greene, Wayne, and other distinguished English and American leaders, took part in this memorable conflict. The Americans retreated to Germantown with a loss of 1,200 men, while the British remained in possession of the field, with a loss of about 800.

Wilmington.—HOTELS:—*Indian Queen.*

Wilmington, the most important town on railway from Philadelphia to Baltimore, is situated between the Brandywine River and Christiana Creek, one mile above their junction, and in the midst of one of the finest agricultural districts in the Middle States. It is built on ground gradually rising to the height of 112 feet above tide-water, and is regularly laid out, with broad streets crossing each other at right angles. Since 1840, both its business and population have much increased: at that time it contained about 8,000 inhabitants, and now the population numbers about 21,000. On the Brandywine River are some of the finest flouring mills in the United States, to which vessels can come drawing eight feet of water. It contains also ship and steamboat yards, a foundry for the manufacture of patent car-wheels, which are used all over the country, and a number of large manufacturing establishments of various kinds. It is the seat of a Catholic College, and is generally distinguished for its Academies and Boarding-schools. It is connected with New Castle, Dover, and Seaford, by railway; and via Downingstown with the Columbia Railroad, from Philadelphia to Columbia.

Havre-de-Grace, on the Philadelphia and Baltimore railway route, is in Maryland, at the head of the Chesapeake Bay, on the Susquehanna River, 36 miles north-east of Baltimore. The cars cross the river by a steam-ferry, sometimes passing in winter upon the ice, as in 1851-2. Havre-de-Grace is quite an old town. It is the southern terminus of the tide-water canal.

PHILADELPHIA WILMINGTON AND BALTIMORE RAILROAD.—This Road extends from Philadelphia to Baltimore, 97 miles. It is the great thoroughfare between the two cities, and during the winter months, the only travelled route.

Leaving the dépôt in the city, the route passes through the suburbs to the Schuylkill at "Gray's Ferry," which it crosses on a substantial bridge, and thence passes onward via Chester, Wilmington, Delaware, Newport, and Elkton, crossing the Susquehanna where it empties into the Chesapeake Bay at Havre-de-Grace: thence, 37 miles beyond to Baltimore.

There is another railroad route between the two cities, but it is never passed over by travellers wishing to go direct between Philadelphia, Baltimore, and the South. Tourists, whose time is not limited, and who are desirous of varying the route of travel, will find that over the Columbia Railroad to the Susquehanna River, thence to York, and from thence to Baltimore, a very pleasant excursion. Distance, 153 miles. *Fare*, $5.00.

PENNSYLVANIA.

PENNSYLVANIA is, in point of population, the second State in the Union, and in all respects one of the most important and interesting. A very singular fact in her history—singular because it has no parallel in the annals of any other member of the American Confederacy—is, that her territory was settled without war or bloodshed. The doctrines of peace and good-will, taught by William Penn and his quiet-loving associates, when they pitched their tents upon the sunny banks of the Delaware, long served, happily, as a charm over the savage natures of their Indian neighbors.

We have no record of battle and siege in the story of this State, from the time of the first settlement at Philadelphia, in 1682, until the date of the French and Indian war in 1755. During this year the famous defeat of Braddock, in which Washington, then in his early youth, distinguished himself, occurred at Pittsburg. In 1763, the massacre of the Conestoga Indians took place in Lancaster County. In 1767, the southern boundary of the State, which has since become famous as Mason and Dixon's line, was made. This line is the proverbial division between the Northern, or Free, and the Southern, or Slave-holding States.

Pennsylvania is memorable in the annals of the American Revolution, in which she played a conspicuous part. Upon her soil occurred the important battles of Brandywine and Germantown (1777).

The traveller will seek, here, also, for the scenes of those celebrated events, the massacres of Wyoming and Paoli. Valley Forge was the chief head-quarters of General Washington, and is made yet more interesting by the memory of the sufferings there of the patriot army during its winter encampment in 1777 and 1778. Philadelphia was the national capital until 1789—a period of nearly ten years—and here the earliest American Congresses assembled. The memorable revolt called the Whiskey Insurrection, happened in Pennsylvania in 1794. This disaffection was bloodless and without sequence, as all disloyalty must ever be in the Keystone State.

Among the great men whom Pennsylvania has given to the Republic, we may cite the honored names of Franklin (though born in Boston), Robert Morris, Fulton, Rush, and Rittenhouse. Mr. Buchanan, the present President, is a citizen of this State.

The landscape of Pennsylvania is extremely diversified and beautiful. One-fourth of her great area of 46,000 square miles is occupied by mountain ranges, sometimes reaching an elevation of 2,000 feet. These hills, links of the great Alleghany chain, run generally from north-east to south-west, through the eastern, central, and southern portions of the State. The spur of this hill-range is called South Mountain, where it rises on the Delaware, below Easton. Next, as we go westward, come the Kittatinny, or Blue Mountains, and the Broad Mountains, south of the North Branch of the Susquehanna. Across this river is the Tuscarora. South of the Juniata are the Sideling Hills, and, lastly, come the Alleghanies, dividing the Atlantic slope from the great Mississippi Valley region. West of the Alleghanies, the only hill-ranges in the State are the minor ones called the Laurel and the Chestnut ridges. This belt of mountains extends over a breadth of 200 miles, enclosing numberless fertile valleys, many charming waters, and the greatest coal fields and iron deposits in the Union.

RIVERS.—Pennsylvania cannot boast the marvellous lake scenery of the Empire State; indeed, she has no lakes, if we except the great Erie waters which wash the shore of the north-west corner of the State. For this want, however, the charms of her many picturesque rivers well atone.

PENNSYLVANIA.

The Susquehanna, the largest river of Pennsylvania, and one of the most beautiful in America, crosses the entire breadth of the State, flowing 400 miles in many a winding bout, through mountain gorges, rocky cliffs, and broad cultivated meadows. See Susquehanna River.

The Juniata is the chief affluent of the Susquehanna. It comes in from the acclivities of the Alleghanies in the west, through a mountain and valley country of great natural delight. See Juniata River.

The Delaware flows 300 miles from its sources in the Catskill Mountains to the Delaware Bay, forming the boundary between Pennsylvania and New Jersey, and afterwards between New Jersey and Delaware. It is one of the chief features of the varied scenery of the New York and Erie Railway, which follows its banks for 90 miles. (See N. Y. and Erie R. R.) Lower down, its passage through the mountains forms that great natural wonder of the State, the Delaware Water Gap. The rocky cliffs here rise perpendicularly to a height of nearly 1,200 feet. (See Delaware Water-Gap.) The navigation of the Delaware is interrupted at Trenton, N. J., by falls and rapids. Philadelphia is on this river, about 40 miles above its entrance into Delaware Bay. The river was named in honor of Lord De La Ware, who visited the bay in 1610.

The Lehigh is a rapid and most picturesque stream. Its course is from the mountain coal districts, through the famous passage of the Lehigh Water Gap below Mauch Chunk, to the Delaware at Easton. Its length is about 90 miles.

The Schuylkill flows 120 miles from the coal regions north, and enters the Delaware five miles below Philadelphia. We shall review it as we call at the towns and places of interest upon its banks.

The Alleghany and the Monongahela Rivers—one 300 and the other 200 miles in length—unite at Pittsburg and from the Ohio. The Youghiogheny is a tributary of the Monongahela.

Philadelphia and Vicinity.*—HOTELS: —Philadelphia is abundantly supplied with excellent hotels of all grades.

The *Continental*, first opened during the present season, is among the largest and most fashionable. It is in Chestnut street, between Eighth and Ninth streets.

The *Girard House*, a first class establishment, is in Chestnut street, directly opposite the *Continental*.

The *La Pierre*, one of the most elegant houses

* See "New Jersey" for routes to New York, and "Delaware" for routes to Baltimore.

in the city, is on the west side of Broad near Chestnut street.

The *Washington Hotel*, in the same vicinage; the *St. Lawrence*, the *American*, the *Merchants'*, the *St. Louis*, and many others.

This great city is, in extent and population, the second in the Union. Its people number about half a million—as many as any of the capitals of Europe (London and Paris only excepted) can show. It was settled in 1682 by a colony of English Quakers, under the guidance of William Penn. The sobriquet of the City of Brotherly Love, which it now bears, was given to it by Penn himself. No striking events mark its history down to the days of the Revolution, and its part in that great drama was more peaceful than warlike. The first Congress assembled here, and subsequent Congresses, during the continuance of the war. The Declaration of Independence was signed and issued here, July 4th, 1776. The Convention which formed the Constitution of the Republic assembled here, May, 1787.— Here resided the first President of the United States, and here, too, Congress continued to meet until about 1797. The city was in possession of the British troops from September, 1777, to June 11, 1778, a result of the unfortunate battles of Brandywine and Germantown.

Philadelphia lies between the Delaware and Schuylkill rivers, six miles above their junction, and nearly 100 miles (by the Delaware River and Bay) from the Atlantic. The site of the city is so low and level, that it does not make a very impressive appearance from any approach. But the elegance and symmetry and neatness of its streets—the high cultivation of all its rural corners, and the picturesque character of the higher suburban land to the northward, fully compensate for this want. The most thronged portion of the city is near the apex of an angle formed by the approach of the two rivers, between which it is built. Streets extend from river to river, and are crossed by other streets at right angles.

PUBLIC SQUARES.—*Washington Square*, a little south-west of the State House, is finely ornamented with trees and gravelled walks, is surrounded by a handsome iron railing, with four principal entrances, and is kept in excellent order. *Independence Square*, in the rear of the State House, is enclosed by a solid brick wall, rising three or four feet above the adjacent streets, surmounted by an iron railing. The entire area is laid off in walks and grass-plots, shaded with majestic trees. It was within this enclosure that the Declaration of Independence was first promulgated, and at the present day it is frequently used as a place of meeting for political and other purposes. *Franklin Square*, between Race and Vine, and Sixth and Franklin streets, is an attractive promenade, with a fountain in its centre, surrounded by a marble basin; it is embellished with a great variety of trees. *Penn Square* is at the intersection of Broad and Market streets, now divided into four parts by cutting Market and Broad streets through it; *Logan Square* is between Race and Vine streets; and *Rittenhouse Square*, between Walnut and Locust streets.

PUBLIC BUILDINGS.—The State House fronts on Chestnut street, and including the wings, which are of modern construction, occupies the entire block, extending from Fifth to Sixth streets. In a room in this building, on the 4th of July, 1776, the Declaration of Independence was adopted by Congress, and publicly proclaimed from the steps on the same day. The room presents now the same appearance it did on that eventful day, in furniture and interior decorations. This chamber is situated on the first floor, at the eastern end of the original building, and can be seen by visitors on application to the person in charge of the State House. In the Hall of Independence is a wooden statue of Washington, and some pictures. Visitors may overlook the city and its surroundings admirably from the cupola of this building.

The *Girard College* is situated on the Ridge Road, in a north-west direction from the city proper, about two miles from the State House. It was founded by the late Stephen Girard, a native of France, who died in 1831, and bequeathed $2,000,000, for the purpose of erecting suitable buildings for the education of orphans.

The commanding site of the edifice occupies an area of about 45 acres, left for the purpose by the founder of the institution. The central, or college building, is 218 feet long, 160 broad, and 97 high, and is a very noble marble structure of the Corinthian order. The other buildings, six in number, surround the main edifice.

The *Merchants' Exchange*, situated between Dock, Walnut, and Third streets, is of white marble. It is a beautiful structure, and of its kind, one of the finest in the country.

The *United States Mint* is in Chestnut street, below Broad street, and fronts on the former street 122 feet. It is built of white marble, in the style of a Grecian Ionic temple, and comprises several distinct apartments. Coining is among the most interesting and attractive of processes, to those who have never witnessed such operations. Visitors are admitted during the morning of each day, until one o'clock, on application to the proper officers.

The *Custom House*, formerly the United States Bank, is located in Chestnut street, between Fourth and Fifth streets. It is a chaste specimen of the Doric order of architecture, after the Parthenon at Athens, with the omission of the colonnades at the sides. It was commenced in 1819, and completed in about five years, at a cost of half a million of dollars.

The *United States Navy Yard* is located in Front street, below Prime, and contains within its limits about 12 acres. It is enclosed on three sides by a high and substantial brick wall; the east side fronts on and is open to the Delaware River. Its entrance is in Front street. The Yard contains every preparation necessary for building vessels of war, and has marine barracks, with quarters for the officers.

Many of the bank edifices of Philadelphia are very elegant, and imposing, built of marble and other rich material.

The CHURCHES of the city are about 300 in number, of all denominations, and new ones are continually making their appearance.

The Catholic *Church of St. Peter and St. Paul*, on Logan Square, is built of red stone, in the Roman style. It is crowned with a dome 210 feet high.

The *Church of St. Mark's* (Episcopal), is a beautiful edifice of light-red sand-stone, with a tower and steeple of admirable grace.

Christ's Church, with its soaring spire, is a very interesting object in its ancient and quaint aspect.

The Church of Calvary (Presbyterian), and the Baptist Church in Broad and Arch streets, are also of sand-stone, with imposing towers and spires. We may also mention among the churches of the greatest architectural interest: St. Stephen's (Episcopal), the Catholic Church of the Assumption, St. Jude's, the Presbyterian Churches, upon Arch and Eighteenth streets, and upon Arch and Tenth streets; the Church of the Nativity, the Baptist Churches on Chestnut and Fifth streets. In the towers of St. Peter's, St. Stephen's, and of Christ Church, there are chimes of bells.

The American Baptist Publication Society is located in Arch street; the Presbyterian Board of Publication is on Chestnut street. Besides these religious associations, there are the American Sunday School Union, the Pennsylvania and the Philadelphia Bible Societies, and the Female and the Friends' Bible Societies, with numerous others.

BENEVOLENT INSTITUTIONS. — The county Almshouse, situated on the west side of the Schuylkill, opposite South street, is an immense structure, consisting of four main buildings, covering and enclosing about 10 acres of ground, and fronting on the Schuylkill River. The site is much elevated above the bank of the river, and commands a fine view of the city and surrounding country. The *Pennsylvania Hospital*, in Pine street, between Eighth and Ninth streets, is an admirable institution. It contains an anatomical museum, and a library of more than 8,000 volumes. In the rear of the lot fronting on Spruce street, is a small building which contains West's celebrated picture of Christ Healing the Sick, presented to this institution by its author. The *United States Marine Hospital or Naval Asylum* has a handsome situation on the east bank of the Schuylkill below South street. It is for the use of invalid seamen, and officers disabled in the service.—*The Pennsylvania Institution for the Deaf and Dumb* is situated on the corner of Broad and Pine streets, having extensive buildings adapted for the purposes of the establishment.—The *Pennsylvania Institution for the Instruction of the Blind* is situated in Race street, corner of Twenty-first street.

ART SOCIETIES. The *Pennsylvania Academy of Fine Arts*, an old and most important institution, has a fine building, with a noble suite of galleries upon Chestnut street, between Tenth and Eleventh streets. It possesses a very valuable and permanent collection of pictures, and makes an annual exhibition of new works. Among its old pictures, are West's Death on the Pale Horse, and Alston's Dead Man Restored. No citizen or stranger should neglect to visit these galleries.

LITERARY AND SCIENTIFIC INSTITUTIONS. The *American Philosophical Society* was founded in 1743, principally through the exertions of Dr. Franklin; its hall is situated in South Fifth street, below Chestnut, and in the rear of the State House. In addition to its library of 15,000 volumes of valuable works, the society has a fine collection of minerals and fossils, ancient relics, and other interesting objects. Strangers are admitted to the hall on application to the librarian. —The *Philadelphia Library* is situated in Fifth street, below Chestnut, on the north corner of Liberty street. It was founded in 1731 by the influence of Dr. Franklin. This institution, together with the Logunian, which occupies the same building, possesses about 65,000 volumes.— The *Athenæum* in Sixth below Walnut street, contains the periodical journals of the day, and a library consisting of several thousand volumes. The rooms are open every day and evening (Sundays excepted) throughout the year.

Strangers are admitted gratuitously for one month, on introduction by a member.—The *Franklin Institute* was incorporated in 1824; it is situated in Seventh street, below Market. Its members are very numerous, composed of manufacturers, artists, mechanics, and persons friendly to the mechanic arts. The annual exhibitions of this institute never fail to attract a large number of visitors. It has a library of about 6,000 volumes, and an extensive reading room, where most of the periodicals of the day may be found. Strangers are admitted to the rooms on application to the actuary.—The *Academy of Natural Sciences*, incorporated in 1817, has a well-selected library of about 14,000 volumes, besides an extensive collection of objects in natural history. Its splendid hall is in Broad street, between Chestnut and Walnut. It is open to visitors every Saturday afternoon.—The *Mercantile Library*, situated on the corner of Fifth and Library streets, was founded in 1822, for the purpose of diffusing mercantile knowledge.—The *Apprentices' Library*, corner of Fifth and Arch streets, consists of about 14,000 volumes, and is open to the youth of both sexes.—The *Historical Society of Pennsylvania*, in the Athenæum buildings, was founded for the purpose of diffusing a knowledge of local history, especially in relation to the State of Pennsylvania. It has caused to be published a large amount of information on subjects connected with the early history of the State, and is now actively engaged in similar pursuits.—The *Friends' Library* in Race street, below Fifth, has about 3,000 volumes, which are loaned, free of charge, to persons who come suitably recommended.—There are several excellent libraries in the different Wards of Philadelphia, which are conducted on the most liberal principles.

MEDICAL INSTITUTIONS. The *University of Pennsylvania*, which is an admirable institution, is situated on the west side of Ninth street, between Market and Chestnut. It was founded in 1791, by the union of the old University and College of Philadelphia.—*Jefferson Medical College* is situated in Tenth street, below Chestnut; it was originally connected with the college at Canonsburg, but it is now an independent institution. The number of pupils averages about 300 annually. The anatomical museum of this institution is open to visitors.—*Pennsylvania Medical College*, in Ninth street, below Locust, is a flourishing institution of recent origin; the first lectures having been delivered in the winter of 1839-'40.—The *College of Physicians* is an old institution, having existed before the Revolution. It is one of the principal sources from which proceeds the pharmacopœia of the United States. —The *Philadelphia College of Pharmacy*, in Zane street, above Seventh, was the first regularly organized institution of its kind in the country. Its objects are to impart appropriate instruction, to examine drugs, and to cultivate a taste for the sciences. The medical schools of Philadelphia are famous, all the Union through, and the students who flock to them every winter may be numbered by thousands.

PRISONS. The *Eastern Penitentiary*, in the north-west part of the city, is situated on Coates street, corner of Twenty-first street, and south of Girard College. It covers about 10 acres of ground, is surrounded by a wall 30 feet high, and in architecture resembles a baronial castle of the middle ages. It is constructed on the principle of strictly solitary confinement in separate cells, and is admirably calculated for the security, the health, and so far as consistent with its objects, the comfort of its occupants. — The *County Prison*, situated on Passyunk Road, below Federal street, is a spacious Gothic building, presenting an imposing appearance. It is appropriated to the confinement of persons awaiting trial, or those who are sentenced for short periods. The *Debtor's Prison*, adjoining the above on the north, is constructed of red sandstone, in a style of massive Egyptian architecture.—The *House of Refuge* is situated in Parish street, between Twenty-third and Twenty-fourth streets, and at Bush Hill is the *House of Correction*.

CEMETERIES. The beautiful cemetery of *Laurel Hill* is situated on the Ridge road, three and a half miles north-west of the city, and on the east bank of the Schuylkill, which is elevated about ninety feet above the river. It contains about 20 acres, the surface of which is undulating, prettily diversified by hill and dale, and adorned with a number of beautiful trees. The irregularity of the ground, together with the foliage, shrubs, and fragrant flowers, which here abound —the finely-sculptured and appropriate monuments—with an extensive and diversified view, make the whole scene highly impressive. On entering the gate, the first object that presents itself to the gaze of the visitor is an excellent piece of statuary, representing Sir Walter Scott conversing with Old Mortality, executed in sandstone by the celebrated Thom. The chapel, which is situated on high ground to the right of the entrance, is a beautiful Gothic building, illuminated by an immense window of stained glass. *Monument Cemetery*, another beautiful enclosure, is situated on Broad street, in the vicinity of Turner's Lane, in the north part of Philadelphia, and about three miles from the State House. It was opened in 1838, and now contains many handsome tombs.—*Ronaldson's Cemetery*, in Shippen street, between Ninth and Tenth, occupying an entire square, and surrounded by an iron railing, is very beautiful. It formerly belonged to Mr. James Ronaldson, from whom it takes its name, who divided it into lots, and disposed of it for its present purposes. —The *Woodlands Cemetery*, beyond the Schuylkill, in West Philadelphia, is a new burial ground, which promises in due course of time to rival even *Laurel Hill* in landscape and monumental attractions.

PLACES OF AMUSEMENT. The *Academy of Music or Opera House*, on Broad and Locust streets, is a grand establishment, with a front of 140 feet, and a flank of 238. The first story is of brown stone, and the rest of pressed brick with brown-stone dressings. The auditorium will seat 3000 persons. The *Walnut Street Theatre* is at the corner of Walnut and Ninth streets. *Arch Street Theatre* is in Arch street, above Sixth. The *Musical Fund Hall* is in Locust street, between Eighth and Ninth streets. The *City Museum*, Callowhill, below Fifth; *Welch's National Circus*, Walnut street, above Eighth; *Concert Hall*, Chestnut, below Thirteenth; *National Hall*, Market street, below Thirteenth; *Sansom Street Hall*, Sansom, above Sixth; the *Assembly Buildings*, Chestnut and Tenth streets.

The *Markets* of Philadelphia are worthy of especial notice, in their great extent and admirable appointment.

Cars and stages to all parts of the city and suburbs, are easily to be found.

THE VICINITY OF PHILADELPHIA.—Laurel Hill and other cemeteries, and the Girard College, we have already mentioned.

Camden, N. J., is opposite Philadelphia. The Camden and Amboy and the Camden and Atlantic R. R. terminate here.

The Fairmount Water Works, which supply the city bountifully, are on the east bank of the Schuylkill, about two miles north-west from the heart of the city, occupying an area of 30 acres, a large part of which consists of the "mount," an eminence 100 feet above tide water in the river below, and about 60 feet above the most elevated ground in the city. The top is divided into four reservoirs, capable of containing 22,000,000 gallons, one of which, is divided into three sections for the purpose of filtration. The whole is surrounded by a beautiful gravel-walk, from which may be had a fine view of the city. The reservoirs contain an area of over six acres; they are 12 feet deep, lined with stone and paved with brick, laid in a bed of clay, in strong lime cement, and made water-tight. The power necessary for forcing the water into the reservoirs, is obtained by throwing a dam across the Schuylkill; and by means of wheels moved by the water, which work forcing pumps, the water of the river is raised to the reservoirs on the top of the "mount." The dam is 1,600 feet long, and the race upwards of 400 feet long and 90 wide, cut in solid rock. The mill-house is of stone, 238 feet long, and 56 wide, and capable of containing eight wheels, and each pump will raise about 1,250,000 gallons in 24 hours.— The Spring Garden Water-works are situated on the Schuylkill, a short distance above Fairmount.

The Falls of the Schuylkill are about four miles above the city, on the river of that name. Since the erection of the dam at Fairmount, the falls have entirely disappeared.— From the city to the falls, however, is a very pleasant drive; and they might be reached in a return visit to the Wissahickon.

The Schuylkill Viaduct, three miles north-west from the city, is 980 feet in length, and crossed by the Reading Railroad.

Wissahickon Creek, a stream remarkable for its romantic and beautiful scenery, falls into the Schuylkill about six miles above the city.— It has a regular succession of cascades, which in the aggregate amount to about 700 feet. Its banks, for the most part, are elevated and precipitous, covered with a dense forest, and diversified by moss-covered rocks of every variety. The banks of the beautiful Wissahickon afford one of the most delightful rides in the vicinity of Philadelphia, and are a great resort for the citizens, picnic parties, and Sunday Schools.

Manayunk, seven miles from the centre of the city, is a large manufacturing place. It is indebted for its existence to the water created by the improvement of the Schuylkill, which serves the double purpose of rendering the stream navigable, and of supplying hydraulic power to the numerous factories of the village.

Germantown is included in the 22d Ward of the city. It is some half a dozen miles from the municipal heart, and may be reached every fifteen minutes by city railroads and steam cars.

Greenwich Point, about three miles below the city, and *Gloucester Point,* directly opposite, are favorite places of resort during the summer season. Steamboats run many times daily from Philadelphia. Fare to the former place, 5 cents —to the latter, 6¼ cents.

Bedford Springs.—Hotels:—

*Routes.—*To Huntington by the Central Pennsylvania Railroad, 202 miles from Philadelphia; from Harrisburg, 92 miles; and from Huntington by the Huntington and Broad Top Railroad to Hopewell, 18 miles by stage. Through from Baltimore, Philadelphia, or Petersburg, to the Springs, same day.

Bedford Springs, about one mile from the town of Bedford, Pa., have long been celebrated for their highly medicinal virtues. The Springs are chalybeate, iron, and sulphur, from the former of which is taken the famous "Bedford Water." The hotel accommodations are excellent, and the pleasure grounds and promenades very attractive.

Doubling Gap White Sulphur Springs.— Hotels:—

*Routes.—*By the Cumberland Valley Railroad to Newville; 30 miles from Harrisburg; 22 miles from Cumberland, and thence by stage.

Valley Forge, the memorable head-quarters of General Washington during the winter of 1777, is 23 miles from Philadelphia, on the railway to Reading. The old head-quarters is still standing near the railroad, from whence it can be seen.

Pottstown, 37 miles from Philadelphia, is prettily situated on the left bank of the Schuylkill. The houses, which are built principally upon one broad street, are surrounded by fine gardens and elegant shade trees. The scenery of the surrounding hills is very fine, especially in the fall of the year, when the foliage is tinged with a variety of rich autumnal tints. The Reading Railroad passes through one of its streets, and crosses the Manatawny on a lattice bridge, 1,071 feet in length.

Reading, 58 miles from Philadelphia by railway, is a pleasant place for a summer home, upon the banks of the Schuylkill River.

Port Clinton is 78 miles from Philadelphia, on the Reading Railroad. It is an agreeable place at the mouth of the Little Schuylkill.

Schuylkill Haven, also on the banks of the Schuylkill, in the midst of a very interesting landscape region. A branch road comes in here from the great coal districts. From Philadelphia, 89 miles by Reading Railroad.

Pottsville, the terminus of the Philadelphia and Reading route, is 93 miles from Philadelphia. It is upon the edge of the coal basin, in the gap by which the Schuylkill comes through Sharp's Mountain.

Allentown, 51 miles from Philadelphia, is upon the railroad from Easton, Pa., to Mauch Chunk. It is built upon high ground, near the Lehigh river, at the junction of Jordan and Little Lehigh creeks. The mineral springs here are highly prized by those who have tried the efficacy of their waters. A visit to "Big Rock," 1,000 feet in elevation, a short distance from the village, will amply repay the tourist, by the extent and richness of the scene there spread out before him in every direction.

Bethlehem is upon the Lehigh, 52 miles from Philadelphia, and 11 miles from Easton. May be reached from New York and Philadelphia by Railway. Routes to Easton, and thence 12 miles by Lehigh Valley Railroad to Mauch Chunk. It is the principal seat of the United Brethren, or Moravians, in the United States, and was originally settled under Count Zinzendorf, in 1741. The village contains a large stone church of Gothic architecture, 142 feet long and 68 feet wide, and capable of seating 2,000 persons. It has also a Moravian Seminary of very high reputation.

Nazareth, another pretty Moravian village, is situated 10 miles north from Bethlehem, and 7 miles north-west from Easton.

Mauch Chunk, is in the midst of the great Pennsylvania coal regions, 43 miles from Easton by railway, and 100 miles from Harrisburg, the State capital. It is upon the Lehigh, in one of its wildest and most romantic passages. Mount Pisgah, a short distance north, rises 1,000 feet along the river. A railway has been constructed, 9 miles, to *Summit Hill*, down which the coal-laden cars come by the force of their own gravity. We are here in the vicinage of the beautiful scenery of the Susquehanna River, which we shall visit in another chapter.

Easton.—HOTELS:—*American.*

From New York 75 miles by N. J. Central R. R. From Philadelphia 80 miles by Belvidere, Delaware & Flemington R. R. Pop., 12,000.

PHILADELPHIA TO PITTSBURG AND THE WEST.

BY THE PENNSYLVANIA RAILWAY.

This route is one of the great highways from the Atlantic to the Mississippi States. The Pennsylvania Central Road, with some completing links, extends 355 miles from the city of Philadelphia, through the entire length of Pennsylvania, to the Ohio River at Pittsburg, connecting there with routes for all parts of the South-west, West, and the North-west. Through trains (15 hours to Pittsburg) run morning, noon, and night. Philadelphia station, south-east corner of Eleventh and Market streets; entrance on Eleventh street.

Lancaster.—HOTELS:—*Michael's (Grapes) Hotel: City Hotel: Red Lion.*

Lancaster, a city of more than 18,000 inhabitants, is upon the Pennsylvania Central Railroad, near the Conestoga creek. It was at one time the principal inland town of Pennsylvania, and was the seat of the State government from 1799 to 1812. In population it now ranks as the fourth in the State. It is pleasantly situated in the centre of a very rich agricultural region, well-built, and has many fine edifices, public and private. It is the seat of Marshall College, organized in 1853, in union with the old establishment of Franklin College, which was founded in 1787. Fulton Hall, an edifice for the use of public assemblies, is a noteworthy structure here, as are some of the score of churches. The oldest turnpike road in the United States terminates here, 62 miles from Philadelphia. One of the sources of the prosperity of Lancaster is the navigation of the Conestoga, in a series of nine locks and slack water pools, 18 miles in length from the town of Safe Harbor in the Susquehanna, at the mouth of the Conestoga. With the help of Tide-Water Canal to Port Deposit, a navigable communication is opened to Baltimore.

Wheatland, the seat of the Hon. James Buchanan, ex-President of the United States, is a few miles from Lancaster.

Harrisburg.—HOTELS:—*Coverly's: Herr's.*

Harrisburg, the capital of Pennsylvania, is upon the east bank of the Susquehanna, 106 miles from Philadelphia. From the dome of the State House, a fine view is obtained of the wide and winding river, its beautiful islands, its interminable bridges, and the surrounding ranges of the Kittatinny Mountains. The Cumberland Valley road diverges at Harrisburg for Chambersburg, 52 miles distant, and the Dauphin and Susquehanna, 59 miles to Auburn, on the Philadelphia and Reading R. R. The north Central Road is to Baltimore, Md., 85 miles: the Columbia Branch to Columbia; also the Lebanon Valley R. R. for Reading, 54 miles.

About 14 miles beyond Harrisburg, the route crosses and leaves the Susquehanna River, and thenceforward follows the banks of the Juniata for about 100 miles to the eastern base of the Alleghanies, the canal keeping the road and river company most of the way—of the Juniata part of the route we shall speak directly—sending the traveller on, if he is in haste, to Pittsburg, over the Alleghanies, by the help of the wonderful specimens of the power of the engineer's art, which will interest him on the way: the tunnel, 3,612 feet long, in which he will pass through the Alleghany Mountains, 2,200 feet above the sea; the great inclined planes of the Portage Railroad, and other marvels of art and of nature.

PENNSYLVANIA.

Pittsburg, Pa.—HOTELS:—The *Monongahela House*.

Pittsburg, Pa., is upon the Ohio river, at the confluence of the Alleghany and the Monongahela. It is situated in a district extremely rich in mineral wealth, and the enterprise of the people has been directed to the development of its resources, with an energy and success seldom paralleled. The city of Pittsburg enjoys, from its situation, admirable commercial facilities, and has become the centre of an extensive commerce with the Western States; while its vicinity to inexhaustible iron and coal mines has raised it to great distinction as a manufacturing place.

The city was laid out in 1765, on the site of Fort Du Quesne, subsequently changed to Fort Pitt. It is situated on a triangular point, at the confluence of the Alleghany and Monongahela rivers, which here form the Ohio. Pittsburg is connected with the left bank of the Monongahela by a bridge 1,500 feet long, which was erected at a cost of $102,000 dollars. Four bridges cross the Alleghany river, connecting Pittsburg with Alleghany City.

There are several places in the vicinity of Pittsburg, which, as they may be considered parts of one great manufacturing and commercial city, are entitled to a notice here. *Alleghany City*, opposite to Pittsburg, on the other side of the Alleghany river, is the most important of them. The elegant residences of many persons doing business in Pittsburg may be seen here, occupying commanding situations. Here is located the *Western Theological Seminary of the Presbyterian Church*, an institution founded by the General Assembly in 1825, and established in this town in 1827. Situated on a lofty, insulated ridge, 100 feet above the Alleghany, it affords a magnificent prospect. The *Theological Seminary of the Associated Reformed Church*, established in 1826, and the *Alleghany Theological Institute*, organized in 1840 by the Synod of the Reformed Presbyterian Church, are also located here. The *Western Penitentiary* is an immense building in the ancient Norman style, situated on a plain on the western border of Alleghany City. It was completed in 1827, at a cost of $183,000. The *United States Arsenal* is located at Lawrenceville, a small but pretty village two and a half miles above Pittsburg, on the left bank of the Alleghany river.

Birmingham is another considerable suburb of Pittsburg, lying about a mile from the centre of the city, on the south side of the Monongahela, and connected with Pittsburg by a bridge 1,500 feet long, and by a ferry. It has important manufactories of glass and iron.

Manchester is two miles below Pittsburg, on the Ohio. The U. S. Marine Hospital is yet below.

It is usual to speak of extensive manufactories as being in Pittsburg, though they are not within the limits of the city proper, but are distributed over a circle of five miles' radius from the courthouse on Grant's Hill. This space includes the cities of Pittsburg and Alleghany, the boroughs of Birmingham and Lawrenceville, and a number of towns and villages, the manufacturing establishments in which have their warehouses in Pittsburg, and may consequently be deemed, from the close connection of their general interests and business operations, a part of the city. There are within the above compass about eighty places of religious worship, and a population of not less than 100,000.

The stranger in Pittsburg will derive both pleasure and instruction by a visit to some of its great manufacturing establishments, particularly those of glass and iron. During the summer season Pittsburg is an immense thoroughfare, large numbers of travellers and emigrants passing through it on their way westward. The population of Pittsburg is about 111,000.

The Juniata. This beautiful river, whose course is closely followed so many miles by the Pennsylvania Railroad and Canal, rises in the south central part of the Keystone State, and flowing eastward falls into the Susquehanna about 14 miles above Harrisburg.

The landscape of the Juniata is in the highest degree picturesque, and many romantic summer haunts will be found, by and by, among its valleys; though at present very little tarry is made in the region, from its attractions being unknown, and the comforts of the traveller being as yet unprovided for. The mountain background, as we look continually across the river from the cars, is often strikingly bold and beautiful. This clause refers to a scene in the upper part of the river, near Water street, a point which the railroad leaves some miles to the south, or left. The Little Juniata, which with the Frankstown branch forms the main river, is a stream of wild romantic beauty. The entire length of the Juniata, as well as its branches, is estimated at nearly 150 miles, and its entire course is through a region of mountains, in which iron ore is abundant, and of fertile limestone valleys.

THE COAL REGION.

From Philadelphia.

The *Philadelphia and Reading Railway* extends 93 miles from Philadelphia to Pottsville, in the heart of the *great Coal regions* of the State. It passes through Valley Forge, Reading, Auburn and other places, for which see Index.

The *Catawissa, Williamsport and Erie Railway*, connects Philadelphia with the Erie Railway at Elmira, N. Y., and by other routes from that point with Niagara Falls and all the lines from New York to the great West and Northwest. It leads to the coal beds of Pennsylvania at Catawissa on the Susquehanna, and thence up the west branch of that river to Williamsport. The entire passage of this road is amidst natural scenes of great variety and beauty.

The North Pennsylvania Railroad extends 33 miles, to Doylestown, Pa.

The Belvidere, Delaware and Flemington Railroad extends, via Easton, Pa. (50 miles), to Belvidere, 64 miles.

THE SUSQUEHANNA AND ITS VICINAGE.

We will now look at the chief scenes and places of interest in Pennsylvania, lying upon and about the great Susquehanna River and its tributaries, and at the railways, canals and other highways of travel which communicate with and intersect that part of the State.

The Susquehanna is the greatest of the rivers of Pennsylvania, traversing as it does its entire breadth from north to south, and in its most interesting and most important regions. It lies about midway between the centre and the eastern boundary of the State, and flows in a zig-zag course, now south-east and now south-west, and so on over and over, following very much the windings of the Delaware, which separates the State from New Jersey. The Pennsylvania Canal accompanies it in all its course, from Wyoming on the north to the Chesapeake Bay on the south. All the great railroads intersect or approach its waters at some point or other, and the richest coal lands of the State lie contiguous to the borders.

The Susquehanna, in its main branch, rises in Otsego Lake, in the S. E. central part of New York, and pursues a very tortuous but generally south-west course. This main, or North, or East Branch, as it is severally called, when it reaches the central part of Pennsylvania—after a journey of 250 miles—is joined at Northumberland by the West Branch, which comes in 200 miles from the declivities of the Alleghanies. The course of this arm of the river is nearly eastward, and, as with the North Branch, through a country abounding with coal, and other valuable products. It is also followed by a canal, for more than a hundred miles up.

The route of the New York and Erie Railway is upon or near the banks of the Susquehanna in southern New York, and occasionally across the Pennsylvania line, for 50 miles, first touching the river near the Cascade Bridge, nearly 200 miles from New York, passing the cities of Binghamton and Owego, and finally losing sight of it just beyond Barton, some 250 miles from the metropolis. The tourist, seeking the picturesque regions of the river, from New York, may take the Erie Route, 201 miles to Great Bend, and thence southward by the Delaware, Lackawanna and Western Road, via Scranton, and stage to Wilkesbarre, in the valley of Wyoming. This railway continues on to Lehigh and Easton (Delaware Water Gap) and Elizabethport, back to New York.

The Catawissa, Williamsport and Erie Railway connects Philadelphia with Catawissa, in a beautiful part of the main arm of the Susquehanna below the Wyoming region, and with Williamsport, in the finest part of the West Branch, continuing on through Elmira, N. Y., to the Falls of Niagara. From Philadelphia, via *Port Clinton*, on the Reading Railroad, to Catawissa 145 miles, to Williamsport 197 miles. By this route passengers may go through from Philadelphia to Buffalo in 16 hours, to Niagara Falls in 18 hours, to Detroit in 26 hours, to Chicago in 36 hours, to St. Louis in 48 hours. Day express from Philadelphia breakfasts at Port Clinton and dines at Williamsport.

The Great Pennsylvania Railroad, via Pittsburg to the West, follows the Susquehanna from the vicinity of Harrisburg some 14 miles up to the mouth of the Juniata.

The Northern Central Road, from Baltimore, touches the Susquehanna at Harrisburg, 85 miles distant, where it connects with the Pennsylvania Railroad for Pittsburg.

A branch road from Harrisburg follows the river down 28 miles to Columbia.

A pleasant route from Philadelphia or New York to the Valley of Wyoming, is by railway from either city to Easton, near the Delaware Water Gap, thence by the Lehigh Valley Road to the coal regions at Mauch Chunk, and thence to Wilkesbarre.

The entire length of the Susquehanna (or Crooked River) is about 500 miles, and the country which it traverses is of every aspect, in turn, from the gentlest pastoral air to the wildest humors of the stern mountain pass. The region most sought, and deservedly so, by the tourist in quest of landscape beauties, is that around and below the Valley of Wyoming. From this point down many miles to Northumberland, where the West Branch comes in, the scenery is everywhere strikingly fine at brief intervals; but the best and boldest mountain passes extend from five to ten miles below the southern outlet of Wyoming, around Nanticoke and Shickshinney. This is the region *par excellence* for the study of the artist. Portions, also, of the West Branch—though not yet very much visited—are remarkably fine.

The Valley of Wyoming and Wilkesbarre.—At Wilkesbarre, in the heart of the Wyoming Valley, there is (near the river) a most excellent hotel. The village is beautifully placed upon a plain twenty feet above the river. Prospect Rock, three miles distant, overlooks the Valley most charmingly.

"Wyoming," says Mr. Minor, in a pleasant history of this vicinity, "though now generally cleared and cultivated, yet to protect the soil from floods, a fringe of trees is left along each bank of the river—the sycamore, the elm, and more especially the black walnut—while here and there, scattered through the fields, a huge shell-bark yields its summer shade to the weary laborers, and its autumn fruit to the black or gray squirrel or the rival plough-boys. Pure streams of water come leaping from the mountains, imparting health and pleasure in their course, all of them abounding with the delicious trout. Along these brooks, and in the swales scattered through the uplands, grow the wild-plum and the butternut; while, wherever the hand of the white man has spared it, the native grape may be gathered in unlimited profusion."

* " Wyoming is a classic and a household name. At our earliest intelligence it takes its place in our hearts as the label of a treasured packet of absorbing history and winning romance. It is the key which unlocks the thrilling recollections of some of the most tragical scenes in our national history, and some of the sweetest imaginations of the poet. Every fancy makes a Mecca of Wyoming. Thus sings Halleck—

When life was in its bud and blossoming,
And waters gushing from the fountain spring
Of pure enthusiast thought, dimm'd my young eyes,
As by the poet borne on unseen wing,
I breathed in fancy 'neath thy cloudless skies,
The summer's air, and heard her echoed harmonies.'

"The pen of Campbell and the pencil of Turner have taken their loftiest and most unbridled flights in praise of Wyoming, and though they have changed, they have not flattered its beauties.

'Nature hath made thee lovelier than the power
Even of Campbell's pen hath pictured.'

" Again, Halleck says of the mythical Gertrude, the fair Spirit of Wyoming, and of the real maidens of the land—

'But Gertrude, in her loveliness and bloom,
Hath many a model here; for woman's eye
In court or cottage, whereso'er her home,
Hath a heart-spell too holy and too high
To be o'erprais'd, even by her worshipper—Poesy!'"

The terrible Battle of Wyoming—which has been so often the theme of the pencil and the pen, occurred on July 3d, 1778. Few of the ill-fated people escaped. Prisoners were grouped around large stones, and were murdered with the tomahawk, amidst yells and incantations of fiendish triumph. One of these stones of inhuman sacrifice may yet be seen in the Valley. It is called Queen Esther's Rock, and lies near the old river bank, some three miles above Fort Forty. The village of Wilkesbarre was burned at this time, and its inhabitants were killed or taken prisoners, or scattered in the surrounding forests.

The site of Fort Forty is across the river from Wilkesbarre, past the opposite village of Kingston, and nearly west of Troy, five miles and a half distant. At this spot, where the slain were buried, there now stands a monument commemorative of the great disaster. It is an obelisk 62¼ feet high, made of granite blocks hewn in the neighborhood. The names of those who fell and of those who were in the battle and survived, are engraved upon marble tablets set in the base of the monument. This praiseworthy work was done by the exertions of the ladies of Wyoming.

Nanticoke and West Nanticoke are little coal villages, at the southern extremity of the Wyoming Valley, where as we have already intimated occur some of the boldest passages of the scenery of the Susquehanna. This point, as others upon the banks of the river below, must be reached by stage, or by the slow and heavily laden canal boats, for railways do not yet traverse the way; and neither are there any better accommodations than those of ordinary village and wayside inns: at least not until we reach Catawissa or Northumberland, where the West Branch comes in. A beautiful view of Wyoming is seen looking northward from the hills on the east side of the river, near Nanticoke; and the scenes below, from the banks of the river and the canal, are most varied and delightful. The coal mines of this neighborhood may easily be penetrated, and with ample remuneration for the venture.

Jessup's is a very cosy, lone inn, upon the west shore, two or three miles below Nanticoke, from whence are seen striking pictures of the river and its strong mountain banks both above

---

* The Author, in Harpers' Magazine for October, 1853, vol. vii., p. 615.

and below; the hills in all this vicinity are impressively bold and lofty, making the comparatively narrow channel of the river seem yet narrower, and *italicizing* the quiet beauty of the many verdant islands which stud the waters here.

Shickshinney and Wapwollopen, yet below, are little places, still in the midst of a rugged hill and valley country. Back of Wapwollopen, on the east shore, is the barren peak of its namesake mountain, and the wild waters of Wapwollopen Creek.

Catawissa is on the line of the railways from Philadelphia for Williamsport, on the West Branch, and thence to Elmira and Niagara. It is connected also by railway with the coal district of Mauch Chunk. The scenery of this vicinity is of great variety and beauty. From the hill-tops—for Catawissa is buried between picturesque hills—remarkable pictures of the winding of the river, and its ever-present companion, the canal, are to be seen—now at the base of grand mural precipices, and, anon, though little verdant intervales.

Northumberland. — HOTELS : — *Central Hotel.*

The west branch of the Susquehanna unites here with the main, or north arm, and the village, the pleasantest of all the region round, is built upon the point formed by the confluence of the two waters. The quiet, cultivated air of Northumberland, and its excellent hotel, will be very likely to detain the not over-hurried traveller awhile.

Sunbury is a prosperous town across the river. The Sunbury and Erie R. R. connects here with the route from Philadelphia to Williamsport and Elmira, and with the Philadelphia and Sunbury route.

Williamsport.—HOTELS:—

Williamsport is the principal town upon the west branch of the Susquehanna. It is a pleasant place, delightfully situated, and much in vogue as a summer resort. The west branch of the canal passes here; and here, too, the railway routes from Philadelphia and from Niagara Falls meet. The river landscape between Williamsport and Northumberland presents in its long extent many charming passages.

Liverpool is a lively little town upon the Susquehanna and the Pennsylvania Canal, below Northumberland, and 29 miles above Harrisburg.

The *Juniata* River comes into the Susquehanna 14 miles above Harrisburg. See Juniata in "Pennsylvania R. R. route."

Harrisburg, the capital of Pennsylvania. See "Pennsylvania R. R. route."

Columbia, Pa.—HOTELS :—

The western terminus of the Philadelphia and Columbia R. R. is on the left bank of the Susquehanna, 28 miles below Harrisburg, and 12 west of Lancaster. A part of the town occupies the slope of a hill, which rises gently from the river and the business part of the town lies along the level bank of the river. The scenery from the hills in the vicinity is highly pleasing. The broad river, studded with numerous islands and rocks, crossed by a long and splendid bridge, and bounded on every side by lofty hills, makes a brilliant display. The junction here of the State railroad from Philadelphia with the main line of the canal, the railroad to York, 12 miles long, and the Tide-water Canal to Maryland, renders Columbia a busy place. The main current of travel, which formerly passed through here, has been diverted by the construction of the Harrisburg and Lancaster Railroad; but the emigrant travel still goes by way of Columbia. A fine bridge crosses the Susquehanna, more than a mile in length.

York, Pa., is ten miles south-west of the Susquehanna, upon the Codorus Creek, 28 miles S. S. E. of Harrisburg, 48 miles from Baltimore, and 92 from Philadelphia. With all these cities, and with yet other points, it is connected by railways. The Baltimore and Susquehanna R. R. unites at York with the York and Cumberland, and with the York and Wrightsville Railroads. The Continental Congress met here in 1777, during the occupation of Philadelphia by the British troops.

Port Deposit is in Maryland, on the east bank of the Susquehanna, at the lowest falls, and five miles from its entrance into the Chesapeake Bay. Fifty millions of feet of lumber are annually floated down the great river, and received at Port Deposit. There are extensive quarries of granite in the neighborhood.

Havre de Grace is at the head of the Chesapeake Bay, at the mouth of the Susquehanna, 36 miles northeast of Baltimore. It is upon the line of the railway from Philadelphia to Baltimore. See that route.

Carlisle, Pa., is a beautiful and interesting town, with a population of 6,000, on the line of the Cumberland Valley R. R., 18 miles below Harrisburg, and 125 miles west of Philadelphia. It lies in the limestone valley country, between the Kittatinny and the South Mountains. Dickinson College (Methodist), which is located in Carlisle, is one of the most venerable and esteemed institutions in Pennsylvania. It was founded in 1783. Carlisle is connected by the Cumberland Valley road with Harrisburg, on the one hand, and with Hagerstown, in Maryland, on the other. General Washington's headquarters were here in 1794, at the time of the Whiskey Rebellion. Some years before, Major André was a prisoner of war in Carlisle.

The Delaware Water Gap.—Hotels:—
Kittatinny House.

The bold passage of the Delaware River, called the Water Gap, is easily and speedily accessible from the cities and vicinage of New York and Philadelphia, and a pleasanter excursion for a day or two cannot be well made. The Delaware River rises on the western declivity of the Catskills, in two streams, which meet at the village of Hancock, a station on the New York and Erie R. R. At Port Jervis (Erie R. R.), after journeying 70 miles, it meets the Kittatinny or Shawangunk Mountain, and next breaks through the bold ridge at the Water Gap. At this great pass the cliffs rise perpendicularly, from 1,000 to 1,200 feet, and the river rushes through the grand gorge in magnificent style. It afterwards crosses the South Mountain, not far below Easton (from which point the Gap is generally approached); next falls over the primitive ledge at Trenton, N. J., grows by and by into a large navigable river, skirts the wharves of the city of Philadelphia, and is lost, 100 miles below, in the Delaware Bay. The whole length of this fine river, from the mountains to the bay, is 300 miles.

From New York, take the New Jersey Central road to Easton, Pa., or go from Philadelphia to Easton in the vicinity of the Water Gap, and thence by other railways. From Great Bend on the Erie Railway, take the Delaware, Lackawanna, and Western Road to the Water Gap.

MARYLAND.

The first settlement in Maryland was made by Leonard Calvert, brother of Lord Baltimore, in 1634, at St. Mary's. It was one of the earliest of the colonies to grant entire freedom of religious faith; virtually, though not, as is often written and said, by formal legal enactments.

Maryland was not the theatre of any of the great battles of the Revolution; but some important scenes of the war of 1812 took place within her borders. The territory of the State was at that period twice invaded by the British troops. They were bravely met and repulsed at the battle of North Point, in the Chesapeake (see battle of North Point), September 13, 1814.

The country which now forms the State of Maryland, was granted to Lord Baltimore by Charles I., and was named in honor of Henrietta Maria, Queen of that monarch. Maryland is one of the most northern of the slave-holding States, and the most southern of the group distinguished as the Middle States. It is one of the original thirteen.

The area of the State is 10,210 square miles, a portion of which is covered by the waters of the Chesapeake Bay, which extends within its jurisdiction 120 miles northward. The country upon both the eastern and western shores of the Bay is generally level and sandy. The long narrow strip which extends westward is a lofty region, crossed by several ridges of the Alleghanies. These ranges, with their intervening valleys, afford charming landscape passages to the traveller, on the route of the Baltimore and Ohio Railway, and make that highway one of the most attractive of the many leading from the eastern cities to the great West. The hill-region of Maryland abounds in rich mineral deposits. The coal lands, though not very great in area, are extremely productive. Copper mines are worked in Frederick and Carroll Counties.

Besides the culture of all the grains, fruits, vegetables, and other products of the Northern States, Maryland grows large quantities of tobacco. The State ranks, in the production of this staple, as third in the Union, and, measuring by population, as second.

The Potomac River forms the boundary line between Maryland and Virginia. Along its passage of 350 miles, from the mountains to the Chesapeake Bay, there is much beautiful and varied scenery. The landscape at its confluence with the Shenandoah, near Harper's Ferry, Virginia, has long been famous among the chief picturesque wonders of America. (See Harper's Ferry.) The Falls of the Potomac, about 14 miles above Georgetown, D. C., will repay a visit. The principal cascade is between 30 and 40 feet perpendicular pitch, and the rocky cliffs on the Virginia side of the river have a very imposing air.

The Patapsco River flows 80 miles from the north part of the State to the Chesapeake Bay, which it enters after passing Baltimore, and 12 miles below that city. It is navigable as far as Baltimore for large merchant ships. It is a rapid stream, and is much utilized as a water-power. The Baltimore and Ohio Railway is built along the whole extent of the western branch of the river.

The Susquehanna River enters the north-east corner of the State, not far from its *debouche* into the Chesapeake.

The Elk, Choptank, Chester, Nanticoke, and Pocomoke, smaller rivers, are all more or less navigable.

The Northern Central Railway extends northward from Baltimore 178 miles to Williamsport, Pa., passing through York, Harrisburg, etc. From Williamsport the route is extended 78 miles to E'mira on the N. Y. and Erie R. R., by the Williamsport and Elmira road. Connects at York with the Wrightville, York and Cumberland R. R.; at Bridgeport junction, with the Cumberland Valley road; at Harrisburg, with the Penn. Cent. and Lebanon Valley; at Dauphin, with the Dauphin and Susquehanna; and at Sunbury, with the Sunbury and Erie.

Baltimore from Philadelphia.

By Philadelphia, Wilmington, and Baltimore Railway, 97 miles, via Wilmington (Del.) and Havre de Grace (Md.), or by the Newcastle and Frenchtown route. See chapter on Delaware.

BALTIMORE AND VICINITY.

Baltimore.—HOTELS :—*Barnum's ; Eutaw House ; Gilmor House; Maltby House.*

Baltimore, one of the four great eastern cities, with a population of over 212,000, is imposingly situated upon the Patapsco River, 12 miles from its entrance into the Chesapeake Bay, and about 200 miles, by these waters, from the sea. Built, as it is, upon hill-slopes and terraces, its appearance is, perhaps, more picturesque than that of any other city in the Union. Striking, indeed, is the unlooked-for scene, gazing from the water upward, through the climbing streets, capped at their tops by soaring spire and dome, in whose midst, and above all, soars the proud crest of the famous monument to Washington ; and hardly less attractive is the picture as the eye looks downward from these elevated points upon the busy city and its surrounding lands and waters.

The present site of Baltimore was chosen in 1729, and its name was bestowed upon it, in 1745, in honor of Lord Baltimore. In 1780 it became a port of entry, with the accompaniments of custom-house, naval officers, etc. In 1782 the first pavements were laid on Baltimore street, the chief avenue of the city at that period, as at the present time. In the same year the first regular communication with Philadelphia was established, through a line of stage-coaches. The corporate character of the city began in 1797 only. The population at this date was only a few thousands, increased by the year 1854 to more than 200,000. The next census will undoubtedly show still greater numbers ; and so, each succeeding enumeration—for the natural advantages of the city promise it ever-increasing progress.

The *Washington Monument*, chief among the structures of this kind, from which Baltimore has won the name of the Monumental City, is a very graceful work, standing upon a terrace 100 feet above the water, in Mt. Vernon Place, at the intersection of Charles and Monument streets. Its base is 50 feet square and 20 feet high, supporting a Doric shaft 176½ feet in height, which is still surmounted by a colossal statue of Washington, 16 feet high. The total elevation is thus 312½ feet above the level of the river. It is built with brick, cased with white marble, and cost $200,000. The ascent is made by a winding stairway within.

Battle Monument, erected to the memory of those who fell defending the city in September, 1814, is at the corner of Calvert and Fayette streets. The square sub-base on which the pedestal or column rests, rises 20 feet from the ground, with an Egyptian door on each front, on which are appropriate inscriptions and representations, in basso-relievo, of some of the incidents of the battle. The column rises 18 feet above the base. This, which is of marble, in the form of a Roman fasces, is encircled by bands, on which are inscribed, in sculptured letters, the names of those whose patriotic achievements it serves to commemorate. It is surmounted by a female figure in marble, emblematic of the City of Baltimore. The whole height of the monument is 52 feet.

Armistead Monument, near the City Spring, is merely a tablet, sunken in a subterranean niche. It was erected to the memory of Col. George Armistead, the commander at Fort Henry, in 1814, through whose intrepidity a British fleet of sixteen sail was repulsed, after having bombarded the fort for twenty-four hours. This stone is often spoken of abroad as among the

Railroad
Havre de
Frenchtown

B.

Baltimore
House;
Balti
with
sit
its
20
as
an
an
is
up
the
mi
fame
less
.ong

monumental wonders of Baltimore—to which glory, however, it has no kind of claim. The good people of the city never think of alluding to it.

PUBLIC BUILDINGS. The *Exchange*, in Gay street, is a large and elegant structure, with a façade of 240 feet. The building has colonnades of six Ionic columns on its east and west sides, the shafts of which are single blocks of fine Italian marble of admirable workmanship. The whole is surmounted by an immense dome, the apex of which is 115 feet above the street. The *Custom House* occupies the first story of the south wing of the Exchange, fronting on Lombard street.

In the north-east is the Merchants' Bank, while the Rotunda is used for the *City Post Office*. The *Reading Room* is a fine apartment, 50 feet square. The *Maryland Institute*, on Baltimore street, has a depth of 355 feet. The first story of this immense building is occupied as a market.

Carroll Hall, at the corner of Baltimore and Calvert streets, contains spacious lecture and exhibition rooms. The railroad depot is an extensive and admirable building.

CHURCHES. The most imposing structure of this class is the *Catholic Cathedral*, corner of Cathedral and Mulberry streets. It is built of granite, in the form of a cross, and is 190 feet long, 177 broad, at the arms of the cross, and 127 feet high, from the floor to the top of the cross that surmounts the dome. The building is well lighted by windows in the dome which are concealed from the view of persons below. At the west end rise two tall towers, crowned with Saracenic cupolas, resembling the minarets of a Mohammedan mosque. This church has the largest organ in the United States, having 6,000 pipes and 36 stops. It is ornamented with two excellent paintings—one, "The Descent from the Cross," was presented by Louis XVI.; the other, "St. Louis burying his officers and soldiers slain before Tunis," was presented by Charles X., of France. The *Unitarian Church*, at the intersection of North Charles and Franklin streets, ranks next to the above in architectural beauty. This edifice is 108 feet long and 78 wide. In front is a colonnade, consisting of four Tuscan columns and two pilasters, which form the arcades. Above, extending around the pediment, is a cornice decorated with emblematic figures and inscriptions. From the portico, the entrance is by bronze doors, in imitation of the Vatican at Rome—three conducting to the body of the building, and two to the galleries.

The Catholics, who are a numerous and wealthy part of the community, have various other elegant church edifices, among which may be mentioned that of *St. Alphonsus*, at the corner of Saratoga and Parker streets, which has a spire of 200 feet; and that of *St. Vincent de Paul*, in Front street. *Grace.Church*, Episcopal, corner of Monument and Park streets, is a superb specimen of the Gothic, in red sandstone. Close by is another Episcopal church, also Gothic, built of gray sandstone. *St. Paul's Church* (Episcopal), at the corner of Charles and Saratoga streets, is a pleasing example of the Norman style.

The *Unitarian Church*, Charles and Franklin streets, has a dome of 55 feet in diameter, which is supported by four arches, each of 33 feet span. The *First Presbyterian Church* is situated at the corner of North and Fayette streets. The total number of churches in Baltimore is some 125.

The city is well provided, too, with educational, benevolent, and literary institutions. The *University of Maryland* is at the intersection of Green and Lombard streets; the Medical Department was founded in the year 1807; the *College of Loyola* is at the corner of Madison and Calvert streets. The *Athenæum*, which is at the corner of Saratoga and St. Paul streets, is occupied by the *Mercantile Library Association*, the *Baltimore Library*, and the *Maryland Hist. Society*. It is in the gallery of the Historical Society that the annual exhibitions of pictures are held. The *St. Mary's College*, a Roman Catholic theological institution, is at the corner of Franklin and Green streets. *M'Kim's Free School* was founded by the liberality of the late Isaac M'Kim. The *Maryland Hospital for the Insane* occupies an eminence in the eastern part of the city. *Mount Hope Hospital*, conducted by the Sisters of Charity, is in Madison street. Near the University, in Lombard street, is the *Baltimore Infirmary*. It is controlled by the regents of the University. In the western part of the city is the *Aged Widows' Home*, a new and elegant edifice. There are also two Orphan Asylums, a House of Refuge, and Almshouses.

Theatres. The *Holiday street* is in Holiday near Fayette street; the *Front street* theatre and circus are in Old Town, Front street, near Jay; the *Museum* is at the corner of Baltimore and Calvert streets. Like "Barnum's" in New York, it serves to gratify the juvenile dramatic taste; *Carroll Hall* is at the corner of Baltimore and Calvert streets.

Green Mount Cemetery is a charming rural spot, about a mile and a half from Battle Monument; entrance at the junction of Belvidere street and York Avenue.

Druid Hill Park is a noble pleasaunce of 550 acres, lately converted from private to pub-

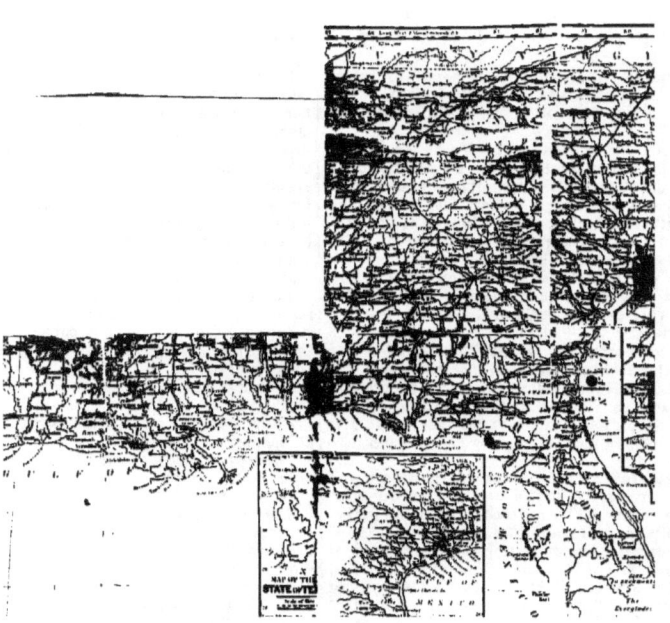

lic use. It abounds in venerable trees and beautiful shrubbery.

North Point, at the mouth of the Patapsco, was the scene of a memorable battle, September 12th, 1814, between the Americans, under General Striker, and the British, under General Ross, in which the former were defeated, and the latter lost three commanders. On the following day, September 13th, Fort McHenry was bombarded for twenty-four hours, by sixteen ships and a land force of 1,200 men. The assailants were repulsed, and the fortress left in the possession of its defenders.

This engagement at North Point and Fort McHenry is duly celebrated in Baltimore on each recurring anniversary, and the Battle Monument was erected in commemoration thereof.

Ellicott's Mills, 15 miles from Baltimore, on the Baltimore and Ohio Railroad, is an exceedingly picturesque little place, in a bold, rocky passage of the Patapsco. (See Baltimore and Ohio Railway.)

Harper's Ferry, Va., and its wonderful scenery, may easily be reached, in a few hours by railway from Baltimore.

The **Thomas Viaduct**, a magnificent granite structure, 360 feet long, 65 feet high, with many arches, is nine miles from the city, on the railway to Washington, where it branches off from the Baltimore and Ohio road.

Annapolis.—HOTELS:—*Mrs. Green and Walton's.*

Annapolis, the capital of Maryland, is a place of the greatest interest, from its antiquity, and its many historical associations. It is reached from Baltimore by the railway from Washington City as far as Annapolis Junction, and thence 21 miles by the Annapolis and Elk Ridge Branch. The city is upon the Severn River, two miles from the Chesapeake Bay. The State House is an interesting edifice. Here is the seat of St. John's College, founded in 1784 by an endowment from the State and by the munificence of individual citizens. At Annapolis, also, is located the United States Naval Academy, established in 1845. Annapolis was founded in 1649. It was first called Providence, next Anne Arundel Town, and lastly, when it received a city charter in 1708, Annapolis, in honor of Queen Anne.

Many important events occurred in Annapolis during the period of the Revolution; and here, at the close of the conflict, occurred the memorable scene of Washington's resignation of his commission. A fine picture of this incident, by Edwin White, has been recently placed in the chamber where it occurred.

Frederick.—HOTELS:—*City Hotel.*

Frederick, one of the largest towns of Maryland, after Baltimore, is reached from that city by the Baltimore and Ohio Railway, 59 miles on the main trunk of that route to Monocacy, and thence three miles by a branch road. Frederick, with some 8,000 inhabitants, is the third city in the State, in population, and in wealth and commercial importance ranks as the second. Some popular Catholic educational establishments are located here.

Hagerstown. — HOTELS: — *Washington House.*

Hagerstown, with a population of about 4,000, is a prosperous place, 26 miles north-west of Frederick, from which it may be easily reached by stage. The Cumberland Valley Railway, at present in operation from Harrisburg, Pa., to Chambersburg, is to be extended to this point; also the Westminster branch of the route from Baltimore to Harrisburg, now terminating at Westminster.

Cumberland.—HOTELS:—*St. Nicholas Hotel.*

Cumberland is on the Potomac River, the Chesapeake and Ohio Canal, and the line of the Baltimore and Ohio Railway, 179 miles from Baltimore City. This is one of the largest and most prosperous towns in Maryland. It is in the mountain region of the narrow strip which forms the western part of the State. For an account of the landscape attractions hereabouts, and of other places and objects of interest in Maryland, see description of the Baltimore and Ohio Railway.

THE CHESAPEAKE BAY.

The Chesapeake is the great highway from Baltimore to the sea. It is the largest bay in the United States, its length being about 200 miles, with a breadth varying from four to forty miles. Its depth permits the passage of the largest ships, nearly to the mouth of the Susquehanna, at the upper extremity. Its shores are profusely indented with arms or estuaries of the oddest shapes, and with the mouths of tributary rivers and creeks.

The Eastern Shore of Maryland and of Virginia.—The waters of the Chesapeake cut off a large portion of Maryland, and lower down a little slice of Virginia on the east, known as the Eastern shore of Maryland and of Virginia. These districts, in the aside position which they thus occupy out of the great current of the national life, invite the traveller by their unique specialities of social habit and character. As railway enterprise, city-lot mania, and other "general orders" of the day, by which the thought and manner of the country is dragooned into universal uniform, and hurried along at forced march, have not yet entered these by-places; there may still be found in them, intact, the feeling, opinion, and life of the "Old Dominion" of a century ago—genuine "first families," with awful pedigrees, hung up in the weather-stained halls of antediluvian homes—manorial homes, with big doors ever open, and surrounded with lordly acres, and attended by retinues of hereditary dependents, which the slave population maintains. Here is yet preserved the old exploded idea, that the present hour, as well as the future, is worth the caring for, and life is considered a thing to be enjoyed, not in anticipation alone, but as it passes, day by day.

Let the care-worn and wearied slip into one of the unnoticed way-steamers of the Great Bay—let him land lazily at ancient Accomac, or thereabouts, and forget a little while the wrinkling perplexities of cabinets and commerce, in the quiet pleasures of simple domestic life within doors, and the genial recreations to which he will be bidden without.

Wild Fowl of the Chesapeake.—These waters, with their tributary streams, are the most famous resort in the United States for every species of aquatic game. Birds of all feathers are drawn hither in marvellous numbers by the abundance of food found on the great flats or shoals along the shores and upon the river inlets.

"Above, around, in numerous flocks are seen,
Long lines of ducks o'er this their fav'rite scene."

"There is," says Dr. Lewis, in his American Sportsman, "no place in our wide extent of country, where wild-fowl shooting is followed with so much ardor as on the Chesapeake Bay and its tributaries, not only by those who make a comfortable living from the business, but also by gentlemen, who resort to these waters from all parts of the adjoining States to participate in the enjoyments of this far-famed ducking ground. All species of wild-fowl come here in numbers beyond credence, and it is really necessary for a stranger to visit the region, if he wishes to form a just idea of the wonderful multitudes and numberless varieties of ducks that darken these waters, and hover in interminable flocks over these famed feeding-grounds. It is not, however, the variety or extraordinary numbers of ducks on the Chesapeake that particularly attracts the steps of so many shooters to these parts, as there are other rivers and streams equally accessible where wild-fowl also abound. But the great magnet that makes the shores the centre of attraction, is the presence of the far-famed CANVASS-BACK, that here alone acquires its peculiar delicacy of flavor, while feeding upon the shores and flats of these waters. It is in quest of these noble ducks that so many repair annually to the waters of the Chesapeake and its numerous tributaries, regardless of the myriads of other ducks that are seen around on every side. The shooter taxes all his energies for the destruction of this one species alone, regarding all others with contempt, as hardly worthy of powder and shot."

"The canvass-backs," says Dr. Sharpless of Philadelphia, in a paper contributed to Audubon's Birds of America, "pass up and down the bay, from river to river, in their morning and evening flights, giving, at certain localities, great opportunities for destruction. They pursue, even in their short passages, very much the order of their migratory movements, flying in a line or baseless triangle; and when the wind blows on the points which may lie in their course, the sportsman has great chance of success. These points or courses of the ducks are materially affected by the winds; for they avoid, if possible, an approach to the shore; but when a strong breeze sets them on to these projections of the land, they are compelled to pass within shot, and often over the land itself.

"In the Susquehanna and Elk rivers there are few of these points for shooting, and there success depends on approaching them while on their feeding-grounds. After leaving the eastern point at the mouth of the Susquehanna and Turkey Point, the western side of the Elk River, which are both moderately good for flying shooting, the first place of much celebrity is the Narrows, between Spesutic Island and the western shore. These Narrows are about three miles in length, and from three to five hundred yards in breadth.

"By the middle of November, the canvass-backs, in particular, begin to feed in this passage, and the entrance and outlet, as well as many intermediate spots, become very successful stations. A few miles down the western shore is Taylor's Island, which is situated at the mouth of the Rumney and Abbey Island at the mouth of Bush River, which are both celebrated for ducks, as well as for swans and geese. These are the most northerly points where large fowl are met with, and projecting out between deep coves, where immense numbers of these birds feed, they possess great advantages. The south point of Bush River, Legoe's Point, and Robbin's and Pickett's Points, near Gunpowder River, are fruitful localities. Immediately at the mouth of this river is situated Carroll's Island, which has long been known as a great shooting ground, and is in the rentage of a company at a high rate. Maxwell's Point, as well as some others up other rivers, and even further down the bay, are good places, but less celebrated than those mentioned. Most of these places are let out as shooting-grounds for companies and individuals, and are esteemed so valuable that intruders are severely treated.

A newspaper correspondent of the past winter, in speaking of the commercial value of the aquatic game of the Chesapeake, says that at Norfolk (which is the great depot of the trade, from whence all the country, far and wide, is furnished), he saw at one house no less than thirty-one barrels, the product of one week's shooting at one spot alone, on Long Island, Back Bay.

Dangers of the Sport.—"Notwithstanding the apparent facilities that are offered of success, the amusement of duck-shooting," says Dr. Sharpless, heretofore quoted, "is probably one of the most exposing to cold and wet; and those who undertake its enjoyment without a courage 'screwed to the sticking-point,' will soon discover that 'to one good a thousand ills oppose.' It is, indeed, no parlor sport; for, after creeping through mud and mire, often for hundreds of yards, to be at last disappointed, and stand exposed on points to the 'pelting rain or more than freezing cold,' for hours, without even the promise of a shot—would try the patience of even Franklin's 'glorious nibbler.' It is, however, replete with excitement and charm. To one who can enter on the pleasure with a system formed for polar cold, and a spirit to endure the weary toil of many a stormy day, it will yield a harvest of health and delight that the roamer of the woods can rarely enjoy."

Voyage down the Chesapeake.—From Baltimore to Norfolk, Virginia, at the lower extremity of the Chesapeake, is a pleasant journey. Good steamers make it daily. It is a charming route, also, to Richmond, turning at or near Norfolk, into the mouth of the James River, and following the many devious miles of those winding and picturesque waters.

The points of chief interest seen in the passage of the Bay, are the embouchure of the Patapsco River and the battle-ground of North Point, near Baltimore, and referred to in our mention of that city. The Bodkin, three miles distant; the harbor of Annapolis, 15 miles still below; and, in the distance, the dome of the venerable capitol in which "Washington, the great and good, set the seal to his sincerity, and finished the edifice of his glory, by voluntarily surrendering his conquering sword to the civil authority of his country." At the lower end of the bay are the famous fortifications of Old Point Comfort and the Rip Raps, protecting the entrance to Hampton Roads and James River. See chapter on Virginia for Norfolk and the James River.

BALTIMORE AND OHIO R. R.

In extent, commercial importance, and pictorial attraction, this great route is one of the most important and interesting in America. It unites the city of Baltimore with the waters and valley of the Ohio, at Wheeling, 397 miles away, making one of the pleasantest and speediest of the great highways from the Atlantic to the Mississippi States. Its whole course is through a region of the highest picturesque variety and beauty, and it is itself a work of the highest artistic achievement in the continual and extraordinary display of skill which the singular difficulties of the way have called forth. It claims, too, especial consideration, and reflects the greatest honor upon the State of Maryland and its beautiful metropolis of Baltimore—as the first railway in America which was built by an incorporated company, and with the assistance of the public purse.

The corner-stone of the road was laid at the very early period in the history of railways of July 4, 1828, and on the 30th of August, 1830, the first section was opened by steam-power, 14 miles, from Baltimore to Ellicott's Mills. The trial of the first engine was made on the 25th of August of that year. On the 1st of June, 1853, the entire route, of nearly 400 miles, was completed, and on the 10th of January, a formal opening of the road was made by a through excursion, with great public fêtes and rejoicings.

The following picturesque description of the journey to the West by this noble highway, is from the pen of William Prescott Smith, Esq., of Baltimore. Its graphic interest will easily excuse its length.

Leaving the city, we cross the **Carrollton Viaduct,** a fine bridge of dressed granite, with an arch of 80 feet span, over Gwynn's Falls; after which, the road soon reaches the long and deep excavation under the Washington Turnpike, which is carried over the railroad by the Jackson Bridge. Less than a mile farther the "deep cut" is encountered, famous for its difficulties in the early history of the road. It is 76 feet in extreme depth, and nearly half a mile in length. Beyond this, the road crosses the deep ravine of Robert's Run, and, skirting the ore banks of the old Baltimore Iron Company, now covered by a dense forest of cedar trees, comes to the long and deep embankment over the valley of Gadsby's run, and the heavy cut through Vinegar Hill immediately following it.

The **Relay House**, eight miles from the inner station, is then reached, where, as the name imports, there was a change of horses during the period in which those animals furnished the motive power of the road.

At this point the open country of sand and clay ends, and the region of rock begins at the entrance to the gorge of the Patapsco River. In entering this defile, you have a fine view of **The Thomas Viaduct** (named after the first President of the Company), a noble granite structure of eight elliptic arches, each of about sixty chord, spanning the stream at a height of sixty-six feet above the bed, and of a total length of some seven hundred feet. This bridge belongs to the Washington Branch Road, which departs from the main line at this place. The pretty village of Elkridge Landing is in sight, and upon the surrounding heights are seen a number of pleasant country seats.

The road now pursues its devious course up the river, passing the Avalon Iron Works, a mile beyond the Relay House, and coming, in a couple of miles farther, to the **Patterson Viaduct**, a fine granite bridge of two arches of fifty-five, and two of twenty feet span. This bridge crosses the river at the Ilchester Mill, situated at a very rugged part of the ravine. The Thistle Cotton Factory appears immediately beyond, and soon after Gray's Cotton Factory, and then the well-known and flourishing town of **Ellicott's Mills**, fourteen miles from Baltimore, covering the bottom and slopes of the steep hills with dwellings, and their tops with churches and other public edifices. The Frederick Turnpike road passes through the town here, and is crossed by the railroad upon the **Oliver Viaduct**, a handsome stone bridge of three arches of twenty feet span. Just beyond this bridge is the Tarpeian Rock, a bold insulated mass of granite, between which and the body of the cliff the railroad edges its way.

The road soon after comes in sight of the Elysville Factory buildings, where it crosses the river upon a new viaduct of three iron spans, each of one hundred and ten feet, and almost immediately recrosses it upon one of three spans, each of one hundred feet in width. From thence it follows the various windings of the stream to the Forks, twenty-five miles from Baltimore. Passing the Marriottsville limestone quarries, the road crosses the Patapsco by an iron bridge fifty feet span, and dashes through a sharp spur of the hill by a tunnel four hundred feet long in mica slate rock. After passing one or two rocky hills at Hood's Mill, it leaves the granite region and enters upon the gentle slopes of the slate hills, among which the river meanders until we reach the foot of **Parr's Ridge**, dividing the waters of the Patapsco from those of the Potomac.

From the summit of the ridge at the Mount Airy Station, forty-four miles from Baltimore, is a noble view westward across the Fredericktown Valley, and as far as the Catoctin Mountain, some fifteen miles distant. The road thence descends the valley of Bush Creek, a stream of moderate curves and gentle slopes, with a few exceptions, where it breaks through some ranges of trap rocks, which interpose themselves among the softer shales. The Monrovia and Ijamsville Stations, are passed at Bush Creek. The slates terminate at the Monocacy River, and the limestone of the Fredericktown Valley commences. That river is crossed by a bridge of three timber spans one hundred and ten feet each, and elevated about forty feet above its bed. At this point, fifty-seven miles from Baltimore, the Frederick Branch, of three miles in length, leaves the Main Road and terminates at the city of that name, the centre of one of the most fertile, populous, and wealthy sections of Maryland.

From the Monocacy to the Point of Rocks, the road having escaped from the narrow winding valleys to which it has thus far been confined, bounds away over the beautiful champaign country lying between that river and the Catoctin Mountain.

The Point of Rocks is formed by the bold profile of the Catoctin Mountain, against the base of which the Potomac River runs on the Maryland side, the mountain towering up on the opposite, Virginia, shore, forming the other barrier of the pass. The railroad turns the promontory by an abrupt curve, and is partly cut out of the rocky precipice on the right, and partly supported on the inner side of the canal on the left by a stone wall of considerable length. Two miles further, another cliff occurs, accompanied by more excavation and walling. From hence the ground becomes comparatively smooth, and the railroad, leaving the immediate margin of the river to the canal, runs along the base of the gently sloping hills, passing the villages of Berlin and Knoxville, and reaching the Weverton Factories, in the pass of the South Mountain.

From this point to **Harper's Ferry**, the road lies along the foot of a precipice for the greater part of the distance of three miles, the last of which is immediately under the lofty cliffs of Elk Mountain, forming the north side of this noted pass. The **Shenandoah River** enters the Potomac immediately below the bridge over the latter, and their united currents rush rapidly over the broad ledges of rock which stretch across their bed. The length of the bridge is about nine hundred feet, and at its western end it divides into two, the left-hand branch connecting with the Winchester and Potomac Railroad, which passes directly up the Shenandoah, and the right hand carrying the Main Road, by a strong curve in that direction, up the Potomac. The bridge consists of six arches of one hundred and thirty, and one arch of about seventy-five feet span, over the river, and an arch of about one hundred feet span over the canal; all of which are of timber and iron, and covered in, except the western arch connected with the Winchester and Potomac Railroad, which is entirely of iron, excepting the floor. This viaduct is not so remarkable for its length as for its peculiar structure, the two ends of it being curved in opposite directions, and bifurcated at the western extremity. Harper's Ferry and all its fine points of scenery are too well known to need description here. The precipitous mountains which rise from the water's edge leave little level ground on the river margin, and all of that is occupied by the United States Armory buildings. Hence the Baltimore and Ohio Railroad has been obliged to build itself a road in the river bed for upwards of half a mile, along the outer boundary of the Government works, upon a trestle-work resting, on the side next the river, upon an insulated wall of masonry, and upon the other side upon strong iron columns placed upon the retaining wall of the Armory grounds. After passing the uppermost building, the road runs along upon the outer bank of the canal which brings the water of the river to the works, and soon crosses this canal by a stone and timber bridge one hundred and fifty feet span. Thence the road passes up the river on the inner side of the canal, and opposite the dam at its head, about one and three-quarters of a mile from the mouth of the Shenandoah, pierces a projecting rock by a tunnel or gallery of eighty feet in length.

The view down the river through this perforation is singularly picturesque, and presents the pass through the mountain at the confluence of the rivers in one of its most remarkable aspects. A short distance above the tunnel, where the river sweeps gradually round to the eastward in the broad smooth sheet of water created by the dam, the railroad leaves the Potomac and passes up the ravine of **Elk Branch**, which presents itself at this point in a favorable direction. This ravine, at first narrow and serpentine, becomes wider and more direct, until it almost loses itself in the rolling table land which characterizes the "Valley of Virginia." The head of Elk Branch is reached in about nine miles, and thence the line descends gradually over an undulating champaign country, to the crossing of the "Opequan" Creek, which it passes by a stone and timber viaduct of one hundred and fifty feet span and forty feet above the water surface. Beyond the crossing the road enters the open valley of Tuscarora Creek, which it crosses twice and pursues to the town of Martinsburg, eighteen miles from Harper's Ferry. At Martinsburg the Tuscarora is again bridged twice, the crossing east of the town being made upon a viaduct of ten spans of forty-four feet each, of timber and iron, supported by two abutments and eighteen stone columns in the Doric style, and which have a very agreeable architectural effect. The Company have erected here large engine-houses and work-shops, and have made it one of their principal stations for the shelter and repair of their machinery.

Westward from Martinsburg the route for eight miles continues its course over the open country, alternately ascending and descending, until it strikes the foot of the North Mountain, and crossing it by a long excavation, sixty-three feet deep, in slate rock, through a depression therein, passes out of the Valley, having traversed its entire breadth upon a line twenty-six miles in length. The soil of the valley is limestone, with slight exceptions, and of great fertility. On leaving these rich and well-tilled lands, we enter a poor and thinly-settled district, covered chiefly with a forest in which stunted pine prevails. The route encounters a heavy excavation and embankment for four or five miles from the North Mountain, and crosses Back Creek upon a stone viaduct of a single arch of eighty feet span and fifty-four feet above the stream. The view across, and of the Potomac Valley, is magnificent as you approach the bridge, and extends as far as the distant mountain range of Sideling Hill, 25 miles to the west. The immediate margin of the river is reached at a point opposite Fort Frederick on the Maryland side, an ancient stronghold, erected a hundred years ago, and still in pretty good preservation.

From this point, thirty miles from Harper's Ferry, the route follows the Virginia shore of the river upon bottom lands, interrupted only by the rocky bluffs opposite Licking Creek, for ten miles to Hancock. The only considerable stream crossed in this distance is Sleepy Creek, which is passed by a viaduct of two spans of one hundred and ten feet each.

The route from Hancock to Cumberland pursues the margin of the Potomac River, with four exceptions. The first occurs at *Doe Gulley*, eighteen miles above Hancock, where, by a tunnel of 1,200 feet in length, a bend of the river is cut off, and a distance of nearly four miles saved. The second is at the Paw Paw Ridge, where a distance of nearly two miles is saved by a tunnel of 250 feet in length. The third and fourth are within six miles of Cumberland, where two bends are cut across by the route with a considerable lessening of distance.

In advancing westward from Hancock, the line passes along the western base of Warm Spring Ridge, approaching within a couple of miles of the **Berkeley Springs**, which are at the eastern foot of that ridge. It then sweeps around the termination of the Cacapon Mountain, opposite the remarkable and insulated eminence called the "Round Top." Thence the road proceeds to the crossing of the Great Cacapon River, nine and a half miles above Hancock, which is crossed by a bridge about 300 feet in length. Within the next mile it passes dam No. 6 of the Chesapeake and Ohio Canal, and soon after, it enters the gap of Sideling Hill.

The next point of interest reached is the **Tunnel at Doe Gulley**. The approaches to this formidable work are very imposing, as for several miles above and below the tunnel, they cause the road to occupy a high level on the slopes of the river hills, and thus afford an extensive view of the grand mountain scenery around.

The Paw Paw Ridge Tunnel is next reached, thirty miles from Hancock, and twenty-five miles below Cumberland. This tunnel is through a soft slate rock, and is curved horizontally with a radius of 750 feet.

The viaduct over Little Cacapon Creek is 143 feet long. About five and a half miles further on, the south branch of the Potomac is crossed on a bridge 400 feet long.

Some two miles above is a fine straight line, over the widely expanded flats opposite the ancient village of Old Town, in Maryland. These are the finest bottom lands on the river, and from the upper end of them is obtained the first view of the **Knobly Mountain**, that remarkable range which lies in a line with the town of Cumberland, and is so singularly diversified by a profile which makes it appear like a succession of artificial mounds. Dan's Mountain towers over it, forming a fine back-ground to the view. Soon after, the route passes the high cliffs known by the name of Kelly's Rocks, where there has been very heavy excavation.

Patterson's Creek, eight miles from Cumberland, is next reached. Immediately below this stream is a lofty mural precipice of limestone and sandstone rock, singularly perforated in some of the ledges by openings which look like Gothic loopholes. The valley of this creek is very straight and bordered by beautiful flats. The viaduct over the stream is 150 feet long. Less than two miles above, and six miles from Cumberland, the north branch of the Potomac is crossed by a viaduct 700 feet long, and rising in a succession of steps—embracing also a crossing of the Chesapeake and Ohio Canal. This extensive bridge carries us out of Virginia, and lands us once more into Old Maryland which we left at Harper's Ferry, and kept out of for a distance of ninety-one miles.

The route thence to Cumberland is across two bends of the river, between which the stream of Evett's Creek is crossed by a viaduct of 100 feet span.

The entrance to the town of **Cumberland** is beautiful, and displays the noble amphitheatre in which it lies to great advantage, the gap of Will's Mountain, westward of the town, being a justly prominent feature of the view.

The brick and stone viaduct over Will's Creek, at Cumberland, is entitled to particular notice. It consists of 14 elliptical arches of 50 feet span and 13 feet rise, and is a well built and handsome structure.

From Cumberland to Piedmont, 28 miles, the scenery is remarkably picturesque, perhaps more so than upon any other section of the road of similar length. For the first 22-miles, to the mouth of New Creek, the Knobly Mountain bounds the valley of the North Branch of the Potomac on the left, and Will's and Dan's Mountains on the right; thence to Piedmont, the river lies in the gap which it has cut through the latter mountain.

The following points may be specially noticed:

The general direction of the road is south-west, for 22 miles, to the mouth of New Creek.

The cliffs, which occur at intervals during the first 10 miles.

The wide bottom lands, extending for the next four miles, with some remarkably bold and beautiful mountain peaks in view.

The high rocky bluffs along Fort Hill, and the grand mural precipice opposite to them, on the Virginia shore, immediately below the "Black Oak Bottom," a celebrated farm embracing 500 acres in a single plain, between mountains of great height.

The Chimney Hole Rock, at the termination of Fort Hill, a singular crag, through the base of which the Railroad Company have driven a tunnel under the road to answer the purpose of a bridge for several streams entering the river at that point.

The crossing of the Potomac, from the Maryland to the Virginia shore, 21 miles from Cumberland, where the railroad, after passing through a long and deep excavation, spans the river by a bridge of timber and iron, on stone abutments and a pier. The view at this point, both up and down the river, is very fine. The bridge is a noble structure, roofed and weather-boarded. It has two spans of 100 feet each, making the total length 320 feet.

The Bull's Head Rock, a mile beyond this point; the railroad, having cut through the *neck*, has left the *head* standing, a bold block of rock breasting the river, which dashes hard against it. Immediately on the other side of the cut made by the railroad through the neck, rises a conical hill of great height. The mouth of New Creek, where there is a beautiful plain of a mile or more in length, and opposite to which is the long promontory of Pine Hill, terminating in Queen's Cliff, on the Maryland side of the river. The profile and pass of Dan's Mountain is seen in bold relief to the north-west, to which direction the road now changes its course. The road skirts the foot of Thunder Hill, and winds along the river margin, bounded by Dan's Mountain and its steep spurs, for seven miles, up to Piedmont. The current of the river is much more rapid here than below, and islands are more frequent.

Piedmont, a flat of limited extent, opposite the small but ancient village of Westernport, at the mouth of George's Creek.

West of Piedmont the road ascends 17 miles by a grade, of which 11 miles is at the rate of 116 feet per mile, to the Altamont Summit. The points worthy of notice in this distance are—

The stone viaduct of three arches, of 56 feet span, over the Potomac River, where the road recrosses into Maryland. It is a substantial and handsome structure, and elevated 50 feet above the water. The road then winds, for five miles, up the valley of Savage River, passing the Everett Tunnell, of 300 feet in length, and 32 miles from Cumberland. This tunnel is secured by a brick arch. The winding of the road up the mountain side, along Savage River, gradually increases its elevation until it attains a height of 200 feet above the water, and placing us far above the tops of the trees growing in the valley, or rather deep ravine, on our right, presents a grand view.

The mouth of Crab-Tree Creek, where the road turns the flank of the Great Back-Bone Mountain—from this point, the view up Savage River to the north and Crab-Tree Creek to the south-west, is magnificent; the latter presenting a vista of several miles up a deep gorge gradually growing narrower; the former a bird's-eye view of a deep, winding trough bounded by mountain ridges of great elevation.

Three miles up Crab-Tree Creek is an excavation 108 feet deep, through a rocky spur of the mountain.

Altamont, the culminating point of the line, at a height of 2,626 feet above tide water at Baltimore—the dividing ridge between the Potomac and Ohio waters—is passed by a long open cut of upwards of 30 feet in depth. The Great Back-Bone Mountain, now passed, towers up on the left hand, and is seen at every opening in that direction.

The Glades, which reach from Altamont to Cranberry Summit, 19 miles, are beautiful natural meadows, lying along the upper waters of the Youghiogheny River, and its numerous tributaries, divided by ridges generally of moderate elevation and gentle slope, with fine ranges of mountains in the back-ground.

The crossing of the great **Youghiogheny River** is by a viaduct of timber and iron—a single arch of 180 feet span resting on stone abutments. The site of this fine structure is wild; the river running here in a woody gorge.

The crossing of the Maryland and Virginia boundary line is 60 miles from Cumberland.

The Falls of Snowy Creek, where the road, after skirting a beautiful glade, enters a savage-looking pass through a deep forest of hemlocks and laurel thickets, the stream dashing over large rocks and washing the side of the road but a few feet below its level.

The descent of 11 miles to Cheat River presents a rapid succession of very heavy excavations and embankments, and two tunnels, viz., the McGuire Tunnel of 500, and the Rodemer Tunnel of 400 feet in length, secured by the most durable arches of stone and brick. There is also a stone and iron viaduct over Salt Lick Creek 50 feet span and 50 feet high. The creek passes through a dense forest of fir trees in its approach to the river.

Cheat River is a dark rapid mountain stream, whose waters are of a curious coffee-colored hue, owing, it is said, to its rising in forests of laurel and black spruce on the highest mountain levels of that country. This stream is crossed by a viaduct consisting of two arches 180 and 130 feet span, of timber and iron on stone abutments and pier.

The ascent of the Cheat River Hill comes next. This is decidedly the most imposing section of the whole line—the difficulties encountered in the four miles west of the crossing of the river being quite appalling. The road, winding up the slope of Laurel Hill and its spurs, with the river on the right hand, first crosses the ravine of Kyer's Run 76 feet deep, by a solid embankment; then, after bold cutting, along a steep, rocky hill-side, it reaches Buckeye Hollow, the depth of which is 108 feet below the road level, and 400 feet across at that level; some more side cutting in rock ensues, and the passage of two or three coves in the hill-side, when we come to Tray Run, and cross it 150 feet above its original bed by an iron trestling 600 feet long at the road level. Both these deep chasms have solid walls of masonry built across them, the foundations of which are on the solid rock, 120 and 180 feet respectively below the road height. They are crossed on elegant cast-iron viaducts.

After passing these two tremendous clefts in the mountain side, the road winds along a precipitous slope with heavy cutting, filling, and walling, to Buckhorn Branch, a wide and deep cove on the western flank of the mountain. This is crossed by a solid embankment and retaining wall 90 feet high at its most elevated point. Some half-mile further, after more heavy cuts and fills, the road at length leaves the declivity of the river, which, where we see it for the last time, lies 500 feet below us, and turns westward through a low gap, which admits it by a moderate cutting, followed soon, however, by a deep and long one through Cassidy's Summit Ridge to the table land of the country bordering Cheat River on the west. Here, at 80 miles from Cumberland, we enter the great western coal field, having passed out of the Cumberland field at 35 miles from that place. The intermediate space, although without coal, will be readily supplied from the adjacent coal basins.

Descending somewhat from Cassidy's Ridge, and passing by a high embankment over the Brushy Fork of Pringle's Run, the line soon reaches the Kingwood Tunnel, of 4,100 feet in length, the longest finished tunnel in America.

Leaving Kingwood Tunnel, the line for 5 miles descends along a steep hill-side to the flats of Raccoon Creek, at Newburgh. In this distance it lies high above the valley, and crosses a branch of it with an embankment 100 feet in elevation. There are two other heavy fills further on. Two miles west of the Kingwood Tunnel is Murray's Tunnel, 250 feet long, a regular and beautiful semi-circular arch cut out of a fine solid sandstone rock, overlaying a vein of coal six feet thick, which is seen on the floor of the tunnel.

From Newburg, westward, the route pursues the valleys of Raccoon and Three Forks Creeks, which present no features of difficulty to the Grafton Station, 101 miles from Cumberland, at the Tygart's Valley River, where the railroad to Parkersburg diverges from that to Wheeling. The distance to these two places (which are 90 miles apart on the Ohio River) is nearly equal, being 104 miles to the former, and 99 to the latter.

Fetterman, a promising-looking village, two miles west of the last point, and 103¼ miles from Cumberland. Here the turnpike to Parkersburg and Marietta crosses the river. The route from Fetterman to Fairmont has but one very striking feature. The Tygart's Valley River, whose margin it follows, is a beautiful and winding stream, of gentle current, except at the Falls, where the river descends, principally by three or four perpendicular pitches, some 70 feet in about a mile. A mile and a half above Fairmont the Tygart's Valley River and the West Fork River unite to form the Monongahela—the first being the larger of the two confluents.

A quarter of a mile below their junction, the railroad crosses the **Monongahela**, upon a viaduct 650 feet long and 39 feet above low water surface. The lofty and massive abutments of this bridge support an iron superstructure of three arches of 200 feet span each, and which formed the *largest iron bridge in America*.

The road, a mile and a half below Fairmont, leaves the valley of the beautiful Monongahela, and ascends the winding and picturesque ravine of Buffalo Creek, a stream some twenty-five miles in length. The creek is first crossed five miles west of Fairmont, and again at two points a short distance apart, and about nine miles further west.

About eleven miles beyond Fairmont we pass the small hamlet of Farmington, and seven or eight miles further is the thriving village of "Mannington," at the mouth of Piles' Fork of Buffalo. There is a beautiful flat here on both sides of the stream, affording room for a town of some size, and surrounded by hills of a most agreeable aspect. Thence to the head of Piles' Fork, the road traverses at first a narrow and serpentine gorge, with five bridges at different points, after which it courses with more gentle curvatures along a wider and moderately winding valley, with meadow land of one or two hundred yards broad on one or other margin. Numerous tributaries open out pretty vistas on either hand. This part of the valley, in its summer dress, is singularly beautiful. After reaching its head at Glover's Gap, 28 miles beyond Fairmont, the road passes the ridge by deep cuts, and a tunnel 350 feet long, of curious shape, forming a sort of Moorish arch in its roof. From this summit, (which divides the waters of the Monongahela from those of the Ohio,) the line descends by Church's Fork of Fish Creek— a valley of the same general features with the one just passed on the eastern side of the ridge.

The road now becomes winding, and in the next four miles we cross the creek on bridges eight times. We also pass Cole's Tunnel, 112 feet, Eaton's Tunnel, 170 feet, and Marten's Tunnel, 180 feet long—the first a low-browed opening, which looks as if it would knock off the smokepipe of the engine; the next a regular arched roof, and the third a tall narrow slit in the rock, originally lined with timbers lofty enough to be taken for a church steeple.

The Littleton Station is reached just beyond, and Board Tree Tunnel is soon at hand.

Leaving Board Tree Tunnel, the line descends along the hill-side of the North Fork of Fish Creek, crossing ravines and spurs by deep fillings and cuttings, and reaching the level of the flats bordering the Creek at Bell's Mill; soon after which it crosses the creek and ascends Hart's Run and Four Mile Run to the Welling Tunnel, 50 miles west of Fairmont, and 28 from Wheeling. This tunnel is 1,250 feet long, and pierces the ridge between Fish Creek and Grave Creek. It is through slate rock like the Board Tree Tunnel, and is substantially arched with brick and stone.

From the Welling Tunnel the line pursues the valley of Grave Creek 17 miles to its mouth at the Flats of Grave Creek on the Ohio River, 11 miles below Wheeling. The first five miles of the ravine of Grave Creek is of gentle curvature and open aspect, like the others already mentioned. Afterwards it becomes very sinuous, and the stream requires to be bridged eight times. There are also several deep cuts through sharp ridges in the bends of the creek, and one tunnel 400 feet long at Sheppard's, 19 miles from Wheeling.

The approach to the bank of the **Ohio River**, at the village of Moundsville, is very beautiful. The line emerging from the defile of Grave Creek, passes straight over the "flats" which border the river, and forming a vast rolling plain, in the middle of which looms up the "great Indian mound," eighty feet high and two hundred feet broad at its base. There is also the separate village of Elizabethtown, half a mile from the river bank, the mound standing between two towns and looking down upon them both. The "flats" embrace an area of some 4,000 acres, about three-fourths of which lie on the Virginia, and the remaining fourth on the Ohio side of the river.

About three miles up the river from Moundsville, the "flats" terminate, and the road passes for a mile along rocky narrows washed by the river, after which it runs over wide, rich, and beautiful bottom lands, all the way to Wheeling.

THE DISTRICT OF COLUMBIA.

THE District of Columbia is a *sui generis* tract, neither State nor Territory, but set apart, *pro bono publico*, as the seat of the Federal Government. It was ceded to the United States for this purpose by Maryland. It occupies an area of sixty square miles. Originally its measure was one hundred square miles, the additional forty coming from Virginia. This part of the cession, however, was retroceded in the year 1846. It includes the city of Alexandria, a few miles below the metropolis, on the Potomac. The present cities of the District are Washington, the National Capital, and Georgetown, close by. Maryland lies upon all sides, except the southwest, where it is separated from Virginia by the Potomac River.

The District of Columbia is governed directly by the Congress of the United States, and its inhabitants have no representation, and no voice in the Federal elections.

Route from Baltimore to Washington City.

The railway from Baltimore to Washington, 40 miles, is over the Baltimore and Ohio route, to the Relay House, nine miles, and thence by the Washington Branch Road, 31 miles. Immediately upon leaving the route of the Baltimore and Ohio Road, the traveller passes over the Thomas Viaduct, a grand structure across the valley of the Patuxent. The Branch route for Annapolis, the capital of Maryland, leaves the Washington road at the Annapolis Junction, 18 miles. Passing Bladensburg, and approaching within a few miles of the city, the grand walls and domes of the Capitol, upon its lofty terrace height, make a magnificent feature in the landscape. The terminus of the road is near the foot of the Capitol Hill.

Washington City.

HOTELS:— *Willard's*, *Brown's*, and the *National;* all spacious first-class establishments.

Washington City, the political capital of the United States, is in the District of Columbia, near the banks of the Potomac River. It is 40 miles distant from Baltimore, 136 from Philadelphia, and 226 from New York, with which cities, as well as with all the chief towns of the Union, it is connected by railway. When the original plan of Washington shall be realized in its full growth to the proportions it was designed to reach—as may yet happen—it will be in its own right, and without the aid of its official position, one of the great cities of the Union. It would be difficult to invent a more magnificent scheme than that of the founder of Washington, or to find a location more eligible for its successful execution. Its easy access from the sea gives it every facility for commercial greatness, and its varied topography almost compels picturesque effect and beauty.

The site was chosen by Washington himself, and it was he who laid the corner-stone of the Capitol. This was on the 18th of September, 1793, seven years before the seat of government was removed thither from Philadelphia.

The scene from the lofty position of the dome of the Capitol, or from the high terrace upon which this magnificent edifice stands, is one of unrivalled beauty, and it gives the visitor at once and thoroughly, a clear idea of the natural advantages of the region, and of the character, extent, and possibilities of the city. Looking eastward, for the space of a mile or more, over a plain yet scarcely occupied, the eye falls upon the broad and beautiful waters of the Potomac, leading by Alexandria and the groves of Mount Vernon, to the sea. Turning westward, it overlooks the city as it at present exists, upon the great highway of Pennsylvania Avenue, to the edifices of the State and Treasury Departments and the President's House, the avenue dropping toward its centre, as a hammock might swing between the two elevated points. Around, on other rising grounds, the various public edifices are seen with fine effect; and, turning again to the left, the view takes in the broad acres of the new national Park, in which are the many unique towers of the Smithsonian Institute, and the soaring shaft of the Washington Monument; off, in the distance, across Rocky Creek, are the ancient-looking walls and roofs of Georgetown.

After a very hasty general peep at the city, the visitor will, of course, turn first to the public edifices, which form its especial attraction.

The Capitol, in its magnitude and in its magnificence of marble and domes, and upon its bold terrace height, will have attracted his curious wonder miles distant, whichever way he may have approached. The corner-stone, as we have said, of this imposing structure was laid by Washington himself, Sept. 18, 1793. In 1814, it was burned by the British, together with the Library of Congress, the President's House, and other public works. In 1818 it was entirely repaired, and in 1851 (July 4), President Fillmore laid the corner-stone of the new buildings, which make the edifice now more than twice its original size. Its whole length is 751 feet, and the area covered is 3½ acres. The surrounding grounds, which are beautifully cultivated and embellished by fountains and statuary, embrace from 25 to 30 acres. The Senate Chamber and the Hall of Representatives of the Congress of the United States, are in the wings of the Capitol, on either side of the central building. The grand rotunda contains eight large pictures, illustrating scenes in American history, painted for the Government by native artists. The edifice is also embellished, both within and without, by many other works of the pencil and of the chisel: chief among them is Greenough's colossal marble statue of Washington, which stands on the broad lawn, before the eastern façade.

The President's Mansion, or the White House, as it is popularly called, is 1¼ miles west of the Capitol, upon a high terrace, at the opposite extremity of Pennsylvania Avenue. The lawns around, containing some twenty acres, drop gradually towards the Potomac River. This elegant but not imposing edifice is built of freestone, painted white. It is two stories high, 170 feet long, and 86 feet deep. On the north point, upon Pennsylvania Avenue, the building has a portico, with four Ionic columns, under which carriages pass. A circular colonnade of six Doric pillars adorns the Potomac front. In the centre of the lawn, across the avenue, on the north, is Clark Mill's bronze equestrian statue of General Jackson, erected in January, 1853. Near the President's Mansion, on the one side, are the very plain buildings of the Navy and the War Departments; and, on the other side, are those of the State and the Treasury Departments.

The Treasury Department is a new and imposing stone structure, 340 feet long and 170 wide. Its total length, when completed, will be 457 feet. The east front, on the bend in the avenue, (made by the intervention of the grounds occupied by the President's Mansion), is embellished by 42 Ionic columns.

The General Post-Office is upon E street, midway between the President's House and the Capitol. It is built of white marble, and its grand dimensions give it an imposing air.

The Patent Office (Department of the Interior) is near the edifice of the General Post Office; when completed it will cover an entire square, and will be one of the largest and most interesting of all the Government structures. Here the visitor may see the models of the countless machines which have grown out of the inventive Yankee brain, and also the cabinets of natural history collected by the exploring expeditions. Here, too, are preserved many most interesting relics of Washington and of Franklin, and the presents of foreign governments.

The Smithsonian Institute is within the area of the New Park, west of the Capitol, and south of Pennsylvania Avenue. This noble institution was endowed by James Smithson, Esq., of England, "for the increase and diffusion of knowledge among men." The edifice is constructed of red sandstone, in the Norman or Romanesque style. Its length is 450 feet; its breadth, 140; and it has nine towers, from 75 to 150 feet high. It contains a lecture-room, capacious enough to hold 2,000 auditors; a museum of natural history, 200 feet in length; a superb laboratory; a library room, large enough for 100,000 volumes; a gallery for pictures and statuary, 120 feet in length.

The *National Monument to Washington* is also within the area of the New Park. The base is to be a circular temple, 250 in diameter, and 100 feet high, upon which there is to be a shaft of 70 feet base, and 500 feet high; the total elevation of the monument being 600 feet. The temple is to contain statues of Revolutionary heroes and relics of Washington. It is to be surrounded by a colonnade of thirty Doric pillars, with suitable entablature and balustrade. Each State contributes a block of native stone or other material, which is to be placed in the interior walls.

The *National Observatory* is located upon the Potomac, and is under the supervision of Lieutenant Maury.

The *Navy Yard*, on the Eastern Branch, about three-fourths of a mile south-east of the Capitol, has an area of 27 acres, enclosed by a substantial brick wall. Within this enclosure, besides houses for the officers, are shops and warehouses, two large ship-houses, and an armory, which, like the rest of the establishment, is kept in the finest order.—The **Navy Magazine** is a large brick structure, situated in the south-eastern section of a plot of 70 acres, the property of the United States, on the Eastern Branch.

The *Congressional or National Cemetery*, is about a mile east of the Capitol, near the Anacostia, or eastern branch. Its situation is high, and commands fine pictures of the surrounding country.

The principal public buildings of the city (not national), are the City Hall in North D street, between Fourth and Fifth streets,; the Columbia College, in the immediate vicinage of the city; the Medical College, and some fifty church edifices.

The residents of Washington number about 61,000; but this estimate is greatly increased during the sittings of Congress, by a very large floating population.

THE VICINITY OF WASHINGTON.

Georgetown.—HOTELS:—

Georgetown is so near as to be almost part and parcel of the Capital. It is at the head of navigation on the Potomac, on high and broken ground. Many elegant mansions, the residences of some of the foreign ministers among them, occupy the "Heights" of the city. Oak Hill Cemetery is a spot of much beauty. An important Catholic College, with both male and female schools, is located here. Population is perhaps nearly 9,000.

The Great Falls of the Potomac a scene of remarkable interest, are 13 miles above Georgetown. The Little Falls are three miles away only. Washington is to be supplied with water by an aqueduct from these Falls.

Alexandria, Va.—HOTELS:—*Newton's Hotel.*

Alexandria, Va., is upon the banks of the Potomac, seven miles below the Capital. It was once within the District of Columbia, but was retroceded to Virginia in 1846, with all the territory of that State which had been before a portion of the national ground. The population of Alexandria is about 11,000.

Bladensburg, a small village in Maryland, on the eastern branch of the Potomac, 6 miles from the Capital, on the Baltimore and Washington Railway, is famous as the Congressional duelling ground.

Mount Vernon, sacred as the home and tomb of Washington, is upon the west bank of the Potomac, 15 miles below the Capital, and eight miles from Alexandria.

The old tomb, which is now fast going to decay, occupies a more picturesque situation than the present one, being upon an elevation in full view of the river. The new tomb into which the remains were removed in 1830, and subsequently placed within a marble sarcophagus, stands in a more retired situation, a short distance from the house. It consists of a plain but solid structure of brick, with an iron gate at its entrance, through the bars of which may be seen two sarcophagi of white marble, side by side, in which slumber in peaceful silence the "Father of his country" and his amiable consort.

This sacred domain remained, since the death of Washington, in the possession of his descendants, until very lately, when it was purchased for the nation for the sum of $200,000, raised by subscriptions, under the auspices of a society of patriotic ladies, who bravely organized themselves as the Ladies' Mount Vernon Association of the Union. Many thousand dollars of the fund came from the lectures and literary-labors of the Hon. Edward Everett.

To reach Mount Vernon from Washington, take the steamer down the Potomac, via Alexandria. Boat leaves tri-weekly, in the morning, returning to the city in the afternoon. Fare, $1.

VIRGINIA.

VIRGINIA,* in the abundance and quality of her political and romantic reminiscences and suggestions, is unquestionably the laureate of our great sisterhood of nations. She was born of the most gallant and creative spirit, and in the most daring and chivalrous days which the world has ever known—the memorable and mighty age of Elizabeth—herself, perhaps, only the hard, ungiving flint, yet majestically striking the light of thought and action from all the dormant genius and power, which came within the range of her influence. Our queenly State grew up a worthy daughter of her great parentage; and in all her history has evinced, as she still perpetuates, its noble spirit. Her whole story is replete with musings for the poet, and philosophy for the historian. What a web of romance may yet be woven from the record of the trials and hair-breadth escapes of her infant life; from the first days of the restless Raleigh, through all the bold exploits of the gallant Smith, the troublesome diplomacy of the wily Opecacanough, the dangerous jealousy of Powhattan, the plots of the treacherous Bacon, to the thrilling drama of the gentle Indian princess. And again, in olden days—in the days of border strife, of bold struggle with the united strategy and cruelty of the French intruder, and the revengeful redskin, she gives us chronicles which, while scarcely yielding in dramatic interest to the incidents of earlier periods, rise higher in the force of moral teachings; while yet again, onward and later, there opens to us the still more thrilling and more lofty story of her mature life, in the proud deeds and grand results of her participation in our eventful Revolution. The be-all and the end-all of that achievement it is not our place now to ask. Much as the world has seen, and much more as it hopes, of mighty consequence, the stupendous effect is not yet felt, nor yet dreamed of, perhaps; but, for what has come, and for what will come, to Virginia, belongs much of the glory—the glory of striking the first blow, by uniting the colonies in resistance to border encroachment; while the last blow, thirty long, struggling years beyond, fell also from her gauntleted hand, when the conquered Cornwallis laid down his shamed sword on the plains of Yorktown. Virginia then led the sounding shout of freedom and empire, which has danced in glad echoes over the Alleghanies, skimmed the vast valleys of the Mississippi and the prairies of the Great West, crossed the snow-clad peaks of the Rocky Mountains, and kissed the far-off floods of the Pacific—a shout which now, more than ever, fills the rejoicing air, and which must grow in grandeur and melody, until it shall exalt and bless the heart of all the earth.

Among the proudest boasts that Virginia may make, is the extraordinary number of great men which she has given to the nation. During half the lifetime of the Republic, its highest office has been conferred upon her sons, who have, in turn, nobly reflected back upon the country the honors they have received. Not only has she been the mother of many and the greatest of our Presidents, but she has reared leaders for our armies and navies, lawgivers for our Senates, judges for our tribunals, apostles for our pulpits, poets for our closets, and painters and sculptors for our highest and most enduring delight. Scanning the map of Middle Virginia, the eye is continually arrested by hallowed shrines—the birth-places, the homes, and the graves of those whom the world has most delighted to honor. Here we pause within the classic groves of Monticello, and look abroad upon the scenes amidst which Jefferson so profoundly studied and taught the world. There, in the little village of Hanover, the burning words of Patrick Henry first awakened the glowing fire of liberty in the bosoms of his countrymen; and here, too, the great Clay was nurtured in that lofty spirit of patriotism, from which sprung his high and devoted public services. Not far off, we may bend again, reverently, over the ashes of Madison and Monroe, of Lee and Wirt, and of a host of others, whom but to mention would be a fatiguing task.

Yet there remains unspoken, though not forgetfully, one other name, the first and greatest—

* The "Romance of American Landscape."

not of Virginia only, not of this wide Republic alone, but of the world itself—a name which may well, and without other laurel, glorify the brow of a nation—the immortal name of WASHINGTON!

It is a pity that Virginia, while no less singularly interesting in her physical than in her moral aspect, has thus far won so little of the attention of our landscapists. Despite the extent and variety of her scenery, from the alluvial plains of the eastern division, through the picturesque hills and dales of the middle region, onward to the summits of the Blue Ridge, with their intervening valleys, and mountain streams, and water-falls—the white cotton umbrella of the artist has scarcely ever been seen to temper its sunshine, except in a few instances of particularly notable interest—as the Natural Bridge and the grand views near Harper's Ferry. The landscape of Virginia is everywhere suggestive; and even in the least varied regions continually rises to the beauty of a fine picture. There are the rich valleys of the James and the Roanoke Rivers, said to resemble, in many of their characteristics, the scenery of the Loire and the Garonne; and, far off among the hills are the rushing and plunging waters of the great Kanawha, and the beetling cliffs of New River.

In the very heart of these natural delights, and superadded to all the political and historical associations at which we have hinted, Virginia attracts us by a wealth of health-giving waters, in the form of mineral springs, in number and nature infinite, where people "go on crutches, looking dismal, and come away on legs, with their faces wreathed in smiles—go with limbs stiffened into pot-hooks-and-hangers, and leave endowed with a good *jointure*—go like shadows, but do not so depart." Magic Waters which, as Peregrine Prolix says, according to popular belief, cure yellow jaundice, white swelling, blue devils, and black plague; scarlet fever, spotted fever, and fever of every kind and color; hydrocephalus, hydrothorax, hydrocele, and hydrophobia; hypochondria and hypocrisy; dyspepsia, diarrhœa, diabetes, and die-of-any-thing; gout, gormandizing, and grogging; liver complaint, colic, and all other diseases and bad habits, *except chewing, smoking, spitting, and swearing.*

For your health or pleasure, dear traveller, we shall conduct you, anon, to these high and mighty shrines of Hygiene, the Virginia Springs.

RAILWAYS IN VIRGINIA.

The links of the Great Northern and Southern route, from Acquia Creek on the Potomac, to Fredericksburg, 15 miles; to Richmond, 60; to Petersburg, 22; to Weldon, N. C., 63.

The Seaboard and Roanoke, in the south-east corner of the State, 80 miles from Portsmouth and Norfolk to Weldon, N. C., on the New York and New Orleans route.

The Richmond and Danville extends 141 miles south-west from Richmond to Danville, on the North Carolina boundary.

The Petersburg and Lynchburg road extends from Lynchburg, 123 miles, to Petersburg, on the Great Northern and Southern line, and thence, 10 miles, to City Point, on James River. It intersects the Richmond and Danville road, about midway, at Burkesville.

The Virginia and Tennessee Railway extends from Lynchburg, 204 miles, to Bristol, thence to Knoxville by the East Tenn. and Virginia R. R.

The Virginia Central Railway extends westward, 95 miles, to Jackson River, through Hanover, Louisa, Gordonsville, Charlottesville, Staunton, Millboro', and other places. Route to the Virginia Springs, Weir's Cave, etc.

The Orange and Alexandria Railway, from Alexandria to Lynchburg, 178 miles. *Stations.—* Alexandria to Springfield, 9 miles; Burke's, 14; Fairfax, 17; Union Mills, 23; *Manassas*, 27 (junction of Manassas Gap Road); Bristoe, 31; Weaversville, 38; Warrenton Junction, 41 (Branch nine miles to Warrenton); Culpepper C. H., 62; Orange C. H., 79; Gordonsville, 88; Lynchburg (via Virginia Central), 170.

Manassas Gap, from Manassas (Orange and Alexandria road) 88 miles to Strasburg.

Winchester and Potomac Railway, 32 miles from Winchester to Harper's Ferry (Baltimore and Ohio Railroad).

Roanoke Valley road, 22 miles from Clarksville to Ridgeway, on the Raleigh and Gaston Railway, N. C.

The North Western Railway (north-west corner of the State) extends from Grafton, on the line of the Baltimore and Ohio Road, to Parkersburg, on the Ohio River.

The Baltimore and Ohio Railway, 397 miles from Baltimore to Wheeling, is partly in Maryland and partly in Virginia. It follows the route of the Potomac River, the dividing line for a long way between the two States. See Maryland for further account of this road.

Richmond.—HOTELS :—The *Exchange* and *Ballard House.*

Richmond, the capital of the "Old Dominion," as Virginia is familiarly called, is in the eastern part of the State, directly on the line of the great railway mail route from New England to New Orleans, through Boston, New York, Philadelphia, Baltimore, Washington, Charleston, and other cities, about 100 miles in a straight line south by west of Washington, from which city it is reached by steamboat down the Potomac River to Acquia Creek, and thence through Fredericksburg by railway, or, more leisurely, from Baltimore and from Philadelphia, every Wednesday and Saturday, and from New York every Saturday, by steamer, outside sea voyage, except from Baltimore, whence the way leads down the Chesapeake Bay, and then (as from New York and Philadelphia) up the James River.

Richmond, as first seen approaching by the river, is a city seated on a hill, says a traveller, and has the imposing aspect of a large and populous capital. It owes this, its first dignity, in some measure to the happy and elevated position of its Capitol, which stands on Shockhoe Hill, and afar off has a handsome and classical appearance; when, however, you approach within criticising distance, it loses some of that enchantment which distance ever lends the view. Though Richmond is not a great capital, it is, nevertheless, a flourishing and interesting city, and now probably contains nearly 33,000 inhabitants, two-thirds of this number being white, and the rest black, free or slave. It has been the scene of some historical events of great dignity and importance. The Capitol stands—we still quote the traveller, whose words we have, with some variation, used in the last sentences—on an elevated plain, near the brow of Shockhoe Hill, and its front looks towards the valley of James River, and over the compact part of Richmond. The view from the portico is extensive, various, and beautiful. It is a Græco-American building, having a portico at one end consisting of a colonnade, entablature, and pediment, whose apicial angle is rather too acute. There are windows on all sides, and doors on the two longer sides, which are reached by high and unsightly double flights of steps placed sidewise, under which are other doors leading to the basement.

Entering by one of the upper doors, an entry leads to a square hall in the centre of the building, surmounted by a dome which transmits light from above. The hall is about forty feet square, and about twenty-five above the floor; has a gallery running round it, in which are nine doors, communicating with various apartments. There are eight niches in the walls, in one of which is a marble bust of La Fayette. Virginia could now, easily and honorably, fill six of the remaining seven. Patrick Henry, Thomas Jefferson, James Madison, James Monroe, John Randolph, and John Marshall, would almost complete the octave.

In the centre of the square hall above described, there is a marble statue of GEORGE WASHINGTON, on which the sculptor's legend reads: "*Fait par Houdon Citoyen Français*, 1788."

The statue is mounted on a rectangular pedestal, four and a half feet high, on one of the larger sides of which is the following honest and affectionate inscription:

"The General Assembly of the Commonwealth of Virginia, have caused this statue to be erected, as a monument of affection and gratitude to

GEORGE WASHINGTON;

who, uniting to the endowments of the Hero the virtues of the Patriot, and exerting both in establishing the Liberties of his country, has rendered his name dear to his Fellow-citizens, and given the world an immortal example of true Glory. Done in the year of

CHRIST

One Thousand Seven Hundred and Eighty-Eight, and in the Year of the Commonwealth the Twelfth."

The simplicity, dignity, and truth of that inscription are worthy of the great original commemorated, and of the young and chivalric State, whose ready gratitude so early erected this lasting monument, and overflowed in language so beautiful and appropriate.

The statue (like the inscription) is an honest Christian statue, and is decently clad in the uniform worn by an American General during the Revolution, and not half covered by the semi-barbarous and pagan toga, with throat uncovered and naked arm, as if prepared for the barber and the bleeder. It is of the size of life, and stands resting on the right foot, having the left somewhat advanced, with the knee bent. The left hand rests on a bundle of fasces, on which hang a military cloak and a small sword, and against which leans a plough. The feet are plunged in whole boots, which are strapped to the knee-buttons of the breeches, which are surmounted by an old-fashioned waistcoat, fortified with deep flaps and most capacious pockets. Military spurs defend the heels, and a capital

VIRGINIA.

pair of Woodstock gloves the hands. The head wears no hat, and has the hair in the fashion of forty years ago, and looks just like his, when he raised his hat in answer to the salutation of some humble fellow-citizen encountered in his morning walk in Chestnut street. The attitude is natural and easy, and the likeness to the great original is strong.

The same generous and patriotic spirit which so early enriched Virginia with this famous work of Houdon, has recently added to the art treasures of the State a fine statue of Clay, and the grand bronze sculptures by the late eminent American artist, Crawford, erected upon the noble terrace-height surrounding the Capitol. This magnificent contribution of Virginia to the art of the country is a colossal equestrian statue of General Washington, elevated upon a grand pedestal or base, embellished with historic scenes in bas-relief, and supported at each angle by statues of other illustrious sons of the State.

Besides the Capitol, the City Hall, the Penitentiary, the Custom-House, are note-worthy edifices. The City Hall is an elegant structure, at an angle of the Capitol Square; the Penitentiary has a façade of 300 feet, near the river, in the west suburb of the city. The estimated cost of the Custom House is nearly half a million of dollars.

Among the churches of Richmond, over 30 in number, is some architectural skill worth observing. The Monumental Church (Episcopal) stands where once stood the Theatre, so disastrously burned in 1811, at the sad sacrifice of the life of the Governor of the State, and more than sixty others of the ill-fated audience in the building at the time.

Richmond College, here, was founded by the Baptists in 1832. *St. Vincent's College* is under the direction of the Catholics. The Medical Department of Hampden and Sydney College, established in 1838, occupies an attractive building of Egyptian architecture.

The Rapids or Falls of James River, which extend six miles above the city, and have a descent of 80 feet, afford valuable water-power. The navigation of the river is opened above the city by the assistance of a canal which overcomes the rapids. The city is supplied with water from the river by means of forcing-pumps, which furnish three reservoirs, of 1,000,000 of gallons each. Richmond is connected by three bridges with Manchester and Spring Hill.

The railway system, of which Richmond has become the centre, is adding daily to its extent and wealth. The Richmond and Petersburg Railroad is its first route southward on the great Northern and Southern transit, and the line from Fredericksburg the first to the northward. The Central Railway, after leaving the city, makes a long excursion to the north, and then turns westward into the interior of the State—the region of the famous Springs. It is completed already 195 miles to Jackson's River, via Gordonsville, Charlottesville, and Staunton, and it is to be continued through the State to Guyandotte on the Ohio River. This is a fine route, from the West or from the East to the Springs, to Weir's Cave, and other wonders of Virginia.

The Richmond and Danville Railway runs south-west 141 miles, to the upper boundary of North Carolina; and connects with the railways of Tennessee, Georgia, and the Carolinas. The city is connected with Lynchburg by railway, and thence by the Virginia and Tennessee line (through all the southern part of the State) with Knoxville, Tennessee. This last route is completed (as we write) 204 miles from Lynchburg south-west to Bristol.

The James River and the Kanawha Canal extend westward between 200 and 300 miles.

Richmond is the great depot for the famous tobacco product of Virginia. It makes also large exports of wheat and flour.

Norfolk.—HOTELS:—*National Hotel.*

Norfolk, Portsmouth, and Gosport Routes.—From Baltimore daily, by steamboat down Chesapeake Bay. From Richmond, by steamer down James River. From Weldon, N. C. (on the great Northern and Southern Railway route), by the Seaboard and Roanoke Railway 80 miles, and from Philadelphia and New York direct by regular lines of steamers on the Atlantic Ocean.

Norfolk is upon the Elizabeth River, eight miles from Hampton Roads and 32 miles from the ocean. It has some 17,000 inhabitants, and is, after Richmond, the most populous city in Virginia. A canal comes in here through the Dismal Swamp. Norfolk is a very pleasant town, irregularly built upon a level plain.

The harbor is large, safe, and easily accessible, defended at its entrance by Forts Monroe and Calhoun. It is a great market for wild fowl, oysters, poultry and vegetables.

The city was laid out in 1705. In 1776 it was burnt by the British. In 1855 it was visited by the yellow fever, which carried off several hundreds of its inhabitants.

Portsmouth, directly opposite Norfolk, is a naval depot of the United States. In the Navy Yard upwards of 1,000 men are employed. The building occupied by the U. S. Naval Hospital is an imposing-looking affair of brick, stuccoed. The Seaboard and Roanoke Railway comes in at Portsmouth from Weldon, N. C. Ferry-boats ply between the town and Norfolk. Gosport lies just below. The United States Dry Dock at this suburb, is a work of great extent and interest.

Petersburg.—Hotels :—*Jarratt's Hotel.*

Petersburg, the third town in Virginia in population (about 16,000), is a port of entry situated on the Appomattox River, distant by railway from Richmond 22 miles; from Fredericksburg, 82 miles; from Washington City, 152 miles. It is on the great route from New York to Charleston and New Orleans. The South Side Railway comes in here from Lynchburg, 133 miles distant. The Appomattox Road connects Petersburg with City Point, 10 miles away, at the entrance of the Appomattox into the James River. The romantic ruins of the old church of Blandford are within the limits of this borough.

Alexandria is upon the Potomac River, seven miles below Washington City by steam ferry. The Orange and Alexandria Railway extends hence 88 miles to Gordonsville, and the Manassus Gap Road 88 miles to Strasburg. This city, until retroceded to Virginia by the general government, was included in the District of Columbia.

Harper's Ferry.—Hotels :—

Harper's Ferry is on the line of the Baltimore and Ohio Railway, 82 miles from Baltimore City. The Winchester and Potomac Railway connects it with Winchester, 32 miles distant. This place, famous for its beautiful scenery, is at the confluence of the Potomac and the Shenandoah Rivers. This meeting of the waters is immediately after their passage through a gap of the Blue Ridge, which was thought by Jefferson to be "one of the most stupendous scenes in nature, and well worth a voyage across the Atlantic to witness." The place was formerly called Shenandoah Falls. A National Armory, employing several hundred hands, is located here. See Baltimore and Ohio Railway.

Winchester.—Hotels :—

Winchester is 114 miles from Baltimore and 32 from Harper's Ferry by the Baltimore and Ohio and the Winchester and Potomac Railways. It is in the midst of a pleasant and picturesque country in the north-eastern part of the State, west of the Blue Ridge and within the limits of the great valley of Virginia.

The Ruins of Jamestown.
This spot, in its history one of the most romantic on the continent, is upon an island near the north bank of the James River, 32 miles above its entrance into the Chesapeake Bay, passed in the voyage from Baltimore to Richmond. The traveller must not neglect the opportunity to recall its varied story of the early colonial adventures and sufferings—the gallantry of Captain John Smith, and the gentle devotion of the Indian Princess Pocahontas. Its revolutionary history, too, may be recalled with interest; its battle-fields of 1781—and many other memorable localities and material remembrances of all the traditionary past—in the ruins of its ancient church and other relics which time has not yet quite obliterated.

Fredericksburg — Hotels : — *Commerce Hotel.*

Fredericksburg is on the line of railways from New York to New Orleans, 70 miles below Washington City and 60 miles above Richmond. It is an agreeable, ancient-looking town, situated in a fertile valley on the banks of the Rappahannock River.

The Birthplace of Washington.
It was in the vicinity of Fredericksburg that Washington was born, and here he passed his early years; and here, too, repose, *beneath an unfinished monument,* the remains of his honored mother.

The birthplace of the Father of his country is about half a mile from the junction of Pope's Creek with the Potomac, in Westmoreland county. It is upon the "Wakefield estate," now in the possession of John E. Wilson, Esq. The house in which the great patriot was born was destroyed before the Revolution. It was a plain Virginia farm-house of the better class, with four rooms, and an enormous chimney, *on the outside,* at each end. The spot where it stood is now marked by a slab of freestone, which was deposited by George W. P. Custis, Esq., in the presence of other gentlemen, in June, 1815. "Desirous," says Mr. Custis, in a letter on the subject to Mr. Lossing, "of making the ceremonial of depositing the stone as imposing as circumstances would permit, we enveloped it in the 'Star-Spangled Banner' of our country, and it was borne to its resting-place in the arms of the descendants of four revolutionary patriots and soldiers. * * We gathered together the bricks of the ancient chimney, which once formed the hearth around which Washington, in his infancy, had played—and constructed a rude kind of pedestal, on which we reverently placed the First Stone, commending it to the respect and protection of the American people in general, and of those of Westmoreland in particular." On the tablet is this simple inscription—"Here, the 11th of February (o. s), 1732, George Washington was Born."

The Tomb of the Mother of Washington. The remains of the mother of Washington repose in the immediate vicinage of Fredericksburg, on the spot which she herself, years before her death, selected for her grave, and to which she was wont to retire for private and devotional thought. It is marked by an unfinished yet still imposing monument. The corner-stone of this sacred structure was laid by Andrew Jackson, President of the United States at the time, on the 7th of May, 1833, in the presence of a grand concourse, and with most solemn ceremonial. After the lapse of almost a quarter of a century the monument remains still unfinished.

The mother of Washington resided, during the latter part of her life, in Fredericksburg, near the spot where she now lies buried. The house of her abode, occupied of late days by Richard Stirling, Esq., is on the corner of Charles and Lewis streets. It was here that her last but memorable interview with her illustrious son took place, when she was bowed down with age and disease.

Hanover Court House, memorable as the scene of Patrick Henry's early triumphs, and as the birthplace of Henry Clay, is near the Pamunky River, and 20 miles above Richmond. The Great Southern Railway passes close by.

The Birthplace of Henry Clay is between three and four miles from Hanover Court House, on the right of the turnpike road to Richmond. The flat, piny region, in which it is situated, is called the Slashes of Hanover; hence the popular *sobriquet* familiarly applied to the great statesman, of the "Millboy of the Slashes." The house is a little, one-story, frame building, with dormer windows, and a large *outside* chimney (after the universal fashion of Southern country-houses) at each gable. In this humble tenement the Immortal Senator was born, in 1777.

Patrick Henry's Early Triumphs at Hanover. Hanover awakens pleasant memories in its stories of the patriotic ardor of Patrick Henry assembling his volunteers and marching to Williamsburg to demand the restoration of the powder which Lord Dunmore had removed from the public magazine, or payment therefor—a daring demand, which he succeeded in enforcing, as the Governor, alarmed at the strength of his cortège, which grew as he went along to 150 in number, sent out the Receiver-General with authority to compromise the matter. The young leader required and obtained the value of the powder, 330 pounds, and sent it to the Treasury, at Williamsburg. This incident happened at Newcastle, once a prosperous village, but now a ruin, with a single house only on its site, a few miles below Hanover Court House, on the Pamunky River.

Williamsburg, the oldest incorporated town in Virginia, and a place of extreme interest in its historical associations, is built upon a plain, between the York and James Rivers, six miles from each. This was the seat of the Colonial Government anterior to the Revolution, and the Capital of the State until 1779. William and Mary College—the oldest educational establishment in the United States, after Harvard University—is located here.

Statue of Lord Botetourt. Of the numerous mementoes of the past which this venerable town contains, the most interesting are to be found in the main street, a broad, pleasantly shaded, and rural-looking avenue. In the centre of the lawn, fronting the edifice of the College, is a mutilated statue of Lord Botetourt, one of the most popular of the old colonial governors. This statue was placed in its present position in 1797.

Palace of Lord Dunmore. The remains of this ancient building, the home of the last of the royal governors of Virginia, is at the head of a pleasant broad court, extending from the Main street in front of the City Hall. It was built of brick. The centre edifice was accidentally destroyed by fire, while occupied by the French soldiers, just after the surrender of Cornwallis at Yorktown. Here the stately old Governor lived, or attempted to live, in royal splendor. All that now remains of his pomp are the two little out-buildings or wings of his palace, yet to be seen by the visitor at Williamsburg.

Brenton Church, a venerable edifice of the early part of the last century, stands on the public square, near Palace street or Court. It is a cruciform building, surmounted by a steeple.

The Old Magazine. On the same area as Brenton Church is an old magazine, an octagonal edifice, built during the administration of Governor Spottswood.

The Old Capitol stood on the site of the present Court House, on the Square, opposite the Magazine. It was destroyed by fire in 1832. A few of the old arches lie yet around half buried in the greensward. It was in the "Old Capitol" that the Burgesses of Virginia were assembled, when Patrick Henry, the youngest member of that body, presented the series of bold resolutions which led to his famous speech—"Cæsar had his Brutus, Charles the First his Cromwell, and George the Third"—concluded by those master-words of raillery, when the excited assembly interrupted him with the cry of "Treason! treason!"—"may profit by their example. If this be treason, make the most of it!"

VIRGINIA. 155

The *Apollo Room of the Raleigh Tavern* is an apartment in another time-honored old building of Williamsburg, in which the House of Burgesses assembled to consider the Revolutionary movements, which were then passing in Massachusetts. This assembly had just been dissolved by the Royal Governor, in consequence of its passage of acts in opposition to those of the Lords and Commons of England just before received. The Queen's Rangers, commanded by Lieutenant-Colonel Simcoe, entered Williamsburg, driving out the Virginia militia, on the stormy night of April 19th, 1781.

The thoughtful traveller will delight himself by recalling other incidents in the history of the localities we have presented to his notice, and in following the course of the great train of events which resulted from or were connected with them.

William and Mary College, at Williamsburg, as we have said already, the oldest establishment of the kind in the Union, after Harvard University, was founded in 1692. Its library numbers between 5,000 and 6,000 volumes, and it has usually in attendance from 150 to 200 students.

The Eastern Lunatic Asylum of Virginia is at Williamsburg.

Yorktown—memorable as the scene of that closing event in the American Revolution, the surrender of the British army, under Lord Cornwallis—is upon the York River, 11 miles from its entrance into Chesapeake Bay, 70 miles east-south-east of Richmond, and about 12 miles from Williamsburg. It was formerly a flourishing town, but is now reduced to the character of a "Deserted Village," with only forty or fifty houses, all told.

All the region of Williamsburg, Yorktown, and the surrounding country of Eastern Virginia, so thickly strewn with memories of thrilling and eventful incidents in the history of the American Revolution, is for the most part a peaceful, level, pastoral land, of piny woods and grassy meadows; yet the village of Yorktown is built upon a high bluff, on either side of which is a deep ravine.

At the time of the famous siege, in 1781, the town contained about sixty houses. In 1814 it was desolated again by fire, and has never since recovered its former activity.

Remains of the Intrenchments, cast up by the British on the south and east sides of the town, are yet to be seen. These mounds vary from 12 to 16 feet in height, and extend, in broken lines, from the river bank to the sloping grounds back of the village.

Cornwallis's Cave is an excavation in the bluff, upon which the village stands, reputed to have been made and used by Lord Cornwallis as a council chamber during the siege. It is exhibited with this character for a small fee. A quarter of a mile below this cave there is another, which there is good reason to believe really was thus occupied by the English commander.

Siege of Yorktown. On the 1st of September, 1781, portions of the British army proceeded up the York River, from the Chesapeake Bay; and, on the 22d, Lord Cornwallis, with his entire force of 7,000 men, arrived, and began his fortifications. He constructed a line of works entirely around the village, and across the peninsula of Gloucester in its rear; besides some field-works some distance off. He was speedily met by the American and French troops, which came in to the number of 12,000; and was at the same time blockaded by the French ships at the mouth of the river. The final result was a general engagement, which resulted, on the 17th of October, in a request from Lord Cornwallis for a cessation of hostilities, and in the total surrender of his army, on the morning of the 19th.

The precise spot, at Yorktown, where the scene of the surrender of the British arms and standards took place, will be pointed out to the inquiring visitor. This great event is the theme of one of Colonel Trumbull's pictures in the Rotunda of the Capitol at Washington.

Lynchburg.—Hotels :—

Lynchburg is a prosperous place on the James River, by railway from Richmond 124 miles, from Petersburg 133 miles. The Virginia and Tennessee Railway, extending hence westward to the railways of Tennessee and other western and south-western routes, is in operation at present 204 miles to Bristol. The James River and Kanawha Canal, following the course of the river from Richmond, passes Lynchburg on its way to Buchanan and Covington.

Lynchburg is on the route to and in the immediate vicinity of the Spring region. The Natural Bridge and the peaks of Otter are here easily accessible.

Lexington.—Hotels :—

Lexington is charmingly situated in the mountain and spring region of Western Virginia, 35 miles north-west of Lynchburg, and 159 from Richmond, by railway to Lynchburg, and thence by stage. Washington College, here, was founded in 1798, and was endowed by General Washington. The Virginia Military Institute was established here, by the State Legislature, in 1838-9.

In July, 1856, a copy in bronze of Houdon's Statue of Washington, in the Capitol at Richmond, was erected here, with gay inaugural ceremonies and fêtes. Lexington was commenced in 1778, and the present population is about 2,000.

Charlottesville.—Hotels :—

Charlottesville, famous as the seat of the University of Virginia, and for its vicinage to Monticello, the home and tomb of Thomas Jefferson, is in the east-central part of the State, 97 miles from Richmond by the Central Railway, and 119 miles from Washington City (via Alexandria) by the Orange and Alexandria and the Virginia Central Railways. The Central Road continues from Charlottesville, via Staunton, into the spring, mountain, and cave region.

The University of Virginia, one of the most distinguished of the colleges of the United States, is situated about a mile west of the village of Charlottesville. It is built (Cyclopædia of American Literature) on moderately elevated ground, and forms a striking feature in a beautiful landscape. On the south-west it is shut in by little mountains, beyond which, a few miles distant, rise the broken, and occasionally steep and rugged, but not elevated ridges, the characteristic feature of which is expressed by the name of Ragged Mountains. To the north-west the Blue Ridge, some 20 miles off, presents its deep-colored outline, stretching to the north-east and looking down upon the mountain-like hills that here and there rise from the plain without its western base. To the east, the eye rests upon the low range of mountains that bounds the view as far as the vision can extend north-eastward and south-westward along its slopes, except where it is interrupted directly to the east by a hilly but fertile plain, through which the Rivanna, with its discolored stream, flows by the base of Monticello. To the south, the view reaches far away until the horizon meets the plain, embracing a region lying between the mountains on either hand, and covered with forests interspersed with spots of cultivated land. The University of Virginia was founded in 1819, by Thomas Jefferson, and so great was his interest in its success, and his estimate of its importance, that in his epitaph, found among his papers, he ranks his share in its foundation third among the achievements and honors of his life—the authorship of the Declaration of Independence being the first, and of the Virginia Statute for Religious Freedom the second. The University is endowed and controlled by the State.

Monticello.—Hotels :—

Monticello, once the beautiful home and now the tomb of Jefferson, is about four miles west of Charlottesville. "This venerated mansion," says Mr. Lossing, in his Feld Book of the Revolution, "is yet standing, though somewhat dilapidated and deprived of its former beauty by neglect. The furniture of its distinguished owner is nearly all gone, except a few pictures and mirrors: otherwise the interior of the house is the same as when Jefferson died. It is upon an eminence, with many aspen trees around it, and commands a view of the Blue Ridge for 150 miles on one side, and on the other one of the most beautiful and extensive landscapes in the world. Wirt, writing of the interior arrangements of the house during Mr. Jefferson's lifetime, records that, in the spacious and lofty hall which opens to the visitor on entering, 'he marks no tawdry and unmeaning ornaments; but before, on the right, on the left, all around, the eye is struck and gratified by objects of science and taste, so classed and arranged as to produce their finest effect. On one side specimens of sculpture, set out in such order as to exhibit at a *coup d'œil* the historic progress of that art, from the first rude attempts of the Aborigines of our country, up to that exquisite and finished bust of the great patriot himself, from the master hand of Carrachi. On the other side, the visitor sees displayed a vast collection of the specimens of the Indian art, their paintings, weapons, ornaments, and manufactures; on another, an array of fossil productions of our country, mineral and animal; the petrified remains of those colossal monsters which once trod our forests, and are no more; and a variegated display of the branching honors of those monarchs of the waste that still people the wilds of the American Continent! In a large saloon were exquisite productions of the painter's art, and from its windows opened a view of the surrounding country such as no painter could imitate. There were, too, medallions and engravings in great profusion.' Monticello was a point of great attraction to the learned of all lands, when travelling in this country, while Mr. Jefferson lived? His writings made him favorably known as a scholar, and his public position made him honored by the nations. The remains of Mr. Jefferson lie in a small family cemetery by the side of the winding road leading to Monticello."

Staunton.—Hotels :—

Staunton is upon a small branch of the Shenandoah, on the Virginia Central Railway, 120 miles west-north-west of Richmond. It has long been a point of rendezvous for tourists to the Spring Region, hard by, though the railway now takes the traveller thither, yet nearer. It is from Staunton that we reach the famous Wier's Cave, 18 miles north-eastward. Staunton is a pretty and prosperous village, with a population of between 2,000 and 3,000. It is the seat of the Western Lunatic Asylum, and of the Virginia Institution for the Deaf, Dumb and Blind.

Wheeling.—HOTELS:—

Wheeling, famous as the Western terminus of the Baltimore and Ohio Railway (from Baltimore City, 397 miles), is on the Ohio River, on both sides of the Wheeling Creek. It is 92 miles below Pittsburg, Pa., and 350 from Cincinnati. The city is built in a glen between bold hills. It is the most important place in Western Virginia in population, trade, and manufactures. Railway lines from the Western States meet the Baltimore and Ohio route at Wheeling.

THE VIRGINIA SPRINGS.

WIER'S CAVE—THE NATURAL BRIDGE—THE PEAKS OF OTTER—AND OTHER SCENES.

Routes to the Springs. *From Baltimore* to Washington, 38 miles—to Alexandria, by steamboat on the Potomac, or by stage, 7 miles—to Gordonsville, by the Orange and Alexandria Railway, 88 miles—thence to the present terminus of the Virginia Central Road, and thence by stage. From Baltimore to Harper's Ferry and Winchester by railway, and thence by railway and stage; a pleasant but not the most expeditious way. Or from Washington by the Potomac River and Fredericksburg Railway 130 miles to Richmond, thence by the Virginia Central Railway, as far as it at present extends, through Charlottesville, Staunton, Goshen, etc. Approaching from the South, travellers should diverge from Richmond.

From Richmond one may also go by railway to Lynchburg, and thence via Lexington, the Natural Bridge, etc.

From the West, passengers should leave the Ohio River and route at Guyandotte, thence by stage.

For still other routes, and to points without the Central Spring region, see each in Index, or further on in this chapter.

The White Sulphur Springs.—Not knowing which of the several routes our traveller may desire to follow, we shall, instead of journeying in any prescribed line from Spring to Spring, jump at once to that central and most famous point, the White Sulphur.

The favorite Spa is on Howard's Creek, in Greenbrier County, directly on the edge of the Great Western Valley, and near the base of the Alleghany range of mountains, which rise at all points in picturesque and winning beauty. Kate's Mountain, which recalls some heroic exploits of an Indian maiden of long ago, is one fine point in the scene, southward; while the Greenbrier Hills lie two miles away, towards the west, and the lofty Alleghany towers up majestically, half a dozen miles off, on the north and east.

The White Sulphur is in the heart of the celebrated group of Western Virginia Springs, with the Hot Spring, 38 miles distant, on the north; the Sweet Spring, 17 miles to the eastward; the Salt and the Red Springs, 24 and 41 miles, respectively, on the south; and the Blue Spring, 22 miles away, on the west.

The vicinage of the White Sulphur is as grateful in natural attractions as the waters are admirable in medicinal value. Its *locale* is a charming valley, environed, like that of Rasselas, by soaring hills, and the summer home in its midst has all the conveniences and luxuries for a veritable Castle of Indolence. Fifty acres, perhaps, are occupied with lawns and walks, and the cabins and cottages of the guests, built in rows around the public apartments, the dining-room, the ball-room, etc., give the place quite a merry, happy village air. There is Alabama Row, Louisiana, Paradise, Baltimore, and Virginia Rows, Georgia, Wolf and Bachelor's Rows, Broadway, the Colonnade, Virginia Lawn, the Spring, and other specialities. The cottages are built of wood, brick, and of logs, one story high; and, altogether, the social arrangement and spirit here, as at all the surrounding Springs, has a pleasant, quiet, home sentiment, very much more desirable than the metropolitan temper of more accessible and more thronged resorts.

It is said that the site of these Springs was once the favorite hunting-ground of the Shawnees, a tradition supported by the remains found in various parts of the valley, in the shape of implements of the chase and ancient graves.

It is not known precisely at what period the Spring was discovered. Though the Indians, undoubtedly, knew its virtues, there is no record of its being used by the whites until 1778, when Mrs. Anderson, wife of one of the early settlers, was borne hither on a litter, from her house fifteen miles off, for the relief of a rheumatic affection. Her recovery, from the employment of the water in bathing and drinking, was noised abroad, and in succeeding years other visitors came, pitching tents near the Spring in the absence of all kind of accommodation. Log-cabins were first erected on the spot in 1784-6, and the place began to assume something of its present aspect about 1820. Since then, it has been yearly improved, until it is capable of pleasantly housing some 1,500 guests.

The Spring bubbles up from the earth in the lowest part of the valley, and is covered by a pavilion, formed of 12 Ionic columns, supporting a dome, crowned by a statue of the *buxom lassie* Hygeia.

The Spring is at an elevation of 2,000 feet above tide-water. Its temperature is 62° Fahrenheit, and is uniform through all seasons. It yields about 30 gallons per minute, and the supply is neither diminished in dry weather, nor increased by the longest rains.

We shall not occupy our little space with the record of the analysis of the water here, or elsewhere, as the visitor may easily inform himself in that respect on the spot. One of its most marked properties, says the waggish Peregrine Prolix, whom we have already quoted, and may perchance again—is a strong infusion of fashion, an animal substance, the quality of which cannot be precisely ascertained, but is supposed to contribute greatly to its efficacy. This esteemed and magic ingredient, when submitted to the ordeal of analysis, is found always to vanish in smoke. There are less erudite, though not more merry doctors about than our most sage friend Prolix.

The Salt Sulphur Springs, three in number, are about twenty-four miles from the White Sulphur, near Union, the capital of Monroe County. Like the White Sulphur, they are beautifully nestled in the lap of mountain ranges. The Springs are near the eastern base of Swope's Mountain. On the north, the Alleghany rises, while Peter's Mountain extends on the south and east.

The Salt Sulphur was discovered in 1805, by Mr. Irwin Benson, while boring for salt water, which he was led to hope for by the fact that the spot had formerly been a favorite "lick" for deer and buffalo. The hotels and cottages at the Salt Sulphur will accommodate some 400 guests. Every reasonable want may be satisfied here, whether it regards the interior creature comforts, or the exterior seekings for the beautiful and curious in physical nature. If one is artistical, he may sketch forever; or if he is geological, or botanical, or conchological, he may fossilize, or herbariumize, or cabinetize, to all eternity.

The Blue Sulphur Spring, in Greenbrier County, is another sweet valley nook, 22 miles west of the White Sulphur, 32 north by east from the Red Sulphur, and 13 from Lewisburg. It is upon the turnpike road to Guyandotte, on the Ohio.

The Blue Sulphur, 25 miles from the White Sulphur, was known long ago, first as a "lick," frequented by vast herds of deer and buffalo from the neighboring forests of Sewell's Mountain. Its geographical position is within the magic hill-circle of the great group of the Western Springs, enjoying all the healthful climates of that most salubrious of regions.

There is, besides the cabins, a large brick hotel here, 180 feet long and three stories high, to which is added, on either side, a wing of stories, and 190 feet façade, with piazzas through the entire length. The fountain is in the centre of a charming lawn, and is crowned, as usual, with a temple-shaped edifice. Here, as in the homes of all the sulphureous sisterhood of this region, the guest will find most hospitable care for all his wants—kind and liberal provision in all things being the common law of the land.

The Red Sulphur Springs, in the southern portion of Monroe County, are 42 miles below the White Sulphur, 17 from the Salt, 32 from the Blue, and 39 from the Sweet. * The approach to the Springs is beautifully romantic and picturesque. Wending his way around a high mountain, the weary traveller is for a moment charmed out of his fatigue by the sudden view of his resting-place, some hundreds of feet immediately beneath him. Continuing the circuitous descent, he at length reaches a ravine, which conducts him, after a few rugged steps, to the entrance of a verdant glen, surrounded on all sides by lofty mountains. The south end of this enchanting vale, which is the widest portion of it, is about two hundred feet in width. Its course is nearly north for about one hundred and fifty yards, when it begins gradually to contract, and changes its direction to the north-west and west, until it terminates in a narrow point. This beautifully secluded Tempe is the chosen site of the village. The north-west portion is occupied by stables, carriage-houses, and shops of various sorts; the southern portion, just at the base of the east and west mountains, is that upon which stand the various edifices for the accommodation of visitors. These buildings are spacious, and conveniently arranged, while the promenades, which are neatly enclosed by a white railing, are beautifully embellished, and shaded from the mid-day sun by indigenes of the forest, the large, umbrageous sugar-maple. The Spring is situated at the south-west point of the valley, and the water is collected into two white marble fountains, over which is thrown a substantial cover.

These Springs have been known and distinguished as a watering-place for near fifty years. The improvements at the place are extensive and well-designed, combining elegance with

* Huntt.

VIRGINIA.

comfort, and are sufficient for the accommodation of 350 persons.

The water of the Spring is clear and cool, its temperature being 54° Fahrenheit.

The Sweet Springs are in the eastern part of Monroe County, 17 miles south-east of the White Sulphur and 22 from the Salt Sulphur. They have been known longer than any other mineral waters in Virginia, having been discovered as early as 1764. So long ago as 1774, these waters were analyzed by Bishop Madison, then the president of William and Mary College.

The Sweet Springs lie in a lovely valley, five miles in length, and between a mile and half a mile broad. The Alleghany Mountain bounds this *Tempe* on the north, and the Sweet Spring Mountain rises on the south. The hotel here is of noble extent, with its grand length of 250 feet, and its dining-hall of 160 feet. The Sweet Springs is one of the gayest places in this wide valley of mineral fountains; and a visit hither is usually the crowning excursion of the Spring season, the jolly breakdown of the ball.

The Red Sweet Springs are a mile only from the Sweet Springs just mentioned, on the way to the White Sulphur. This water is chalybeate, and a powerful agent in cases requiring a tonic treatment. The landscape here is most agreeable. A mile and a half from the Sweet Springs are the admired Beaver-dam Falls.

THERMAL WATERS.

The Warm Springs are in Bath County, about 170 miles, nearly west, on the great Spring Route, from Staunton, or points further west on the Virginia Central Railway by the Hot and White Sulphur to the Ohio River, at Guyandotte. They are situated in a delightful valley, between lofty hill ranges. Fine views are opened all about on the Warm Spring Mountain. From the "Gap," where the road crosses, and from "the Rock," 2,700 feet above tide water, the display is deservedly famous.

Hot Springs. Five miles removed from the Warm Springs (Bath County) at the intersection of two narrow valleys, are the Hot Springs. The scenery here, though very agreeable, as is that of all the region round, is not especially striking. The accommodations for guests, however, are as admirable as elsewhere; and the waters are not less serviceable.

The Bath Alum Springs are at the eastern base of the Warm Spring Mountain, five miles east of the Warm Springs, 47 miles east of the White Sulphur, and 45 west of Staunton. The valley of the Bath Alum is a cosy glen of 1,000 acres, shut in, upon the east, by McClung's Ridge; on the south-east, by Shayer's Mountain; on the west, by the Piney Ridge; and on the south-west by Little Piney.

The Rockbridge Alum Springs are in Rockbridge County, on the main road from Lexington to the Warm Springs, 17 miles from the first and 22 from the second point. The valley in which they are found lies below the North Mountain on the east, and the Mill Mountain on the west.

The Fauquier White Sulphur, in Fauquier County, are 56 miles only from Washington, and about 40 from Fredericksburg. Take the Orange and Alexandria Railway from Alexandria, 41 miles, to Warrenton Junction, thence, nine miles, by Branch to Warrenton, and you are close by. Take the Virginia Central Railway from Richmond, thence deflect at Gordonsville, on the Orange and Alexandria road, to Warrenton Junction; thence, as before, to Warrenton, by Branch road. Distance from Richmond 132 miles.

Jordan's White Sulphur. These Springs are in Frederick County, five miles from Winchester, and one mile and a half from Stephenson Depot, on the Winchester and Potomac Railway. Distance from Harper's Ferry 28½, from Baltimore (Baltimore and Ohio Railway to Harper's Ferry) 116¼ miles.

The Shannondale Springs are in Jefferson County, five miles and a half from Charleston, a point on the railway from Harper's Ferry to Winchester. Distance to Charleston from Harper's Ferry, 10 miles; from Baltimore (Baltimore and Ohio Railway) 92 miles.

The Berkeley Springs, in Bath, Morgan County, are two miles and a half from Sir John's Depot, a point on the Baltimore and Ohio Railway, 130 miles west of Baltimore. This is a very ancient and distinguished resort, esteemed and frequented by Washington before the Revolution. Strother's Hotel is a house to linger at as long as possible. O'Farrell's Hotel is another and a good house here.

The Capon Springs are 23 miles south-west of Winchester, at the base of the North Mountain. Take the Baltimore and Ohio Railway, from Baltimore to Harper's Ferry, 82 miles; thence, the Potomac and Winchester, 32 miles, to Winchester; thence by stage; or take the Orange and Alexandria road, from Alexandria, 27 miles, to Manassus Station and the Manassus Gap Railway, 61 miles more, to Strasburg.

Healing Springs, Bath County. These thermal waters lie in a pleasant valley of eight or ten miles extent, between the Warm Spring Mountain on the east, and the Valley Mountain on the west. In the neighborhood is the fin-

cascade, from which this locality of the southern group of the Healing Springs (here particularly referred to) derives its name of Falling Spring Valley.

Dibrell's Spring is on the main road from Lynchburg to the White Sulphur, 19 miles west, by a direct road from the Natural Bridge, or 28 miles thence, via Buchanan. It is in the extreme north-western part of Botetourt County, 30 miles east of the Alleghanies, and at the base of Gordon Mountain.

Rawley's Springs are in Rockingham County, upon the southern slope of the North Mountain, 12 miles east of Harrisburg, and 120 miles north-east of the White Sulphur.

Grayson's Sulphur are west of the Blue Ridge, in Carroll County, 20 miles south of Wytheville, on the New River—a region of remarkable natural beauty.

The Alleghany Springs are in Montgomery County, on the south fork of the Roanoke River, 10 miles east of Christiansburg, on the Virginia and Tennessee Railway. From Richmond to Christiansburg, 210 miles west; from Lynchburgh, 86 miles.

Pulaski Alum Spring, in Pulaski County, north-west, on Little Water Creek, 10 miles from Newbern, and seven miles, in a direct line, from the Virginia and Tennessee Railway.

New London Alum is in the County of Campbell, 10 miles south-west of Lynchburg. (See Lynchburg, for route thither.)

The Huguenot Springs is a watering-place in Powhattan County, 17 miles from Richmond. Take the Richmond and Danville Railway to the Springs Station, about 10 miles, thence by good omnibuses or stages. A pleasant excursion from Richmond.

There are many other mineral fountains in Virginia discovered and undiscovered. We have mentioned in our catalogue all of much resort; and many more than the visitor can explore in one short summer.

The Natural Bridge is in Rockbridge County, in Western Virginia, 63 miles from the White Sulphur Springs. From Washington, the traveller hither may take the Orange and Alexandria Railway to Gordonsville; on the Virginia Central and the Central to Millboro'; and thence by stage. From Richmond or other points by railway to Lynchburg; and thence by canal packet thirty-five miles to the bridge. In the pleasant book of Virginia Letters, upon which we have already freely drawn, Peregrine Prolix thus records the story of his visit to the Natural Bridge:

"Every body in this vicinity will tell you that the distance from Lexington to the Natural Bridge is 12 miles; but the shortest route is 14 miles, six of which being supposed to be impassable in consequence of the superabundance of rain. The driver of my hack, by name Oliver (a *melanthrope* of great skill in his art), pursued a route three miles longer. Not being aware of the inconceivable badness of the road, and being naturally averse to early rising, I did not leave Lexington until nine o'clock. Oliver soon horrified me by turning into the road we travelled last evening, and informing me we must pursue it for six miles, and then take a cross-road for three miles to get into the direct route. This was bad news; for in a region of bad roads, the cross-roads are the worst, and are as bad as the cross women. And, indeed, until within two miles of the bridge, the road is so pre-eminently abominable, that it has won to itself the title of purgatory, and like that uncomfortable place, when once in, it requires much whipping to get you out.

"Notwithstanding the difficulties of mud and mire, rut and rock, hill and hollow, the skilful Oliver landed me safe at the house near the bridge at two P. M. A melanthropic guide conducted me immediately down a winding rocky path to the bottom of the deep chasm, in which flows the little stream called Cedar Creek, and across the top of which, from brink to brink, there still extends an enormous rocky stratum, that time and gravity have moulded into a graceful arch. The bed of Cedar Creek is more than two hundred feet below the surface of the plain, and the sides of the enormous chasm, at the bottom of which the water flows, are composed of solid rock, maintaining a position almost perpendicular. These adamantine walls do not seem to me to be water-worn, but suggested the idea of an enormous cavern, that in remote ages may have been covered for miles by the continuation of that stratum of which all that now remains is the arch of the Natural Bridge. I do verily believe that this stupendous object is the *ruin of a cave*, one of those antres vast, in which our limestone regions abound, and which perhaps existed previous to the upheaving of our continent, and was tenanted by Naiads, Tritons, and other worthies of the deep.

"The first sensation of the beholder is one of double astonishment; first, at the absolute sublimity of the scene; next, at the total inadequacy of the descriptions he has read, and the pictures he has seen, to produce in his mind the faintest idea of the reality. The great height gives the arch an air of grace and lightness that must be seen to be felt, and the power of speech is for a moment lost in contemplating the immense dimensions of the surrounding objects. The middle of the arch is forty-five feet in perpendicular thickness, which increases to sixty at its juncture with the vast abutments. Its top, which is covered with soil supporting shrubs of various sizes, is two hundred and ten feet high. It is sixty feet wide, and its span is almost ninety feet. Across the top passes a public road, and being in the same plane with the neighboring country, you may cross it in a coach without being aware of the interesting pass. There are several forest trees of large dimensions growing near the edge of the creek directly under the arch, which do not nearly reach its lowest part.

"The most imposing view is from about sixty yards below the bridge, close to the edge of the creek; from that position the arch appears thinner, lighter, and loftier. From the edge of the creek at some distance above the bridge, you look at the thicker side of the arch, which from this point of view approaches somewhat to the Gothic. A little above the bridge, on the western side of the creek, the wall of rock is broken into buttress-like masses, which rise almost perpendicularly to a height of nearly two hundred and fifty feet, terminating in separate pinnacles which overlook the bridge. It requires a strong head (perchance a thick skull) to stand on one of these narrow eminences and look into the yawning gulf below.

"When you are exactly under the arch and cast your glances upwards, the space appears immense; and the symmetry of the ellipsoidal concave formed by the arch and the gigantic walls from which it springs, is wonderfully pleasing. From this position the views in both directions are sublime and striking from the immense height of the rocky walls, stretching away in various curves, covered in some places by the drapery of the forest, green and graceful, and in others without a bramble or a bush, bare and blue.

"Reader, do not allow the coolness of the neighbors, or the heat of the weather, or the badness of the roads, or the goodness of your equipage, or the inertia of your disposition, or the gravity of your baggage, or the levity of your purse, or the nolition of your womankind, or any other creature of any other kind, to prevent you from going to see the Natural Bridge; you never saw its like before and never will you look upon its like again."

The Peaks of Otter. These famous mountain heights are in the same region as the Natural Bridge. They lie in the county of Bedford, 10 miles from the village of Liberty, and 35 miles from Lynchburg—railway from Richmond to Lynchburg, and thence by stage.

The summits of the Peaks of Otter are about two miles apart. The northern mountain rises 4,200 feet above the plain, and 5,307 above the sea. It is the southern or conical peak which is most often ascended.

"After riding about a quarter of a mile," says a visitor to these peaks, "we came to the point beyond which horses cannot be taken, and dismounting our steeds, commenced ascending on foot · the way was very steep, and the day so warm that we had to halt often to take breath. As we approached the summit, the trees were all of a dwarfish growth, and twisted and gnarled by the storms of that high region. There were also a few blackberry bushes, bearing their fruit long after the season had passed below. A few minutes longer brought us to where the trees ceased to grow; but a huge mass of rocks, piled wildly on top of each other, finished the termination of the peak. Our path lay for some distance around the base of it, and under the overhanging battlements, and rather descending for a while, until it led to a part of the pile which could with some effort be scaled. There was no ladder, nor any artificial steps, and the only means of ascent was by climbing over the successive rocks. We soon stood upon the wild platform of one of nature's most magnificent observatories, isolated and apparently above all things else terrestrial, and looking down upon and over a beautiful, variegated, and at the same time grand, wild, wonderful, and almost boundless panorama. Indeed, it was literally boundless, for there was a considerable haze resting upon some parts of the 'world below,' so that, in the distant horizon, the earth and sky seemed insensibly to mingle with each other. I had been there before. I remember, when a boy of little more than ten years old, to have been taken to that spot, and how my unpractised nerves forsook me at the sublimity of the scene.

"On this day it was as new as ever; as wild, wonderful and sublime as if I had never before looked from those isolated rocks, or stood on that awful summit. On one side, towards Eastern Virginia, lay a comparatively level country in the distance, bearing strong resemblance to the ocean; on the other hand were ranges of high mountains, interspersed with cultivated spots, and then terminating in piles of mountains, following in successive ranges, until they were lost also in the haze. Above and below, the Blue Ridge and Alleghanies run off in long lines; sometimes relieved by knolls and peaks, and in one place above us making a graceful curve, and then again running off in a different line of direction.

"Very near us stood the rounded top of the other peak, looking like a sullen sentinel for its neighbor.

"We paused in silence for a time. We were

there almost cut off from the world below, standing where it was fearful even to look down. It was more hazy than at the time of my last visit, but not so much so as to destroy the interest of the scene.

"There was almost a sense of pain at the stillness which seemed to reign. We could hear the flappings of the wings of the hawks and the buzzards, as they seemed to be gathering a new impetus after sailing through one of their circles in the air below us. North of us, and on the other side of the Valley of Virginia, were the mountains near Lexington, just as seen from that beautiful village—the Jump, North, and House mountains succeeding each other. They were familiar with a thousand associations of our childhood, seeming mysteriously, when away from the spot, to bring my early home before me—not in imagination such as had often haunted me when I first left to find another in the world, but in substantial reality. Further on down the valley, and at a great distance, was the top of a large mountain, which was thought to be the Great North Mountain, away down in Shenandoah County. I am afraid to say how far off. Intermediate between these mountains, and extending opposite and far above us, was the Valley of Virginia, with its numerous and highly cultivated farms. Across this valley, and in the distance, lay the remote ranges of the Alleghany, and mountains about, and, I suppose, beyond, the White Sulphur Springs. Nearer us, and separating Eastern and Western Virginia, was the Blue Ridge, more than ever showing the propriety of its cognomen of the 'back bone,' and on which we could distinctly see two zigzag turnpikes, the one leading to Fincastle and the other to Buchanan, and over which latter we had travelled a few days before. With the spyglass we could distinguish the houses in the village of Fincastle, some twenty-five or thirty miles off, and the road leading to the town. Turning towards the direction of our morning's ride, we had beneath us Bedford County, with its smaller mountains, farms and farmhouses, the beautiful village of Liberty, the county roads, and occasionally a mill-pond, reflecting the sun like a sheet of polished silver. The houses on the hill at Lynchburg, twenty-five or thirty miles distant, are distinctly visible on a clear day, and also Willis' mountain, away down in Buckingham County. The tourist may take a carriage from Liberty or at Buchanan, to the Peaks. A fine well-graded turnpike leads thence and a good hotel is at the base of the mountain.

Weir's Cave. This wonderful place, scarcely inferior in its mysterious grandeur to the celebrated Mammoth Cave of Kentucky, is in the north-eastern corner of Augusta County, Va., 17 miles from Staunton (on the Central Railway), 16 miles from Waynesboro', 8 miles from Mount Sidney, 14 miles from Harrisburg, and 32 from Charlottesville and the University of Virginia.—Take the Central Railway from Richmond, or the Orange and Alexandria from Washington City to Gordonsville and the Central Railway onward to Staunton: thence by stage 17 miles to the Cave.

Weir's Cave (sometimes written *Weyer's*), was named after Bernard Weyer, who discovered it in 1804, while in chase of a wild animal who fled thither for escape. Many of the countless apartments in this grand subterranean castle are of exquisite beauty—others again are magnificent in their grand extent. Washington Hall, the largest chamber, is no less than 250 feet in length. A traveller visiting the cave on the occasion of an annual illumination, thus describes this noble apartment:

"There is a fine sheet of rock-work running up the centre of this room, and giving it the aspect of two separate and noble galleries, till you look above, where you observe the partition rises only twenty feet towards the roof, and leaves the fine arch expanding over your head untouched. There is a beautiful connection here standing out in the room, which certainly has the form and drapery of a gigantic statue; it bears the name of the Nation's Hero; and the whole place is filled with these projections—appearances which excite the imagination by suggesting resemblances, and leaving them unfinished. The general effect, too, was perhaps indescribable. The fine perspective of this room, four times the length of an ordinary church; the numerous tapers, when near you, so encumbered by deep shadows as to give only a dim, religious light, and when at a distance, appearing in their various attitudes like twinkling stars on a deep, dark heaven; the amazing vaulted roof spread over you, with its carved and knotted surface, to which the streaming lights below in vain endeavored to convey their radiance; together with the impression that you had made so deep an entrance, and were so entirely cut off from the living world and ordinary things, produces an effect which, perhaps, the mind can receive but once, and will retain for ever."

"Weir's Cave," says the same writer, "is, in my judgment, one of the great natural wonders of this new world, and for its eminence in its own class, deserves to be ranked with the Natural Bridge and Niagara, while it is far less known than either. Its dimensions, by the most direct course, are more than 1,600 feet, and by the more winding paths twice that length; and its objects are remarkable for their variety, formation and beauty. In both respects, it will, I think, compare, without injury to itself, with the celebrated Grotto of Antiparos. For myself, I acknowledge the spectacle to have been most interesting; but to be so, it must be illuminated as on this occasion. I had thought that this circumstance might give the whole a toyish effect; but the influence of 2,000 or 3,000 lights on these immense caverns is only such as to reveal the objects, without disturbing the solemn and sublime obscurity which sleeps on every thing. Scarcely any scenes can awaken so many passions at once, and so deeply. Curiosity, apprehension, terror, surprise, admiration, and delight, by turns and together arrest and possess you. I have had before, from other objects, one simple impression made with greater power; but I never had so many impressions made, and with so much power, before. If the interesting and the awful are the elements of the sublime, here sublimity reigns, as in her own domain, in darkness, silence, and deeps profound."

Madison's Cave is within a few hundred yards of Weir's. It is thus described Mr. Jefferson, in his "Notes on Virginia:"

"It is on the north side of the Blue Ridge, near the intersection of the Rockingham and Augusta line with the south fork of the southern river Shenandoah. It is in a hill of about 200 feet perpendicular height, the ascent of which on one side is so steep, that you may pitch a biscuit from its summit into the river which washes its base. The entrance of the cave is in this side, about two-thirds of the way up. It extends into the earth about 300 feet, branching into subordinate caverns, sometimes ascending a little, but more generally descending, and at length terminates in two different places at basins of water of unknown extent, and which I should judge to be nearly on a level with the water of the river; however, I do not think they are formed by refluent water from that, because they are never turbid; because they do not rise and fall in correspondence with that in times of flood, or of drought, and because the water is always cool. It is probably one of the many reservoirs with which the interior parts of the earth are supposed to abound. The vault of this cave is of solid limestone, from 20 to 40 or 50 feet high, through which water is continually percolating. This, trickling down the sides of the cave, has encrusted them over in the form of elegant drapery; and dripping from the top of the vault generates on that, and on the base below, stalactites of a conical form, some of which have met and formed massive columns."

The Blowing Cave is on the stage road between the Rockbridge and the Bath Alum Springs, one mile west of the village of Milboro. It is in a high ledge near the bank of the Cow Pasture River. The entrance of the Cave is semi-circular and about four feet high, elevated 30 or 40 feet above the road below. When the internal and external atmosphere is the same, there is no perceptible current issuing from it. In intense hot weather, the air comes out with so much force as to prostrate the weeds at the entrance. In intense cold weather, the air draws in. There is a *Flowing and Ebbing Spring* on the same stream with the Blowing Cave, which supplies water-power for a grist-mill, a distillery and a tannery. It flows irregularly. When it commences, the water bursts out in a body, as if let loose from a dam. Mr. Jefferson called this a Syphon Fountain. There are two others of the kind in Virginia—one in Brooks Gap, in Rockingham County, and the other near the mouth of the North Holston.

The Hawk's Nest, sometimes called *Marshall's Pillar*, is on New River, in Fayette County, a few rods only from the road leading from Guyandotte, on the Ohio, to the White Sulphur Springs—96 miles from Guyandotte, and 64 miles from the Springs. It is an immense pillar of rock, with a vertical height of 1,000 feet above the bed of the river. Thus writes a foreign tourist of this impressive picture:

"You leave the road by a little by-path, and after pursuing it for a short distance, the whole scene suddenly breaks upon you. But how shall we describe it? The great charm of the whole is connected with the point of sight, which is the finest imaginable. You come suddenly to a spot which is called the Hawk's Nest. It projects on the scene, and is so small as to give standing only to some half dozen persons. It has on its head an old picturesque pine; and it breaks away at your feet abruptly and in perpendicular lines, to a depth of more than 1,000 feet. On this standing, which, by its elevated and detached character, affects you like the monument, the forest rises above and around you. Beneath and before you is spread a lovely valley. A peaceful river glides down it, reflecting, like a mirror all the lights of heaven, washes the foot of the rocks on which you are standing, and then winds away into another valley at your right. The trees of the wood, in all their variety, stand out on the verdant bottoms, with their heads in the sun, and casting their shadows at your feet, but so diminished as to look more like the pictures of the things than the things themselves. The green hills rise on either hand and all around, and give completeness and beauty to the scene; and beyond these appears the gray outline of the more distant mountains, bestowing grandeur to what was supremely beautiful. It is exquisite. It conveys to you the idea of per-feet solitude. The hand of man, the foot of man, seem never to have touched that valley. To you, though placed in the midst of it, it seems altogether inaccessible. You long to stroll along the margin of those sweet waters, and repose under the shadows of those beautiful trees; but it looks impossible. It is solitude, but of a most soothing, not of an appalling character, where sorrow might learn to forget her griefs, and folly begin to be wise and happy."

The Ice Mountain is a remarkable natural curiosity, in the county of Hampshire. It is upon the North River (eastern bank), 26 miles north-west of Winchester. May be reached from Baltimore by Baltimore and Ohio Railway to Harper's Ferry, by railway thence to Winchester, from Winchester by stage.

The Ice Mountain rises 500 feet above the river. "The west side, for a quarter of a mile," says Mr. Howe, in his history of Virginia, "is covered with a mass of loose stone, of a light color, which reaches down to the bank of the river. By removing the loose stone, fine *crystal ice* can always be found in the warmest days of summer. It has been discovered even as late as the 15th of September; but never in October, although it may exist throughout the entire year, and be found, if the rocks were excavated to a sufficient depth. The body of rocks where the ice is found is subject to the full rays of the sun, from nine o'clock in the morning until sunset. The sun does not have the effect of melting the ice as much as continued rains. At the base of the mountain is a spring of water, colder by many degrees than spring water generally is."

The Salt Pond is a charming lake, on the summit of Salt Pond Mountain, one of the highest peaks of the Alleghany. It is in Giles County, 10 miles east of Parisburg and five miles from the Hygeian Springs. This Pond (we again quote from Mr. Howe) "is about a mile long, and one-third of a mile wide. At its termination it is dammed by a huge pile of rocks, over which it runs; but which once passed through the fissures only. In the spring and summer of 1804, immense quantities of leaves and other rubbish washed in and filled up the fissures, since which it has risen full 25 feet. Previous to that time, it was fed by a fine large spring at its head; that then disappeared, and several small springs now flow into it at its upper end. When first known, it was the resort of vast numbers of elk, buffalo, deer, and other wild animals, for drink." The waters of this Pond have not, despite its name, any saline taste; on the contrary it abounds in fine fresh trout and other fish.

Caudy's Castle (Howe) "was so named from having been the retreat of an early settler, when pursued by the Indians. It is the fragment of a mountain, in the shape of a half cone, with a very narrow base, which rises from the banks of the Capon to the height of about 500 feet, and presents a sublime and majestic appearance. *Caudy's Castle*, as also the *Tea Table*, and the *Hanging Rocks*, mentioned below, may all be visited from the Capon Springs.

"**The Tea Table** is about ten miles from Caudy's Castle, in a deep rugged glen, three or four miles east of the Capon. It is about four feet in height, and the same in diameter. From the top issues a clear stream of water, which flows over the brim on all sides, and forms a fountain of exquisite beauty.

"**The Hanging Rocks** are about four miles north of Romney. There the Wappatomka River has cut its way through the mountain of about 500 feet in height. The boldness of the rocks and the wildness of the scene surprise the beholder.

"A bloody battle," says tradition, "was once fought at the Hanging Rocks, between contending parties of the Catawba and Delaware Indians, and it is believed that several hundred of the latter were slaughtered. Indeed, the signs now to be seen at this place exhibit striking evidence of the fact. There is a row of Indian graves between the rocks and public road, along the margin of the river, of from 60 to 70 yards in length. It is believed that very few of the Delawares escaped."

NORTH CAROLINA.

NORTH CAROLINA.—Much less romantic interest possesses the public mind, though not justly, in regard to this State, than almost any other of the Old Thirteen.

The history of the region does not, to be sure, present many very brilliant points, although attempts to colonize it were made at a very early day—as long ago as 1585-9, and by Sir Walter Raleigh—and though the people were engaged, like their neighbors, in bloody struggles with the Indian tribes. Yet the State did memorable service in the Revolution, and especially in being the first publicly and solemnly to renounce allegiance to the British crown, which she did in the famous

Mecklenburg Declaration of Independence, May 20th, 1775—more than a year before the similar formal assertion of the other States.

In picturesque attraction, the State is popularly considered to be wholly destitute; an impression which results from an erroneous estimate of her topography, which travellers in the course of years have made, from the uninteresting forest travel in the eastern portion, traversed by the great railway thoroughfare from the Northern to the Southern States; the only highway until within very late years, and to this day the only one very much in use.

The Pine, or Eastern part of North Carolina, stretching sixty miles inland, is a vast plain, sandy, and overrun with interminable forests of pine. Yet this wilderness is not without points and impressions of interest to the tourist, more particularly when it is broken, as it often is, by great stretches of dank marsh, sometimes opening into mystical-looking lakes, as on the little Dismal Swamp, lying between Pamlico and Albemarle Sounds, and in the Great Dismal Swamp, which the State shares with Virginia. Then in these woods we may watch the process of the gathering of the sap of the pines, for those famous staple manufactures, " tar, pitch, and turpentine."

The coast, too, of North Carolina is one of the most celebrated on the western borders of the Atlantic—the one most watched and feared by mariners and all voyagers, that upon which the dreaded capes Hatteras and Lookout and Fear are found.

While the innumerable bays and shoals and islands are thus cautiously avoided by the passing mariners, they are as eagerly sought by the fisherman and the sportsman. Immense quantities of shad and herring and other fish are taken here, and the estuaries of the rivers and the bays are among the favorite resorts of wild fowl of every species; making this coast scarcely less attractive to the sportsman, than is the Chesapeake Bay and the shore of Long Island.

The interior of the State is a rude, hilly country, which, though it is not at present, may yet be, softened into the blooming beauty of New England. Beyond to the westward, lies the great mountain district, which, when it comes to be better known, as the railways now approaching it from all sides promise that it soon will be, will place the State in public estimation among the most strikingly picturesque portions of the Union. Two great ridges of the Alleghanies traverse this grand region, some of their peaks rising to the noblest heights, and one of them reaching a greater altitude than any summit east of the Rocky Mountains. Wild brooks innumerable and of the richest beauty, water-falls of wonderful delight, and valleys lovely enough for loveliest dreams, are seen in this yet almost unknown land. We shall lead our traveller thither anon; after a little longer glimpse at the general characteristics of the country; at the facilities for locomotion which are at command, and after a brief visit to places and scenes in the eastern and middle sections of the State.

Mineral products of great variety and value are found in North Carolina, as in the neighboring mountain districts of South Carolina and Georgia. Until the discovery of the auriferous lands of California, this was the most abundant gold tract in the United States. The mines here of this monarch of metals have been profitably worked for many years. At the branch mint at Charlotte, in the mining region, gold was coined, between and including the years 1838 and 1853, to the value of no less than $3,790,033; the highest annual product being $396,734, in the year 1852.

The copper lands of the State, says Professor Jackson, are unparalleled in richness. Coal, too, both bituminous and anthracite, is found here in great abundance, and of the finest quality. Iron ore also exists throughout the mountain districts. Limestone and Freestone may be had in inexhaustible supply. Marl is abundant in all the counties on the coast, and silver, lead, manganese, salt, and gypsum have been discovered.

The rivers of North Carolina have no very marked picturesque character, except the mountain streams in the west, where, besides other charming waters, the shores of the Beach Road for forty miles, are unsurpassed in bold and changeful beauty. The greater number of the rivers run from 200 to 400 miles, in a south-east direction through the State to the Atlantic. A few small streams empty into the Tennessee. The Roanoke and the Chowan extend from Virginia to Albemarle

NORTH CAROLINA.

Sound. The Cape Fear River traverses the State and enters the sea near the southern extremity of the State. Travellers by the old steamer route from Wilmington to Charleston, will remember the passage of this river from the former place, 25 miles to its mouth at Smithville.

"*Quel beau pays!*" exclaimed a visitor from Guadaloupe, as he entered the stream from the sea, and looked out upon its white sandy shores luxuriant with the trailing foliage of the live-oak.

"'Quel beau pays!'" echoed the captain of the incoming barque, in surprise; "do you, just from the grand mountains and valleys of Guadaloupe, call this miserable flat region a beautiful country?"

"For that very reason, mon ami. It is exactly because I have so long seen only mountains and valleys that these beautifully wooded plains, so new to my sight, and in such direct contrast with all I have ever gazed upon before, charm me so much. Mon Dieu, quel beau pays!"

The reader will understand our anecdote according to the teachings of his own experience.

The Neuse and the famous Tar Rivers come from the north to Pamlico Sound. The Yadkin and the Catawba enter South Carolina, and are there called, one the Great Pedee, and the other the Wateree. These and the other rivers of this State are so greatly obstructed at their mouths by sand banks, and above by rapids and falls, that their waters are not navigable for any great distance, or by any other than small craft. Vessels drawing ten or twelve feet of water ascend the Cape Fear River as far as Wilmington, and steamboats yet beyond to Fayetteville. Steamboats sail up the Neuse 120 miles, to Waynesboro', up the Tar 100 miles, to Tarborough, the Roanoke 120 miles, to Halifax, and up the Chowan 75 miles.

RAILWAYS.—The Wilmington and Weldon road, 162 miles long, traverses the entire breadth of the State, in the eastern portion, from Weldon through Halifax, Brattleborough, Rocky Mount, Joyners, Wilson, Nahunta, Goldsborough Mt., Mount Olive, Faison's, Strickland's, Teachey's, Washington, and Bordeaux, to Wilmington. It is a link in the great mail route from the Northern to the Southern cities. Railways also diverge from the above line to Raleigh. The Raleigh and Gaston road from Weldon, 97 miles, and the North Carolina road, from Goldsborough, 48 miles. This road continues on from Raleigh, through Hillsborough, Graham, Greensborough, Lexington, Salisbury, and intermediate stations, to Charlotte, 175 miles beyond Raleigh. At Charlotte it unites with the railway system of South Carolina.

The Raleigh and Gaston railway extends (with connecting links) from Weldon, on the Great Northern and Southern mail route, 97 miles, to Raleigh.

The Roanoke Valley road deflects from the Raleigh and Gaston, and unites with the Virginia routes.

The North Carolina and N. C. Central Railways extend from Goldsborough, on the Great Northern and Southern route, to Raleigh, 48 miles, and thence north-west via Hillsboro', Graham, etc., to Greenborough; thence southwardly to Charlotte, uniting with the South Carolina railways. Distance from Raleigh to Charlotte 175 miles.

Other routes are now in progress, which will traverse all the western parts of the State, and unite the eastern and middle districts, at many points, with the railways of Tennessee and the Great West.

Raleigh.—HOTELS:—

Raleigh, from New York, by the Great Southern line of railway, through Philadelphia, Baltimore, Washington, and Richmond, to Weldon, N. C., thence by the Raleigh and Gaston Railway. Distance from Washington, 286 miles; from Weldon, 97 miles. From Charleston, S. C., by the great mail route, to Goldsborough, N. C., on the Wilmington and Weldon link; thence by the North Carolina Central Railway.

Raleigh, the capital of North Carolina, is situated a little north-east of the centre of the State, near the Neuse River. It is a pleasant little city, on a high and healthful position. Union Square is an open area of ten acres, occupying a centre, on the sides of which are the principal streets. The State House, which is on this square, is one of the most imposing of the Capitols of the United States. It is built of granite, after the model of the Parthenon, with massive columns and a grand dome. The former State House was destroyed by fire in 1831, and with it the celebrated statue of Washington by Canova. The State Lunatic Asylum is here, and the North Carolina Institution for the Deaf and Dumb. Population, 5,000.

Wilmington.—Hotels :—

Wilmington, the largest, and the chief commercial city of North Carolina, is in the southeastern extremity of the State, upon the Cape Fear River, 34 miles from the sea. Reached from New York, Philadelphia, Baltimore, etc., by the Great Southern route, upon which it is a prominent point. Travellers from Charleston and New Orleans formerly took the steamer here for a coast voyage as far as Charleston; now the route is continued by the Manchester and Wilmington railway to Kingsville, on the Columbia Branch of the South Carolina road. A more direct way to Charleston will be opened by the North-eastern railway, to deflect at a convenient point from the Wilmington and Manchester line. This is a busy place, full of manufacturing and commercial life. It offers, however, no very great attractions to the traveller in quest of the picturesque, though it played a part in the drama of the Revolution. Major Craig took possession of the town in January, 1781, and occupied it until the surrender of Cornwallis. Population about 11,000.

Newbern.—Hotels :—

Newbern, a pleasant, old town of about 5,000 inhabitants, is at the confluence of the Neuse and the Trent Rivers, midway on the Atlantic line of the State, 50 miles above Pamlico Sound. It is on the line of the Atlantic and North Carolina Railway, which extends from Goldsborough, on the Wilmington and Weldon link of the main southern route (N. Carolina Road) to Morehead City, opposite Beaufort. Distance from Goldsborough, 59 miles, from Morehead City, 34 miles.

Fayetteville.—Hotels :—

Fayetteville is a thriving place of some 8,000 people. It is at the head of navigation, on Cape Fear River, 60 miles south of Raleigh, and 100 miles above Wilmington. Reached at present on plank roads from Raleigh, and from the Wilmington and Weldon railway.

Charlotte.—Hotels :—

Charlotte is one of the chief towns in the western part of North Carolina. Reached from Raleigh by the North Carolina railway, 175 miles, and from Charleston and Columbia, S. C., by the South Carolina and Columbia Branch, and the Charlotte and South Carolina railways; from Columbia, 109 miles; from Charleston, 237 miles. A plank road, 120 miles long, connects this town with Fayetteville.

Charlotte is in the midst of the gold region of the State, and is the seat of a United States Branch Mint. Some interesting historical memories are awakened at Charlotte. It was here that the patriots of Mecklenburg County assembled in convention, in 1775, and boldly passed a series of resolutions, declaring themselves independent of the British Crown; thus anticipating by a year the immortal Declaration of '76. The British troops occupied the town in 1780, and for a little while it was the head-quarters of the American forces. Here General Greene took command of the Southern army from General Gates, fifty days after the departure of Cornwallis.

Battle of Guilford Court House.—The scene of this interesting event in the history of the American Revolution, is in the County of Guilford, in the north-western part of the State.

THE MOUNTAIN REGION.

No section of the United States is richer in beautiful landscape than is all the western part of North Carolina, traversed by some of the noblest spurs of the Blue Ridge. Turn, here, which way you will, every varying point presents a picture of new and wonderful charm.

Black Mountain, 20 miles north-east of Ashville, rises to the magnificent height of 6,476 feet, and is thus the loftiest peak east of the Rocky Mountain ranges. The scene from its crown is of surpassing grandeur.

The Swannanoa Gap is a magnificent mountain pass, between Ashville and Morgantown. The *Falls of the Catawba* are hard by.

The Hickory-Nut Gap is another grand clove on the giant hills, rich in wonderful pictures of precipices and cascades.

Pilot Mountain, in Burke County, is a bold peak, almost isolated in the midst of a comparatively level region. In the olden time it was the landmark of the Indians in their forest wanderings; hence its present name.

The Hawk's Bill, in Burke County, is a stupendous projecting cliff, looking down 1,500 feet upon the waters of a rushing river.

The Table Rock, a few miles below the Hawk's Bill, rises cone-shaped, 2,500 feet above the valley of the Catawba River.

The Ginger Cake Rock, also in Burke County, is a singular pile, upon the summit of the Ginger Cake mountain. It is a natural stone structure, in the form of an inverted pyramid, 29 feet in height. It is crowned with a slab 32 feet long and two feet thick, which projects half its length beyond the edge of the pyramid upon which it is so strangely poised. Though seeming just ready to fall, nothing could be more secure. A fine view down the dark ravine below is commanded at this point.

The French Broad River, in its wild mountain course of 40 miles, or more, from Ashville to the Tennessee line, abounds in admirable scenes. It is a rapid stream, and in all its course lies deep down in mountain gorges—now foaming over its rocky pathway, and now sleeping, sullen and dark, at the base of huge precipitous cliffs. A fine highway follows its banks, and often trepasses upon its waters, as it is crowded by the jealous overhanging cliffs. Near the Tennessee boundary, and close by the Warm Springs, this road lies in the shadow of the bold mountain precipices known as the Painted Rocks and the Chimneys. The Painted Rocks have a perpendicular elevation of between 200 and 300 feet. Their name comes from the Indian pictures yet to be seen upon them. The chimneys are lofty cliffs, broken at their summits into detached piles of rocks, bearing much the likeness of colossal chimneys, a fancy greatly improved by the fire-place-looking recesses at their base, and which serve as turnouts in the narrow causeway. The picture embracing the angle in the river, beyond the Chimney Rocks, is especially fine.

The Indian name of the French Broad is Tselica. Under this title, Mr. Simms has woven into beautiful verse a charming legend of the river. "The tradition of the Cherokees," he says, "asserts the existence of a siren in the French Broad, who implores the hunter to the stream, and strangles him in her embrace, or so infects him with some mortal disease, that he invariably perishes."

The Warm Springs, across the river from the vicinage of the Painted Rocks, is a very pleasant and popular summer resort. The excellent hotel here occupies a fine plateau, very grateful to the sight, in its contrast with the rugged character of the wild landscape all around.

Route.—To reach the mountain region of North Carolina, from the north, follow the great southern route from Washington via the Orange and Alexandria, the Virginia and Tennessee, and the East Tennessee and Virginia Railways, via Lynchburg. From Charleston, S. C., take the South Carolina railways to Spartanburg, and thence by stage to Ashville; or railway lines through from Charleston, via Columbia, S. C., and Charlotte, N. C., to Salisbury, on the North Carolina Central route, and thence, as before, by stage to Morgantown and Ashville.

SOUTH CAROLINA.

SOUTH CAROLINA is one of the most interesting States in the Union, in its legendary and historic story, in its social characteristics, and in its physical aspect.

Upon its settlement by the English, in 1670, John Locke, the famous philosopher, framed a Constitution for the young Colony, after the pattern of that of Plato's Model Republic. Later (1690) the native poetic humor of the people received a new prompting from the influx of French Huguenots, driven from their own land by the Revocation of the Edict of Nantz. This chivalric spirit was fostered by the wars which they shared with the Georgians, under Oglethorpe, against the Spaniards in Florida, and by the gallant struggles in which they were perpetually involved with the Yemassee and other of their Indian neighbors. Next came the long and painful trial of the Revolution, in which these resolute people were among the first and most ardent to take up arms in the cause of right—the most persistent and self-sacrificing in the prosecution of the contest, under every rebuff, and the last to leave the bloody and devastating fight—a story now told undeniably and gloriously everywhere through her romantic territory, upon the battle-fields, from the mountains to the sea.

The generous temper, from which all this brave history grew, has been ever since nourished and developed by the social circumstances of the people; the kindly and benign influences of a pastoral or agricultural life, cementing, endearing, and perpetuating, through a thousand links, family love, associations, attainments, and possessions. These characteristics have been yet further brought out by the climate, by the physical nature of their home, and by the domestic dependence of one portion of the community, and the ennobling effect of the consciousness of power and the obligations it imposes upon the other.

The physique of the Palmetto State is exceedingly varied. Here, on the sea-board and the

SOUTH CAROLINA.

south, broad savannas and deep, dank lagunes, covered with teeming fields of rice, and fruithful in a thousand changes of tropical vegetation; in the middle districts great undulating meadows, overspread with the luxuriant maize, or white with snowy carpetings of cotton; and, again, to the northward, bold mountain ranges, lovely valleys, and matchless waterfalls.

> "The sunny land, the sunny land, where Nature has displayed
> Her fairest works, with lavish land, in hill, and vale, and glade;
> Her streams flow on in melody, through fair and fruitful plains;
> And, from the mountains to the sea, with beauty plenty reigns!"

Railways. The South Carolina Railway traverses the lower portion of the State, 137 miles from Charleston to Augusta, Georgia. There are many villages but no important towns on this route, excepting Aikin, a semi-watering place, 17 miles from Augusta.

The North Eastern Railway will extend north from Charleston to Florence, where it will tap the great highway, from Boston to New Orleans, which now leaves Charleston to the eastward.

The Cheraw and Darlington extends 40 miles to Cheraw from Florence terminus of the North-eastern road from Charleston, on the Wilmington and Manchester.

The Columbia Branch extends 66 miles from Branchville midway, on the South-eastern road to Columbia, the Capital of the State.

The North Eastern Railway, from Charleston 102 miles to Florence link of great route from Charleston to New York.

The Charleston and Savannah Railway is completed from Charleston to Coosawatchie, 61 miles, and will soon be opened through.

The Wilmington and Manchester extends 172 miles from Kingsville, Columbia Branch of South Carolina road. STATIONS—Kingsville to Wateree, Junction, 9 miles (Camden Branch Road diverges here); Manchester, 15; Sumterville, 25; Maysville, 34; Lynchburg, 43; Timmonsville, 52; Florence, 64 (North Eastern Road for Charleston and the Cheraw and Darlington, for Cheraw, diverge here); Mar's Bluff's, 70; Pee Dee, 76; Marion, 85; Mullen's, 92; Nichol's, 99; Fine Bluff, 108; Grist's, 118; Whitesville, 127; Flemington, 137; Maxwell's, 144; Brinkley's, 154; Wilmington, 171 miles.

The Camden Branch extends 37 miles from Kingsville—Columbia Branch of the South Carolina Road. STATIONS—Kingsville to Clarkson's, 4 miles; Manchester Junction, 9; Middleton, 11; Claremont, 18; Hopkins, 28; Camden, 37 miles.

The Charlotte and South Carolina Railway extends northward, through the mountain region, 105 miles, from Columbia to Charlotte, North Carolina. The principal places passed are Winnsboro' and Chester. At Chester a railway diverges for Yorkville.

The Greenville and Columbia Railway extends northwest, via Newberry C. H., 143 miles from Columbia to Greenville, with branches and connecting ines to Spartanburg. Laurensville, Abbeville, and Anderson.

The Spartanburg and Union Railway deflects at Alson, from the Greenville and Columbia Railway, 55 miles north-west of Columbia. When finished to Spartanburg it will be 67 miles long.

The Laurens Road extends 32 miles from Newberry C. H. (Greenville and Columbia Railway) to Laurensville.

The Abbeville Branch of the Greenville and Columbia Road deflects at Cokesbury, 19 miles to Abbeville.

The Anderson Branch (Greenville and Columbia) deflects at Belton, 10 miles to Anderson. From this point and from Spartanburg other roads are in progress to connect with the railway routes of North Carolina and Tennessee.

Charleston from New York.

From New York daily, by railway, to Philadelphia, Baltimore, Washington City, Fredericksburg, and Richmond, Va., Welden and Wilmington, N. C.; thence by Wilmington and Manchester Railway to Kingsville, on the Columbia Branch of the South Carolina Road; or more directly by the newer route—via North Eastern Railway, which deflects from the Wilmington and Manchester road at Florence.

The pleasantest mode of travel, however, from New York to Charleston or Savannah, is by the fine line of steamships, which make the voyage in some 60 hours twice a week, leaving New York (pier No. 4) every Wednesday and Saturday. The fare is about the same whether by sea or land; meals in the latter case adding so much to the cost.

From New Orleans to Charleston. Steamers daily to Mobile and to Montgomery, Alabama; thence, by railway, to Atlanta; thence, by Georgia Road, to Augusta; thence, by South Carolina Road, to Charleston.

From Savannah to Charleston. Steamers every Monday, Thursday, and Friday.

8

Charleston.—HOTELS:—The hotels are numerous and among the most stately edifices in the city. They are usually kept in a style which will rank with any in the country. Among the most conspicuous of these are the "Charleston Hotel," the "Mills House," the "Calder House," the "Pavilion," and the Planters' Hotel." The charges at these houses range from $1 50 to $2 50 *per diem.* The "Charleston Hotel," the "Mills House," and "Pavilion Hotel," are particularly good specimens of Charleston architecture.

Charleston, the metropolis of South Carolina, is picturesquely situated at the confluence of the Ashley and Cooper Rivers, which combine to form its harbor. This harbor is deep and spacious, drawing 17 feet of water. The *coup d'œil* is noble, broad, imposing, and highly picturesque. Though the grounds are low, hardly more than 12 feet above high water, the effect is good; and the city, like Venice, seems, at a little distance, to be absolutely rising out of the sea. The bay is almost completely landlocked, making the harborage and roadstead as secure as they are ample. The adjuncts contribute to form a *tout ensemble* of much beauty. Directly at the entrance of the city stands Castle Pinckney, a fortress which covers an ancient shoal. On the sea-line rises Fort Moultrie, famous as Fort Sullivan, in beating off, and nearly destroying, the British fleet, under Sir Peter Parker, in 1776. On the eastern extremity of the same island (Sullivan's) on which Fort Moultrie stands, you may trace the outline of the fortress, which, under Colonel Thompson, with 700 Carolina rifles, defeated Sir Henry Clinton at the very moment when Moultrie drove Sir Peter Parker away from the South. Within the harbor you are arrested by the imposing battlements of Fortress Sumter, which covers the channel with a formidable array of cannon. This fort, with that of Moultrie, constitute the chief defences of the place upon the sea. On James Island you are shown the ruins of old Fort Johnson. On the opposite headlands of the Haddrill you may trace the old lines which helped in the defence of the city, eighty years ago, but which are now mostly covered by the smart village of Mount Pleasant. These points, north, east, and south, with the city lying west of them, bound the harbor, leaving an ample circuit of bay—coursing over which, from south to north, the eye gladly pursues the long stretch of Cooper River, the Etiwando of the Red man, along the banks of which, for many miles, the sight is refreshed by noble rice-fields, and in many places by fine old structures of the ancient and present gentry.

Steamers ply up this river, and return the same day; affording a good bird's-eye view of the settlements, along a very picturesque shore line, on either hand. It was up this river that Mr. Webster distinguished himself by shooting an alligator, or rather shooting *at* him—the alligator diving at the shot, and leaving the matter sufficiently doubtful to enable an old lawyer and politician to make a plausible case of it.

Standing on James Island, or on the battlements of Fort Sumter, the eye notes the broad stream of the Ashley, winding from west of the city, round its southernmost point, to mingle in with the waters of the Cooper. The Ashley was anciently a region of great wealth and magnificence. It is still a river of very imposing aspects—broad, capacious, with banks of green, through which you may still behold some antique and noble edifices. Within the harbor, if you can appropriate a couple of days, you may find them agreeably employed, especially in the summer months, by a trip to Fort Sumter, to James Island, to Mount Pleasant, and Sullivan's Island. The two latter places are favorite and healthy retreats for the citizens of Charleston in midsummer. The "Mount Pleasant Hotel" is ample, cool, and well kept, with the usual adjuncts of bowling and billiard-saloons. The forests in the immediate neighborhood afford fine drives and picturesque rambles. You pass in twenty minutes from Mount Pleasant to Sullivan's Island. The Moultrie House, at this place, is one of the finest watering-places in the southern country. The sea-bathing is secure; the beach—one of the most capacious—affords hard drives, along the line of breakers, for nearly three miles, to the eastern end of the island, where the sea, angrily struggling with shoals to press into the estuaries behind Sullivan and Long Islands, keeps up a perpetual and not unpleasant roar—exhibiting its passions in a way to inspire no terror.

Charleston was originally founded about 1670. It was subsequently laid out on a plan furnished from England, which was then considered of very magnificent scale; but the streets were narrow, though regularly laid out, and no provision was made for public squares. In this respect the city is still very deficient. But the general style of building, which gives to each private dwelling a large court of its own, with trees and verandahs, renders the want of public squares less sensibly felt. Originally built of wood, and ravaged by frequent fires, Charleston has become, in a large degree, a city of brick. Its public buildings are some of them antique as well as noble edifices. St. Michael's Church, the State House (now employed for the Courts of Justice), and the Old Custom House, are all solid

SOUTH CAROLINA. 171

and imposing structures, raised during the colonial period. St. Michael's Tower is held in great admiration among the Charlestonians. The Custom House has a traditional character, as distinguished by the British in the Revolution as the prison-house of the patriots. It was in this building that Hayne, the martyr, was kept in bonds; and hence he was led out to execution. The New Custom House, of marble, is making rapid progress, and promises to be one of the finest specimens of American architecture. The several churches of St. Philip (Episcopal), St. Finsbar (Catholic), Citadel Square (Baptist), Central (Presbyterian), are all fine edifices; the towers of St. Finsbar, of St. Philip, and the Baptist, rising more than 200 feet.

Among the objects of public curiosity is the Orphan Asylum—a magnificent structure, of great capacity. It generally contains from 150 to 250 orphans, the numbers of both sexes being nearly equal. The Military Academy (citadel) is a State institution. One-half of its 180 members are beneficiary. The plan of education is borrowed, in part, from the system at West Point, and in part from the *Polytechnique School of France*. Its graduates are among the most distinguished and successful, perhaps, of all our colleges, and are more thoroughly grounded in the *useful* pursuits than any other. To examine these two institutions will afford the stranger very grateful employment for a day.

The environs of the city afford a variety of very pleasant *drives*. The Battery, which is the Charleston *Prado, Plaza, Alameda, Carrousel*, is of great resort on pleasant afternoons; thronged with carriages and pedestrians. Its gardens are, on such occasions, crowded with happy children. But take a coach and drive to the *Magnolia Cemetery*—a beautiful "city of the silent"—the Greenwood and Mount Auburn of Charleston. You will find this a lovely retreat; well laid out—mingled woods and waters—looking out on the Cooper, whose streams find their way into its pretty lakelets, over which the live oak hangs its Druid mosses. From this scene drive across the Ashley River; cross this broad stream, here a mile in width, and find yourself at once in the *country*, among cotton plantations and lovely farmsteads. If you have time, continue your drive a few miles farther, to the "Old Parish Church of St. Andrew," one of the most antique churches built by the early settlers under the Anglican *régime*.

The great avenue from Charleston into the country was pronounced by *Archdale*, one of the Lords Proprietors, such an avenue as no prince of Europe could boast. This was due to the noble oaks and magnolias, the myrtles and the jessamines, which lined it on either hand, making it a covered way, embowered in shade, grateful in green, venerable with moss, and giving out a perpetual fragrance from a world of summer flowers.

Returning to the city, you will find yourself interested in numerous public buildings and institutions, all of which are of interest to the traveller, who is either studious or simply curious. Charleston is especially rich in her public charities:—the South Carolina, Fellowship, Hibernian, Hebrew, German, and a variety besides all of whom have large endowments and fine buildings. She has a Literary and a Medical College in prosperous exercise. The College Library contains some 10,000 volumes: the Charleston Library, some 30,000; the Apprentices', 12,000. The College Museum is second to none in the United States.

The commerce of Charleston, once equal to that of any city on the Atlantic, has undergone many fluctuations. It is now reviving, and gradually increasing in extent and profit. She is slowly building up a marine of her own. Her chief exports are rice, cotton, tobacco, lumber, tar, pitch, and turpentine. Her farms now contribute their *spring* supplies to New York and other Northern cities. The quantity of rice raised within the State, and exported through Charleston, exceeds that of any other State and city; and the enterprise of her merchants and citizens, in the construction of railways to the Appalachian Mountains, is adding largely to her importance as a *depôt*, and place of trade and transit for the great interior of the West. She has steam lines to New York, Philadelphia, Baltimore, Havana, and Florida. Her population is now estimated at 65,000 inhabitants, of whom 20,000 are slaves.

We have indicated Fort Moultrie as a spot distinguished by one of the greatest battles of the Revolution; but the chronicles of Charleston show, besides, a long series of gallant struggles with powerful enemies. She has been threatened by the Red men, who, in formidable alliance, brought down their numerous tribes to her very gates. She has been assailed by fleets of the Spaniards and the French. Her colonial existence was one long struggle with the Spaniards and the savages. In the revolutionary contest she took a first and most distinguished part *against* the Crown; was thrice assailed by the British, and only succumbed finally to their arms, after a leaguer of two months, and when half the city was in ruins, and the people were suffering from famine. She has contributed some of the most able and patriotic men to the Republic in arts, arms, statesmanship, science, and literature. She is the birthplace of Christopher Gadsden, William Moultrie, Charles

Cotesworth and Thomas Pinckney, Henry Middleton, Arthur Middleton, Thomas Lynch, John and Edward Rutledge, William Lowndes, Joel R. Poinsett, Stephen Elliott, Hugh Legaré, Holbrook, Haynes (R. Y.), and scores besides, who have left honorable memorials, national as well as sectional, of which she may be justly proud, and to which the Confederacy itself is happy to do honor. The descendants of these great men still survive, and serve to give character to society, and to add to the attractions of the city. Let the traveller, if he can, give a week to Charleston, and he will find its scenery, its society, its characteristics, quite sufficient to exercise his curiosity and thoughts during that period; but if he can appropriate two days only, we have shown him how these may be profitably spent.

The Seaboard and Lowland towns, villages, and plantations, may be reached by the steamboats which ply between Charleston and Savannah, or by stage or carriage from the line of the railway. The traveller will not see them in their own peculiar beauty, because the climate in summer time, when the wonderful tropical vegetation covers the rank earth, is not to be braved by the unacclimated. The planters themselves, indeed, remove with their families, at this season, either to the uplands or to the little sandy pine-covered elevations with which the country is dotted. The negroes, alone, can bear the summer airs of the lowlands without ill results. In the winter, however, life may easily be made enjoyable in the villages here, under the balmiest and most healthful of temperatures, and in the midst of genial and refined society.

Beaufort, in the extreme southern part of the State, 10 miles inland, on two great arms of the sea, is a pleasant little village, where one might winter quietly and healthfully. The steamers (inland route) from Charleston to Savannah call here.

The Lowlands of Carolina.—The journey on the South Carolina railway will give the traveller some inkling of the lowland features of the southern landscape, though not in its strongest or most interesting character. Since much of the way is through extensive pine forests, which makes the rhyming sneer bestowed upon this part of the country not altogether inapt:

"Where to the North, pine trees in prospect rise;
Where to the East, pine trees assail the skies;
Where to the West, pine trees obstruct the view;
Where to the South, pine trees forever grew!"

But a second glimpse will reveal, amidst all these "pine trees," the towering cypress, with its foliage of fringe and its garlands of moss—the waxen bay-leaf, the rank laurel, and the clustering ivy; and, if you are watchful, you may catch, in the rapid transit of the cars through the swamps, glimpses of almost interminable cathedral aisles of cypress and vine, sweeping through the deeper parts of the boundless lagoons. But a railroad glimpse, and especially at the speed with which you travel here, is quite insufficient for reasonable observation. At Woodlands, a mile only south of Midway, the centre of the road, lives the distinguished poet and novelist, Simms; and, as he is always upon hospitable thoughts intent, we will pay him a flying visit, not doubting of our welcome. Yonder, in that wide and spreading lawn, stands our author's mansion—an old-fashioned brick structure, with massive and strange portico. The ranks of orange-trees and live oak which sentinel his castle, are the objects of his tenderest care—true and ardent lover of nature as he is. Mr. Simms has a particular fondness for the especial grape-vine, depending in such fantastic and numberless festoons from the limbs of yon venerable tree. He has immortalized it in his song; and, as it is a good specimen of its class—a class numerous in the South—we will pay it an humble tribute in our prose. It is strong-limbed as a giant—and, but for the grace with which it clings to the old forest-king, would seem to be rather struggling with him for his sceptre, than loyally and lovingly suing for his protection. The vine drops its festoons, one beneath the other, in such a manner that half a dozen persons may find a cozy seat, each over his fellow, for a merry swing. On a dreamy summer eve, you may vacillate, in these rustic couches, to your heart's content, one arm thrown round the vine will secure you in your seat, while the hand may hold the favorite book, and the other pluck the delicious clusters of grapes, which, as you swing, encircle your head like the wreath upon the brow of Bacchus. If the rays of the setting sun be hot, then the rich and impenetrable canopy of foliage above you will not prove ungrateful.

A stroll over Mr. Simms' plantation will give you a pleasant inkling of almost every feature of the Southern lowlands, in natural scenery, social life, and the character and position of the slave population. You may sleep sweetly and soundly within his hospitable walls, secure of a happy day on the morrow, whether the rain holds you prisoner within doors, or the glad sunshine drags you abroad. He will give you a true Southern breakfast, at a very comfortable hour, and then furnish you abundant sources of amusement in his well-stocked library, or suffer you to seek it elsewhere, as your fancy listeth. At dinner, you shall not lack good cheer, for either the physical or the intellectual man, and then you may take a pleasant stroll to the quiet banks of the Edisto

—watch the raft-men floating lazily down the stream, and interpret as you will the windings and echoes of their boat-horns—or you may muse in the shaded bowers of Turtle Cove, or any of the many other inlets and bayous of the stream. Go where you may, you must not fail to peep into the dark and solemn swamps. You may traverse their waters on wild bridges of decayed and fallen trees; you may dream of knight and troubadour, as your eye wanders through the gothic passages of cypress, interlacing their branches, and bearing the ever-dependent moss, which hangs mournfully, as if weeping over the desolation and death which brood within the fatal precincts. If you fear not to startle the wild-fowl, to disturb the serpent, or to encounter the alligator, you may enter your skiff, and, sailing *through* the openings in the base of the cypress, you may penetrate at pleasure, amidst bush and brake, into the mystic chambers of these poisonous halls. Mr. Simms has beautifully described these solemn scenes in his "Southern Passages and Pictures:"

"'Tis a wild spot, and hath a gloomy look;
The bird sings never merrily in the trees,
And the young leaves seem blighted. A rank growth
Spreads poisonously round, with power to taint,
With blistering dews, the thoughtless hand that dares
To penetrate the covert. Cypresses
Crowd on the dark, wet earth; and stretched at length,
The cayman—a fit dweller in such home—
Slumbers, half buried in the sedgy grass,
Beside the green ooze where he shelters him.
A whooping crane erects his skeleton form,
And shrieks in flight. Two summer-ducks aroused
To apprehension, as they hear his cry,
Dash up from the lagoon, with marvellous haste,
Following his guidance. Mostly taught by these,
And started by our rapid, near approach,
The steel-jawed monster, from his grassy bed,
Crawls slowly to his slimy, green abode,
Which straight receives him. You behold him now,
His ridgy back uprising as he speeds,
In silence, to the centre of the stream,
Whence his head peers alone."

* * * * *

Rambling, once upon a time, through the negro quarters of Mr. Simms' plantation, we amused ourself in studying the varied characters of the slaves, as shown in the style of their cabins, the order in which they kept them, the taste displayed in their gardens, etc.; for every man has all the material and time at his command to make himself and his family as comfortable as he pleases. The huts of some bore as happy an air as one might desire; neat palings enclosed them; the gardens were full of flowers, and blooming vines clambered over the doors and windows. Others, again, had been suffered by the idle occupants to fall into sad decay; no evidence of taste or industry was to be seen in their hingeless doors, their fallen fences, or their weed-grown gardens. These lazy fellows were accustomed even to cut down the shade trees which had been kindly planted before their homes, rather than walk a few yards further for other and even better fuel. The more industrious of the negroes here, as elsewhere, employ their leisure hours, which are abundant, in the culture of vegetables and in raising fowls, which they sell to their masters, and thus supply themselves with the means to purchase many little luxuries of life. For necessaries they have no concern, since they are amply and generously provided with all which they can require. Others who will not thus work for their pin-money, are dependent upon the kindness of their masters, or more frequently upon their ingenuity at thieving. Many of them sell to their master in the morning the produce they have stolen from him the previous night. At least, they all manage to keep their purses filled; and we were assured that not one, had he occasion or desire to visit Charleston or Augusta, but could readily produce the means to defray his expenses. One old woman was pointed out to us, who had several times left the plantation with permission to remain away as long as she pleased; yet, although her absences were sometimes of long continuance, she was too wise not to return to a certain and good home. Wander how and whither she would, in due time her heart would join the burden of the song:

"Oh! carry me back to old Virginny,
To old Virginny's shore!"

While once visiting some friends in Carolina, we had the pleasure of witnessing the bridal festivities of one of the servants of the family, a girl of some eighteen years. The occasion was one of those pleasant things which long hold place in the memory. For days previous, the young ladies of the household gayly busied themselves in kind preparations for the event; in instructions to the bride, in the preparation of her white muslin robe, of her head-dress, and other portions of her toilet, in writing her notes of invitation to her sable friends—Mr. Sambo Smith or Miss Clara Brown, according to the baptismals of their respective masters, whose names the negroes of the South always assume. In our quality of artist, we had the pleasure to expend our water-colors in wreaths of roses, and pictures of cupids, hearts, and darts, and so on, upon the icings of the cakes which the young ladies had prepared for the bridal feast; and who knows but that our *chef d'œuvres* were consumed by ebony lips on that memorable night! The ceremony took place in the cabin of the bride, and in presence of the whites; and then followed revelry, feasting, and dancing upon the lawn, much to the delight of the happy pair and their dark friends, and scarcely less to the pleasure of the bride's kind mistresses and of all of us who witnessed their sports from the parlor windows. By the way, when you journey in the South, line your pockets with tobacco, dispense it generously to the darkies, and they are your friends for life.

As we have said, Woodlands and its vicinage will enlighten you as to the *genus* of the scenery of all the lowlands of the South. This *genus*, however, you will find, as you ramble from the seaboard towards the interior, subdivided into many species, each widely varying from the other. Upon the seaboard, and its many lovely and luxuriant islands, you will find the *beau ideal* of Southern soil, climate, vegetation, architecture, and character. Here abound those lovely inlets and bays, which make up the absence of the lake scenery of the North. These bayous and lakelets are covered with the rankest tropical vegetation; they abound in every species of wild-fowl—birds of the most gorgeous plumage, songsters of the sweetest notes—the mocking-bird and the nightingale, the robin, and a host of other equally celebrated warblers. Here, the foliage is so dense and rich, in form and color, that a poor imagination will readily people the spot with elves and sprites; and there, again, so dark and solemn are the caverns, overshadowed by the impenetrable roofs of leaves, that you may readily interpret the screech of the owl, the groan of the bull-frog, and the hiss of the serpent into the unearthly wail of damned spirits. These are fitting haunts for the sad and contemplative mind at the witching hour of night.

Here, the rice plantations abound. Many of them are of great extent, some of the planters employing several hundred slaves. The white population is thus necessarily thin yet opulent. The cabins of the negroes on these extensive domains, surrounding the mansion and its many outbuildings of the proprietor, give to every settlement the aspect of a large and thriving village. There is something peculiarly fascinating in this species of softened feudal life. The slaves are for the most part warmly attached to their masters, and they watch over their interests as they would their own. Indeed, they consider themselves part and parcel of their master's family. They bear his name, they share his bounty; and their fortune depends wholly upon his. Through life they have every comfort; the family physician attends them when sick, and in their old age and imbecility they are well protected. They glory in their master's success and happiness; their pride is in exact proportion to the rank of the family they serve; and, whatever that may be, they still cherish a haughty and self-satisfied contempt for "poor white folks."

"Go 'way, Sambo," we once heard one of these jovial lads exclaim to another, whose ill-fortune it was to serve a less opulent planter than himself; "go 'way, Sambo, your massa only got fifty niggers; my massa got hundred." And he pulled up his shirt-collar, and marched pompously off with the step and air of a millionnaire.

The masters, themselves, descended from an old chevalier stock, and, accustomed through many generations to the seclusion of country life, and that life under Southern skies, and surrounded with all the appliances of wealth and homage, have acquired an ease, a grace, a generosity, and largeness of character, incompatible with the daily routine of the petty occupations, stratagems, and struggles of modern commercial and metropolitan life, be it in the South or the North.

Where the swamps and bayous do not extend, the country, still flat, is mostly of a rich sandy soil, which deeply tinges the waters of all the rivers from the Atlantic to the Mississippi. This is the grand characteristic of the southern portions of all the Gulf States. The rivers, as they extend towards the interior, are lined with high sandy bluffs, which, still further northward, give place, in their turn, to mountain ledges and granite walls. These streams, from the Mississippi to the Alabama, the Chattahoochee and the Savannah, to the smaller rivers of Carolina and Florida, are filled with sandy islands, ever changing their position and form. Frequently high freshets occur in them, completely altering their channels, and bearing away the produce of whole plantations, from the cotton bale to the family domicile, and the century-aged tree which shaded it. In crossing the smaller watercourses of the South, we have often observed marks of the extent of a freshet upon high trees, at an elevation of 50 or 60 feet above our head. They are sometimes an excessive bore to the hurried traveller, holding him water-bound for days together, and invariably in places where, of all others, he does not love to tarry.

We happened to be in Augusta years ago, during a great rise in the waters of the Savannah. In the course of some few hours, the river had extended its limits throughout the city, and over the plain for miles in every direction. It was a novel and beautiful sight to gaze from your balcony upon this unlooked-for Venice. Boats were sailing in every direction through the streets—even the ponderous crafts of the Savannah, capable of holding fifty or sixty men. We observed the pretty vessel of the "Augusta

Boat Club," dashing up Broad street and under the hotel windows, with the crew in full dress, music sounding, and gay banners waving upon the air! A ferry was established to pick up passengers at their doors or windows, and convey them to the base of the Sand-hills, a summer retreat, some three miles to the northward. The cross streets leading from the river were washed away to the depth of many feet, and for days afterwards passengers were transported across them in flats and bateaux.

From these freshets, with the innumerable stagnant pools which they leave, together with the miasma arising from immense quantities of decaying vegetable matter, spring many of the local fevers and diseases of the South. In Augusta, the yellow fever followed the great freshet, and carried off, during the brief space of a few weeks, nearly three hundred of the inhabitants. This terrible scourge had not previously visited the city for eighteen years, and has not since returned.

Georgetown, one of the oldest settlements in South Carolina, is about 15 miles from the sea on Winyaw Bay, near the junction of the Pedee, Black, and Waccamaw Rivers. Some revolutionary memories are awakened here. In 1780 the vicinage was the scene of a skirmish between American and British troops, and in 1781 it was taken from the enemy by General Marion, and the military works destroyed.

Columbia.—HOTELS :—The *Congaree*.

Columbia, the capital of South Carolina, is 128 miles from Charleston, by the South Carolina Railway and the Columbia Branch. It is connected by railway with the great route from New York to New Orleans, with Augusta, Georgia, and with Camden, Cheraw, and most of the interior and mountain villages of the State. It is a beautiful city, situated on the bluffs of the Congaree, a few miles below the charming falls of that river. It is famous for its delightfully shaded streets, its wonderful flower gardens, and the model plantations in its vicinity. Nothing can be more inviting than the walks and drives in the neighborhood. The South Carolina College, located here, is a prosperous institution, with from 150 to 200 students. The new capitol building of granite, now in progress, will be a noble edifice, costing about three millions of dollars.

The college and State libraries are large and choice. The lunatic asylum is an object of great interest. Here, also, is the theological college of the Presbyterian Church, and a Roman Catholic establishment. The population of Columbia is about 9,000.

Camden.—HOTELS :—*Mansion House*.

Camden is 83 miles north-east of Columbia, with which it is connected by railway, though with such considerable detours as to increase the distance to 52 miles. A direct line is in progress. Distance from Charleston by railway, 140 miles. It is on the Wateree River, navigable to this point by steamboats. Camden is a place of great historic interest. A battle was fought near by, August, 1780, between the Americans, under General Gates, and the British, under Lord Cornwallis; and another in April, 1781, between General Greene and Lord Rawdon. The scene of the latter struggle is the south-eastern slope of Hobkirk's Hill, now called Kirkwood, a beautiful summer suburb of the old town. Upon the Green, in front of the Presbyterian Church, on De Kalb street, there is a monument over the grave of Baron De Kalb, who fell in the battle of August, 1780, at Camden. The corner stone was laid in 1825, by La Fayette. The head-quarters of Cornwallis, to be seen here, is a fine old building in ruins. On the Market House, there is a well-executed metallic effigy, 10 feet high, of King Haiglar, a most famous chieftain of the Catawbas. Mr. Simms has made this Indian King the theme of one of his fine legends.

Fort Motte, an important Revolutionary relic, is upon high terrace ground, near the Bull's Head Neck, on the Congaree, just above its meeting with the Wateree, 33 miles below Columbia, and *en route* thence from Charleston.

Cheraw, near the northern line of the State, is at present 207 miles from Charleston, and 129 miles from Columbia by railway. The northeastern railway, now partly in operation, will open a direct and much nearer route from the former city, and a direct road from the latter is in contemplation. Cheraw is on the Great Pedee River, at the head of steam navigation.

Orangeburg is on the line of the South Carolina (Columbia branch) Railway, 97 miles from Charleston, and 49 from Columbia. It is a spot of historic interest, near the banks of the Edisto River. It formed a link in the chain of military posts established by the British after the fall of Charleston. Among the old relics here, are some remains of the works erected by Rawdon, near the Edisto, and the old Court House, which bears traces, in the shape of bullet marks, of the assault made by Sumter in 1781.

Eutaw Springs. This interesting spot, the scene of the famous battle of Eutaw, is about 40 miles below Orangeburg, and 60 miles north-west of Charleston.

THE MOUNTAIN VILLAGES AND SCENERY OF SOUTH CAROLINA.

The northern districts of South Carolina, form, with the neighboring hill-region of Georgia, and the western portion of North Carolina, one of the most interesting chapters in the great volume of American landscape beauty and wonder. In mountain surprises, picturesque valley nooks, and delicious waterfalls, this region is nowhere surpassed in all the Union. Beautiful and healthful villages, with high social attractions, afford most agreeable homes and head-quarters to the hunter of the picturesque. These villages are favorite summer resorts of the people of the lowlands of the State; and their elegant mansions and villas are every year more and more embellishing all the vicinage.

Greenville. — HOTELS : — *Mansion House; Goodlett House.*

Greenville, in the north-west corner of the State, lies at the threshold of the chief beauties of this region, and gives ready access to all the rest. It is distant by railway—from Charleston, 271 miles—from Columbia, 128 miles. The village is beautifully situated on Reedy River, near its source, and at the foot of the Saluda Mountain. It is one of the most popular summer resorts in the up-country of Carolina, being in the immediate vicinity of the Table Mountain, the White Water, and the Slicking Falls, the Jocassee, and Saluda Valleys, the Keowee River, Paris Mountain, Cæsar's Head, and numerous other bold peaks of the Blue Ridge.

The Table Mountain is in Pickens District, in the north-west corner of South Carolina, about 20 miles above the village of Greenville. It is one of the most remarkable of the natural wonders of the State, rising as it does 4,300 feet above the sea, with a long extent on one side of perpendicular cliffs, 1,000 feet in height. The view of these grand and lofty rocky ledges is exceedingly fine from the quiet glens of the valley of the cove below, and not less imposing is the splendid amphitheatre of hill-tops seen from its crown. The record of one of our own journeys to this interesting locality stands thus in our note-book:

Approaching the broad perpendicular side of the mountain, at its base, we came upon it suddenly; a right-angled descent in our path revealed one of the most charming *coup d'œils* I ever enjoyed. In the foreground lay, in pastoral beauty, the sweet valley of the cove, diversified with greensward and cultivated land, and embellished with a most picturesque and orthodox log-cabin. In the middle ground, rose from the bosom of the vale, a line of mountains, robed in richest verdure, upon which, as a crowning-point, the mighty Rock displayed its towering front. Besides these magic features, were others of winning beauty. Turning the eye, the Bald Mountain, Cæsar's Head, and other chains were visible. The ear, too, detected, though unseen, the infantile murmurings of the Saluda River, as it swept through the valley, from its source, a few miles north of the great Rock, and between it and the adjoining space of the Alleghanies. The Stool Mountain to the left, and near the Rock, forms a prominent feature in the picture. I was told a pretty Indian legend, substantiating the former existence of an aboriginal brobdignagdian, whose colossal person and lordly appetite could be satisfied with no humbler seat than the "Stool" in question, and no less a board than the noble "Table." Hence the names. Such accommodations would suit well for the statue of the prince, into which the Grecian sculptor was assigned the trifling task of cutting the Athenian Acropolis! The Rock, of course, derives its name from its resemblance in form to the table. This resemblance, however, is only general. It is a solid mass, oblong in form. The northern front perpendicular, and over half a mile in extent. The eastern is considerably inclined. The southern admits of easy ascent. On the north, the elevation of the rock is about 1,100 feet, gradually declining towards the western verge. The entire elevation above the level of the sea is 4,300 feet.

A long and toilsome ramble over hill and dale, led us to the foot of the rock at the usual place of ascent, on the eastern façade. On the way, we encountered near the rock a little lake, more properly called the "Pool." It was environed with straggling and massive pieces of stone, that had fallen at various times from above. Probably crumbs, that escaped at the orgies of the before-mentioned ideal lord of the domain. The ascent is made by means of flights of wooden steps, secured to the rock. Of these steps, we counted about 130. They are substantially built, and with the assistance of the rail or banister, the passage is safe and tolerably easy. From the summit we enjoyed a wide-spread and most enchanting panorama.

Among the many mountains seen from our eyry station, was the commanding form of Cæsar's Head. It is the highest in the vicinity, and well deserving a visit. Across this valley was the distant gleam of the Fall of Slicking; its long line of sparkling spray heightened much the beauty of the scene. The Stool Mountain, which is prominent from the valley below, here dwindles to its proper height.

The top of the rock, which is comparatively level, is of great extent. In many places the surface is stony, in others alluvial and covered with noble trees. Near the centre, the remains of a hut exist; a building erected as a kitchen to a hotel, which it was once contemplated to erect on the rock. Though the enterprise was given up, it is not at all impracticable. The 50 or 60 acres of tillable land might furnish provisions, while for water, there is a spring, of the most grateful purity and coolness, near the middle of the isolated and elevated demesne.

The Falls of Slicking are in the mountain glens, on the opposite side of the valley, at the base of the Table Rock.

Leaving the cabin at the base of the Saluda Mountain, the tourist in his ascent, soon finds himself following the windings of the river. After the passage of about one-quarter of a mile he reaches the "Trunk," so called from its being the point of junction of two different branches of the river or creek; the distance between these streams as you continue to ascend, gradually increases, and when near the summit they are widely separated; they bear one name, and abound, each, in cascades. The right-hand branch is the more picturesque, and is the one by which the visitor is usually conducted.

The "Trunk" is decidedly the gem of all the locales, and for that reason many forbear visiting it as they *ascend* the mountain, philosophically leaving it until they have surveyed the lesser beauties. Such shall be our course now. Following then the right or south branch of the stream, the traveller is now lost amid the forest trees, and now reaching a spread of table-land, sees at his feet a tranquil stream; above him sport the feathery waters; below, wave the tops of giant trees; and beyond arises, in majestic grandeur, the Table Rock, surrounded by numerous attendant peaks. Again he is hidden in the thick foliage, and again and again he reaches the rocky terrace with its basin and its cascade, and its mountain distance, each view improved by the increased elevation. Near the summit is such a terrace as I describe, with a perpendicular fall of considerable extent. From this point is a charming view of the neighboring mountains of Cæsar's Head, Bald Mountain, the Pinnacle Rock, and other spurs. This site is second only to the "Trunk," to which we now return.

At the "Trunk"—a scene of remarkable charms, where one may linger long unweariedly—the two streams fall perpendicularly some 70 feet, mingling in one in the basin below. This basin is easily accessible, and nowhere is there a more secluded or more wildly picturesque spot. Save when in his meridian, the sun's rays seldom violate its solitude. On one side are the two cascades leaping in snowy masses from rock to rock, and on others are mighty bulwarks of venerable stone, here and there studded with the adventurous shrub, or overhung with rich foliage.

Pendleton is an agreeable little village, on Eighteen Mile Creek, Anderson District, in the mountain region of the north-west corner of South Carolina. The South Carolina Railway and its branches approach a few miles below at Anderson Court House, thus very nearly connecting it with Charleston, Columbia, Greenville, and most of the middle towns of the State. It is interesting, from its vicinage to much picturesque scenery, and to Fort Hill, once the home of Calhoun.

Fort Hill, once the residence of the statesman John C. Calhoun, is a few miles only from the village of Pendleton. It is a plain but comfortable building of wood, with piazzas and other fittings and arrangements, after the usual fashion of southern country houses. Here Mr. Calhoun lived in his months of release from the toils of public life, venerated by the humblest and highest of his neighbors for his noble and gentle private virtues and graces, no less than he was honored abroad for his unrivalled genius as a statesman and orator.

Walhalla, a flourishing German settlement, is in this region.

Pickens Court House is a few hours' ride, on horseback or carriage, north of Pendleton and west of Greenville. It is within excursion distance of the Keowee River, the Valley of Jocasse, the Cataract of the White Water, and other interesting scenes.

The Keowee, a beautiful mountain stream, in Pickens District, S. C., with the Tugaloo River, forms the Savannah. The road to the Valley of Jocasse lies along its banks.

"I have been where the tides roll by,
Of mighty rivers deep and wide,
On every wave and argosy—
And cities builded on each side:
Where the low din of commerce fills
The ear with strife that never stills.

"Yet not to me have scenes like these,
Such charms as thine, oh peerless stream!
Not cities proud my eye can please—
Not argosies so rich I deem—
As thy cloud-vested hills that rise—
And forests looming to the skies!"

The Keowee region is full of romantic memories of the Cherokee wars.

The Jocasse Valley, in Pickens District, near the northern line of the State, is one of the most charmingly secluded little nooks in the world, environed as it is on every side, except that through which the Keowee steals out, by grand mountain ridges. The chief charm of Jocasse is, that it is small enough to be felt and enjoyed all at once, as its entire area is not too much for one comfortable picture. It is such a valley as painters delight in.

The White Water Cataracts are an hour or two's tramp yet north of Jocasse. Their chief beauty is in their picturesque lines and in the variety and boldness of the mountain landscape all around: though they would still maintain their claims to the universal admiration, for their extent alone, even were the accessory scenes far less beautiful than they are. The number of visitors here is increasing year by year, and the time is approaching when this and the thousand other marvels of nature in the Southern States will win tourists from the North, as the White Mountains and the Catskills, and Trenton and Lake George now attract pilgrims from the South.

Adjoining this most attractive region of South Carolina, and easily accessible therefrom, are the many beautiful scenes of the western portion of North Carolina, of which we have already spoken, and of Tallulah, and Toccoa, and Yonah, and Nacoochee, and numerous other lovely spots in the hill-region of Georgia, which we have yet to visit.

Spartanburg.—HOTELS :—

Spartanburg is connected with Charleston by railway via Columbia, and Union—distance 220 miles. The village of Spartanburg is in the midst of a mineral region, famous for its gold and iron. Here, too, are some celebrated limestone springs. The place is the seat of a University, endowed by Benjamin Wofford, and controlled by the Methodists; also of a prosperous Female College. A distinguished Asylum for the Deaf, Dumb, and Blind is located here. Within the limits of this district is the memorable revolutionary battle-field of the Cowpens.

The Battle-field of the Cowpens (January 17, 1781) is on the hill-range called the Thickety Mountain. In the olden time the cattle were suffered to graze upon the scene of the contest—from whence its name. Without reviewing the incidents in detail of the important fight of the Cowpens, we will remind the reader that it was a brave one, resulting in the defeat and retreat of the British under Tarleton, with a loss of 10 officers and 90 privates killed, and 23 officers and 500 privates taken prisoners. The American loss was about 70, of whom only 12 were killed.

Yorkville.—HOTELS :—*Jasper Stowe's Hotel.*

Yorkville, midway on the upper boundary of South Carolina, is in the heart of its beautiful mountain scenery, and is, besides, the particular point from whence the tourist may the most easily and speedily reach the scenes of the historic events, which so heighten the pleasure of travel in all this region—every plantation telling a thrilling tale of its own—for during the last three years of the war of the Revolution, there was unceasing struggle here between the partisan bands of the patriots and the British troops.

Route. Yorkville is 212 miles from Charleston by the South Carolina Railway and the Columbia Branch to Columbia, thence by the Charlotte and South Carolina Railway to Chester, and thence by the King's Mountain Railway to Yorkville. A line of railways comes in at Chester, just below Yorkville, from Weldon and Goldsboro', N. C. (on the great Northern and Southern route), via Raleigh and Charlotte, N. C. This is a pleasant access from New York, via the mountain region of North Carolina, to that of South Carolina and Georgia.

The village of Yorkville is situated upon an elevated plain on the dividing ridge between the Catawba and the Broad Rivers. In the vicinage there are some valuable sulphur and magnesia waters, to add to the attractions of winning scenery and romantic story which the region so abundantly offers to the tourist.

King's Mountain Battle-field lies about 12 miles north-east of Yorkville, about a mile and a half south of the North Carolina line. The King's Mountain range extends about sixteen miles southward, sending out lateral spurs in various directions. The scene of the memorable battle fought in this region is six miles from the summit of the hill. A simple monument to the memory of Ferguson and others marks the spot, and on the right there is a large tulip tree, upon which it is said ten tories were hanged.

The story of the eventful battle of King's Mountain is thus told in the words of General Gates: "On receiving intelligence," he says in his report, "that Major Ferguson had advanced up as high as Gilbert Town, in Rutherford County, and threatened to cross the mountains to the western waters, Col. William Campbell with 400 men from Washington County, Virginia, Colonel Isaac Shelby with 240 men from Sullivan County, N. C., and Lieut. Colonel John Sevier with 240 men of Washington County, N. C., assembled at Watanga, on the 25th of September (1780), where they were joined by Col. Charles McDowell with 160 men from the counties of Burke and Rutherford, having fled before the enemy to the western waters. We began our march on the 26th, and on the 30th we were joined by Col. Cleaveland, on the Catawba

River, with 350 men from the counties of Wilkes and Surry. No one officer having properly a right to the command in chief, on the 1st of October we despatched an express to Major General Gates, informing him of our situation, and requesting him to send a general officer to take command of the whole.

"In the mean time Col. Campbell was chosen to act as commandant, until such general officer should arrive. We marched to the *Cowpens* on Broad River, in South Carolina, where we were joined by Col. James Williams, with 400 men, on the evening of the 6th of October, who informed us that the enemy lay encamped somewhere near the Cherokee Ford of Broad River, about 30 miles distant from us. By a council of principal officers it was then thought advisable to pursue the enemy that night with 900 of the best horsemen, and have the weak horses and footmen to follow us as fast as possible. We began our march with 900 of the best men about 8 o'clock the same evening, and marching all night, came up with the enemy about 3 o'clock, P. M., of the 7th, who lay encamped on the top of King's Mountain, 12 miles north of the Cherokee Ford, in the confidence that they would not be forced from so advantageous a pass. Previous to the attack, on our march the following disposition was made: Col. Shelby's regiment formed a column in the centre on the left, Col. Campbell's regiment another on the right, while part of Colonel Cleaveland's regiment, headed in front by Major Joseph Winston and Colonel Sevier formed a large column on the right wing. The other part of Cleaveland's regiment, headed by Colonel Cleaveland himself, and Col. Williams' regiment composed the left wing. In this order we advanced, and got within a quarter of a mile of the enemy before we were discovered. Col. Shelby's and Col. Campbell's regiments began the attack, and kept up a fire on the enemy while the right and left wings were advancing to surround them, which was done in about five minutes, and the fire became general all around. The engagement lasted an hour and few minutes, the greater part of which time a heavy and incessant fire was kept up on both sides. Our men in some parts where the regulars fought, were obliged to give way a distance, two or three times, but rallied and returned with additional ardor to the attack. The troops upon the right having gained the summit of the eminence, obliged the enemy to retreat along the top of the ridge to where Col. Cleaveland commanded, and were there stopped by his brave men. A flag of truce was immediately hoisted by Captain Depeyster, the commanding officer (Major Ferguson having been killed a little before), for a surrender. Our fire immediately ceased, and the enemy laid down their arms (the greater part of them charged) and surrendered themselves prisoners at discretion. It appears from their own provision returns for that day, found in their camp, that their whole force consisted of 1,125 men. * * Total loss of the British, 1,105 men, killed, wounded, or made prisoners."

"No battle during the war," says Mr. Lossing, in his Field Book where we find the preceding report of the struggle at King's Mountain, "was more obstinately contested than this: for the Americans were greatly exasperated by the cruelties of the Tories, and to the latter it was a question of life and death. It was with difficulty that the Americans, remembering Tarleton's cruelty at Buford's defeat, could be restrained from slaughter, even after quarter was asked. In addition to the loss of men on the part of the enemy mentioned in the report, the Americans took from them 1,500 stand of arms. The loss of the Americans in killed was only twenty, but they had a great number wounded." Battle fought Oct. 7, 1780.

Crowder's Knob, the highest peak of King's Mountain, is about 3,000 feet above the level of the sea.

The Mountain Gap, near the Cherokee Ford, the Great Falls of the Catawba, and Rocky Mount, the scene of another of the partisan struggles, and Hanging Rock, where Sumter fought a desperate fight, are other interesting scenes and localities of this hill-region of Carolina.

FLORIDA.

FLORIDA is much visited when cold winter winds and snows prevail, by those who love mild and balmy atmospheres, and especially by invalids in quest of health-restoring climates. The villages of St. Augustine, Jacksonville, Pilatka and neighboring places, which are those most particularly sought, are near the Atlantic coast, in the extreme north-eastern part of the State.

They may be speedily and pleasantly reached by steamers from Charleston and Savannah, as we shall show, after a very hasty peep at the specialties in the history and character of the region.

The shrine of the life and health-giving Goddess, Hygeia, was sought under the southern skies of Florida centuries ago, as it is to-day. Ponce de Leon came here in 1512, hoping to find the fabled fountain of perpetual youth and strength. He was not so fortunate, though thousands of others have since been, in a grateful degree.

After the brave De Leon, came Narvaez, more unlucky still, for when he had resolutely penetrated to the interior with his four hundred gallant followers, no man ever heard of him or of them again.

De Soto followed in 1539, with a not much happier reward, for though he subdued the savages and took possession of their land, it was only to leave it again and to pass on. Battle and strife have, with intervals of quiet, so characterized Florida, almost to the present day, that its name would seem but irony did it really refer, as is generally supposed, to the floral vegetation of the soil, instead of to the simple happening of the discovery of the country on *Pascua Florida* or Palm Sunday.

The earliest settlements in Florida were made by the French, but they were driven out by the Spaniards, who established themselves securely at St. Augustine in 1565, many years before any other settlement was made on the western shores of the Atlantic. Before the Revolution, Florida warred with the English Colonies of Carolina and Georgia, and passed into British possession in 1763. It was reconquered by Spain in 1781, and from that period until within very late years, it has been the field of Indian occupation and warfare. The reconquest by Spain in 1781, was confirmed in 1783, and in 1821 that power ceded the country to the United States. Its territorial organization was made in 1822, and its admission into the Union as a State occurred March 3, 1845. A sanguinary war was waged from 1834 to 1842, between the troops of the United States and the Indian occupants, the Seminoles, led by their famous chief Osceola. Since that period the savages have been removed to other territory, excepting some remnants still in possession of the impenetrable swamps and jungles of the lower portions of the State.

Florida is the grand peninsula forming the extreme South-eastern part of the United States. Its entire area eastward lies upon the Atlantic, and the Gulf of Mexico washes almost the whole of the western side. Georgia and Alabama are upon the north. The country is for the most part level, being nowhere more than 250 or 300 feet above the sea. "The southern part of the peninsula," says Mr. De Bow, in his "Resources of the South and West," "is covered with a large sheet of water called the Everglades—an immense area, filled with islands, which it is supposed may be reclaimed by drainage. The central portion of the State is somewhat elevated, the highest point being about 171 feet above the ocean, and gradually declining towards the coast on either side. The country between the Suwanee and the Chattahoochee is elevated and hilly, and the western region is level. The lands of Florida, Mr. De Bow continues, "are almost *sui generis*, very curiously distributed, and may be designated as high hummock, low hummock, swamp, savannas, and the different qualities of fine land. High hummock is usually timbered with live and other oaks, with magnolia, laurel, etc., and is considered the best description of land for general purposes. Low hummock, timbered with live and water oak, is subject to overflows, but when drained is preferred for sugar. Savannas, on the margins of streams and in detached bodies are usually very rich and alluvious, yielding largely in dry seasons, but needing, at other times, ditching and dyking. Marsh savannas, on the borders of tide streams, are very valuable, when reclaimed, for rice or sugar-cane.

The swampy island-filled lake called the Everglades is covered with a dense jungle of vines and evergreens, pines and palmettos. It lies south of Okeechobee, and is 160 miles long and 60 broad. Its depth varies from one to six feet. A rank tall grass springs from the vegetable deposits at the bottom, and rising above the surface of the water, gives the lake the deceitful air of a beautiful verdant lawn. The soil is well adapted, it is thought, to the production of the plantain and the banana.

In the interior of Florida there is a chain of lakes, of which the extreme southern link is Lake Okechobee, nearly 20 miles in length. Many of these waters are extremely picturesque in their own unique beauty of wild and rank tropical vegetation.

The rivers of the State are numerous, and, like the lakes, present everywhere to the eye of the stranger very novel attractions, in the abundance and variety of the trees and shrubs and vines which line all their shores and bayous. The largest of the many rivers is the Appalachicola, which crosses the western arm of the State to the Gulf of Mexico. The St. Mary's is the boundary on the extreme northern corner, Georgia being upon the opposite bank. Its waters fall into the Atlantic, as do those of the St. Johns river, in the same section of the State.

The St. Johns River is the point to which we purpose to direct the more particular attention of the tourist at this time, not for its own beauties' sake—for it is but a straggling, sluggish stream, possessing no very salient picturesque attractions—but as the access to the famous winter and invalid resorts of Florida, the villages of St. Augustine, Jacksonville, Pilatka, and other places.

Route to St. Augustine, etc. Two fine steamers leave Charleston, S. C., and three leave Savannah, Geo., every week for Pilatka, on the St. Johns River. Fare from either place to Piccolata (18 miles from St. Augustine), $8. From Piccolata to Augustine (3½ hours' stage) $1 to $2. Charleston steamers sometimes visit Augustine direct.

The steamer Darlington leaves Jacksonville every Saturday morning for Enterprise, the present limit of steamboat navigation on the St. Johns, stopping at Pilatka over Sunday, resuming her voyage Monday morning, and arriving at Enterprise that (Monday) night. Returning, leaves Enterprise Wednesdays. Fare, $6.

The St. Johns River comes from a marshy tract in the central part of the peninsula, flowing first north-west to the mouth of the Ochlawaha, and thence about northward to Jacksonville, and finally eastward to the Atlantic. It is navigated by steamboats only to Pilatka, though vessels drawing eight feet of water may pass up 107 miles, to Lake George. The entire length of the river is 200 miles. The country which it traverses is covered chiefly with dank cypress swamps and desolate pine barrens.

Jacksonville.—HOTELS:—The only good hotel is the *Judson House.*

Jacksonville, 25 miles from the mouth of the St. Johns, is the most important point on the river. It is a flourishing, busy town of from 1,000 to 1,500 inhabitants, has numerous saw-mills, and considerable commerce. Many invalids remain here, and seek no further.

The next in order, frequented by strangers, is *Fleming's Island* (47 miles up), situated at the confluence of Black Creek with the St. Johns. It is a quiet, home-like, and pleasant place, not infested by low company.

Of **Middleburgh,** 16 miles up Black Creek, report speaks favorably. It has been but recently resorted to by invalids. It consists of a few houses only.

Magnolia Mills (56 miles up the river), a large, solitary hotel, on the west bank of the St. Johns, is kept by Dr. Benedict, a northern physician, of established reputation. Good rooms and good entertainment may be expected there.

Next comes **Piccolata** (60 miles up), a village of but one house, where passengers for St. Augustine, 18 miles east, can generally get a tolerable night's lodging, when desired.

Pilatka.—HOTELS:—*Spear's House.*

Pilatka, on the west bank, 25 miles, or two hours, further south, is a new and thriving town, deriving considerable trade from the fertile back country. Here are two or three more or less tolerable places of entertainment. Passengers for Orange Springs and Ocala take stage here.

Welaka, on the east bank, is a new settlement. Every attention is shown to strangers by its gentlemanly proprietor—110 miles up the St. Johns.

Enterprise, also on the east bank, on Lake Monroe, and the *ultima thule* of steamboat adventure, boasts a new, large, commodious, and well-kept hotel. The hunting and fishing are good in the vicinity—180 miles up the river.

Thirty miles east from Enterprise, on the sea-coast, and four miles from Mosquito Inlet, is *New Smyrna,* consisting of two houses. Reached by mail-wagon, once a week. Mr. Sheldon entertains company, and ensures them capital sport.

FLORIDA.

Mail boat leaves here for Indian River every second week.

St. Augustine.—HOTELS:—St. Augustine is well furnished with hotels and boarding-houses, and there is unusually ample and comfortable accommodation for all comers. The principal hotels are the *Magnolia*, Buffington, proprietor—a well-built, well-kept, and well-furnished resort—and the *Planters'*, Mrs. Loring, lessee, a popular house.

First-class boarding-houses are kept by Mrs. Reid, Mrs. Fazio, and Miss Mather. There are also others of less note.

The hotel prices are $1 50 and $2 a day; $9 and $10 per week; fire extra. The boarding-house charges are less, being from $6 upward.

Visitors, unless more than ordinarily difficult and exacting, will find the tables satisfactorily furnished; admirably so, considering the isolation of the place, and its remoteness from markets and commercial cities. The winter fare consists of groceries and butter from the north; delicious fish and oysters, beef, game, poultry, venison, duck, wild turkey, and occasionally green turtle; green peas and salads are rarely lacking, even in mid-winter; game birds are abundant, such as quail, snipe, etc.

St. Augustine is built along the seaward side of a narrow ridge of land, situated between salt marsh and estuary half a mile from the beach, two miles from the ocean, in sight of the bar and light-house, and within hearing of the surf. The soil is sandy loam and decomposed shell, and is very productive. Approaching by a bridge and causeway crossing the St. Sebastian River and marsh, we enter a well-shaded avenue, flanked by gardens and orange groves, which leads directly to the centre of the quaint old city. Here is the public square, a neat enclosure of some two acres, facing which, on either side, stand the Court House, the Market and wharf, the Protestant Episcopal Church—a plain building, in the pointed style, handsomely furnished—and, immediately opposite, the venerable Roman Catholic Church, a striking edifice of seemingly great antiquity, but built only about eighty years ago. It is of the periwig pattern, and in the worst possible taste. One of its bells bears date 1682. Connected with this church is a small convent and school.

A minute's walk brings us to the sea-wall or breakwater, a broad line of massive masonry, built about 1840 by order of Government, at great cost, for the protection of the city, but whose chief use is that of affording to the inhabitants the pleasantest promenade in fine weather. This wall extends half a mile southward to the now deserted barracks and magazine, and as far northward to Fort Marion, formerly St. Mark, a picturesque and decayed fortress, which once commanded the whole harbor, looming up out of the flat landscape, grand as a Moorish castle, and forming the most conspicuous and interesting relic of the Spanish occupation.

Parallel to this sea-wall, run north and south, with short intersections, the three principal streets or lanes, long, narrow, without pavement or sidewalk, irregularly built up with "dumpy" but substantial houses, rather dingy and antediluvian, mostly of stone, or with the lower stories stone and the upper of wood. They have invariably the chimneys outside, and are ornamented with projecting balconies and latticed verandas, from which the gay paint has long since faded, being all toned and weather-stained into one sombre gray hue, which, in keeping with the surroundings, is the joint result of age, neglect, sun, and saline air. Every house is separated from its neighbor by more or less of garden plot, ill protected by broken fence and crumbling wall, wherein they raise two or more crops of vegetables every year, figs in perfection, and roses in unmeasured abundance.

Augustine is sometimes styled the "Ancient City," and is, indeed, the oldest in the United States. Its appearance is in strict keeping with its venerable age, seen in the unequivocal marks of decay and decrepitation. Perhaps the friable nature of the common building material contributes to this ruinous appearance, all the older houses being constructed of a stratified concrete of minute shell and sand called "coquina," in blocks conveniently obtained, easily worked, hardening by exposure, but abrading and crumbling in course of time. And yet this material seems everlasting; for the old stuff of dilapidated buildings, and houses disused by diminution of population, forms, by refacing, the excellent material for new. Coquina houses, however, are invariably dark, and always damp in winter, on which account frame dwellings, although not so cool summer houses, are much preferred by the innovating Yankees. But the Minorcan, or sub-Spanish population, still adhere to their traditions, and refuse to be reformed. They build for the summer time—the longest season—and wisely build, when they do build, the same solid, squat, low-doored, narrow-windowed, disagreeably-dark and rheumatically-damp dwellings as ever. Visitors, however, in choosing winter quarters, will do well to prefer those hotels which are of frame, and have a cheerful sunny exposure.

Northerners seeking in Florida a milder climate and permanent winter residence, have generally preferred St. Augustine. And with the best reason. The proximity of the Gulf Stream renders it warmer in winter and cooler in summer than the settlements on the St. Johns River. It is at present the most southern habitable place on the eastern coast; and it has peculiar advantages over all other towns in East Florida—in its churches, its company, and its comforts. Good society may always be had there; the citizens are hospitable, and among the visitors are always some agreeable persons, cultivated and distinguished.

Visitors begin to arrive about the holidays, and the first "stranger" is looked for with as much anxiety as the first Connecticut shad. From the middle of March until the middle of April is the height of the season, and then the hotels are crowded. Then, too, the city is gay. Everybody is sociable, idle, happy, *sans souci*. Pleasure parties you meet at every turn, groups on every corner, bathing in the sweet air that flows through shady streets from yon blue rushing sea. Deliciously fresh and mild is the atmosphere during the first spring heats. Then the soft south wind fills the senses with a voluptuous languor, and the evening land breeze comes laden with the fragrance of orange blossoms and the breath of roses. A moonlight walk upon the sea wall suggests the Mediterranean, and the illusion is heightened by the accents of a foreign tongue.

The effect of these happy climatic and social conditions is very noticeable. The most morose tempers seem to lose their acerbity, and even the despairing invalid catches the contagion of cheerfulness.

Two-thirds of the population of Augustine (amounting to 1,300 whites) are of Spanish origin, and still speak the Spanish language. The women are pretty, modest, dark-eyed brunettes; dress neatly in gay colors, are skillful at needlework, and good housewives. The men exhibit equally characteristic traits of race and nationality. The people are generally poor. There are no manufactures. The town produces little, and exports nothing—its chief support, since the loss of its orange groves, being derived from Government offices, receipts from strangers, and the hire of slaves. It has one saw-mill, rarely running. It has a bathing-house, for the prevention of sickness, and three good physicians and a dentist to cure it. Perhaps no city in the Union is healthier than Augustine.

St. Marys.—Hotels:—*Rail Road Hotel.*

St. Marys may be included in this region, though it lies in the State of Georgia, yet still near the north-east line of Florida. It is upon the St. Marys River, nine miles from the sea. The village is a pleasant one, and the healthfulness of its climate makes it deservedly a place of invalid resort.

Tallahassee.—Hotels:—*City Hotel.*

Tallahassee, the capital of Florida, is a pleasant city, of some 1,400 inhabitants, in the centre of the northern and most populous part of the State, near the head of the Gulf of Mexico. It is connected by railroad, 26 miles, with St. Marks, near the Gulf. It is regularly built upon a somewhat elevated site. Some of its public edifices are highly respectable, but do not call for any especial remark.

Chief among the attractions of Tallahassee are the many beautiful springs found in the vicinity. Ten miles from the city is a famous fountain, called *Wachulla*. It is an immense limestone basin, as yet unfathomed in the centre, with waters as transparent as crystal.

St. Marks is on St. Marks River, near the Gulf of Mexico, and 26 miles from Tallahassee, by railroad.

From Pensacola to Tallahassee, Flo.—To La Grange (on Choctawhatchie Bay), by steamboat, 65 miles; by stage to Holmes Valley, 25; Oakey Hill, 42; Marianna, 66; Chattahoochee, 90; Quincy, 108; Salubrity, 117; Tallahassee, 130.

From Jacksonville to Tallahassee, Flo.—To the White Sulphur Spring, 82 miles. This curious spring rises in a basin ten feet deep and thirty in diameter; it discharges a quantity of water, and after running a course of about 100 feet, enters the Suwanee River. The waters have been found very beneficial in cases of consumption, rheumatism, and a variety of other complaints. Visitors will find ample accommodation here. From the mineral spring to Madison, 35 miles; Lipona, 73; Tallahassee, 98—or 180 miles from Jacksonville.

Appalachicola.—Hotels:—

Appalachicola is at the entrance of the river of the same name into the gulf of Mexico, through the Appalachicola Bay. It is easily accessible by the river and the Gulf, and is a place of large cotton shipments. It is 135 miles south-west of Tallahassee.

Pensacola.—Hotels:—*Bedell House, Winter's House, St. Mary's Hall.*

Pensacola is upon the Pensacola Bay, in the extreme south-west corner of the State, 10 miles from the Gulf of Mexico and 64 east of Mobile. The harbor here is one of the safest on the Florida coast, which is not remarkable for safe harbors. It is well sheltered by St. Rosa Island, and is defended by Forts Pickens, McCrea, and

Barrancas. The population of Pensacola is about 2,000.

Route from Pensacola to Mobile, Ala.—To Blakely, 50; Mobile, 64 miles.

Tampa is on Tampa, formerly Espiritu Santo Bay, which opens on the Gulf of Mexico, near the centre of the western coast of Florida.

Key West City is upon the island of Key West, off the southern extremity of the peninsula, occupying the important post of key to the Gulf passage. It was first settled in 1822, and is now the most populous city of Florida, having a population of about 3,000. It is a military station of the United States. Some 30,000 bushels of salt are annually made at Key West by solar evaporation. Great quantities of sponges, too, are found and exported; but the chief business of the island accrues from the salvages upon the wrecks cast upon the coast. Forty or fifty vessels are every year lost in the vicinity, by which the island profits to the amount of $200,000. The Marine Hospital here, 100 feet long, is a noteworthy building. Fort Taylor, a strong and costly post, defends the harbor. The Charleston and Havana steamers touch at Key West once a week. There is no other reliable mode of access.

A railway now extends from Fernandina on the Atlantic coast, south-westerly across the peninsula, to Cedar Keys on the Gulf of Mexico. Stage lines diverge to various points in the interior. The Pensacola and Georgia Railway will cross the upper part of the State from Jackson west to Tallahassee. This route is at present in operation 25 miles from Tallahassee to Monticello. Other lines will soon connect Tallahassee with Pensacola, and with Savannah, Macon, &c.

GEORGIA.

This great State possesses unrivalled sources of prosperity and wealth, and though they are as yet only in the dawn of development, the traveller will not hesitate to predict for her a glorious future, when he notes the spirit of activity, enterprise, and progress, which so markedly distinguishes her from other portions of the South. While Nature is here everywhere most prodigal in means, man is earnest in improving them. With the will and energy of northern enterprise, utilizing the advantages of a southern soil, who can cipher out the grand result?

Georgia was settled the latest of the "Original" Thirteen States of the Union. She derived her name with her charter from George II., June 9th, 1732. Her first colony was planted by General Oglethorpe, on the spot where the city of Savannah now stands, in 1773; sixty-three years after the settlement of South Carolina, and a century behind most of the original colonies. Three years after the arrival of Oglethorpe, Ebenezer was planted by the Germans, 25 miles up the Savannah River. Darien, on the sea, was commenced about the same time by a party of Scotch Highlanders. Among the early troubles of the colony was a war with the Spaniards in Florida, each party in turn invading the territory of the other.

The people of Georgia took a vigorous part in the Revolution; and the State was in possession of the British a portion of that time. The city of Savannah was taken by them, December 29th, 1778. A bold attempt was made by the combined American and French forces to recapture it, but failed, with the loss to the allies of 1,100 men. The Great Cherokee Country, in the upper part of the State, came into the full possession of the whites in 1838, when the Indians were removed to new homes beyond the Mississippi.

The sea-coast of Georgia, extending about eighty miles, is very similar in character to that of the Carolinas, being lined with fertile islands cut off from the main land by narrow lagoons or sounds. The famous sea-island cotton is grown here; and wild fowl are abundant in all varieties. Upon the main, rice plantations flourish, with all the semi-tropical vegetation and fruit which we have seen in the ocean districts of South Carolina.

Passing northward to the central regions of the State, the cotton fields greet our eyes at every step, until the surface of the country becomes more and more broken and hilly, and, at last, verges upon the great hill-region traversed by the Appalachian or Alleghany Mountains. These great ranges occupy all the northern counties, and present to the charmed eye of the tourist, scenes of beauty and sublimity not surpassed in any section of the Union.

Rivers.—There are many fine rivers in Georgia; but, as with the water-courses of the South generally, they are often muddy, and their only beauty is in the rank vegetation of their shores, with here and there a bold sandy bluff.

The **Savannah** divides the States of Georgia and South Carolina, through half their length. Its course, exclusive of its branches, is about 450 miles. The cities of Augusta and Savannah are upon its banks, and it enters the Atlantic 18 miles below the latter place. From June to November it is navigable for large vessels as far as Savannah, and for steamboats up to Augusta, 230 miles. The river voyage between these points is a very pleasant one, presenting to the eye of the stranger many picturesque novelties, in the cotton fields which lie along the banks, through the upper part of the passage; and in the rich rice plantations below. Approaching Savannah, the tourist will be particularly delighted with the mystic glens of the wild swamp reaches, and with the luxuriant groves of live-oak which shadow the ancient-looking manors of the planters. A few miles above the city of Savannah, he may visit the spot where Whitney invented and first used his wonderful cotton-gin. Whitney was a Yankee schoolmaster of an inquiring turn of mind, and it was during his intervals of rest from pedagogical rule, that he grew impatient of the slow process of picking the cotton-seed from the fibres with the fingers, and set himself to work so effectually to remedy the difficulty. A noble monument should mark the place, and commemorate the achievement; but alas! we live in an irreverent or a forgetful age and country.

The alligator is often seen sunning himself on the shores of the lower waters of the Savannah, being abundant in the contiguous swamps. They are dangerous reptiles to deal with, especially when in ill humor. We once saw a large specimen of this genus, who had swallowed, as his "post-mortem" discovered, a bottle of brandy and a certificate of membership in a Methodist church. The coroner's inquiry asked after the owners of the articles, but inference, only, answered the question.

"When our canoe," says Sir Charles Lyell, in his record of travels in this region, "had proceeded into brackish water, where the river banks consisted of marsh land, covered with a tall, reed-like grass, we came close to an alligator, about nine feet long, basking in the sun. Had the day been warmer, he would not have allowed us to approach so near to him; for these reptiles are much shyer than formerly, since they have learned to dread the avenging rifle of the planter, whose stray hogs and sporting dogs they often devour. About ten years ago, Mr. Cooper tells us he saw two hundred of them together in St. Mary's River, extremely fearless. The oldest and largest individuals on the Altamaha have been killed, and they are now rarely twelve feet long, and never exceed sixteen and a half feet. As almost all of them have been in their winter retreats ever since the frost of last month, I was glad that we had surprised one in his native haunts, and seen him plunge into the water by the side of our boat. When I first read Bartram's account of alligators more than twenty feet long, and how they attacked his boat and bellowed like bulls, and made a sound like distant thunder, I suspected him of exaggeration; but all my inquiries here and in Louisiana convinced me that he may be depended upon. His account of the nests which they build in the marshes is perfectly correct. They resemble haycocks, about four feet high, and five feet in diameter at their bases, being constructed with mud, grass, and herbage. First they deposit one layer of eggs on a floor of mortar, and having covered this with a second stratum of mud and herbage eight inches thick, lay another set of eggs upon that, and so on to the top, there being commonly from one hundred to two hundred eggs in a nest. With their tails they then beat down round the nest the dense grass and reeds, five feet high, to prevent the approach of unseen enemies. The female watches her eggs until they are all hatched by the heat of the sun, and then takes her brood under her care, defending them, and providing for their subsistence. Dr. Luzenberger, of New Orleans, told me that he once packed up one of these nests, with the eggs, in a box for the Museum of St. Petersburgh, but was recommended, before he closed it, to see that there was no danger of any of the eggs being hatched on the voyage. On opening one, a young alligator walked out, and was soon after followed by all the rest, about one hundred, which he fed in his house, where they went up and down the stairs, whining and barking like young puppies. They ate voraciously yet their growth was so slow as to confirm him in the common opinion, that individuals which have attained the largest size are of very great age; though whether they live for three centuries, as some pretend, must be decided by future observations."

The **Oconee** rises in the gold lands of the mountain districts of Georgia, and traverses the State until it meets the Ogeechee, and with that river reaches the sea under the name of the Altamaha. Milledgeville, the capital of Georgia, is upon the Oconee, 300 miles from the ocean; and Athens, one of the most beautiful places in the State, and the seat of the University of Georgia, is also passed by its waters. Small steamboats may ascend the Oconee as far as Milledgeville; but now, with the more speedy travel by railway, there is little need of them.

The **Ockmulgee** is navigable for small steamboats to Macon.

The **Flint River**, in the western part of the State, passes by Lanier, Oglethorpe, and Albany, and uniting with the Chattahoochee, at the south-west extremity of the State, forms the Appalachicola. The length of the Flint River is about 300 miles. Its navigable waters extend 250 miles, from the Gulf of Mexico to Albany.

The **Chattahoochee** is one of the largest and most interesting rivers of Georgia. It pursues a devious way through the gold region westward from the mountains in the north eastern part of the State, and makes the lower half of the dividing line between Georgia and Alabama. At the point where it enters Florida, it is joined by the Flint River, and the united waters are thenceforward called the Appalachicola. The Chattahoochee is navigable for large steamboats as far up as Columbus, 350 miles from the Gulf of Mexico. The principal towns on this river besides Columbus, are Eufaula, West Point, and Fort Gaines.

Just above Columbus there are some picturesque rapids in the Chattahoochee, overlooked by a fine rocky bluff, famous in story as the "Lover's Leap." The scene would be a gem in regions the most renowned for natural beauty. On the left, the river pursues its downward course to the city, in a straight line. Its flow is rapid and wild, broken by rocks, over which the water frets and foams in angry surges. The bed of the stream is that of a deep ravine, its walls lofty and irregular cliffs, covered to their verge with majestic forest growth. From this point the city of Columbus is but partially visible. The village of Girard and the surrounding hills on the Alabama side, form a distinct and beautiful background to the picture. The fine bridge which spans the river at Columbus, and the steamboats which bear the exchanges of wealth over the waters, are dimly seen through the mist which clothes the Falls of Coweta.

Railway Routes. The *Georgia Railway* extends, in a westerly direction, 171 miles from Augusta to Atlanta, passing through Belair, Berzelia, Dearing, Thomson, Camak, Cumming, Crawfordville, Union Point, Greensboro', Oconee, Buckhead, Madison, Rutledge, Social Circle, Covington, Conyer's, Lithonia, Stone Mountain, and Decatur. A branch line, 10 miles long, extends from Camak to Warrenton, the capitol of Warren County; another of 18 miles from Cumming to Washington, the capital of Wilkes County; another from Union Point to Athens, the capital of Clarke County. The road (the Georgia) connects at Augusta with the South Carolina road for Charleston. The Augusta and Waynesboro' extends 53 miles to Millen, a station on the Central road, from Savannah to Macon Stations: Waynesboro', Thomas and Lumpkin.

The Western and Atlantic Road extends from the Georgia Railway at Atlanta, 138 miles, northward to Chattanooga, Tennessee. *Stations*—Atlanta to Vining's, 8 miles; Marietta, 20; Acworth, 35; Allatoona, 40; Cartersville, 47 · Cass, 52; Kingston, 59; Adairsville, 69; Calhoun, 78; Resaca, 84; Tilton, 91; Dalton, 100; Tunnell Hill, 107; Ringgold, 115; Johnson, 120; Chickamauga, 128; Boyce, 133; Chattanooga, 138 miles. This road is continued (from Dalton) by the East Tennessee and Georgia, to Knoxville, Tennessee.

The *Rome Railway* deflects from the Western and Atlantic at Kingston, and extends 20 miles to Rome.

The Atlanta and Lagrange Road extends from the Georgia Road at Atlanta, 87 miles to West Point, from whence it is continued by other routes to Montgomery, Alabama. *Stations*—Atlanta to East Point, 6 miles; Fairburn, 18; Palmetto, 25; Powell's, Newnan, 40; Grantville, 52; Hogansville, 59; LAGRANGE, 72; Long Cane, 78; West Point, 87 miles.

The Central Railway extends 191 miles from Savannah to Macon. *Stations*—Savannah to Eden, 20 miles; Guyton, 30; Egypt, 40; Armenia, 46; Halcyondale, 50; Ogeechee, 62; Scarboro', 71; Millen, 79 (branch road 53 miles to Augusta); Cushingville, 83; Birdsville, 90; Midville, 94; Holcomb, 108; Speir's Turnout, 112; Davisboro', 123; Tennille, 136; Oconee, 146; Emmett, 153; Kingston, 160; Gordon, 171 (branch to Milledgeville and Eatonton); Griswoldville, 182; Macon, 191 miles.

Milledgeville and Eatonton Branch of Central Road. *Stations*—Gordon to Wolsey, 9 miles; Milledgeville, 18; Dennis, 29; Eatonton, 38 miles.

Macon and Western extends 101 miles from Macon to Atlanta, terminus of Georgia railway. *Stations*—Macon, Junction, Howard's, 6 miles; Crawford's, 13 miles; Smarr's, 19; Forsyth, 24;

GEORGIA. 187

Collier's, 30; Goggin's, Barnesville, 40; Milner's, 47; Thornton's, Griffen, 58; Fayette, 65; Lovejoy's, Jonesboro', 79; Rough and Ready, 90; East Pond, 95; Atlanta, 101 miles.

The Muscogee Railway extends from Macon, terminus of Central road, 99 miles, to Columbus, with Branch to Americus. *Stations*—Macon to Echeconnee, 17 miles; Mule Creek, 21; Fort Valley, 28 (Americus Branch); Everett's, 35; Reynolds', 41; Butler, 50; Columbus, 99 miles.

The South-Western (or Americus Branch of Muscogee) station as above, from Macon to Fort Valley, 28 miles; thence to Marshallville, 7; Winchester, 9; Oglethorpe, 21; Anderson, 30; Americus, 41; Sumter, 51; Albany, 76 miles.

The Savannah, Albany and Gulf Road will connect Savannah and Tallahassee. It extends at present from Savannah 107 miles to McDonald's Station, from which point a line of stages runs to Thomasville, 110 miles, passing through Mill Town and Troupville, and connecting with stages for Tallahassee and other places in Florida. Various deflections from the routes we have here named are in progress or in contemplation in Georgia, all confirming the reputation of the State as the Southern leader in this great field of human enterprise and progress.

Savannah.—Hotels:—The principal hotels, and they are most excellent ones, are the *Pulaski House* in Johnson or Monument Square, the *City Hotel* in Bay street, and the *Screven House*. They are all eligibly and pleasantly situated in the heart of the city.

Savannah, the largest city of Georgia, with a population of about 16,000 whites and 12,000 blacks, is upon the south bank of the Savannah river, 18 miles from the sea. Its site is a sandy terrace, some forty feet above low water mark. It is regularly built, with streets so wide and so unpaved—so densely shaded with trees, and so full of little parks, that but for the extent and elegance of its public edifices, it might seem to be an overgrown village, or a score of villages rolled into one. There are no less than twenty-four little green squares scattered through the city, and most of the streets are lined with the fragrant flowering China tree, or the Pride of India, while some of them, as Broad and Bay streets, have each four grand rows of trees, there being a double carriage-way, with broad walks on the outsides, and a promenade between.

Among the public buildings of note in Savannah are the new Custom House, the City Exchange, Court House and Theatre, the State Arsenal, the Armory, the Oglethorpe and the St. Andrew's Halls, the Lyceum, the Market House, and the Chatham Academy. The St. John's (Episcopal) Church, and the Independent Presbyterian Church, are striking edifices. The city has, besides, a dozen other Protestant and some Catholic churches, and a Jewish Synagogue. The State Historical Society has a fine Library. The public Library has over 7,000 volumes. There are also other literary associations and reading-rooms. The principal charitable institutions of the city are the Orphan Asylum, the Hibernian and Seaman's Friend Societies, the Georgia Infirmary, the Savannah hospital, the Union and the Widow's Societies, and the Savannah Free School.

In Johnson or Monument Square, opposite the Pulaski House, there is a fine Doric Obelisk erected to the memories of Greene and Pulaski, the corner stone of which was laid by Lafayette during his visit in 1825. It is a marble shaft, 53 feet in height. The base of the pedestal is 10 ft. 4 in. by 6 ft. 8 in., and its elevation is about 12 feet. The needle which surmounts the pedestal is 37 feet high. Another and very elegant structure has since been built in Chippewa Square, to the memory of Pulaski. This general fell gallantly during an attack upon the city, while it was occupied by the British in the year 1779.

The vicinage of Savannah, though flat, is exceedingly picturesque along the many pleasant drives, and by the banks of the river and its tributary brooks, leading everywhere through noble avenues of the live oaks, the bay, the magnolias, the orange and a hundred other beautiful evergreen trees, shrubs and vines.

The *Cemetery of Bonaventure*, close by, is a wonderful place. It was originally a private estate, laid out in broad avenues, which cross each other. These avenues are now grand forest aisles, lined with live oaks of immense size; their dense leafage mingling overhead, and the huge lateral branches trailing upon the ground with their own and the superadded weight of the heavy festoons of the pendant Spanish moss. A more beautiful or more solemn home for the dead than in the shades of these green forest aisles, cannot be well imagined. The endless cypress groves of the "silent cities" by the Bosphorus, are not more impressive than the intricate web of these still forest walks.

Bonaventure has thus been sketched by starlight:

"Along a corridor I tread,
High over-arched by ancient trees,
Where, like a tapestry o'erhead,
The gray moss floats upon the breeze:
A wavy breeze which kissed to-day
Tallulah's falls of flashing foam,
And sported in Toccoa's spray—
Brings music from its mountain home.

GEORGIA.

> "The clouds are floating o'er the sky,
> And cast at times a fitful gloom,—
> As o'er our hearts dark memories fly,
> Cast deeper shades on Tatnall's tomb;
> While glimmering onward to the sea,
> With scarce a rippling wave at play,
> A line of silver through the lea,
> The river stretches far away." *

Savannah was founded by General Oglethorpe in 1732. It was occupied in 1778 by the British, and came back into the possession of the Americans in 1783. But few Revolutionary remains are now to be seen, the city having overgrown most of them. Batteries, ramparts, and redoubts have given place to the more pleasant sights of fragrant gardens and shady parks. Mounds and ditches, however, may be traced near the edge of the swamp, south-east of the town. *Jasper's Spring*, the scene of a brave and famous exploit of the war time, may yet be visited. It lies near the Augusta road, two miles and a half from the city westward; the spring is a fountain of purest water, in the midst of a marshy spot, covered with rank shrubbery, at the edge of a forest of oak and pine trees. The interest of the place is in its association only. Sargent Jasper, aided only by one companion, watched by this spring for the passage of an American prisoner, under a British guard of eight men, whom he boldly and successfully assailed, restoring the captive to his country and his friends. In memory of this action, Sargent Jasper's name has been given to one of the public parks of the city.

Savannah is one of the healthiest of the southern cities, and its climate is constantly improving, owing, it is said, to the improved manner of cultivating the great rice lands in the neighborhood. No pleasanter winter home for invalids or others can be found: for, to the balmy climate of the region, and every appliance of physical comfort, there are superadded extraordinary social attractions in the cultivated manners and the hospitable hearts of the people.

Routes from Savannah.—Georgia is famous the Union over for her railroad enterprise. In this respect, at least, she leads all the southern States. Her endless rails traverse her borders, and especially in the central and northern portions, in every direction; linking all her towns and districts to each other, and with all the surrounding States. Between ten and eleven thousand miles of railroad—either finished, or being built—now centre in Savannah, communicating thence, directly or indirectly, with Macon and Columbus, and with Montgomery in Alabama, with Augusta, Atlanta, and onward to Tennessee, etc. Roads, too, are in process of construction, and nearly completed to Charleston and to Pensacola and other points in Florida.

The Central Railroad extends from Savannah, 192 miles, to Macon, with Branch deflecting from Waynesborough to Augusta, and another to Milledgeville. It unites also with the South-Western road, to be extended west to the Chattahoochee river; and from that route by the Muscogee road to Columbus. The Macon and Western links the Central road from Savannah with the Georgia railroad from Augusta at Atlanta; the Western and Atlanta prolongs it thence to Chattanooga in Tennessee, and by other routes to Knoxville. All these and other routes we shall duly follow as we continue our journey through the south and south-west.

Florida is reached at Jacksonsville, St. Augustine, and other places, by regular tri-weekly steamers from Savannah. *See* chapter on Florida.

Augusta.—HOTELS :—*The Planter's*—a first class house.

Augusta, one of the most beautiful cities in Georgia, and the second in population and importance, is on the eastern boundary of the State, upon the banks of the Savannah River, and at the head of its navigable waters, 120 miles N. N. W. from Savannah, and 136 N. W. from Charleston, with both of which cities it has long been connected by railroad. Augusta has now a population of over 13,000, and it is every year greatly increasing. The principal street, parallel with the river, is a noble avenue, in length and breadth. This is the Broadway of the city, wherein all the shopping and promenading are done, and where the banks, and hotels, and markets are to be found. Of late years, Augusta has spread itself greatly over the level lands westward.

A pleasant ride of two or three miles from the heart of the town, brings us to a lofty range of sand-hills, covered with charming summer residences. This high ground is in healthful atmospheres, even when epidemics prevail—as they very rarely do, however—in the city streets below.

There are delightful drives along the banks of the Savannah, particularly below the city; and across the river at Hamburg there are some beautiful wooded and grassy terraces, known as Shultz's Hill, and much resorted to as a pic-nic ground.

Augusta has some fine public buildings and churches. The City Hall, built at a cost of

---

* Bonaventure is upon the Warsaw River, which may be seen gleaming through the forest passages. "Tatnall's tomb," a family vault of the former possessors of the spot, was here, alone, before its adoption as a public cemetery.

$100,000, the Medical College, the Richmond Academy, and the Masonic Hall, are every way creditable to the architectural taste and the liberality of the people. The churches are about fifteen in number. There are also here an arsenal and hospital, and gas works.

The rapid development of the up-country of Georgia, within a few years, has brought down to Augusta, by her railways, great prosperity; and the water power which has been secured by means of a canal, which brings the upper floods of the Savannah River to the city, at an elevation of some forty feet, is enlarging and enriching it by extensive and profitable manufactures. This canal, 9 miles in length, was constructed in 1845.

Routes from Augusta.—To Charleston by the South Carolina Railway; to Savannah by the Central road and the Waynesboro' Branch, and by steamers down the Savannah; to Atlanta by the Georgia Railway, and thence into Alabama or Tennessee by connecting lines; to Macon, Athens, Columbus, and most of the northern towns, by deflecting or intersecting lines of the Georgia road. See Index for the various routes, places, and scenes.

Macon.—Hotels:—The *Lanier House.*

Macon is on the Ockmulgee, 191 miles west-north-west of Savannah, by the Central Railway, of which it is the northern terminus. From Augusta, by the Augusta and Waynesboro', 53 miles, to Millen, on the Central Railway; thence, 112, by the Central road from Savannah. Total distance from Augusta, 165 miles; from Milledgeville (the capital), by railway, 38 miles; from Atlanta, on the Georgia Railway, 101 miles; from Columbus, by the Muscogee and South-western railways, 99 miles. The South-western extends (at present) to Americus, 71 miles from Macon, uniting with the Muscogee for Columbus at Fort Valley. Macon is one of the chief cities in Georgia, in population (about 9,000). It is a prosperous commercial place, and a great cotton mart. The Georgia Female College is located here. Rose Hill Cemetery, on the Ockmulgee, is a pretty rural bit of native woodland. Lamar's Mound is a high rising ground, covered with fine private residences, continued by the pleasant suburban village of Vineville.

Columbus is on the Chattahoochee River, the western boundary of the State; 290 miles from Savannah, by the Central, the South-western, and the Muscogee Railways, via Macon; from Augusta, 264 miles, by the Augusta and Waynesboro', the Central, the South-western, and the Muscogee Railway, or 310 miles by the Georgia Railway to Atlanta, thence by the Atlanta and Lagrange, and the Montgomery and West Point, via Opelica, Alabama; from Macon, by railway, 99 miles; from Atlanta, 189; from Montgomery, Alabama, by railway, 92 miles. Columbus is a handsome commercial city, of some 9,000 inhabitants. Large quantities of cotton are shipped hence for the Gulf of Mexico, via the Chattahoochee. See Chattahoochee River for picturesque scenes in this neighborhood. Girard, Alabama, is connected with Columbus by a fine bridge.

Atlanta.—Hotels:—*Trout House.*

Atlanta is a new and thriving city, at the western terminus of the Georgia Railway. Distance by that route, 171 miles from Augusta; from Macon (railway), 101 miles, and from Savannah (railway), via Macon, 292 miles. The railway routes of Tennessee and of Virginia meet at Atlanta; also railways from Columbus and from Montgomery, Alabama. Atlanta, not many years ago wild forest-land, has already attained to a population of 16,000.

Athens.—Hotels:—*Lanier House.*

Athens is a beautiful up-country town on the Oconee River. From Augusta, by the Georgia Railway, to Union Point, 70 miles; thence by the Athens branch, 43 miles. Total, 113 miles from Savannah, by railway, via Augusta. Athens is the seat of Franklin College, the University of Georgia.

Milledgeville. — Hotels: — *Milledgeville Hotel; McComb's Hotel; Washington Hall.*

Milledgeville, the capital of Georgia, a town of about 3,000 people, is upon the Oconee River, in the midst of a fine cotton-growing region. From Savannah, by the Central Railway, to Gordon, 171 miles, and thence by the Milledgeville and Eatonton, 18 miles. Total, 189 miles. From Augusta, by the Augusta and Waynesboro', to Millen, on the Central road, 53 miles; thence by the Central (as from Savannah). Total distance from Augusta, 163 miles; from Columbus, 135 miles, and from Atlanta, 139 miles.

The Capitol at Milledgeville is a large semi-Gothic structure.

The Oglethorpe University is at Midway, a pretty village on the railway, 1¼ miles below Milledgeville.

The Mountain Region of Georgia.—Throughout all Northern Georgia, the traveller will find a continuation of the charming Blue Ridge landscape, which we have already explored in the contiguous regions of Upper South Carolina, and North Carolina *West.* This picturesque district in the "Pine State" extends from Rabun County, in the north-eastern corner of the State, to Dade, in the extreme north-west, where the summit of the Lookout Mountain oversees the valley of the Tennessee. Here are the famous gold lands, and in the midst of them the Dahlonega branch of the United States Mint.

The most frequented, if not the finest scenes in this neighborhood are in the north-east, as the wonderful Falls of Tallulah and Toccoa, the valley of Nacochee and Mount Yonah in Habersham County, the Cascades of Eastatoia and the great Rabun Gap in Rabun; all within a day's ride of the Table Mountain, Cæsar's Head, Jocasse, the Whitewater Falls, and other wonders of South Carolina. Further west are the Falls of Amicalolah, the Cahutta Mountain, the Dogwood Valley, and Mount Look-out. This was formerly the hunting-ground of the Cherokees; and, indeed, not many years have passed since the final removal of this tribe to new homes beyond the Mississippi.

Clarksville, a pleasant village in Habersham County, is a favorite summer residence of the people of the "Low country" of Georgia, and the point of rendezvous for the exploration of the landscape of the region—the point from whence to reach Tallulah, Toccoa, Nocoochee, etc. From Charleston or Columbia, or other places in South Carolina, follow the railways to Greenville or to Anderson, S. C., and proceed thence by stage, one to two days' ride, to Clarksville; or take the Georgia railways from Augusta to Athens, and thence by stage, one or two days' travel, to Clarksville, passing the Madison Springs, Mount Currahee, and Toccoa.

Toccoa Falls (for route see Clarksville, above), is in the County of Habersham, a few miles from the village of Clarksville.

The late Judge Charlton, describing this famous scene, says:

Several years have passed away since I last stood at the beautiful Fall of the Toccoa. It was one of the delightful summer days peculiar to the climate of Habersham County. The air had all the elasticity of the high region that surrounded us, and the scenery was of a character to elevate our spirits and enliven our fancy.

A narrow passage led us from the road-side to the foot of the Fall. Before us appeared the perpendicular face of rock, resembling a rugged stone wall, and over it,

"The brook came babbling down the mountain's side."

The stream had lost much of its fulness from the recent dry weather, and as it became lashed into fury, by its sudden fall, it resembled a silver ribbon, hung gracefully over the face of the rock, and waving to and fro with the breath of the wind. It reminded me more forcibly than any other scene I had ever beheld, of the poetic descriptions of fairy-land. It is just such a place—as has been often remarked by others—where we might expect the fays and elves to assemble of a moonlight night, to hold their festival on the green bank, whilst the spray, clothed with all the varied colors of the rainbow, formed a halo of glory around their heads. It is, indeed, beautiful, surpassingly beautiful: the tall trees reaching but half-way up the mountain height, the silver cascade foaming o'er the brow of the hill, the troubled waves of the mimic sea beneath, the lulling sound of the falling water, and the call of the mountain birds around you, each and all come with a soothing power upon the heart, which makes it anxious to linger through the long hours of the summer day.

Tearing ourselves away from the enchantment that held us below, we toiled our way up to the top of the Fall, using a path that wound around the mountain. When we reached the summit, we trusted ourselves to such support as a small tree, which overhangs the precipice, could give us, and looked over into the basin beneath. Then, growing bolder as our spirits rose with the excitement of the scene, we divested ourselves of our boots and stockings, and waded into the stream, until we approached within a few feet of the cascade. This can be done with but little danger, as the brook keeps on the even and unruffled tenor of its way until just as it takes its lofty plunge into the abyss below.

The height of the Fall is now 186 feet; formerly it was some feet higher, but a portion of the rock was detached some years ago by the attrition of the water, and its fall has detracted from the perpendicular descent of the stream.

"Beautiful streamlet! onward glide,
In thy destined course to the ocean's tide!
So youth impetuous, longs to be—
Tossed on the waves of manhood's sea:
But weary soon of cloud and blast,
Sighs for the haven its bark hath passed;
And though thou rushest now with glee,
By hill and plain to seek the sea—
No lovelier spot again thou'lt find,
Than that thou leavest here behind;
Where hill and rock 'rebound the call'
Of clear Toccoa's water-fall!"

There are picturesque legends connected with this winsome spot; one of them narrates the story of an Indian chief and his followers, who, bent upon the extermination of the whites, and trusting to the guidance of a woman, was led by her over the precipice, and, of course, perished in their fall.

GEORGIA.

The Cataracts of Tallulah are 12 miles from Clarksville (see route to Clarksville), by a road of very varied beauty. From Toccoa to Tallulah the cut across is five or six miles only. There is a comfortable hotel near the edge of the gorges traversed by this wild mountain stream, and hard by its army of waterfalls.

The Tallulah or *Terrora*, as the Indians more appositely called it, is a small stream, which rushes through an awful chasm in the Blue Ridge, rending it for several miles. The ravine is 1,000 feet in depth, and of a similar width. Its walls are gigantic cliffs of dark granite. The heavy masses piled upon each other in the wildest confusion, sometimes shoot out, overhanging the yawning gulf, and threatening to break from their seemingly frail tenure, and hurl themselves headlong into its dark depths.

Along the rocky and uneven bed of this deep abyss, the infuriated Terrora frets and foams with ever-varying course. Now, it flows in sullen majesty, through a deep and romantic glen, embowered in the foliage of the trees, which here and there spring from the rocky ledges of the chasm walls. Anon, it rushes with accelerated motion, breaking fretfully over protruding rocks, and uttering harsh murmurs, as it verges a precipice,

> "Where, collected all,
> In one impetuous torrent, down the steep
> It thundering shoots, and shakes the country round:
> At first, an azure sheet, it rushes broad;
> Then whitening by degrees as prone it falls,
> And from the loud-resounding rocks below
> Dash'd in a cloud of foam, it sends aloft
> A hoary mist, and forms a ceaseless shower."

The most familiar point of observation is the Pulpit, an immense cliff which projects far into the chasm. From this position, the extent and depth of the fearful ravine, and three of the most romantic of the numerous cataracts are observed. At various other localities fine glimpses down into the deep gorge are afforded, and numerous other paths lead to the bottom of the chasm. At the several cataracts—the *Lodore*, the *Tempesta*, the *Oceana*, the *Serpentine*, and others,—the picture is ever a new and striking one—which the most striking and beautiful, it would be very difficult to determine. The natural recess called the Trysting Rock, once the sequestered meeting-place of Indian lovers, is now a halting-spot for merry groups as they descend the chasm, just below the Lodore cascade. From this point, Lodore is upon the left, up the stream; a huge perpendicular wall of parti-colored rock towers up in front and below; to the right are seen the foaming waters of the Oceana cascade, and the dark glen into which they are surging their maddened way. Tempesta, the Serpentine, and other falls, lie yet below.

The wild grandeur of this mountain gorge, and the variety, number, and magnificence of its cataracts, give it rank with the most imposing waterfall scenery in the Union.

The **Valley of Nacoochee**, or the Evening Star, is said by tradition to have won its name from the story of the hapless love of a beauteous Indian princess, whose sceptre once ruled its solitudes. With or without such associations, it will be remembered with pleasure by all whose fortune it may be to see it. The valley-passages of the South are specialties in the landscape, being often so small and so thoroughly and markedly shut in, that each forms a complete picture, neither more nor less, in itself. The little vale of Jocassee, in South Carolina, is such a scene, and that of Nacoochee is another, and yet finer example.

Nacoochee, like Tallulah and Toccoa, is a matter of a day's excursion from Clarksville.

Mount Yonah looks down into the quiet heart of Nacoochee, lying at its base. If the tourist should stay over-night in the valley, as he will be apt to do, he will take a peep at the mountain panorama to be seen from the summit of old Jonah.

The **Falls of the Eastatoia** are some three or four miles from the village of Clayton, in Rabun, the extreme north-eastern county of Georgia. They lie off the road to the right, in the passage of the Rabun Gap, one of the mountain ways from Georgia into North Carolina. Clayton may be reached easily from Clarksville, the next town southward, or in a ride of 12 miles from the Falls of Tallulah.

The village of Clayton is an out-of-the-way little place, occupying the centre of a valley completely encircled by lofty mountain ranges.

The Eastatoia, or the Rabun Falls, as they are otherwise called, would be a spot of crowded resort, were it in the midst of a more thickly peopled country. The scene is a succession of cascades, noble in volume and character, down the ravined flanks of a rugged mountain height. From the top of one of the highest of the falls, a magnificent view is gained of the valley and waters of the Tennessee, north of the village of Clayton, and the hills which encompass it.

In the neighborhood of Eastatoia, and, indeed, all through Rabun County, the traveller will find everywhere delightful hill, valley, and brook scenery. We once traversed all the region leisurely, and with great pleasure, *en route* from Clarksville to the French Broad River, in North Carolina.

Mountain Accommodations. We

ought, perhaps, to remind the traveller, that when he leaves the frequented routes hereabouts, or anywhere among the Southern hills, he must voyage in his own conveyance, wagon or horseback (the latter the better), stop for the night at any cabin near which the twilight may find him, content himself with such fare as he can get (we won't discourage him by presenting the *carte*), and pay for it moderately when he resumes his journey in the morning.

Union County, lying upon the north-west line of Habersham, is distinguished for natural beauty, and for its objects of antiquarian interest.

The **Track Rock**, in Union, bearing wonderful impressions of the feet of curious animals now extinct, must be seen to be believed.

Pilot Mountain, also in Union, is a noble elevation of some 1,200 feet.

Hiawassee Falls, in the Hiawassee River; there are some beautiful cascades, some of them from 50 to 100 feet in height.

The **Falls of Amicalolah** are in Lumpkin County, south-west of Habersham. They lie some 17 miles west of the village of Dahlonega, near the State road leading to East Tennessee. The name is a compound of two Cherokee words—"Ami," signifying water, and "Calolah," rolling or tumbling; strikingly expressive of the cataract, and affording us another instance of the simplicity and significant force of the names conferred by the untutored sons of the forest.

The visitor will rein up at the nearest farmhouse, and make his way thence, either up the Rattlesnake Hollow to the base of the Falls, or to the summit. The range of mountains to the south and west, as it strikes the eye from the top of the falls, is truly sublime; and the scene is scarcely surpassed in grandeur by any other, even in this country of everlasting hills. The view from the foot embraces, as strictly regards the falls themselves, much more than the view from above, and is therefore, perhaps, the better; both, however, should be obtained in order to form a just conception of the scene; for here we have a succession of cataracts and cascades, the greatest not exceeding 60 feet, but the torrent, in the distance of 400 yards, descending more than as many hundred feet. This creek has its source upon the Blue Ridge, several miles east of the falls; and it winds its way, fringed with wild flowers of the richest dyes, and kissed in autumn by the purple wild-grapes which cluster over its transparent bosom; and so tranquil and mirror-like is its surface, that one will fancy it to be a thing of life, conscious of its proximate fate, rallying all its energies for the startling leap; and he can scarcely forbear moralizing upon the oft-recurring and striking vicissitudes of human life, as illustrated in the brief career of this beautiful streamlet.

From an elevated point, attained in ascending the mountain on the east, Dahlonega, embosomed in its lovely hills, is distinctly visible; several of the principal buildings are distinguishable—among them, the United States Branch Mint.

The Look-out Mountain. On the summit of this beautiful spur, the north-west corner of Georgia and the north-east extremity of Alabama meet on the southern boundary of Tennessee. Almost in the shadow of the Look-out heights lies the busy town of Chattanooga, in Tennessee, on the great railway route from Charleston via the Georgia roads to Knoxville, and thence by the Virginia railways to the north; and on the other hand westward, through Nashville, to the Ohio and the Mississippi. See Chattanooga in the chapter on Tennessee.

The country around the "Look-out" is extremely picturesque; the views all about the mountain itself are admirable, and nothing can exceed in beauty the charming valley of the Tennessee and its waters, as seen from its lofty summit. It is, too, in the immediate vicinage of other remarkable localities, the Dogwood Valley, hard by; Georgia and the Nickajack Cave in Alabama.

The Nickajack Cave. The mouth of this wonderful cavern which has only to be known in order to be famous, is in Alabama, although otherwise it traverses Georgia territory. We leave it, therefore, for our chapter on Alabama.

There are some other mountain and waterfall pictures in Georgia besides those in the upper tier of counties—a few isolated scenes lower down, standing as outposts to the hill-region, as Mount Currahee, the Rock Mountain, and the Falls of Towaliga.

Mount Currahee is on the upper edge of Franklin County, adjoining Habersham, where we have already visited the Falls of Tallulah and Toccoa, Nacoochee and Yonah, and on the stage route from Athens (see route to Clarksville) to those scenes. It is about 16 miles above the village of Cairnesville, and a few miles below the Toccoa cascade.

Mount Currahee, in the midst of mountains, might not be very noticeable; but isolated as it is, and as an appetizer for the feast of wild beauties which the traveller from the lowlands is anticipating, it is always a scene of much interest.

The **Rock Mountain** is a place of great repute and resort in the western part of the State. It is in De Kalb County, where also is Atlanta, the western terminus of the Georgia Railway. It may thus be easily reached by the Georgia Road from Augusta, and all points thereon, and from all places on the many different railways meeting at Atlanta. (See Atlanta.) The precise locality of the Rock or Stone Mountain is at the Stone Mountain station on the Georgia Railway, 15 miles east of Atlanta, and 9 miles east of Decatur, the capital of the county. Accommodations are ample. The mountain stands alone in a comparatively level region. It covers 1,000 acres of surface. Its circumference is about six miles. Its height above the sea 2,230 feet, yet increased by the addition of an observatory.

The western view of the mountain, though perhaps the most beautiful, is not calculated to give the beholder a just conception of its magnitude. To obtain this, he must visit the north and south sides, both at the base and at the summit. Pursuing, for half a mile, a road which winds in an easterly direction along the base of the mountain, the traveller arrives directly opposite its northern front. There the view is exceedingly grand and imposing. This side of the mountain presents an almost uninterrupted surface of rock, rising about 900 feet at its greatest elevation. It extends nearly a mile and a half, gradually declining toward the west, while the eastern termination is abrupt and precipitous. The side is not perpendicular, but exhibits rather a convex face, deeply marked with furrows. During a shower of rain, a thousand waterfalls pour down these channels, and if, as sometimes happens, the sun breaks forth in his splendor, the mimic torrents flash and sparkle in his beams, like the coruscations of countless diamonds.

Near the road is a spring, which, from the beauty of its location, and the delightful coolness of its water, is an agreeable place of resort. It is in a shady dell, and its water gushes up from a deep bed of white and sparkling sand. A more exquisite beverage a pure taste could not desire.

Among the curiosities of the mountain, there are two which are especially deserving of notice. One is the "Cross Roads." There are two crevices or fissures in the rock, which cross each other nearly at right angles. They commence as mere cracks, increasing to the width and depth of five feet at their intersection. They are of different lengths, the longest extending probably 400 feet. These curious passages are covered at their junction by a flat rock, about 20 feet in diameter.

Another is the ruins of a fortification, which once surrounded the crown of the mountain. It is said to have stood entire in 1788. When, or by whom, it was erected is unknown. The Indians say that it was there before the time of their fathers.

The Falls of the Towalaga would be beautiful anywhere, and they are therefore particularly so, occurring as they do in a part of the State not remarkable for its picturesque character. They lie some distance south of the Rock Mountain, and may be easily reached from Forsyth or Griffin, on the line of the railway from Macon to Atlanta.

The river above the falls is about three hundred feet in width, flowing swiftly over a rocky shoal. At its first descent, it is divided by a ledge of rock, and forms two precipitous falls for a distance of fifty feet. The falls are much broken by the uneven surface over which the water flows, and on reaching their rocky basin, are shivered into foam and spray.

From the foot of this fall the stream foams rapidly down its declivitous channel for two hundred feet, and again bounds over a minor precipice in several distinct cascades, which commingle their waters at its base in a cloud of foam.

The Indian Springs are in Butts County, near the Falls of the Towalaga. Stop at Forsyth or Griffin, on the railway between Macon and Atlanta.

The Madison Springs are on the stage route from Athens to the waterfall region of Habersham County, 7 miles from Danielsville, the capital of Madison County. Take the Georgia Railway and Athens branch to Athens—thence by stage.

The Warm Springs, in Merriweather County, are 36 miles by stage from Columbus. A nearer railway point is Lagrange, on the Atlanta and Lagrange Railway, connecting at Atlanta with the Georgia road from Augusta. These springs discharge 1,400 gallons of water per minute, of 90 degrees Fahrenheit.

The Sulphur Springs are 6 miles north of Gainesville, Hall County, in the upper part of the State. The nearest railway point is Athens, on the branch of the great Georgia Road—thence by stage.

The Rowland Springs are about 6 miles from Cartersville, in Cass County. Cartersville is a station on the great railway route from Charleston via Augusta to Tennessee. Western and Atlantic link 47 miles above Atlanta; 91 miles below Chattanooga.

The Red Sulphur Springs, or "*the Vale of Springs*," are at the base of Taylor's Ridge, in Walker County, the north-west corner of the State. Western and Atlantic Railway. In the vicinity is the Look-out Mountain and other beautiful scenes. No less than twenty springs are found here in the space of half a mile,—chalybeate, sulphur, red, white and black, and magnesia.

The Thundering Springs are in Upson

County, in the west central part of the State. Nearest railway station, Forsyth, on the western and Macon route from Macon to Atlanta.

The Powder Springs—sulphur and magnesia—are in Cobb County, accessible from Marietta, 20 miles above Atlanta, on the Western and Atlantic railway.

ALABAMA.

The natural beauties of Alabama, excepting in the peculiar features of the southern lowlands seen near the coast, are not of such marked interest to the tourist as the landscape of some other States. Still we shall lead attention to many objects most noteworthy and enjoyable.

In the upper region are the extreme southern outposts of the great Appallachian hill ranges; but, as if wearied with all their long journey, they here droop their once bold heads and fall to sleep, willing, perhaps, to accept the poetical signification of the name of the new territory into which they now enter—Alabama, *Here we rest.*

While the upper portion of the State is thus rude and hilly, the central falls into fertile prairie reaches. The extreme southern edge for fifty or sixty miles from the gulf is sometimes a sandy, sometimes a rich alluvial plain.

The climate, like most of all the southern line of States, varies from the characteristics of the tropics below, through all the intermediate degrees to the salubrious and invigorating air of the mountain lands above.

The chief agricultural product of Alabama is cotton, of which great staple it yields more than any other State in the Union. Extensive canebrakes once existed, but they have been greatly cleared away. Sugar cane grows on the south-west neck, between Mobile and the Mississippi. Many of the rich alluvial tracts yield rice abundantly. Tobacco, also, is produced. Indian corn, oats, sweet potatoes, buckwheat, barley, flax, and silk, are much cultivated, besides many other grains, fruits, and vegetables, and large supplies of live stock of all descriptions.

Mineral Products. Alabama is rich in great deposits of coal, iron, variegated marbles, limestone, and other mineral treasures. Gold mines, too, have been found and worked. Salt, sulphur, and chalybeate springs abound.

History. It is supposed that Alabama was first visited by white men in 1541, when the gallant troops of De Soto passed through its wildernesses, on their memorable exploring expedition to the great Mississippi. In 1702, a fort was erected in Mobile Bay by a Frenchman named Bienville, and nine years later the present site of the city of Mobile was occupied. At the peace of 1763, this territory passed into the possession of the English, with all the French possessions (except New Orleans) east of the Mississippi. Until 1802 Alabama was included in the domain of Georgia, and after 1802 and up to 1817 it was a part of the Mississippi Territory. At that period it was formed into a distinct government, and was admitted in 1819 into the Union as an independent State.

The Alabama River is a grand navigable stream, formed by the meeting, some ten miles above Montgomery, of the Coosa and the Tallapoosa. About 45 miles above the Mobile it is joined by the Tombigbee, and the united waters are thence known as the Mobile River. The Alabama is navigable for large steamers through its whole course of 460 miles, from the city of Mobile to Wetumpka. It has long been, and still is, a part of the great highway from Boston and New York to New Orleans. It flows through a country of rich cotton fields, broad savanna lands, and dense forest tracts.

The Tombigbee River flows 450 miles from the north-east corner of Mississippi, first to Demopolis, Alabama, where it unites with the Black Warrior, and thence to the Alabama River, about 45 miles above Mobile. Its course is through fertile savanna lands occupied by cotton plantations. Aberdeen, Columbus, Pickensville, Gainesville, and Demopolis, are upon its banks. Large steamboats ascend 366 miles, to Columbus.

The **Black Warrior River** unites at Demopolis with the Tombigbee (see Tombigbee, above). Tuscaloosa, the capital of the State, is upon its banks. To this point large steamboats regularly ascend, 305 miles, from Mobile. The Indian name of this river was Tuscaloosa, and it is still thus sometimes called.

The **Chattahooche** forms a part of the Eastern boundary of the State. See Georgia.

The **Hill-Region**.—The upper part of Alabama is picturesquely broken by the Alleghanies, which end their long journey hereabouts. In the north-east extremity of the State there are many fine landscape passages.

The **Nickajack Cave** enters the Raccoon Mountain a few miles below Chattanooga, Tennessee, and the Lookout Mountain, and immediately finds its way into Georgia. A magnificent rocky arch of some 80 feet span forms the mouth of the cavern, high up in the mountain side. Just beneath is a dainty little lakelet, formed by the waters of a mysterious brook, which comes from the interior of the cave, and disappears some distance from the point of egress, rising again without. How the waters of this singular pond vanish no one knows, any more than how they come; but vanish they do, for some distance, when they are again seen, making their way, like all ordinary mortal waters, toward other streams. The passage of the cave is made in a canoe, on this subterranean and nameless stream, now through immense chambers of grand stalactites, and now through passages so narrow, that to pass, one must crouch down on his back and paddle his way against the walls and roof of the procrustean tunnel. We thus explored the Nickajack some years ago for seven miles, without finding its end or any signs thereof. At that period no traveller had before penetrated so far, and we have not heard of any additional revelations since. This wonderful Avernus was, at one period of long ago, the rendezvous of the band of a certain negro leader, known as Nigger Jack. His mountain head-quarters were thus called "Nigger Jack's" Cave, a patronymic refined at this day into the more romantic name of the Nickajack. Large quantities of saltpetre are found here.

Natural Bridge.—In Walker County there is a remarkable natural Bridge, thought by some travellers to be more curious than the celebrated scene of the same kind in Virginia.

The **Muscle Shoals** are an extensive series of rapids in that part of the Tennessee River which lies in the extreme northern part of the State. The descent of the water here is 100 feet in the course of 20 miles. The neighborhood is a famous resort of wild ducks and geese, which come in great flocks in search of the shell-fish from which the rapids derive their name. Boats cannot pass this part of the Tennessee except at times of very high water. A canal was once built around the shoals, but it has been abandoned and is falling into decay.

Mineral Springs abound in the upper part of Alabama. The Blount Springs, in Blount County, near the Black Warrior River, are much resorted to; and so also the Bladen Springs, in Choctaw County, in the western part of the State, near the line of the Mobile and Ohio Railways. At Tuscumbia a spring issues from a fissure of the limestone rock, discharging 20,000 cubic feet of water per minute. It forms a considerable brook, which enters the Tennessee 2½ miles below. There are valuable sulphur springs in Shelby and Talladega counties. The Shelby Springs are near Columbiana, on the Alabama and Tennessee River Railway.

Huntsville is a beautiful mountain village of Alabama, on the line of the Memphis and Charleston Railway, 150 miles above Tuscaloosa and 116 below Nashville, Tennessee.

Railways.—The Mobile and Ohio extends 261 miles northward to Okolona, at which point a direct line, now partly in operation, will continue it onward to the mouth of the Ohio River.

The Alabama and Florida road will connect Montgomery and Pensacola.

The Mobile and Girard road will traverse the State from Columbus, Ga., to Mobile. Completed from Montgomery 47 miles to Chunnuggee.

The Pensacola and Georgia road will cross the southern line of Alabama from Pensacola to Tallahassee.

The Montgomery and West Point road, 88 miles, connects at West Point with road to Atlanta, Ga.

The Alabama and Tennessee Railroad runs 110 miles from Selma to Talladega. Steamers from Selma to Mobile. Shelby Springs on this route.

The Memphis and Charleston extends from Memphis, on the Mississippi River, along the lower line of Tennessee and the upper line of Alabama, to Chattanooga, 310 miles; thence by the Georgia and Carolina railways to Atlanta, Macon, Augusta, Savannah, Charleston, &c., and in another direction, to Knoxville, and thence to Virginia.

Mobile.—HOTELS :— *Battle House.*

Mobile is in direct railway communication with all the cities of the north and west. Steam-

ALABAMA.

boats connect daily with New Orleans, 165 miles westward. From St. Louis it is reached by steamers on the Mississippi to New Orleans, or steamers to Cairo, and thence by the Mobile and Ohio Railway. From Montgomery, 330 miles above, by steamboats daily on the Alabama River. Distance of Mobile from New York, 1566 miles; time, four to five days; fare, between 40 and 50 dollars.

From Montgomery to Mobile, by Steamboat.

To Washington		12
Lowndesport	10	22
Vernon	9	31
Miller's Ferry	9	40
Benton	14	54
Selma	28	82
Cahawba	16	98
Portland	23	121
Bridgeport	17	138
Canton	4	142
Prairie Bluff	10	152
Prairie Bluff Landing	24	176
Bell's Landing	20	196
Claiborne	22	218
Gosport	7	225
Oliver's Ferry	8	233
French's Landing	9	242
James' Landing	6	248
Tombigbee River	39	287
Fort St. Philip	23	310
MOBILE	21	331

Fare, $10.

From Mobile to New Orleans, by Steamboat.

To Cedar Point, Ala		30
Portersville	12	42
Pascagoula	13	55
Mississippi City	28	83
Cat Island	11	94
East Marianne	11	105
West Marianne	5	110
St. Joseph's Island	5	115
Grand Island	4	119
Lake Borgne	9	128
Fort Coquilles	11	139
Point aux Herbes	7	146
Lakeport (on Lake Pontchartrain)	15	161

By Railroad.

NEW ORLEANS	5	166

Fare, $5.

Mobile was founded by the French, about the year 1700, and was ceded by that nation to England in 1763. In 1780 England surrendered it to Spain, and on the 5th of April, 1813, it was made over by the Spanish government to the United States. It was incorporated as a city in December, 1819. The present population is about 22,000.

The city is pleasantly situated on a broad plain, elevated 15 feet above the highest tides, and has a beautiful prospect of the bay, from which it receives refreshing breezes. Vessels having a draft of more than 8 feet of water cannot come directly to the city, but pass up Spanish River, six miles round a marshy island, into Mobile River, and then drop down to the city. As a cotton mart and a place of export, Mobile ranks next in importance to New Orleans and Charleston. In 1850 the tonnage of this port was upwards of 25,000 tons. The city is supplied with excellent water, brought in iron pipes for a distance of two miles, and thence distributed through the city. This port is defended by Fort Morgan (formerly Fort Bower), situated on a long low, sandy point at the mouth of the bay, opposite to Dauphin Island. A lighthouse is built on Mobile Point, the lantern of which is 55 feet above the level of the sea.

A number of sailing vessels ply regularly between Mobile and New Orleans, and places in the Gulf of Mexico, and the principal cities on the Atlantic coast. Steamboats also keep up a daily communication with New Orleans, via Lake Borgne, and likewise with Montgomery, continuing the route hence to Charleston, S. C., and the East. The Mobile and Ohio Railroad, a most important work for the city of Mobile and the States through which it will pass, is now under active construction, a portion of which is already opened. This road, in connection with its great link, the Illinois Central Railroad, will be one of the greatest works of the age, extending from the Gulf of Mexico to Lake Michigan, and embracing nearly twelve degrees of latitude. *See* Railways of Alabama, *ante*.

Spring Hill College (Roman Catholic) is located here.

Montgomery. — HOTELS :— The *Exchange Hotel* is a large first-class house.

Montgomery, the capital and the second city of Alabama in population and trade, and one of the most prosperous places in the South, is on the Alabama River, 331 miles from Mobile by water. *See* Mobile for routes hence to that city and to New Orleans. Montgomery is connected by railway with the Georgia roads, and is upon the Great Northern and Southern Line from New York to New Orleans. Population about 8,000.

From Montgomery to Tuscaloosa, Ala., by stage. —To Wetumpka, 15; Kingston, 39; Maplesville, 61·; Randolph, 71; Centreville, 85; Scottsville, 93; Mars, 99; *Tuscaloosa*, 123.

Tuscaloosa.—HOTELS :—*Mansion House.*

Tuscaloosa is upon the Black Warrior River, at the head of steamboat navigation, 125 miles by plank road from Montgomery. It is one of the principal towns of Alabama, and was once the capital. It is the seat of the University of Alabama, established, 1831. The State Lunatic Asylum and a United States Land Office are located here also. Population about 4,000.

For route to Montgomery, *see* Montgomery, *ante*.

From Tuscaloosa to Tuscumbia, Ala., by stage.

—To New Lexington, 24; Eldridge, 51; Thorn Hill, 73; Russelville, 103; *Tuscumbia*, 111.

From Tuscaloosa to Huntsville, Ala., by stage.
—To McMath's, 32; Jonesboro', 44; Elyton, 56: Mount Pinson, 70; Blountsville, 96; Oleander, 120; Lacy Springs, 132; Whitesburg, 139; *Huntsville*, 149.

MISSISSIPPI.

MISSISSIPPI, like Alabama, was first visited by Europeans at the time (about 1541) when the Spanish expedition bore the bright banner of De Soto through all the great belt of forest swamps which lies upon the Mexican Gulf—from the palm-covered plains of Florida on the east, to the far-off floods of the mighty "Father of Waters," on the west.

The enmity of the Indians, and other obstacles, prevented any permanent occupation of the new country at this period. In 1682, La Salle descended the Mississippi River, and visited the territory of its present namesake State. Two years after, he set out again for the region, with a resolute band of colonists, but the venture failed before it was fairly begun, various misfortunes preventing his ever reaching his destination. Iberville, a Frenchman, made the third attempt at a settlement, but with no better success than his predecessors met with. A beginning was, however, at length accomplished, by Bienville and a party of Frenchmen. This expedition settled in 1716 at Fort Rosalie, now the city of Natchez. A dozen years later (1728) a terrible massacre of the new comers was made by their jealous Indian neighbors, which checked, but yet did not stay, the "course of empire." "Manifest destiny" was the watchword of America then, even as it is now; and the whites "still lived," despite decapitation. Other sanguinary conflicts with the aborigines took place in 1736, '39, and 52, with the same final result—the defeat and devastation of the Indian tribes, and the triumph of the invading whites.

The territory fell into the possession of the British crown upon the conclusion of the peace of Paris, in 1763. The strength of the new colony was augmented about this period by portions of the dispersed Acadian communities of Nova Scotia; and soon after a stream of colonists stole down from the New England territories, by the way of the Mississippi and the Ohio Rivers. In 1798 the colony was organized as a Territory, Alabama forming a portion thereof. The State history of Mississippi began December 10, 1817.

Much of the area of Mississippi is occupied by swamp and marsh tracts. There is within her territory, between the mouth of the Yazoo River and Memphis in Tennessee, a stretch of this description, covering an area of nearly 7,000 square miles. It is sometimes a few miles broad, and sometimes not less than a hundred. These low portions of the State are subject to inundation at the time of freshets, and great is the cost and care to protect them, as well as all the lands of a similar character lying along the Mississippi. Banks, or levees are built along the river shores to restrain the unruly floods, but sometimes a breach or crevasse, as such rent is called, occurs, and then woful is the damage and great the risk, not only of property but of life.

Where the country is not thus occupied by swampy or marshy stretches, it sweeps away in broad table-lands, shaped into grand terraces, or steps descending from the eastward to the waters of the great river. The steps are formed by two ranges of bluffs, which sometimes extend to the river shores, and halt abruptly in precipices of fifty and even a hundred feet perpendicular height. These bluffs are features of great and novel attraction to the voyager on the Mississippi River.

The climate of Mississippi has the same general characteristics of the other Southern States, passing from the temperatures of the torrid zone, southward, to more temperate airs above—unlike

Alabama, however, and the South-eastern States of Georgia and Carolina, it has no bold mountain lands within its area.

The climate of Mississippi cannot at present be commended for salubrity; though, as the marsh lands become cleared and cultivated, the fatal miasmas which at present taint the air at certain seasons and in particular districts, will decrease—nay, perhaps disappear entirely; and the dread caution, like that over the entrance to Dante's Inferno, may no more require to be written upon any part of her faithful domain. The winters here, and in the neighboring State of Louisiana, have a temperature a few degrees lower than that of the same latitudes near the Atlantic. The fig and the orange grow well in the lower part of the State, and the apple flourishes in the higher hilly regions. Cotton is the great staple of Mississippi, the State being the third in the Union in this product; the second even, the amount of population being the measure. Besides cotton, however, the varied soil yields great supplies of Indian corn, tobacco, hemp, flax, silk, and all species of grains and grasses, besides live-stock of very considerable value.

Mississippi has no very extensive mineral products; or, if she has, they have not as yet been developed. Some gold has been found, but in no important quantity.

Most of the water-courses here are tributaries of the Mississippi. They run, chiefly, in a south-west direction, following the general slope of the country. Some lesser waters, in the eastern sections, find their way to the Gulf of Mexico, as tributaries of the Pearl River, in the centre of the State, and of the Tombigbee and Pascagoula, in Eastern Mississippi and Western Alabama.

The Yazoo River is a deep and narrow stream, and sluggish in its movements. It is nearly 300 miles in length, exclusive of its branches, and is navigable for steamboats in all its course, and at all seasons, from its mouth to its sources. Its way leads through great alluvial plains of extreme fertility, covered everywhere by luxuriant cotton fields. Vicksburg is 12 miles below the union of the Yazoo with the Mississippi.

The **Tallahatchie**, the largest branch of the Yazoo, has a length almost as great as that river, 100 miles of which may be traversed by steamers.

The Big Black River is some 200 miles long. Its course and destiny are the same as that of the Yazoo, as also the character of the country which it traverses.

The Pearl River pursues a devious course from the north-east part of the State, 250 miles, to Lake Borgne, and thence to the Gulf of Mexico. Jackson, the capital of the State, is upon the Pearl River, south-west of the central region. Small boats sometimes ascend the river as far as this place, though the navigation is almost destroyed by the accumulations of sand-bars and drift-wood.

RAILWAYS IN MISSISSIPPI.

The Mobile and Ohio Road extends, first, along the western edge of Alabama, and afterwards near the eastern line of Mississippi, 261 miles northward from the city of Mobile, Alabama, to Okolona, Mississippi.

The Southern Mississippi, part of a line which will cross the centre of the State from east to west, extends at present eastward from Jackson, the capital of the State, about 45 miles to Forest, and westward, 44¼ miles to Vicksburg.

The Mississippi and Tennessee extends southward from *Memphis*, 97 miles, to Granada, from whence it is continued by the Mississippi Central and the New Orleans, Jackson and Great Northern road to New Orleans.

The Mississippi Central, from Jackson, Tenn., 237 miles, south to Canton, Miss. At Jackson it meets the Mobile and Ohio road north from Mobile, and at Canton it is continued southward by the New Orleans and Great Northern line to the Crescent City.

The New Orleans, Jackson, and Great Northern Railway, from New Orleans 206 miles north to Canton, Miss., continued by other railways direct to Memphis, to the Ohio and all northern cities, is in operation, north-west, to the Mississippi boundary. *See* Railways of Louisiana.

Jackson.—HOTELS:—*Bowman House.*

Jackson, the capital of Mississippi, is upon the Pearl River, south-west from the centre of the State. It is connected by railway, 46 miles, with Vicksburg, on the Mississippi River, by railway, 183 miles, with New Orleans. The Southern Mississippi road extends, at present, 44¼ miles east of Jackson, to Forest. It is a chief point on the great railway route lately opened from New Orleans northward. The State Capitol, the Penitentiary, Lunatic Asylum, and a United States Land Office, are here. Population, about 6,000.

Cooper's Well, in Hind County, 12 miles west of Jackson, is noted for the mineral qualities of its waters.

Natchez.—Hotels :—*Mansion House.*

Natchez, on the Mississippi River, 279 miles above New Orleans, is the most populous and commercial place in the State. It is built upon a bluff, 200 feet above the water, overlooking the great cypress swamps of Louisiana. The lower part of the town, where the heavy shipping business is done, is called Natchez-Under-the-Hill. In Seltzertown, near Natchez, there is a remarkable group of ancient mounds, one of which is 35 feet high. Smaller remains of the kind are found yet nearer the town.

The broken and varied character of the country about Natchez is in most agreeable contrast with the flat lands on the opposite side of the river. The streets are wide and regular, and, to a great extent, elegantly built. The public edifices are well constructed, and the private mansions are pleasantly surrounded with trees and gardens. The town is the centre of an extensive trade, continually upon the increase. Steamers come and go with inspiring despatch, and branch railways will soon link the town with all the great routes of the Union.

Vicksburg.—Hotels :—

Vicksburg is upon the Mississippi, 400 miles above New Orleans, and 46 miles, by railway, from Jackson, the capital of the State. Population, about 4,000.

Aberdeen, a town of some 4,000 inhabitants, is upon the Tombigbee River, 165 miles north-east of Jackson, 28 north of Columbus, and 540 from Mobile, by water. Steamboats ply regularly from Mobile.

Columbus.—Hotels :—

Columbus, population about 4,000, is upon the Tombigbee River, 60 miles below Aberdeen, and 145 miles north-east of Jackson. *See* Jackson for route thither. Regular steamboat communication with Mobile.

Holly Springs is 210 miles above Jackson. It is connected by railway (north) with the line from Memphis to Chattanooga, Tennessee. The Chalmers' Institute and St. Thomas' Hall for boys, the Franklin Female College, and the Holly Springs Female Institute, are here.

The vicinage of Holly Springs is remarkable for its natural beauty and its salubrious climate.

The Lauderdale Springs, sulphur and chalybeate, are in Lauderdale County, in the extreme north-west corner of the State.

LOUISIANA.

Louisiana is one of the most interesting States in the Union, from the romantic incidents of its early history, the peculiar features of its landscape, and its unique social character and life.

The traveller, looking upon the face of the Great River, will recall the bright hopes of De Soto, when he, too, so gazed with delighted wonder; then he will muse upon that hapless destiny which gave the gallant explorer a grave beneath the very floods which he was the first to find and enter, with such exultant anticipations. Then he will remember the visit of La Salle to the mouth of the river, in 1691—next, the attempted settlement, in 1699, under the brave lead of Iberville; then comes the enterprise of Crozart, to whom the country was granted by Louis XIV. in 1712; next comes its history from 1717, while in possession of the famous French financier John Law, and his company of rash speculators, with all the incidents of the story of the brilliant but fleeting "Mississippi Bubble;" next the restoration of the territory to the French Crown, its transfer to Spain in 1762, its retrocession to France in 1800, and its final acquisition by the United States in 1803, when this Government purchased it for $11,500,000, and the further payment of certain claims of American citizens against the Government of France. Of the history of the region in its participation in our national trials, and especially of the memorable event of the battle of New Orleans, we shall speak by and by.

Louisiana in no part of its territory reaches a greater elevation than 200 feet above the level of the Gulf of Mexico, while very much of the Southern region is so low that it becomes inundated

LOUISIANA.

at high water. Marshes extend from the coast; then come the low prairie lands which approach the central parts of the State; and above, the country grows broken and hilly, west of the basin of the Mississippi. In the extreme north-west is a marshy tract of 50 miles in length and 6 in breadth, full of small lakes, made by the interlacings of the arms of Red River. It is estimated that an area of between 8,000 and 9,000 square miles, lying respectively upon the Mississippi and Red Rivers, is subject to inundation annually.

About three-fifths of the whole area of the State is alluvial and diluvial; the rest is occupied by the tertiary formation, and contains coal and iron, ochre, salt, gypsum, and marl. In the vicinity of Harrisonburg, near the north-eastern line of the State, and among the freestone hills which rise hereabouts precipitously to a height of 80 and 100 feet, large quartz crystals have been found, and quantities of jasper, agates, cornelians, sardonyx, onyx, feldspar, crystallized gypsum, alumine, chalcedony, lava, meteoric stones, and fossils.

The exhalations from the marshes in the long hot summers affect the atmosphere, and make Louisiana, in much of its territory, dangerous to the acclimated, and quite unapproachable to strangers, at the season when the especial features of the landscape may be seen in all their greatest glory.

Cotton and sugar-cane are the great products of this State. Of the latter staple, it yielded in 1850 nine-tenths of the whole supply raised in the United States.

The bays and lakes, formed by expansions of the rivers in the marsh lands near the coast, make a marked feature in the landscape of Louisiana, as lakes Pontchartrain, Borgne, Maurepas, &c. Some of these waters we shall see again when we reach New Orleans.

Besides the Mississippi and the Red Rivers, of which the reader will find accounts elsewhere in our volume, the streams in Louisiana do not offer very great attractions to the traveller.

Railways.—But little need of railway communication has heretofore been felt in Louisiana, so great are the facilities of travel by water; though the iron roads now in progress, chartered or projected, will traverse the country in all directions, and connect it advantageously with the neighboring States.

The New Orleans, Jackson and Great Northern Railway extends 206 miles north from New Orleans, through Jackson, the capital of Mississippi, to Canton, and thence to Memphis, Tenn., and all points north and east.

The N. O., Opelousas, and Great Western line is to connect New Orleans and Houston. It extends at present to Brashear (80 miles), on Berwick Bay; connects on Mondays with steamers for Galveston and Sabine Pass, and on Wednesdays and Saturdays for Galveston and Indianola. Returning, the trains connect on Mondays, Wednesdays, and Fridays.

The Mexican Gulf R. R. runs from New Orleans to Proctorsville, on Lake Borgne.

The New Orleans, Milneburg and Lake Pontchartrain, and the New Orleans and Carrollton Railways, are short routes from New Orleans.

The West Feliciana Railway extends 26 m. from Bayou Sara to Woodville.

The Clinton and Port Hudson Road, from Port Hudson, 14 m. to Clinton. A road is now in progress from Vicksburg west to Shreveport, where it is to connect with the proposed Southern Pacific route.

New Orleans. — HOTELS :— The *Crescent City* is famous for the extent and style of its hotels, in a land of sumptuous establishments of this kind.

The *St. Charles* is a splendid "institution" on St. Charles street. Destroyed by fire, it was rebuilt by the close of 1852, at a cost of nearly $600,000. The house was leased at the rate of $30,000 per annum until 1855, and at $40,000 since that period. It has accommodation for nearly 1,000 guests.

The *St. Louis Hotel*, another superb palatial establishment, is upon St. Louis street. It holds the same high rank as the St. Charles.

The *City Hotel* (Camp and Common streets) is another magnificent palace-home for the stranger.

The *St. James*, yet another fine house, is now being constructed, and will be ready in the autumn of this year (1860).

Routes. New Orleans was formerly reached from the north by the routes through the Carolinas, Georgia, and Alabama. Now the favorite and most speedy way is to Washington, and thence, by the Virginia and Tennessee railways and the Great Northern road, through Mississippi; or west to Chicago or Cincinnati, and thence south, by rail. Those who prefer water travel may go inland to some point on the Ohio or the Mississippi, as Cincinnati or Cairo, or St. Louis, and find there good steamers to take them in four or five days to the Crescent City. Distance about 1,700 miles.

New Orleans, the metropolis of the Southwestern States, is built within a great bend of the Mississippi River (from whence its name of the Crescent City), 94 miles from its *debouchure* into the Gulf of Mexico. It is distant from New York 1,663 miles, from Philadelphia 1,576, Boston 1,887, Baltimore 1,478, Washington City 1,438, Charleston, S. C., 879, Cincinnati 1,548, St. Louis 1,201, Pittsburg 2,025, Chicago 1,628, and the Falls of St. Anthony 1,993.

The city is built on land gently descending from the river towards a marshy ground in the rear, and from two to four feet below the level of the river at high water mark. It is prevented from overflowing the city by an embankment of earth, termed the *Levee*, which is substantially constructed, for a great distance along the banks of the river. This Levee is 15 feet wide and 4 feet high, and forms a delightful promenade. It is accessible at all times by vessels of the largest description coming from the ocean, and its advantages of communication with the upper country, and the whole valley of the Mississippi, are at once stupendous and unrivalled. It is not an exaggeration to say that, including the tributaries of this noble river, New Orleans has upwards of 17,000 miles of internal navigation, penetrating the most fertile soils, and a great variety of climates; though at present the resources of this immense valley are only partially developed.

This city is the chief cotton mart of the world. Not unfrequently from a thousand to fifteen hundred flat-boats may be seen lying at the Levee, that have floated down the stream hundreds of miles, with the rich produce of the interior country. Steamboats of the largest class may be observed arriving and departing almost hourly; and, except in the summer months, at its wharves may be seen hundreds of ships and other sailing craft, from all quarters of the globe, landing the productions of other climes, and receiving cargoes of cotton, sugar, tobacco, lumber, provisions, &c. Indeed, nothing can present a more busy, bustling scene than exists here in the loading and unloading of vessels and steamers, with hundreds of drays transporting the various and immense products which come hither from the West.

The receipts and exports of cotton from New Orleans exceeded in the years 1859-'60 two millions and a quarter of bales, the value of which may be set down at one hundred millions of dollars. Besides cotton, a vast amount of other products, as sugar, tobacco, flour, pork, etc., are received at New Orleans, and thence sent abroad. The total value of these products for the year ending Sept. 1, 1859, amounted to $172,952,664. Besides its exports, New Orleans has a large import trade of coffee, salt, sugar, iron, drygoods, liquors, etc., the yearly value of which exceeds $17,000,000.

THEATRES.—New Orleans is as amply supplied with public amusements as with public houses. Both are esteemed there as among the first of human considerations.

A superb opera house was erected on Bourbon street in 1859, and has proved to be a great success, being always crowded by the worshippers of the lyric muse. A new edifice has been erected in the place of the old St. Charles Theatre.

The Orleans Theatre.—The representations at this house are in the French language. It is a very popular resort of the large foreign population of the city.

The American Theatre is another of the leading dramatic establishments. There are still many other minor theatres, and places of amusement, in the city.

The edifices of the City Bank, on Toulouse street, of the Canal Bank, on Magazine street, and of the Bank of Louisiana, are note worthy objects.

THE MARKETS.—The stranger here will be much interested by a visit to the markets. St. Mary's, in the Second District, the Washington Market, in the Third District, and the meat market, on the Levee, are all extensive establishments.

COTTON PRESSES.—There are some 20 or more great cotton presses in New Orleans, each occupying usually a whole block to itself. They are well worth inspection. A fine view of the city may be had from the summit of the dome, which surmounts the centre building of the edifice known as the New Orleans Cotton Press; 150,000 bales of cotton are, it is said, annually pressed at this last mentioned establishment.

Churches.—The city possesses many elegant church edifices.

The Church of St. Louis, opposite Jackson Square, makes an imposing appearance. The entrance is flanked on either side by a lofty tower. The present building was erected in 1850, upon the site of the old church, which was pulled down. The Presbyterian Church, opposite Lafayette Square; the Jewish Synagogue (formerly the Canal street Episcopal Church); St. Patrick's Church, on Camp street, and the new Episcopal Church on Canal street, are all fine structures. The spire of St. Patrick's is a striking feature in the picture of the city, as seen from the river approach. There are 40 or more churches in New Orleans, about one-half of which are Roman Catholic. The church of St. Alphonso, on Constance, and of St. Mary's, on Josephine street, are new and elegant edifices.

The Custom House is, after the Capitol at Washington, the largest building in the United States. It covers an area of 87,333 superficial feet, having a front on Canal street of 334 feet, on Custom House street of 252 feet, on the New Levee of 310 feet, and on the Old Levee of 297 feet. Its height is 82 feet. The chief business apartment is 116 long by 90 broad, and has no less than 50 windows. There is, luckily, no window-tax, though, in the United States. This grand edifice is built of granite, from the Quincy quarries of Massachusetts.

The United States Branch Mint is a noble structure at the corner of Esplanade and New Levee streets. It is three stories high 282 feet in length, and 108 feet deep. It has, besides, two wings, each 81 feet long.

The City Hall is a fine Grecian building of marble. It is at the corner of St. Charles and Havia streets, opposite Lafayette Square.

The Odd Fellows' Hall is a large edifice, opposite Lafayette Square, on Camp street; built, 1852.

The Merchants' Exchange is on Royal near Canal street. *The City Post Office* is in the Exchange, also the Merchants' Reading Room.

The streets of New Orleans are wide, well-paved, and are regularly laid out, usually intersecting each other at right angles. The broadest is Canal street, with a width of 190½ feet, with a grass plot, 25 feet wide, extending in the centre through its whole length. The houses are built chiefly of brick, and are usually five or six stories high. The private dwellings in the suburbs are many, of them very charming places, buried in the grateful shadow of tropical leaves—the magnolia, lemon, myrtle, and orange-tree.

Jackson Square, formerly Place d'Armes, covers the centre of the river-front of the Old Town Plot, now the First District. It is a place of favorite resort. Its shell-strewn paths, its beautiful trees and shrubbery, and its statuary, are all agreeable pleasures to enjoy.

Lafayette Square, in the Second District, is another elegant public park, superbly adorned with fine shade trees and shrubbery.

Congo Square is in the rear of the city. Like the other public grounds, it is a delightful place to lounge away a summer evening.

Literary and Charitable Institutions.—The *University of Louisiana* is on Common street, between Baronne and St. Phillipi streets, occupying the whole front of the block. It has a prosperous *Law School* and a *Medical School*. This University was organized in 1849. The Medical College, which stands in the centre of the block, has a façade of 100 feet. This department was established in 1835. It has a large Anatomical Museum and extensive and valuable collections of many kinds. The State made an appropriation of $25,000 towards the purchase of apparatus, drawings, plates, etc., illustrative of the various branches of medical study. This college had in the year 1859 no less than four hundred students. There is also a school of medicine, numbering two hundred students.

The Charity Hospital (in which the medical students of the University enjoy great facilities for practice) is situated on Common street, between St. Mary's and Gironde street.

The United States Naval Hospital is on the opposite side of the river, a little way above Algiers.

Newspapers.—Over 20 newspapers are published in New Orleans, half of which are dailies of deservedly high repute, at home and abroad. Several of them are printed in the French language. The New Orleans Picayune is famous the world over. De Bow's Review, a commercial journal of distinguished ability, is published here.

Water and Gas Works.—The city is supplied with water from the river, raised by steam to an elevated reservoir, and thence distributed through the streets. Some six millions of gallons are used daily. Gas was introduced in 1834—water the same year.

Cemeteries.—Some of these homes of the dead in New Orleans are deserving of particular notice, both from their unique arrangement and for the peculiar modes of interment. Each is enclosed with a brick wall of arched cavities (or ovens, as they are called here), made just large enough to admit a single coffin, and raised, tier upon tier, to a height of about twelve feet, with a thickness of ten. The whole enclosure is divided into plots, with gravel paths intersecting each other at right angles, and is densely covered with tombs, built wholly above ground, and from one to three stories high. This method of sepulchre is adopted from necessity, and burial *under ground* is never attempted, excepting in the Potter's Field, where the stranger without friends, and the poor without money, find an uncertain rest: the water with which the soil is always saturated often lifting the coffin and its contents out of its narrow and shallow cell, to rot with no other covering than the arch of heaven.

New Orleans was named in honor of the Duke

of Orleans, Regent of France, during the minority of Louis XV. It was the place selected for the seat of the monarchy meditated in the treason of Aaron Burr. Great was the alarm of the citizens in January, 1804, at that prospective insurrection.

The Battle of New Orleans.—This memorable battle-ground lies about four miles from the St. Charles Hotel. It is washed by the waters of the great Mississippi, and surrounded by cypress-swamps and cane-brakes. The action took place January 8th, 1815, between the British troops, under General Pakenham, and the Americans, under Jackson, the former suffering a signal defeat. Pakenham was approaching the city by the way of Lakes Borgne and Pontchartrain, at the time of this terrible repulse. His loss in killed and wounded was nearly 3,000, while the Americans had but 7 men killed and 6 wounded. Jackson's troops fought securely and effectively behind improvised defences of cotton-bags, while the enemy was, unluckily for himself, unsheltered and powerless in the open marshy field. This engagement occurred after the signing of the treaty of peace, but, of course, before intelligence of that event had reached the country.

"Next morning, at daylight," says a traveller, of his approach to New Orleans from Mobile, "we found ourselves in Louisiana. We had already entered the large lagoon, called Lake Pontchartrain, by a narrow passage, and, having skirted its southern shore, had reached a point six miles north of New Orleans. Here we disembarked, and entered the cars of a railway built on piles, which conveyed us in less than an hour to the great city, passing over swamps in which the tall cypress, hung with Spanish moss, was flourishing, and below it numerous shrubs just bursting into leaf. In many gardens of the suburbs, the almond and peach trees were in full blossom. In some places the blue-leaved palmetto, and the leaves of a species of iris (*Iris cuprea*), were very abundant. We saw a tavern called the "Elysian Fields Coffee House," and some others with French inscriptions. There were also many houses with porte-cochères, high roofs, and volets, and many lamps suspended from ropes attached to tall posts on each side of the road, as in the French capital. We might, indeed, have fancied that we were approaching Paris, but for the negroes and mulattoes, and the large verandas reminding us that the windows required protection from the sun's heat.

"It was a pleasure to hear the French language spoken, and to have our thoughts recalled to the most civilized parts of Europe, by the aspect of a city forming so great a contrast to the innumerable new towns we had lately beheld."

Our traveller, just quoted, thus writes of the **Markets.** One morning we rose early to visit the market of the First Municipality, and found the air on the bank of the Mississippi filled with mist as dense as a London fog, but of a pure white instead of yellow color. Through this atmosphere the innumerable masts of the ships alongside the wharf were dimly seen. Among other fruits in the market we observed abundance of bananas, and good pine-apples, for 25 cents (or a shilling) each, from the West Indies. There were stalls where hot coffee was selling, in white china cups, reminding us of Paris. Among other articles exposed for sale were brooms made of palmetto leaves, and wagon loads of the dried Spanish moss, or *Tillandsia*. The quantity of this plant hanging from the trees in the swamps surrounding New Orleans, and everywhere on the Delta of the Mississippi, might suffice to stuff all the mattresses in the world. The Indians formerly used it for another purpose—to give porosity or lightness to their building materials. When at Natchez, Dr. Dickeson showed me some bricks dug out of an old Indian mound, in which the tough woody fibre of the *Tillandsia* was still preserved. When passing through the stalls, we were surrounded by a population of negroes, mulattoes, and quadroons, some talking French, others a patois of Spanish and French, others a mixture of French and English, or English translated from French, and with the French accent. They seemed very merry, especially those who were jet-black. Some of the creoles also, both of French and Spanish extraction, like many natives of the south of Europe, were very dark.

Amid this motley group, sprung from so many races, we encountered a young man and woman, arm-in-arm, of fair complexion, evidently Anglo-Saxon, and who looked as if they had recently come from the North. The Indians, Spaniards, and French standing round them, seemed as if placed there to remind us of the successive races whose power in Louisiana had passed away ; while this fair couple were the representatives of a people, whose dominion carries the imagination far into the future. However much the moralist may satirize the spirit of conquest, or the foreigner laugh at some of the vain-glorious boasting about "destiny," none can doubt that from this stock is to spring the people who will supersede every other in the northern, if not also in the southern continent of America—

————"Immota manebunt,
Fata tibi . . .
Romanos rerum dominos."

The Levee. Soon after our arrival we walked to the Levee or raised bank of the Mississippi, and ascending to the top of the high roof of a large steamer, looked down upon the yellow muddy stream, not much broader than the Thames at London. At first we were disappointed that the "Father of waters" did not present a more imposing aspect; but when we had studied and contemplated the Mississippi for many weeks, it left on our mind an impression of grandeur and vastness, far greater than we had conceived before seeing it.

Panorama of the City. We went next, for the sake of obtaining a general view of the city and its environs, to the top of the cupola of the St. Charles Hotel. If the traveller has expected, on first obtaining an extensive view of the environs of this city, to see an unsightly swamp, with scarcely any objects to relieve the monotony of the flat plain, save the winding river and a few lakes, he will be agreeably disappointed. He will admire many a villa and garden in the suburbs, and in the uncultivated space beyond, the effect of uneven and undulating ground is produced by the magnificent growth of cypress and other swamp timber, which have converted what would otherwise have formed the lowest points in the landscape into the appearance of wooded eminences. From the gallery of the cupola we saw the well-proportioned, massive square tower of St. Patrick's Church, recently built for the Irish Catholics, the dome of St. Louis Hotel, and immediately below us that fine bend of the Mississippi, where we had just counted the steamers at the wharf. Here, in a convex curve of the bank, there has been a constant gain of land, so that in the last twenty-five years no less than three streets have been erected, one beyond the other, and all within the line of several large posts of cedar, to which boats were formerly attached. New Orleans was called the Crescent City, because the First Municipality was built along this concave bend of the Mississippi. The river in this part of its course varies in breadth from a mile to three quarters of a mile, and below the city sweeps round a curve for 18 miles, and then returns again to a point within five or six miles of that from which it had set out. Some engineers are of opinion that, as the isthmus thus formed is only occupied by a low marsh, the current will in time cut through it, in which case, the First Municipality will be deserted by the main channel. Even should this happen, the prosperity of a city, which extends continuously for more than six miles along the river, would not be materially affected, for its site has been admirably chosen, although originally determined, in some degree, by chance. The French began their settlements on Lake Pontchartrain, because they found there an easy communication with the Gulf of Mexico. But they fixed the site of their town on that part of the great river which was nearest to the lagoon, so as to command, by this means, the navigation of the interior country.

Pere Antoine's Date Palm. Walking through one of the streets of New Orleans, near the river, immediately north of the Catholic Cathedral, we were surprised to see a fine date palm, 30 feet high, growing in the open air. The tree is seventy or eighty years old, for Père Antoine, a Roman Catholic Priest, who died about twenty years ago, at the age of eighty, told Mr. Bringier that he planted it himself when he was young. In his will he provided that they who succeeded to this lot of ground should forfeit it if they cut down the palm. Wishing to know something of Père Antoine's history, we asked a Catholic creole, who had a great veneration for him, when he died. He said it could never be ascertained, because, after he became very emaciated, he walked the streets like a mummy, and gradually dried up, ceasing at last to move; but his flesh never decayed, or emitted any disagreeable odor.

If the people here wish to adorn their metropolis with a striking ornament, such as the northern cities can never emulate, let them plant in one of their public squares an avenue of these date palms.

Baton Rouge, the capital of Louisiana, is upon the Mississippi, 129 miles above New Orleans. It is built upon the first of the famous bluffs of the Great River seen in ascending its waters. It is thought to be one of the most healthy places in this part of the country. Besides the State Capitol, the city contains a College and a United States Arsenal and Barracks. The name of Baton Rouge is said to have come thus: When the place was first settled, there was growing on the spot a cypress (a tree of a reddish bark) of immense size and great height, denuded of branches. One of the settlers playfully remarked that it would make a handsome cane. From this small jest grew Baton Rouge (red cane).

The Home of Zachary Taylor. Baton Rouge is interesting as having been the home of the military hero, and President of the United States, General Taylor.

THE MISSISSIPPI.

This mighty river was discovered in 1672, yet its true source was not fully determined until its exploration by Schoolcraft, who, in 1832, found that it took its rise in the small lake called Itasca, situated in 47° 10' N. lat., and 94° 54' W. long. from Greenwich. This lake, called by the

French *Lac la Biche*, is a beautiful sheet of water, of an irregular shape, about eight miles in length, situated among hills covered with pine forests, and fed chiefly by springs. It is elevated above 1,500 feet above the ocean, and is at a distance of more than 3,000 miles from the Gulf of Mexico.

The river drains an extent of territory which, for fertility and vastness, is unequalled upon the globe. This territory, termed the "Mississippi Valley," extends from the sources of the Mississippi in the north to the Gulf of Mexico in the south, and from the Alleghany Mountains on the east to the Rocky Mountains on the west. Or, to give its outline more definitely, we will take a position on the Gulf of Mexico, where it empties its accumulated waters, and run a line north-westward to the Rocky Mountains, from whence issue the sources of the Arkansas, Platte, and other smaller streams; from this point, along the Rocky Mountains to the sources of the Yellowstone and Missouri Rivers; around the northern sources of the latter river to the head-quarters of Red River, a branch of the Assinoboin; around the sources of the Mississippi proper, to the head-quarters of the Wisconsin and Illinois Rivers; between the confluents of the lakes, and those of the Ohio, to the extreme source of the Alleghany River; along the dividing line between the sources of streams flowing into the Ohio River, and those flowing towards the Atlantic; between the confluents of the Tennessee, and those streams emptying into Mobile Bay; between the sources discharged into the Mississippi, and those into the Tombigbee and Pearl Rivers; to the mouth of the Mississippi, and from its mouth to the outlet of the Atchafalaya. The whole presenting an outline of more than 6,000 miles, or an area of about 1,210,000 square miles. The Mississippi River is navigable for steamboats, with but partial interruption, as far north as the Falls of St. Anthony, a distance of 2,037 miles; its course, however, is extremely crooked, and not unfrequently a bend occurs from 20 to 30 miles round, while the distance across is not more than a mile or two. In some instances, however, these distances have been shortened by what are termed "cut-offs," which are made by opening a narrow channel across the neck of a bend, when, on admitting the water, the current, running with such velocity, soon forces a channel both wide and deep enough for the largest steamboats to go through. The navigation is frequently rendered dangerous, owing to the mighty volume of water washing away from some projecting point large masses of earth, with its huge trees, which are carried down the stream. Others, again, are often imbedded in the mud, with their tops rising above the water, and not unusually causing the destruction of many a fine craft. These are called, in the phrase of the country, "snags" and "sawyers." The *whirls*, or *eddies*, caused by the striking peculiarities of the river in the uniformity of its meanders, are termed "points" and "bends," which have the precision, in many instances, as though they had been struck by the sweep of a compass. These are so regular, that the flat-boatmen frequently calculate distances by them; instead of the number of miles, they estimate their progress by the number of bends they have passed.

A short distance from its source, the Mississippi becomes a tolerably sized stream; below the Falls of St. Anthony it is half a mile wide, and below the Des Moines rapids it assumes a medial width and character to the mouth of the Missouri. About 15 miles below the mouth of the St. Croix River, the Mississippi expands into a beautiful sheet of water, called *Lake Pepin*, which is 24 miles long, and from two to four miles broad. The islands, which are numerous, and many of them large, have, during the summer season, an aspect of great beauty, possessing a grandeur of vegetation which contributes much to the magnificence of the river. The numerous sand-bars are the resort, during the season, of innumerable swans, geese, and water-fowl. The Upper Mississippi is a beautiful river, more so than the Ohio; its current is more gentle, its water clearer, and it is a third wider. In general it is a mile wide, yet for some distance before commingling its waters with the Missouri it has a much greater width. At the junction of the two streams it is a mile and a half wide. The united stream, flowing from thence to the mouth of the Ohio, has an average width of little more than three-quarters of a mile. On its uniting with the Missouri it loses its distinctive character; it is no longer the gentle, placid stream, with smooth shores and clean sand-bars, but has a furious and boiling current, a turbid and dangerous mass of waters, with jagged and dilapidated shores. Its character of calm magnificence, that so delighted the eye above, is seen no more.

A little below 39°, on the west side, comes in the mighty Missouri, which, being longer, and carrying a greater body of water than the Mississippi, and imparting its own character to the united stream below, some have thought, ought to have given its name to the river from the junction. Between 36° and 37°, on the east side, comes in the magnificent Ohio, called by the French, on its first discovery, *La Belle Rivière;* for a hundred miles above the junction it is as wide as the parent stream.

"No person who descends the Mississippi

river for the first time, receives clear and adequate ideas of its grandeur, and the amount of water it carries. If it be in the spring of the year, when the river, below the mouth of the Ohio, is generally over its banks, although the sheet of water that is making its way to the Gulf is, perhaps, 30 miles wide, yet, finding its way through deep forests and swamps, that conceal all from the eye, no expanse of water is seen but the width that is curved out between the outline of woods on either bank, and it seldom exceeds, and oftener falls short of a mile. But when he sees, in descending from the Falls of St. Anthony, that it swallows up one river after another, with mouths as wide as itself, without affecting its width at all; when he sees it receiving, in succession, the mighty Missouri, the broad Ohio, St. Francis, White, Arkansas, and Red Rivers, all of them of great depth, length, and volume of water; when he sees this mighty river absorbing them all, and retaining a volume apparently unchanged, he begins to estimate rightly the increasing depths of current that must roll on in its deep channel to the sea. Carried out of the Balize, and sailing with a good breeze for hours, he sees nothing on any side but the white and turbid waters of the Mississippi, long after he is out of sight of land."

TABLE OF PLACES ON THE MISSISSIPPI RIVER, WITH THEIR INTERMEDIATE AND GENERAL DISTANCES.

Distances from the Falls of St. Anthony to St. Louis.

To Fort Snelling, Min. St. Peter's River,		7
ST. PAUL	5	12
Lake Pepin, and Maiden's Rock,	60	72
Chippewa River	25	97
La Crosse	89	186
Root River	5	191
Bad Axe River	20	211
Upper Iowa River	9	220
Prairie du Chien	56	276
Fort Crawford	2	278
Wisconsin River	2	280
Prairie la Port	20	300
Cassville	10	310
Peru	20	330
DUBUQUE	8	338
Fever River	17	355
GALENA, Ill., 7 miles up Fever River,		
Belleview, Iowa	7	362
Savannah, Ill	19	381
Charleston, Iowa	2	383
Lyons, Iowa	15	398
New York, Iowa	5	403
Camanche, Iowa	7	410
Albany, Ill	8	418
Parkhurst, Iowa	19	437
Davenport, Iowa, and Rock Island,	13	450
To Bloomington, Iowa	31	471
New Boston, Ill	26	497
Iowa River	1	498
Oquawke, Ill	20	518
BURLINGTON, Io	15	533
Skunk River, Io	7	540
Madison, Io	16	556
Montrose, Io., and NAUVOO, Ill.	10	566
Keokuk	12	578
Des Moines River, and Warsaw, Ill.	4	582
Tully, Mo	18	600
La Grange, Mo	8	608
Quincy, Ill	12	620
Marion City, Mo	8	628
Hannibal, Mo	11	639
Louisiana, Mo	27	666
Clarksville, Mo	13	679
Hamburg, Ill	13	692
Westport, Mo	14	706
Gilead, Ill	15	721
Bailey's Landing, Mo	3	734
Illinois River, Ill	15	749
Grafton, Ill	2	751
Alton, Ill	18	769
Missouri River, Mo	5	774
ST. LOUIS, Mo	18	792

Distances from St. Louis, Mo., to Cairo, and Mouth of the Ohio River.

To Cahokia, Ill		3
Carondelet, or Vide Pouche, Mo.	4	7
Jefferson Barracks, Mo	2	9
Harrison, Mo	20	29
Herculaneum, Mo	2	31
Selma	4	35
Fort Chartres Island	15	50
St. Genevieve, Mo	11	61
Kaskaskia River, Ill	14	75
Chester, Ill	1	76
La Cuarso's Island	14	90
Devil's Bake-oven, and Grand Tower,	15	105
Bainbridge, Mo	17	122
Devil's Island	8	130
Cape Girardeau, Mo	6	136
Commerce	12	148
Dog-tooth Island	11	159
Elk Island	8	167
CAIRO, Ill., AND MOUTH OF OHIO RIVER,	8	175

Distances from the Mouth of the Ohio River to New Orleans.

To Island No. 1		6
Columbus, Ky	12	18
Wolf's Island, or No. 5	1	19
Hickman, Ky	18	37
New Madrid, Mo	42	79
Point Pleasant, Mo	7	86
Little Prairie, Mo	27	113
Needham's Island, and Cut-off	25	138
Bearfield Landing, Ark	3	141
Ashport, Tenn	5	146
Osceola, Ark	12	158
Plum Point	3	161
1st Chickasaw Bluff	5	166
Fulton, Tenn	2	168

VALLEY OF THE OHIO.

To Randolph, Tenn., and 2d Chickasaw Bluff,	10	178
3d Chickasaw Bluff	17	195
Greenock, Ark	30	225
Wolf River, Tenn. Memphis, Tenn.	20	245
Norfolk, Miss	10	255
Commerce, Miss	17	272
Peyton, Miss	31	303
St. Francis River, and Sterling, Ark.	13	316
Helena, Ark	10	326
Yazoo Pass, or Bayou, and Delta, Miss.	10	336
Horse-shoe Bend	8	344
Montgomery's Pt. Ark. Victoria, Miss.	58	402
White River, Ark	4	406
Arkansas River, Napoleon, Ark.	16	422
Bolivar Landing	13	435
Columbia, Ark	53	488
Point Chicot	4	492
Greenville, Miss	4	496
Grand Lake Landing, Ark	40	536
Princeton, Miss	5	541
Bunches Bend and Cut-off	10	551
Lake Providence, La	19	570
Tompkinsville, La	15	585
Campbellsville, La	16	601
Millikinsville, La	10	611
Yazoo River, Miss., and Sparta, La.	8	619
Walnut Hills, Miss	10	629
VICKSBURG, Miss	2	631
Warrenton, Miss	10	641
Palmyra Sett, Miss	15	656
Carthage Landing, La	4	660
Point Pleasant, La	10	670
Big Black River	14	684
Grand Gulf, Miss	2	686
St. Joseph's, La., and Bruinsburg, Miss.	10	696
Rodney, Miss	10	706
NATCHEZ, Miss	41	747
Ellis Cliff, Miss	18	765
Homochitto River, Miss	26	791
Fort Adams	10	801
Red River Island, and Cut-off	11	812
Raccourci Cut-off and Bend	10	822
Bayou Sara, St. Francisville, and Pt. Coupee, La.	30	852
Waterloo, La	6	858
Pt. Hudson, La	5	863
BATON ROGUE, La	25	888
Plaquemine, La	23	911
Bayou la Fourche, and Donaldsonville, La.	34	945
Jefferson College	16	961
Bonnet Quarre Ch	24	985
Red Church, La	16	1001
Carrolton, La	19	1020
Lafayette, La	4	1024
NEW ORLEANS, La	2	1026

THE OHIO RIVER

Is formed by the junction of the Alleghany and Monongahela, the former being navigable for keel-boats as far as Olean, in the State of New York, a distance of about 250 miles; the latter is navigable for steamboats to Brownsville, 60 miles, and by keel-boats upwards of 175 miles. At Pittsburg commences the Ohio, and after running a course of about a thousand miles, unites its waters with those of the Mississippi. No other river of the same length has such a uniform, smooth, and placid current. Its average width is about 2,400 feet, and the descent, in its whole course, is about 400 feet. At Pittsburg it is elevated about 1,150 feet above the ocean. It has no fall, except a rocky rapid of 22½ feet descent at Louisville, around which is a canal 2½ miles long, with locks sufficiently capacious to admit large steamboats, though not of the largest class. During half the year this river has a depth of water allowing of navigation by steamboats of the first class through its whole course. It is, however, subject to extreme elevations and depressions. The average range between high and low water is probably 50 feet. Its lowest stage is in September, and its highest in March. It has been known to rise 12 feet in a night. Various estimates have been made of the rapidity of its current, but owing to its continually varying, it would be difficult to assign any very exact estimate. It has been found, however, according to the different stages of the water, to vary between one and three miles; in its lowest, however, which is in the autumn, a floating substance would probably not advance a mile an hour.

Between Pittsburg and its mouth it is diversified by many considerable islands, some of which are of exquisite beauty; besides a number of tow-heads and sand-bars, which in low stages of the water greatly impede the navigation. The passages between some of the islands and the sand-bars at their head are among the difficulties of the navigation of the Ohio.

In the infancy of the country, every species of water craft was employed in navigating this river, some of which were of the most whimsical and amusing description. The barge, the keel-boat, the Kentucky-flat or family-boat, the pirogue, ferry-boats, gondolas, skiffs, dug-outs, and many others, formerly floated in great numbers down the currents of the Ohio and Mississippi Rivers to their points of destination, at distances sometimes of three thousand miles.

"Whoever has descended this noble river in the spring, when its banks are full, and the beautiful red-bud and *Cornus Florida* deck the declivities of the bluffs, which sometimes rise 300 feet in height, impend over the river, and cast their grand shadows into the transparent waters, and are seen at intervals in its luxuriant bottoms, while the towering sycamore throws its venerable and majestic arms, decked with rich foliage, over the other trees—will readily acknowledge the appropriateness of the French name, '*La Belle Riviere.*'"

Table of places on the Ohio, from Pittsburg to Cincinnati, with their intermediate and general distances:

Place	Dist.	Total
To Middletown, Pa.	11	
Economy, Pa.	8	19
Freedom, Pa.	6	25
Beaver, Pa.	6	20
Georgetown, Pa.	14	44
Liverpool, Ohio.	4	48
Wellsville, Ohio.	4	52
Steubenville, Ohio.	19	71
Wellsburg, Va.	7	78
Warrenton, Ohio.	7	85
Martinsville, Ohio.	8	93
Wheeling, Va. } Bridgeport, Ohio }	1	94
Elizabethtown, Va. } Big Grave Creek, Va. }	13	107
New Martinsville, Va.	19	117
Sisterville, Va.	29	146
Newport, Ohio.	12	158
Marietta, and } Pt. Harmer, O. }	18	176
Vienna, Va.	6	182
Parkersburg, Va. } Belpre, Ohio }	6	188
Blennerhasset's Island	2	190
Hockingsport, Ohio.	11	201
Bellville, Va.	4	205
Murraysville, Va.	5	210
Shade River, Ohio.	1	211
Ravenswood, Va.	11	222
Letartsville, Ohio.	22	244
Pomeroy.	14	258
Coalport, Ohio } Sheffield, Ohio }	1	259
Point Pleasant, Va. Gt. Kanawha River, Va. }	12	271
Gallipolis, O.	4	275
Millersport, O.	24	299
Guyandotte, Va. } Proctoraville, O. }	13	312
Burlington, O.	8	320
Big Sandy River, Va. } Cattlottsburg, Va. }	4	324
Hanging Rock, O.	13	337
Greenupsburg, Ky.	6	343
Wheelersburg, O.	8	351
Portsmouth, O. } Scioto River, O. }	12	363
Rockville, O.	16	379
Vanceburg, Ky.	3	381
Rome, O.	7	389
Concord, Ky.	6	395
Manchester, O.	7	402
Maysville, Ky. } Aberdeen, O. }	12	414
Charleston, Ky.	7	421
Ripley, O.	2	423
Higginsport, O.	7	430
Augusta, Ky.	4	434
Mechanicsburg, Ky.	7	441
Neville, O.	3	444
Moscow.	4	448
Pt. Pleasant, O. } Belmont, Ky. }	4	452
To New Richmond.	5	457
Little Miami River, O.	14	471
Columbia, } Jamestown, Ky. }	1	472
Cincinnati, O. } Newport and Covington, Ky. }	5	477

Distances from Cincinnati to the mouth of the Ohio.

Place	Dist.	Total
To North Bend, O.	16	
Great Miami River, O.	4	20
Lawrence, Ia.	2	22
Petersburg, Ky.	3	25
Aurora, Ia.	2	27
Belleview, Ky.	6	33
Rising Sun, Ia.	3	36
Big Bone Lick Creek, } Hamilton, Ky. }	12	48
Paoriot, Ia.	2	50
Warsaw, Ky.	10	60
Vevay, Ia.	10	70
Kentucky River.	10	80
Madison, Ia.	12	92
Hanover Landing, Ia.	6	98
New London, Ia.	4	102
Westport, Ky.	6	108
Utica, Ia.	15	123
Jeffersonville, Ia.	9	132
Louisville, Ky. and from Pittsburg.	1	133 … 610
Shippingsport, Ky.	2	135
Portland, Ky. } New Albany, Ia. }	1	136
Salt River and } West Point, Ky. }	18	154
Brandenburg, Ky.	18	172
Mockport, Ia.	3	175
Northampton, Ia.	7	182
Amsterdam, Ia.	3	185
Leavenworth, Ia.	8	193
Fredonia, Ia.	5	198
Alton, Ia.	13	211
Concordia, O.	10	221
Rome, Ia., and } Stevensport, Ky. }	11	232
Cloversport, Ky.	10	242
Carmelton, Ia.	13	255
Troy, Ia.	6	261
Lewisport, Ky.	6	267
Rockport, Ia.	12	279
Owensburg, Ky.	9	288
Bon Harbor, Ky.	3	291
Enterprise, Ia.	3	294
Newburg, Ia.	15	309
Green River, Ky.	6	315
Evansville, Ia.	9	324
Hendersonville, Ky.	12	336
Mount Vernon, Ia.	26	362
Uniontown, Ky.	15	377
Wabash River.	5	382
Raleigh, Ky.	6	388
Shawneetown, Ill.	5	393
Caseyville, Ky.	9	402
Cave in Rock, Ill.	14	416
Elizabeth, Ill.	6	422
Golconda, Ill.	23	445
Cumberland River and } Smithland, Ky. }	17	462
Tennessee River and } Paducah, Ky. }	12	474

To Belgrade, Ill...................	8	482
Fort Massac, Ill...................	2	484
Caledonia, Ill.....................	25	509
America, Ill......................	3	512
Trinity, Ill.......................	5	517
CAIRO, Ill. and MOUTH OF THE OHIO RIVER	5	522
and from Pittsburg.................		999

Distances from Pittsburg and Cincinnati.

	F'm Cin.	F'm P'burg.
To St. Louis, Mo...................	697	1174
Falls of St. Anthony..............	1489	1966
Memphis, Tenn....................	767	1244
Vicksburg........................	1153	1630
Natchez..........................	1269	1746
New Orleans......................	1548	2025

TEXAS.

TEXAS, one of the younger of the great family of American States, came into the Union through much tribulation, her history marked with wars and rumors of wars. In the year 1821 the inducements held out to settlers in this region by the Government of Mexico, to whom the territory at that period belonged, caused an immense rush of emigration thither from the United States. This new and hardier population had grown so great by the year 1832, as to quite absorb and destroy the original feeble spirit of the land under Mexican rule, and to embolden the exotic population to seek the freedom and independence there, to which they had been accustomed at home. With both the will and power to accomplish their purpose, they first demanded admission for their State as an independent member of the Mexican confederacy; and that being refused, they declared themselves wholly free of all allegiance whatsoever to that government. This assumption resulted in a war with Mexico, which after various fortunes was determined in favor of the Texans by the total defeat and capture of the Mexican President Santa Anna, at the memorable battle of San Jacinto, April 21st, 1836. The little village of San Jacinto is in Harris County, near the present city of Houston, in Buffalo Bayou, near its entrance into Galveston Bay.

Texas continued to be an independent nation after the battle of San Jacinto, until her admission in 1846, as a member of the great North American Confederacy.

This fresh turn in events, and the disputes which followed, in respect to boundary lines between the new State and the territory of Mexico, were soon followed by the war between that country and the United States. Again, Texas became the scene of battle and bloodshed, enriching her soil with gallant and brave associations. Two of the famous fights in this war, under the sturdy and victorious lead of the American General, Taylor, occurred within the limits of the present State.

The immortal field of Palo Alto is near the southern extremity of Texas, between Point Isabel and Matamoras, 9 miles north-east of the latter town. The battle took place on the 8th of May, 1846. The American troops numbering 2,111, led by General Taylor, had 32 killed and 47 wounded, while the Mexicans, under General Arista, amounting to 6,000 men, had 252 killed. The American loss unhappily included the gallant Major Ringgold.

The battle-field of Resaca de la Palma lies in the south-eastern extremity of the State, near the entrance of the Rio Grande into the Gulf of Mexico. It is in close vicinage with the field of Palo Alto, 4 miles north of Matamoras, on the route to Point Isabel. This gallant engagement occurred on the 9th of May, 1846, the day following the victory of Palo Alto. The Mexicans, to the number of 6,000, under General Arista, were totally defeated by about 2,000 Americans, commanded by General Taylor. The loss of the former was about 500 killed and wounded, besides all their artillery and furniture: that of the latter was 39 killed and 82 wounded.

Though the Lone Star* has since these days of trial gone on prospering and to prosper, she is not yet entirely at peace in all her borders. At the north-west plains of the State the people are still exposed to the murderous incursions of their Indian neighbors, the fierce and warlike Camanches, Apaches, and other tribes.

* The device of the flag of the Republic of Texas.

The Landscape of Texas.—No one of the Southern States has a greater variety of surface than has Texas. Along the coast on the south-east there is a flat reach of from 30 to 60 miles in breadth; next comes a belt of undulating prairie country extending from 150 to 200 miles wide, and this again is succeeded in the west and north-west by a region of bold hills and table-lands. The plateau of Texas, including some portions of New Mexico, extends about 250 miles, from north to south, and 300 miles from the Rio Grande east. The upper part, Llano Estacado or "Staked Plain," is 2,500 feet above the sea. This immense district is totally destitute of trees and shrubbery, excepting, sometimes, the immediate edge of the streams. Even the stunted grasses which the rains call up, soon wither and die. The Colorado, the Brazos and the Red rivers, find their sources here.

The extreme northern part of the State, extending, perhaps, 60 miles or more, is occupied by a portion of the great American desert. The high lands of the west and north-west are yet a wilderness, visited only by a few bold hunters in quest of the buffalo and other wild animals which abound there. The region, though, is said to have an inviting aspect, and to be well watered and fertile.

The Colorado Hills extend in a north and south direction, east of the Colorado River. Between the Colorado and the Rio Grande, and north of the sources of the San Antonio and Nueces rivers, are broken and irregular chains of hills, probably outposts of the great Rocky Mountain ranges. Some of these hills—as the Organ, the Hueco, and the Guadaloupe Mountains—have an elevation of 3,000 feet above the Rio Grande; and the Guadaloupe group rises to that height above the adjacent plains.

Texas abounds in mineral wealth, as might be supposed from her proximity to the rich mining districts of Mexico. Gold and silver lie buried, no doubt, in large supplies in her soil. Indeed, the latter metal has been already found at San Saba and upon the Bidas River. Exciting rumors prevailed for a while, some few years since, of the detection of gold, west of the Colorado River, and between it and the San Saba Mountains. Coal is supposed to exist about 200 miles from the coast, in a belt extending south-west from Trinity River to the Rio Grande. Iron is found in many parts of the State; and copperas, agates, lime, alum, chalcedony, jasper, and red and white sandstone. There are, too, salt-lakes and salt-springs. In a pitch lake, 20 miles from Beaumont, there are deposits of sulphur, nitre, and fire-clay.

The coast of Texas, like that of the borders of all the Southern States on the Atlantic and the Gulf of Mexico, is lined with a chain of low islands, separated from the main land by bays and lagoons. There are the bays of Galveston, Matagorda, Espiritu Santa, Aranzas, Corpus Christi, and Laguna del Madre. These bays are some 30, and some nearly 100 miles in length.

The Rio Grande, or Rio Bravo del Norte, the largest river in Texas, of which it forms the southern boundary, is 1,800 miles in length. It comes from the Rocky Mountains to the Gulf of Mexico. It is a shallow stream, much broken by rapids and sand-bars, though small steamboats ascend its waters 450 miles from the sea, to Kingsbury Rapids. The "Great Indian Crossing" is about 900 miles from its mouth. At this place is the famous ford of the Apaches and the Camanches, when they make their predatory visits into Mexico.

The Colorado River runs from the tablelands in the north-west part of the State 900 miles to Matagorda Bay. Austin City, Bastrop, La Grange, Columbus, and Matagorda are upon its banks. Austin, the capital of the State, at the head of steamboat navigation, is 300 miles from the sea; at Matagorda, at its mouth, many portions of this river are extremely picturesque.

The Brazos is one of the largest of the Texan rivers. It runs from the table-lands of the west to the Gulf of Mexico, 40 miles below Galveston; the direct distance from its source to its mouth is 500 miles, and, by the windings of its channel, 900 miles. It passes by Waco, Washington, Columbia, and Richmond. At high water the Brazos is navigable 300 miles from its mouth, to Washington, and steamboats may ascend 40 miles, to Columbia, at all seasons. Much of its course is through alluvial plains, occupied with sugar and cotton plantations, fields of Indian corn, and forests of red cedar and of live oak.

The Nueces comes, like most of the rivers of Texas, from the table and hill districts of the west, and flows through the State into the Gulf of Mexico. The Nueces follows a very eccentric course of 350 miles to the Nueces Bay. It may be ascended by steamers 100 miles.

The *San Antonio*, the *Guadaloupe*, the *Trinity*, the *Neches*, and the *Sabine*, other chief rivers of Texas, are, in general character, course, and extent, much like those of which we have already spoken more at length.

The Soil of Texas is as varied as its surface and climate, and, for the most part, extremely fertile. The great staple is cotton, which thrives all over the State, and is of very superior quality in the Gulf districts. Sugar may be profitably cultivated in the level regions. Tobacco is raised with ease, and with scarcely less success than in Cuba itself. All the grains and grasses of the north are found here, with every variety of tropical and other fruits and vegetation. The live oak, in many varieties, abounds in the forests, besides the palmetto, cedar, pine, hickory, walnut, ash, pecan, mulberry, elm, sycamore, and cypress.

Wild Animals. There is every opportunity for the adventurous hunter, in the wildernesses and prairies of Texas, where wild animals of many species abound. In the north-west he may find the wild horse, or mustang, and the fierce buffalo. The deer and the antelope, the moose and the mountain-goat, are plentiful—not to mention the jaguars, the pumas, wild-cats, black bears, ocelots, wolves, and foxes, and such smaller game as peccaries, opossums, raccoons, hares, rabbits, and squirrels. A special feature of the wild life here is the prairie-dog, or marmot, dwelling in holes burrowed in the ground. Their numbers are so great that the traveller may sometimes journey for days together without losing sight of them.

Wild Birds are abundant in many varieties, birds of prey and birds of sport. There is the bald-headed eagle and the Mexican eagle, vultures, owls, hawks, wild turkeys, wild geese, prairie hens, canvass-back and other ducks, teal, brandt, pheasants, quails, grouse, woodcocks, pigeons, partridges, snipes, plovers, red-birds, and turtle-doves. By the waters are found, also, the crane, the swan, the pelican, the water turkey, and the king-fisher. The smaller birds are numerous, and among them many of the most brilliant plumage, as the oriole, the paroquet, the cardinal, the whippoorwill, and the sweet-toned mocking-bird. Blackbirds abound, and woodpeckers, blue-jays, starlings, red-birds, swallows, martens, and wrens.

In the rivers and bays there are all the varieties of water life, from alligators to perch, pike, trout, turtles, and oysters.

Snakes and reptiles of all sorts are at home in Texas. Rattlesnakes, moccasins, copperheads, coach-whips, and garden snakes, horned frogs, and lizards, the ugly centipedes and the poisonous tarantula.

The Houston and Texas Central R. R. extends from Houston 70 miles to Navasota. Connects at Houston with steamers for Galveston and New Orleans, and at Hempstead with a daily line of stages for Washington, Chappell Hill, Brenham, La Grange, Austin, New Braunfels, and San Antonio.

The Buffalo, Bayou, Brazos, and Colorado Railway extends from Galveston (steamboat) 65 miles to Harrisburg; to Richmond (railway), 32 miles more; and thence by stages via Columbus and Bastrop. Distance from Galveston to Austin 240 miles.

The Houston Tap and Brazoria R. R., from Houston to Columbia, 50 miles.

Galveston.—HOTELS:—*Island City House.*

Galveston, with a population of 8 or 10 thousand, is yet the largest city and the commercial metropolis of Texas. It is built on an island at the mouth of Galveston bay. The island of Galveston is about 30 miles in length and 3 miles broad. It is a thriving place, and with the spirit of progress, and its advantages as the best harbor on the coast, will no doubt increase rapidly in importance. Galveston is provided with good hotels, a reasonable supply of newspapers, churches, and schools. The Roman Catholic university of St. Marys, the R. C. cathedral, and the Episcopal church are large, noticeable structures of brick, in the Gothic style. There is in the city also a convent of Ursuline Nuns.

The island of Galveston was for a number of years the rendezvous and head-quarters of the famous pirate of the Gulf, Lafitte, until his settlement was broken up in 1821 by Lieutenant Kearney, commanding the United States brig Enterprise.

Railroads and steamers are bringing Galveston within speedy reach of the great country around it. Passengers may now leave the city on the Buffalo, Bayou, Brazos, and Colorado route,— first by steamboat 65 miles to Harrisburg—thence to Richmond by rail, 97 miles, and from Richmond by stage, to Bernard, 115 miles; to Columbus, 145 miles; to Bastrop, 210 miles; and to Austin, 240 miles: through from Galveston to Austin in 60 hours, including 18 hours' rest. Stages go from Columbus, on this route, to Hallettsville, Gonsales, Seguin, and San Antonio, and from Austin to all points of Western and North-western Texas. The whole line of railway between Galveston and Austin is rapidly progressing, and will no doubt soon be completed. Steamers leave Galveston daily for the city of New Orleans.

Houston.—HOTELS:—*Fannin House.*

Houston from New Orleans is by steamer via Galveston. Houston is the second of the Texan cities in commercial importance. Its population is about 6,000. It is situated on the low lands of the coast stretch, upon the Buffalo Bayou, 82 miles north-west of Galveston and 200 miles east-south-east of Austin City. Much of the surrounding country is a treeless savanna, covered with fine pasturage. This is a great *entrepot* for the cotton, sugar, and other products of the adjacent country. Houston was settled in 1836, and was once the capital of Texas. There are excellent hotels here. A railroad, to extend hence to Austin City, is partly in operation. *See* Galveston. Cars leave Houston by the Houston and Texas Central Railway, connecting at Hempstead (50 miles) with daily stage lines for San Antonio, and various other towns in the interior.

Austin.—Hotels :—

Austin, the capital of Texas, is upon the Colorado River, 200 miles by land from its mouth, and 230 miles west-north-west of Galveston. The landscape of the vicinage is strikingly picturesque. The seat of government was established here in 1844. The present population of nearly 4,000 is steadily increasing, and in due time the city will no doubt become a large and prosperous business mart. From New Orleans by steamer to Galveston. For routes thence *see* Galveston.

San Antonio.—Hotels :—*Menger House.*

San Antonio, with a population of about 8,000, is one of the largest towns in Texas. It is in Bexar County, on the San Antonio River, 110 miles south-west of Austin City. Fort Alamo, in the vicinity, contains a United States Arsenal. Many of the residences here are very elegant and beautiful.

Brownsville.—Hotels :—

Brownsville, formerly Fort Brown, is opposite Matamoras, on the Rio Grande, 40 miles from its mouth. It is 300 South of Austin. Brownsville is one of the chief towns of the State, with a population of about 6,000. It was named in honor of Major Brown, who commanded the garrison at the period of the Mexican war. He was mortally wounded by a shell from the enemy's batteries (May 6, 1846) while General Taylor was occupied in opening a communication with Point Isabel. The American army entered Matamoras without opposition after the success of Palo Alto and Resaca de la Palma.

ARKANSAS.

Arkansas is one of the younger States, having been admitted into the Union as late as 1836. It was formerly a part of the territory of Louisiana, and was settled by the French at Arkansas Post, about 1685. Its history has no very marked points, beyond rude frontier contests with the Indian tribes. It is a wild, desolate region of swamps, marshes, and lagoons, for a hundred miles back from the Mississippi River. This great plain is broken at intervals by elevations sometimes thirty miles in circuit. At flood periods, when the land is, as it often is, inundated, these points become temporary islands. Great levees are in process of construction along the banks of the river, by which means much of this vast tract will be converted into valuable land, with a soil of the richest nature. The Ozark Mountains bisect the State unequally. The middle regions, and the district north of the Ozark ranges, have a broken and varied surface.

The climate, soil, vegetation, and products of the lower portion of Arkansas, are all similar to those of the other south-western States; while the hilly regions above have, in all these respects, the more northern characteristics. The southern section is unhealthy, while the uplands are as salubrious as any part of the north-western State.

Productions. The rich, black, alluvion of the river, yields Indian corn in great luxuriance. This product, with cotton, tobacco, rice, many varieties of grain, wool, hops, hemp, flax, and silk, are the staples.

The Forest trees include great quantities of the cotton-wood, gum, ash, and cypress, in the bottom lands; and the usual vegetation of the north in the uplands. The sugar-maple, yielding large supplies of sap, is found here.

Wild animals range the forests and swamps in Arkansas as in Texas; and quails, wild turkeys, geese, and other birds abound. Trout and other fish are plentiful in the rivers and streams.

Minerals. Coal, iron, zinc, lead, gypsum, manganese, salt, and other mineral products exist here. Gold, too, it is said, has been found. "There is," says a writer, "manganese enough in Arkansas to supply the world; in zinc, it exceeds every State except New Jersey; and has more gypsum than all the other States put together; while it is equally well supplied with marble and salt."

Reaching Arkansas, we leave the sea-board, which we have followed almost without intermission thus far, in our rapid tour of the Union, from the St. Lawrence, southward and westward. Arkansas has no seaboard, though the great highway of the Mississippi well supplies this want; laving, as its waters do, nearly all its eastern boundary, and receiving the floods and freights of most of the many great rivers which traverse every part of its wide area.

The **Arkansas River**, rising in the Rocky Mountains, comes from the Indian Territory on the west, and traverses the middle of the State for 500 miles, gathering up in its long course the waters of many tributary streams, and bearing them to the great floods of the Mississippi. The entire length of this river is 2,000 miles. It is navigable for steamers 800 miles. Next to the Missouri, the Arkansas is the largest of the vassals of the "Father of Waters."

The **White River** is 800 miles in length. It is navigable from the Mississippi—into which it debouches, not far from the mouth of the Arkansas—350 miles to the mouth of the Black River, and at some periods of the year 50 miles yet higher up, to Batesville. As along the other rivers of Arkansas, the cypress covers the swamps of the Mississippi vicinage, and gives place to the pine and other vegetation higher up. This stream has numerous large affluents, among them the Big North Fort, Bryant's Fork, the Little North Fork, and Buffalo Fork.

The **St. Francis**, the **Red River**, the Washita, and other waters, bear the same general characteristics as the streams already mentioned. There are no lakes in this State of especial extent or interest.

Railways have not thus far been much needed in Arkansas, with her great facilities of water communication, and her thin population. A route is now in progress from Memphis to Little Rock. This line is completed at this time between the former city and Madison. In some future edition of this work, we shall no doubt be called upon to unravel the iron web of travel here, as, now, in most of the other States of the Union.

Little Rock.—Hotels:—*Anthony House.*

Little Rock—accessible by steamboat from the Mississippi, and from Memphis as far west as Madison, by railway; the remaining portion of the railway route is in progress. Arkansas has as yet no towns of any considerable extent.

Little Rock, the capital, with a population of 3,000 or 4,000, is the largest. It is situated on the top of a rocky bluff, the first of these characteristic precipices which is seen in the ascent of the Arkansas River, 300 miles up. The State House is a handsome, rough-cast brick edifice. The Penitentiary is located here, and there is also a United States Arsenal. Regular communication with points on the Arkansas and the Mississippi Rivers.

Route From Little Rock to Fort Smith and Fort Gibson.—To Lewisburg, 45; Pt. Remove, 52; Dwight, 76; Scotia, 82; Clarksville, 98; Horse-Head, 109; Ozark, 121; Pleasant Hill, 135; Van Buren, 160; Fort Smith, 165; Fort Gibson, 23 miles.

From Little Rock to Batesville, Ark. — To Oakland Grove, 30; Searcy, 50; Batesville, 95 miles.

From Batesville to Hix's Ferry.—To Sulphur Springs, 10; Smithville, 35; Jackson, 50; Hix's Ferry, 80 miles.

From Little Rock to Helena, Ark.—To Big Prairie, 25; Rock Roe, 38; Lawrenceville, 48; Lick Creek, 76; Helena, 91 miles.

From Little Rock to Napoleon, Ark.—To Pine Bluff, 50; Richland, 72; Arkansas Post, 118; Wellington, 133; Napoleon, 148 miles.

From Little Rock to Columbia, Ark.—To Pine Bluff, 50; Bartholomew, 120; Columbia, 145 miles.

From Little Rock to Memphis, Tenn.—To Clarendon, 65; St. Francis, 115; Marion, 145; Mississippi River, 154; Memphis, 155 miles.

From Little Rock to Fulton, Ark.—To Benton, 24; Rockport, 55; Raymond, 80; Greenville, 93; Washington, 129; Fulton and Red River, 144 miles.

The *Hot Springs* are situated a few miles north of the Washita River.

A line of stages runs hence from Little Rock, 53 miles.

Projecting over the Hot Spring Creek there is a point of land from 150 to 200 feet high, forming a steep bank. More than one hundred springs issue hence, in temperature varying from 135° to 160° Fahrenheit. The region is one of very great resort.

Alabaster Mountain. In Pike County, on the Little Missouri River, there is a mountain of alabaster, of fine quality, and white as new-fallen snow.

Natural Bridge. In the neighborhood of the Alabaster Mountain, there is a remarkable natural bridge formation, which is regarded as a very curious and interesting scene.

Van Buren, the most commercial town of Arkansas, is 160 miles west-north-west of Little Rock, within five miles of the Indian Territory. It is pleasantly situated on the Arkansas River.

Batesville, with a population of about 2,000, is upon the White River, 400 miles from its mouth. Small steamers ascend at nearly all seasons. Batesville is distant from Little Rock (see route) 90 miles; from Memphis, Tenn. 115 miles.

Fort Smith is a thriving village on the Arkansas River, 163 miles west-north-west, by land, of Little Rock.

Camden, is upon the Washita River, 110 miles from Little Rock.

Napoleon, 125 miles south-east of Little Rock, is upon the Mississippi River, at the mouth of the Arkansas. It is a busy and thriving place—the seat of a United States Marine Hospital.

Arkansas Post, is upon the Arkansas River, some 50 miles from its mouth. It is an ancient settlement, having been occupied by the French as early as 1685. It was, for many years, the chief depot of the peltries of the country far around.

TENNESSEE.

The territory, which now forms the State of Tennessee, was settled before any other of the lands west of the Alleghanies, Fort Loudon having been built by adventurers from North Carolina as early as 1757. The early history of the country is, like that of the neighboring State of Kentucky, full of the records of bloody struggles with the Indian occupants of the soil.

The little band of pioneers at Fort Loudon, were not, of course, suffered to rest peacefully in their new home; on the contrary, they were all either butchered or driven away. In a few years, though, the axes of the whites again rung through the wild forests, and their cabins dotted the land, gradually clustering into villages and towns. Tennessee was admitted, in 1796, as the sixteenth member of the American Union. She played a very honorable part in the war of 1812.

The landscape of Tennessee is most varied and agreeable, though none of the great natural wonders of the Republic lie within her borders. Her mountain, valley, and river scenery is exceedingly beautiful, and will become famous as it becomes known. The Cumberland Hills, and other ranges of the Appalachian chain, pass through her western area, separating her from North Carolina, and shutting in the valleys of the Holston and other rivers. The height of the mountain ridges and summits here is variously estimated at from 1,500 to 2,000 or more feet. They are most of them covered with a rich forest growth to the top, where the axe and the plough have not changed their native character. The central portion of the State, stretching from the mountains to the Tennessee River, has a broken surface, while beyond, towards the Mississippi, which makes the western boundary, the country is comparatively level.

Many valuable mineral products are found here—coal and iron in great abundance, and rich deposits of copper. Gold, too, has been detected, and silver, lead, zinc, manganese, magnetic-iron ore, gypsum of superior quality, and a great variety of beautiful marbles, slate, nitre, burr-stones, and limestone. Salt and mineral springs, the latter of very valuable character, abound.

The climate here, excepting in the river lowlands, is most agreeable and healthful; exempt alike from the winter severities of the North, and from the summer heats of the South.

Immense quantities of live stock are raised in Tennessee; more, indeed, than in any other part of the Union. It is, too, a vast tobacco, cotton, and corn-growing region. The culture of hemp, buckwheat, rye, oats, barley, maple, sugar, and many other agricultural products occupy the industry and contribute to the wealth of the people.

The Tennessee River enters the State at its south-east extremity, from North Carolina, and forms the chief affluent of the Ohio. Its sources are among the Alleghanies, in Virginia, flowing under the names of the Clinch and the Holston Rivers, until they unite at Kingston, in Tennessee. The first course of the main stream is southwest to Chattanooga, near the point where the States of Tennessee, Georgia, and Alabama meet. From Chattanooga it turns towards the north-west, until the obstruction of the Cumberland Mountains bends its current southward again, and sends it off on a *détour* of 300 miles into Upper Alabama and the north-east corner of the Mississippi. It gets back to Tennessee at this point, and for the second time traverses the entire breadth of the State, crosses Kentucky, and reaches the end of its journey at Paducah, 48 miles from the mouth of the Ohio. The length of the Tennessee proper is about 800 miles; including its longest Branch, the Holston, its waters extend 1,100 miles. The only important obstruction in the navigation of the Tennessee is that great 20 miles stretch of rapids in Alabama, the Muscle Shoals (see Alabama). Steamboats ascend the river nearly 300 miles, to the foot of these rapids, and above to Knoxville, on the Holston, nearly 500 miles. A railway supplies the missing link in the passage of the river caused by the intervention of the rapids. Knoxville and Chattanooga are the principal places in Tennessee passed by this river. In Alabama, Tuscumbia, and Florence; and in Kentucky, Paducah.

The upper waters of the Tennessee, and all that portion of the river in the eastern and middle parts of the State, are extremely beautiful; varied as the landscape is, by wild mountain scenes, and fertile pastoral lands. In the neighborhood of Chattanooga, where the Look-out Mountain lifts its bold crest, the scenery is especially attractive. It would be difficult to find a more charming picture than that from the summit of the Look-out Mountain, over the smiling valley of the Tennessee, and the capricious windings of the river.

The chief rivers of the Tennessee, besides its great namesake, and the two branches from which it is formed—the Holston and the Clinch—are the Hiawasse, from Georgia, the Hatchee, and the Duck River. All the waters of the State are ultimately absorbed by the Mississippi, in its western boundary.

Railways in Tennessee. Nashville and Chattanooga, 151 miles from Nashville, in the North Central part of the State, to Chattanooga, near the Georgia and Alabama lines, connecting with the Georgia and South Carolina Railway system. To be extended north-west to the Ohio River.

Tennessee and Alabama. In operation southward to Columbia; to be extended and connected with routes from Mobile, Alabama, and from New Orleans.

East Tennessee and Georgia. From Knoxville, south-west, 103 miles to Dalton, Georgia, connecting with the railways of that State. To be extended north-east, by the East Tennessee and Virginia, to the railways of Virginia and west, from Knoxville to Nashville.

Memphis and Charleston, 310 miles from Memphis to Chattanooga, partly on the southern borders of extreme Western Tennessee, through the upper part of Mississippi and Alabama, into East Tennessee.

Memphis and Granada, southward, from Memphis to the Mississippi and Louisiana roads.

Besides these routes now in operation many others are being constructed or are proposed.

Nashville.—HOTELS:—*Nashville Hotel.*

Nashville, the capital of Tennessee and the most important town in the commonwealth, is most agreeably situated on the south side of Cumberland River, and at the head of steamboat navigation. The site of the town consists of an entire rock, covered in some places by a thin soil, and elevated from 50 to 175 feet above the river. This place, owing to its healthy location, is the resort of numbers from the lower country during the heat of summer. Numerous steamboats of the first class are owned here, which ply at regular intervals between Nashville and Cincinnati, and other places.

Both the public and private buildings of Nashville are highly creditable to the taste and the liberality of the people; many of the latter are really sumptuous in their character. The capitol, in its bold position 175 feet above the river, and in its elegant and costly architecture, is a very imposing structure. It is built of fine limestone, much like marble, which was quarried on the spot. Its noble dimensions are 240 by 135 feet. Its cost was about $1,000,000. The Lunatic Asylum is a superb affair, and so, too, is the Penitentiary, with its 310 feet façade. Here is the University of Nashville, founded in 1806. Its Medical School has over 100 students. The Mineral Cabinet of the late Dr. Troost is the richest private collection in the United States. A wire Suspension Bridge spans the Cumberland River here. It was built at a cost of $100,000. The city is lighted with gas, and is supplied with water from the river. The population of Nashville was in 1853, about 20,000. The city is 200 miles from the mouth of the Cumberland River, 230 miles east-north-east of Memphis, 206 miles south-west of Lexington, Kentucky, and 684 miles from Washington. Railroads are in progress, which will connect it with Louisville, Kentucky, and thence with all the great railways, north and east—others, which will unite it with all the Atlantic States via Knoxville and the Virginia routes, while it is already in daily and unbroken communication with the Atlantic, via the Nashville and Chattanooga route, connecting with the Georgia railways, Atlanta, Augusta, Savannah, and Charleston, S. C., and with Montgomery, on the great line from New York to New Orleans.

The Hermitage, Home of General Jackson. The traveller, while in this vicinage, will not fail to make a pilgrimage to the spot sacred as the hearthstone of the great General and Statesman, Andrew Jackson.

Memphis.—HOTELS:—*Gayoso House*, corner of Selby and McCall streets.

Memphis is finely situated upon the Tenth Chickasaw Bluff of the Mississippi, at the mouth of the Wolf River. It is in the south-west corner of the State, upon the site of Fort Pickering. The city presents a striking appearance as seen from the water, with its esplanade several hundred feet in width, sweeping along the bluff and covered with large warehouses. It is the chief town on the Mississippi, between New Orleans and St. Louis. Its population amounted in 1853 to over 12.000. Memphis is 781 miles from New Orleans, 126 miles below St. Louis, and 209 miles from Nashville. The Memphis and Charleston railway connects the city via Chattanooga, Tenn., and Atlanta and Augusta, Georgia, with the Atlantic at Savannah, and at Charleston, S. C. A railway to Little Rock, Arkansas, and others to Nashville, are in course of construction.

Knoxville.—HOTELS:—*Lamar House*.

Knoxville is upon the Holston River, four miles from its junction with the French Broad; 185 miles east of Nashville and 204 miles south-east of Lexington, Kentucky. It is connected by the East Tennessee and Georgia railway with all the great routes of Georgia to the Atlantic, and with the highway to New Orleans, via Montgomery and Mobile in Alabama; also by the East Tennessee and Virginia railway, with Richmond, Virginia, and all the great thoroughfares of the country. The great route from Boston to New Orleans will pass through Knoxville. The city is a pleasant and prosperous one, with a population at this time of some 10,000. Formerly it was the capital of the State. The university of East Tennessee, founded in 1807, is here. Here, too, is the largest manufactory of window glass in the Southern States.

Chattanooga.—HOTELS:—*Crutchfield House.*

Chattanooga is upon the Tennessee River, in the southern part of t dary is touched by Al 250 miles from Knoxvi south-east of Nashvill centre, being the termi Chattanooga route, fr upon the Georgia rout and thence through Vi line from Charleston, Memphis. The Tenn two-thirds of the ye small boats, from th Population about 6,000 in the chapter upon (landscape surrounding

Columbia.—HOTE

Columbia (populatic Duck River, 41 miles line of the great railw struction from Nashvi and New Orleans—in endy from Nashville to lege is located here. emy occupies an imp was the home of Mr. F 1844 to the Presidency

Murfreesboro'.—

Murfreesboro' is 30 the railway route via (Augusta, to Charleste town is built in a beau ley. It is the seat of tl tist), established in 1841 capital of Tennessee fr and thriving town.

Jackson is upon th miles below Nashville Ohio Railway passes h

Lebanon, the seat versity, is 30 miles east

Caves and Mou While in Eastern Tenn not fail to see some of t Cumberland Mountain Rock, here, are some s feet of men and animal far from Manchester, fort, enclosed by a w growing, whose age years. This mysterio between two rivers, a acres.

KENTUCKY.

"THE highest phase of Western character," says Mr. Tuckerman, "is doubtless to be found in Kentucky, and in one view best illustrates the American in distinction from European civilization. In the North this is essentially modified by the cosmopolite influence of the seaboard, and in the South by a climate which assimilates her people with those of the same latitudes elsewhere; but in the West, and especially in Kentucky, we find the foundations of social existence laid *by the hunter*—whose love of the woods, equality of condition, habits of sport and agriculture, and distance from conventionalities, combine to nourish independence, strength of mind, candor, and a fresh and genial spirit. The ease and freedom of social intercourse, the abeyance of the passion for gain, and the scope given to the play of character, accordingly developed a race of noble aptitudes; and we can scarcely imagine a more appropriate figure in the foreground of the picture than Daniel Boone, who embodies the honesty, intelligence, and chivalric spirit of the State."

The first visit of Boone to the wilderness of Kentucky was about the year 1769, at which period he and his hardy companions made the earliest settlement at Boonesborough. In 1774, Harrodsburg was begun, and Lexington a year or two afterwards. The pioneers in their western forests met with all the adventure their hearts could desire—more, indeed; for so great was their exposure and suffering, for many long years, from the cruel enmity of the savage populations, that the country came to be known as "the dark and bloody ground." A memorable battle was fought near the Blue Lick Springs, Aug. 19, 1782, between the Kentuckians and the Indians—an unequal and disastrous conflict, in which the colonists were routed, with a loss of sixty men, among them a son of the gallant Boone.

In 1778, Du Quesne, with his Canadian and Indian army, was bravely repulsed at Boonesborough. Kentucky came into the Union in 1792, being the second State admitted after the Revolution.

The physical aspect of Kentucky is one of changing and wonderful beauty, as we shall see in subsequent visits to some of her marvellous natural scenes. The Cumberland Mountains traverse the eastern counties, and a line of hills follows the course of the Ohio River, with meadow stretches between, sometimes ten, and even twenty miles in width. The State is well supplied with coal, iron, and other minerals. Salt and mineral springs of great repute abound.

The chief agricultural staples of this region are hemp, flax, tobacco, and Indian corn: of the first two of these products, a greater quantity is raised here than in any other State. In tobacco, Kentucky is second only to Virginia, and in the product of Indian corn she is behind Ohio alone.

Rivers.—The Ohio River forms the entire northern boundary of Kentucky, and the Mississippi washes all her western shore; thus giving her, with the aid of the many streams which come from the interior of the State into these great highways, the greatest possible facilities for the transportation of her staples to all markets.

The Kentucky River, like most of the streams here, is remarkable for picturesque beauty; its passage, in a course of 200 miles, north-west, to the Ohio, is often through bold limestone ledges, ranged on either side of the narrow dark channel in grand perpendicular cliffs. "Deepen Trenton Falls," says Mr. Willis, "for one or two hundred feet, smooth its cascades into a river, and extend it for thirty miles—*thirty miles* between perpendicular precipices, from three to five hundred feet high, and only a biscuit-toss across at the top—and you have a river of whose remarkable beauty the world is strangely ignorant."

The Cumberland River is one of the largest of the tributaries of the Ohio. It has its source in the Cumberland Mountains, in the south-east corner of the State, and flows 600 miles, making a bend into Tennessee, and then traversing western Kentucky. It is navigable for steamers 200 miles to Nashville, and sometimes to Carthage, while small craft may ascend 300 miles yet higher. About 14 miles from Williamsburg there is a fine fall of 60 feet perpendicular in this river.

The Licking River flows from the Cum-

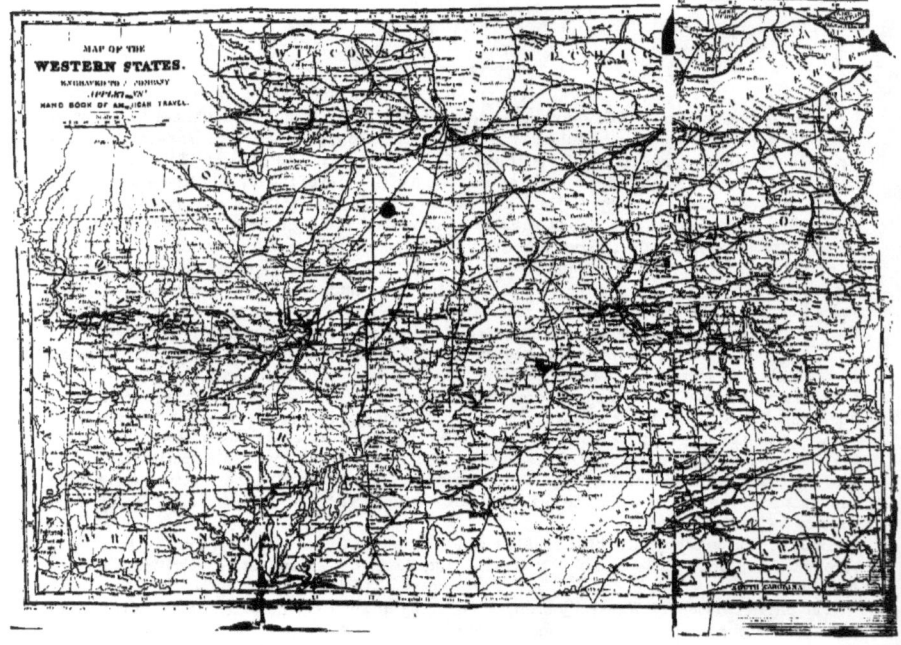

berland Mountains, 200 miles, into the Ohio, opposite Cincinnati. Steamboats may ascend 50 miles to Falmouth. This river varies in width from 50 to 100 yards. Its banks are often lofty and precipitous, covered with huge forest trees. The South Licking and the North Fork are among its tributaries.

Green River is about 300 miles in length. It rises in the eastern section of the State, and flows westward for some 150 miles, through the limestone regions and by the Mammoth Cave, finally entering the Ohio nine miles above Evansville in Indiana. It is navigable in high water, and by the aid of locks and dams, for steamboats, 200 miles to Greensburg.

Salt River, named in token of the Salt Springs which abound in its vicinity, enters the Ohio 22 miles below Louisville. This is the fabled retreat of defeated politicians and other unhappy adventurers.

The Tennessee River rises among the Cumberland Mountains of Eastern Kentucky, and flows 70 miles within the limits of this State. (See Tennessee.)

Railways. Covington and Lexington, 99 miles south from Covington, on the Ohio, opposite Cincinnati, to Lexington, on a branch of the Elkhorn river, via Falmouth.

Louisville and Frankfort and Lexington and Frankfort Railways, 94 miles from Louisville, on the Ohio, to Lexington via Frankfort.

Louisville and Nashville. This route extends 185 miles, forming an important link in the great chain of roads which bind Tennessee so closely with all the northern and southern States of the Union.

Louisville.—HOTELS:—The *National*, on Main street; the *United States*, on Jefferson street.

To reach Louisville from Boston, New York, Philadelphia, Baltimore, and intermediate places, see Cincinnati and routes to that city. From Cincinnati, take the steamer down the Ohio river, 133 miles to Louisville; or take the Ohio and Mississippi Railway 87 miles to Seymour, and thence by the Jeffersonville Railway 59 miles to Jeffersonville, opposite Louisville. *This is the best land route.* Total distance to Louisville from Cincinnati, 146 miles. Louisville may also be reached less directly, from Cincinnati by the Kentucky Railways, the Covington and Lexington, 99 miles to Lexington, and thence by the Louisville and Lexington, via Frankfort, 94 miles; from Cincinnati to Louisville, by this route, 193 miles. Louisville may be reached from Pittsburg, western terminus of the Pennsylvania Railway, from Philadelphia or from Wheeling, western terminus of the Baltimore and Ohio Railway from Baltimore, by steamboat down the Ohio. From St. Louis there is a *direct* Railway communication by the new route of the Ohio and Mississippi road from Cincinnati to St. Louis. The way heretofore has been by routes higher up; the "Terre Haute and Alton" to Indianapolis, and thence by the Jeffersonville Railway.

Louisville, with a population of about 100,000, is the chief city of Kentucky. Its position is at the Falls of the Ohio, where Beargrass Creek enters *La Belle Rivière*. The topography is most agreeable, affording fine views from many parts of the terrace elevation of 70 feet.

The Falls, which are quite picturesque in appearance, may be seen from the town. In high stages of the water they almost entirely disappear, and steamboats pass over them; but when the water is low, the whole width of the river, which is scarcely less than a mile, has the appearance of a great many broken rivers of foam, making their way over the falls. The river is divided by an island, which is now rapidly disappearing. To obviate the obstruction to the navigation caused by the falls, a canal two and a half miles in length has been cut round them, to a place called Shippingsport. It was a work of immense labor, being, for the greater part of its course, cut through the solid rock. The extent of the city river-wards is over two miles. The course of the leading streets is in this direction. They are, for the most part, wide, well paved, and delightfully shaded with noble trees.

The principal public buildings are the Court House, the Medical College, the Male High School, the Female High School, the Custom House, the Blind Asylum, the Cathedral, and several fine churches, of various denominations. The Cave Hill Cemetery is an interesting and much admired spot.

Silver Creek, 4 miles below the city, on the Indiana side (cross by ferry from Portland), is a beautiful rocky stream, and a favorite fishing and pic-nic place of the Louisville ruralizers. There is a small but fair hotel here. Another pleasant excursion is to the mouth of Harrod's Creek, 8 miles up the Ohio. There are, too, famous drives on the Lexington and Bardstown turnpikes, through a beautiful and richly cultivated country. The road along the borders of Bear Grass Creek, Lexington-wards, is very agreeable. The fine forest vegetation, the charming parklike groves, the hemp fields and the *blue grass* pastures, all help to furnish forth delight in the Louisville suburban rides and rambles.

Louisville is connected with Frankfort, 65 miles distant, by Railway, and with Lexington,

94 miles. A railway to Nashville has been recently completed, by which the tourist is put in the immediate neighborhood of the Mammoth Cave.

Lexington.—HOTELS:—The *Phœnix House.*

From Cincinnati, by the Lexington and Covington Railway, 99 miles: from Louisville, by the Louisville and Frankfort and Lexington and Frankfort, 94 miles.

Lexington, upon the Town Fork of the Elkhorn River, is one of the most beautiful and most opulent of the Kentucky cities. In population (about 13,000) it is the second place in the State. The streets are regular, broad, well-paved, well-built, and delightfully shaded. Here is the seat of the *Transylvania University*, the *Law* and *Medical* schools of which are held in high repute. The *University Library* numbers over 14,000 volumes. The *State Lunatic Asylum* occupies a prominent *locale.*

A monument in honor of Henry Clay, whose home of *Ashland* is close by, has been erected in Lexington. This city was once the Capital of Kentucky.

Ashland, the Home of Clay. The old Clay homestead (lately replaced by a new mansion) stood about a mile and a half from Lexington, and the locality is, of course, the chief object of interest in this neighborhood. "Walking slowly and thoughtfully up," says Mr. Greeley, "a noble avenue that leads easterly from Lexington, the traveller finds the road terminating abruptly in front of a modest, spacious, agreeable mansion, only two stories in height, and of no great architectural pretensions. Mr. Clay lived at Ashland between forty and fifty years. The place bore the name when he went to it, probably, as he said himself, on account of the ash timber with which it abounds, and he made it one of the most delightful retreats in all the West. The estate is about 600 acres large, all under the highest cultivation, except some 200 acres of park, which is entirely cleared of underbrush and small trees, and is, to use the words of Lord Morpeth, who stayed at Ashland nearly a week, the nearest approach to an English park of any in this country. It serves for a noble pasture, and here Mr. Clay had some of the finest horses and Durham cattle in America. The larger part of the farm is devoted to wheat, rye, hemp, etc., and the crops look most splendid. Mr. Clay paid great attention to the ornamentation of the land with beautiful shade trees, shrubs, flowers, and fruit orchards. From the road which passes the place on the north-west side, a carriage-way leads up to the house, lined with locusts, cypress, cedar and other fine trees, and the rose, jasmine, and ivy clamber about them, and peep through the grass and the boughs like so many twinkling fairies. The mansion is nearly hidden from the road by the surrounding trees; and is as quiet and secluded, save to the throng of pilgrims continually pouring thither, as though it were a wilderness. After the death of Mr. Clay, the estate of Ashland was sold at public auction, but was purchased by James B. Clay, the great statesman's eldest son, and so the honored and beloved little homestead remains yet, happily, in the family possession. Let it be sacredly and forever preserved."

Covington.—HOTELS:—*Magnolia House.*

Covington, opposite Cincinnati (see routes from all points to Cincinnati), is one of the principal cities of Kentucky, with a population of about 14,000. It is upon the Ohio, immediately below the point where the Licking River comes in. Across the Licking is the suburban town of Newport. Steam ferries unite it with Cincinnati, and the great suspension bridge (see Cincinnati) will soon make a yet better means of communication thence. Covington is built upon a broad and beautiful plain, very much after the topography of the great Ohio city opposite, to which, indeed, it may be regarded as suburban. This is the seat of the *Western Theological College*, a prosperous and richly-endowed institution. There are here large manufactories of cotton, hemp, silk, and tobacco. The place, too, like Cincinnati, is greatly addicted to the salting and packing of pork and beef.

Newport.—HOTELS:—*Barlow's Hotel.*

Newport, across the Licking River from Covington, has a population of about 9,000. Like the neighboring cities of Covington and Cincinnati, to which it owes its prosperity, it is delightfully and advantageously situated. It will probably soon absorb the large adjoining villages of Jamestown and Brooklyn.

Frankfort.—HOTELS:—*Capitol House.*

Frankfort, the capital of Kentucky, is situated on the east bank of the Kentucky River, 60 miles above its entrance into the Ohio. The site of the town is a deep valley, surrounded by precipitous hills. The river flows in deep limestone banks; the quarries of which yield a fine stone, or marble, of which many of the houses are built. The heights on the north-east afford fine peeps at the beautiful scenery of the Kentucky waters. The State Capitol occupies an eminence, midway between the river and upper end of the valley. It is a fine structure, built of marble, quarried in the neighborhood. Here, too, is the State Penitentiary. The town is connected with the village of South Frankfort, across the river, by a good bridge. Population some 6,000. Distance from Louisville, by the Louisville and Frankfort Railway, 65 miles; from Lexington, by the Frankfort and Lexington Railway, 29 miles; from Cincinnati, by railways via Lexington, 128 miles.

Maysville.—HOTELS :—*Lee House.*

Maysville (population some 8,000), is upon the Ohio River, 60 miles above Cincinnati, and 60 miles north-east of Lexington, from Cincinnati by steamboat. At Portsmouth, Ohio, some 50 miles above, on the Ohio River, railway lines come in from all parts of the country, north and east. Maysville is upon Limestone Creek, whose name it formerly bore. The position of the town is in the midst of a varied hill-landscape. It is, in business and population, the fourth city of Kentucky, and its greatest hemp mart. This is the entrepôt for the merchandise and produce imported and exported by the north west section of the State. Railways will soon connect it with other points.

Paducah (population nearly 4,000), is upon the Ohio, just below the mouth of the Tennessee, 340 miles from Louisville; 473 miles from Cincinnati. Paducah bears the name of an Indian chief who once lived in the neighborhood.

Harrodsburg, a town of over 3,000 people, and the oldest settlement in Kentucky, is upon an eminence near Salt River, about 30 miles below Frankfort and Lexington. The first cabin ever built in the State was erected here by Captain James Harrod, in 1774. Here is the seat of *Bacon College*, and of a *Military Academy*. The greatest attraction, however, of Harrodsburg, is its celebrated mineral springs, which make it the most famous summer resort of all the country round.

Harrodsburg Springs. *See Harrodsburg, ante.* This is one of the most fashionable watering places of Kentucky, and is, in the crowded season of July and August, "the grand field of tournament for Western flirtation, and the gathering point for politicians out of harness, and for such wealthy Westerners and Southerners as like to spend their money on the side of the Alleghanies that slopes toward home." The hotel here with all its surroundings and appointments, is most admirable. Dr. Graham, the liberal proprietor, has already expended more than $800,000 upon the embellishment of the place and so expended it, that it all sensibly contributes to the comforts and enjoyments of his guests.

Knob Lick is an interesting spot, within excursion distance of the Harrodsburg Springs—15 miles distant. The Knobs or hillocks here are from 100 to 200 feet high, more or less conical, some of them insulated, others connected by crumbling isthmuses; the whole forming a group of barren conoidal eminences, which are finely contrasted with the deep verdure of the surrounding plain.

The **Devil's Pulpit** is a wonderful rock and ravine; a passage in the bold landscape in the Kentucky River, accessible from Harrodsburg in a twenty-mile excursion.

The **Blue Lick Springs** is a watering place of high repute, on the Licking River, in Nicholas County. Easily reached by stage from Paris, a station on the Covington and Lexington Railway; 19 miles from Lexington; 80 miles from Covington, opposite Cincinnati. These springs contain soda, magnesia, lime, sulphuretted hydrogen, and carbonic acid, in combination with muriates and sulphates.

Drennon Springs (black and salt sulphur), are upon the banks of the Kentucky, in Henry County. They may be reached by steamboat from Louisville.

Poplar Mountain Springs are upon the Poplar Mountain top, in Clinton County, four miles from Albany. The scenery in this vicinage is of remarkable beauty. Upon Indian Creek, not far from the springs, there is a fine waterfall, of 90 feet perpendicular descent.

The **White Sulphur Springs** are in Grayson County, four miles from Litchfield. They are very numerous within a small area.

The **Tar and Breckenridge White Sulphur Springs** are in Breckenridge County, four miles from Cloverport. They are readily accessible from the Ohio River. The Breckenridge coal is found in this vicinity.

The **Tar and Sulphur Springs** are upon Green River, in Davies County, near the "Old Vernon Settlements." There are other springs of reputation in this vicinity.

The **Esculapia Springs**, Chalybeate, and White Sulphur, are in a beautiful valley of Lewis County.

The **Fox and the Phillips' Springs** are in the abundant spring region of Fleming County.

The **Lettonian Springs** (sulphur), are upon the Bank Lick Road, near the Ohio River, and about four miles from Covington. This is a pleasant excursion point from Cincinnati.

The **Parroquet Springs** are near Sheppardsville, in Bullitt County.

KENTUCKY.

The **Sink Holes of Kentucky.** Of these curious cavities or depressions in the surface of the ground, known as sinks, remarkable examples are found in Kentucky. Sinking Creek in Breckenridge County suddenly disappears, and is not seen again within a distance of half a dozen miles. Near Munfordsville, in Hart County, there is a strange spring connected with a millpond, the waters of which overflow the dam every twenty-four hours, rising 12 or 15 inches, and receding to their ordinary level with the precision of the tides. Six miles east of the same town, there is a hole, in form like an inverted cone, which is 70 feet in diameter at the surface, and but 10 or 12 feet across, at a depth of 25 or 30 feet. Stones cast into this pit, give no indication of touching the bottom. There is yet another extraordinary sink in this neighborhood, on the top of an elevation, called Frenchman's Knob. It has been descended by means of a rope, 275 feet, but without finding bottom.

Natural Bridge. There is an extraordinary natural Bridge in the romantic county of Christian. It makes a grand span of 70 feet, and is 30 feet high.

Dismal Rock is a frowning precipice, 160 feet high, in Edmonson County.

Cumberland Gap. This passage of the Cumberland River through the mountains, in Knox County, is an imposing scene. The waters make their way between huge cliffs, 1,300 feet in height.

Waterfalls. Besides the cascades of the Indian Creek, near Poplar Mountain, of which we have already made mention, there are numerous beautiful waterfalls among the hills of Kentucky. The Kentick Creek in Cumberland County, presents some fine pictures of this kind. The traveller must not overlook, either, if his time serves for the exploration, the Rock House in Cumberland; the Indian Rock in Edmonson; Pilot Rock in Christian; and the Flat and the Anvil Rocks in Union County.

The **Mounds and Fortifications,** which are numerous in Kentucky, afford employment enough for the antiquarian tourist. In Allen County, 17 miles from Bowling Green, there is a wall of solid limestone, 200 yards in length, 40 feet high; at its base, 30 feet thick, and at its summit, 6 feet. It crosses a neck formed of a curve in Drake's Creek, and shuts in a peninsula of about 200 acres, elevated 100 feet above the river. Upon the crown of this eminence, an area of three acres is surrounded by a wall and ditch, making the place a fortress of immense strength. Other strange ancient works, older than tradition, may be found in Warren, Spencer, Boone, La Rue, Montgomery, Barren, and Bourbon Counties.

The **Big Bone Licks** of Boone County exhibit the great bones of the Mastodon, and other extinct animals. Curious fossil remains are found in Bourbon County. Impressions of the feet of men and of animals may be seen in a rock near Morganfield, in Union County.

The **Mammoth Cave.** Many and varied as are the natural beauties and wonders in Kentucky, the most strange and magnificent of them all remains yet to be seen in the weird halls and chambers of the famous Mammoth Cave.

Route.—Tourists from the Eastern cities will leave Louisville by the Louisville and Nashville railway, and stop at Cave City, nine miles from the Cave. From Cave City the distance is traversed by a good carriage road. Steamers ply on the Green River from Louisville to within the distance of a mile only of the cave.

The Mammoth Cave is in Edmonson County, south of the centre of the State.

Cave Hotel is in the near vicinity of the grand Plutonian halls, but 200 feet, indeed, from the gloomy portals. The journey through these stupendous vaults and passages is long and toilsome, despite the marvels which everywhere beguile the way. As it takes days to see these wonderful scenes, so it would require many pages to describe them, which compels us to be content with the briefest catalogue of the chief points of interest.

After exploring the ante-chambers and the Audubon Avenue, which is a mile in length, 50 or 60 feet high, and as many wide, we return and pass through the vestibule for a second time, entering the main cave or Grand Gallery, a mighty tunnel of many miles extent. The Kentucky Cliffs passed, we descend some 20 feet to the Church. This is a grand apartment, 100 feet in diameter, with a roof formed of one solid, seamless rock, suspended 63 feet overhead. Nature has supplied these solemn halls with a natural pulpit, and a recess where a mighty organ and a countless choir could be placed. Religious services have been performed in the dim, religious light of torches, under this magnificent roof. The *Gothic Avenue* is reached by a *détour* from the main cave, and a descent of some 30 feet. It is two miles in length, 40 feet wide, and 15 feet high. This place was once called the Haunted Chamber. Louisa's Bower, Vulcan's Furnace, and the new and old Register Rooms, are now passed in succession. The Gothic Chapel rivals all the marvels of the highest and nicest art, in the strength, beauty, and proportions of its grand columns, and its exquisite ornamentation. The Devil's Arm Chair is a large stalagmite pillar, in the centre of which is a spacious seat, grand enough for the gods. After passing numerous other stalactites and stalagmites, we look, in succession, at Napoleon's Breastwork, the Elephant's Head, and the Lover's Leap. This last scene is a large pointed rock, more than 90 feet above the floor, and projecting into a grand rotunda.

Just below the Lover's Leap, a *détour* may be made to the lower branch of the Gothic Avenue, at the entrance of which we may see an immense flat rock, called Gatewood's Dining Room; and to the right, a beautiful basin of water, named the Cooling Tub. Beyond is Flint Pit. Still pursuing our *détour*, we pass, one after the other, Napoleon's Dome, the Cinder Banks, the Crystal Pool, the Salts Cave, and a wonderful place, still beyond, called Annetti's Dome, through a crevice of which a waterfall comes.

Re-entering the main cave or the Grand Avenue, we arrive, soon, at the Ball Room, where nature has provided every necessary fitting of gallery and orchestra. Willie's Spring has its pleasant story, which will delight the wondering visitor until he is called upon for astonishment at the sight of the great rock, known as the Giant's Coffin.

Here begin the incrustations, ever varied in form and character, which are so much the delight of all visitors. The Giant's Coffin passed, we sweep round with the Great Bend. Opposite is the Sick Room. Hereabouts there is a row of cabins for consumptive patients.

The *Star Chamber* is a splendid hall, with perpendicular arches on each side, and a flat roof. The side rocks are of a light color, and are strongly relieved against the dark ceiling, which is covered with countless sparkling substances, resembling stars.

The *Cross Room* has a ceiling of 170 feet span, and yet not a single pillar to uphold it. The Black Chambers contain ruins which remind us of old baronial castle walls and towers. Through the Big Chimneys we ascend into an upper room, about the size of the main cave. Here are heard the plaintive whispers of a distant waterfall; as we come nearer, the sound swells into a grand roar, and we are close to the cataract. To enter the place called the Solitary Chambers, by the way of the Humble Chute, we have to crawl upon our hands and knees for 15 or 18 feet beneath a low arch. Here is the Fairy Grotto, the character of which admirably realizes the promise of its name. The Chief City or Temple, is an immense vault, two acres in area, covered by a solid rocky dome, 120 feet high. Other localities, in the direct passage of the cave, as in some of the many *détours*, are appropriately named the Stoops of Time, the Covered Pit, the Side Saddle, and the Bottomless Pit, the Labyrinth, the Dead Sea, the Bandit's Hall, and the River Styx, and the Rocky Mountains.

No more serious accident, it is said, than an occasional stumble, has ever been known to occur. Colds, instead of being contracted, are more often cured by the visit. Nowhere is the air in the slightest degree impure. So free is the cave from reptiles of every kind, that St. Patrick might be supposed to have exerted his fabled annihilating power in its favor. Combustion is everywhere perfect. No decomposition is met with. The waters of the springs and rivers of the cave are habitually fresh and pure. The temperature is equable at all seasons at 59° Fahrenheit.

Thus, no one need, through any apprehension, deny himself the novel delight of a ramble along the 226 avenues, under the 47 domes, by the 8 cataracts, the 23 pits, and the "thousand and one" marvellous scenes and objects of this magnificent and most matchless cave.

The **Richardson or Diamond Cave** (recently discovered) lies on the way from the railway station to the Mammoth Cave; 1¼ miles from the former, and 5 miles from the latter. It is said to be in its surprising attractions second only to its famous neighbor.

OHIO.

Ohio is one of the largest and most important of the great Western States, and the third in the Republic in population and wealth. It extends over an area 200 miles in length, and 195 miles in breadth. On its northern limits are Michigan and Lake Erie; Pennsylvania and Virginia encompass it eastward. The waters of the Ohio separate it from Kentucky on the south, and westward is the State of Indiana.

The central portions of Ohio are, for the most part, level lands, with here and there, more especially towards the north, tracts of marsh. In the north-west there is an extensive stretch of very fertile country, called the Black Swamp, much of which is yet covered with forest. Some prairies are seen in these middle and northern parts of the State. Huge boulders are found hereabouts, as upon all the plains of the West, but where they came from, or how, nobody knows. North of the middle of the State there is a range of highlands which apportion the waters for the Ohio on the south, and for Lake Erie on the north, the former recipient getting the lion's share. A second ridge interrupts the Ohio slope near the middle of the State, and thence, all the rest of the way southward, the country is broken and hilly, terminating, often, upon the waters of the Ohio, in abrupt and lofty banks.

The great bituminous coal veins of Pennsylvania, Virginia, and Kentucky, extend into Ohio, supplying her well with this valuable product. Of iron, also, she possesses ample stores.

The Ohio River forms most of the eastern and all of the southern boundary of the State, and is the recipient of the other principal streams of the region. See index for description of the Ohio in previous pages.

The Muskingum River is formed of the Tuscarawas and the Walhonding, which rise in the upper part of the State and meet at Coshocton. From this point the course of the Muskingum is nearly south east, 110 miles to the Ohio, at Marietta. Steamboats reach Dresden, 95 miles up.

The Scioto River receives its main affluent at Columbus, and flows thence nearly south to the Ohio at Portsmouth. Its passage is about 200 miles, through a fertile valley region. The route of the Ohio and Erie Canal is near the Scioto, below, for a distance of 90 miles.

The Miami River flows 150 miles from the northwest central part of the State, past Troy, Dayton, and Hamilton, to the Ohio, 20 miles below Cincinnati. It is a rapid and picturesque stream, traversing a very populous and productive valley tract. Its course is followed for 70 miles by the Miami Canal.

In the upper part of Ohio are the Maumee, the Sandusky, the Huron, the Cuyahoga, and other smaller rivers, which find their way to Lake Erie.

Lake Erie forms about 150 miles of the north and north-eastern boundary of Ohio.

Though there are many scenes of quiet beauty on the rivers and in the valleys of Ohio, yet the State possesses no landscape of any considerable fame; no celebrated and accepted shrines for Nature's devotees and pilgrims. There are, however, some objects of curious antiquarian interest—remarkable earth-works, which have for many long years attracted attention and inquiry. These mounds are scattered all over the country. There are some examples existing at Circleville. Another very remarkable one is found at Marietta; this mound is 30 feet high, and is surrounded by an elliptical wall, 230 by 215 feet. In Warren County is Fort Ancient, which has about 4 miles of embankment from 18 to 20 feet high. In Ross County are Clark's Works, 2,800 feet long and 1,800 broad, enclosing some smaller works and mounds.

A subterranean Lake is supposed to exist at Bryan, in Williams County, as water, when bored for, is found at a depth of 40 or 50 feet, at all times and in great abundance; and fish, too, sometimes coming up with it.

Ohio owes her wonderful prosperity—her almost marvellous growth, in the period of half a century, from a wild forest tract to the proud rank she now holds among the greatest of the great American States—mainly to the rich capabilities of her generous soil and climate. Nearly all her vast territory is available for agricultural uses. In the amount of her products of wool and of Indian corn, she has no peer in all the land—while she is exceeded by only one other State in her growth of wheat, barley, cheese, and live-stock; by only two States in the value of her orchards, oats, potatoes, buckwheat, grasses, hay, maple sugar, and butter. Tobacco also is one of her staples, and among other articles which she yields abundantly, are hops, wine, hemp, silk, honey, beeswax, molasses, sweet potatoes, and a great variety of fruits. Her vines, which are known and esteemed everywhere, have yielded, in the vicinity of Cincinnati alone, half a million of gallons of wine in a year.

In the forests and woodlands are found the oak, the sugar and other maples, the hickory, the sycamore, poplar, ash, and beech—the pawpaw, the buckeye (Ohio is called the Buckeye State), the dogwood, and many other trees.

Railways. If Ohio were famous for nothing else, her railways would immortalize her name. The very best way to catalogue these iron roads here would be to say, that no matter between what two given points you may desire to pass, you will be sure to find a locomotive to drag you. In round terms, several thousand miles of railway are in operation in this State, with yet many other routes in progress. Ohio, and her neighbors, Indiana and Illinois, form the great triumvirate of locomotive States. Looking upon the map, no one would attempt the vain labor to unravel the intricate web which the restless spider Travel has woven all over this region. "*Ironing done here*" seems to be the sign of the land, as it was over Punch's map of the world, during the railway mania in England. Happily, many as are the roads, they are not too many, but all contribute to the prosperity and glory of the country, near and afar off.

Ohio, in the number and population of her cities and towns, exceeds all the States of the West. To Cincinnati, her chief commercial metropolis, her peers have conceded the royal title of "the Queen City." New Orleans alone, in all the vast valley of the Mississippi, surpasses it.

Cincinnati.—HOTELS:—The *Burnet House* is very pleasantly and centrally located on Third and Vine streets; the *Spencer House*, near the Landing; *Broadway Hotel*, near the River and Landing; *Walnut Street House*, Walnut and Gano streets; *Gibson's House*, Walnut street, near Fourth.

ROUTES:—From *New York*. By Hudson River or the Harlem Railway to Albany, and thence by the Central Railroad to Buffalo, or by the N. Y. and Erie Railroad to Dunkirk or Buffalo, 459 miles; from Dunkirk, or Buffalo, above (N. Y.), via Erie (Penn.), by the Cleveland and Erie Railroad, along the shore of Lake Erie to Cleveland (Ohio), 142 miles; Cleveland and Columbus Railroad, 135 miles, to Columbus; Little Miami Railroad, 120 miles, to Cincinnati. Total distance from New York, 856 miles.

From *Philadelphia*. By Pennsylvania Railroad, 355 miles, to Pittsburg (Pa.), 187 miles to Crestline, 60 miles to Columbus, 120 miles to Cincinnati. Total, 722 miles.

From *Baltimore*. Baltimore and Ohio Railroad, 397 miles, to Wheeling (Va.), Central Ohio, via Zanesville, to Columbus, 141 miles; Little Miami Railroad, 120 miles, to Cincinnati. Total, 658.

From *St. Louis*. Ohio and Mississippi Railway.

From *New Orleans*. Mississippi and Ohio River Steamers, or by Railway.

"The Queen City of the West," as Cincinnati is called, is the largest capital of the Mississippi region, and, with its population of 250,000, it is the fifth in extent and importance in all the Union. Its central position on the Ohio River has made it a receiving and distributing depot for all the wide and rich country tributary to those great waters. The city is delightfully situated in a valley of three miles extent, enclosed by a well-defined *cordon* of hills, reaching, by gentle ascent, an elevation above the river of some 400 feet. These high points command imposing views of the city and its surroundings, far and near.

The chief portion of Cincinnati lies upon two plateaus or terraces, the first 50 feet above low-water mark, and the second 108 feet. The upper plain slopes gradually, for a mile, to the foot of Mount Auburn—a range of limestone hills, charmingly embellished with villas and vineyards. The city occupies the river shore for more than three miles, and its area is rapidly extending in every direction. The central and commercial quarter is well and compactly built. The streets are mostly of good width, well paved and well lighted with gas. The principal thoroughfares are Broadway, Main, Pearl, and Fourth streets. Main street, the great business highway, five and a half miles long, traverses the city from the Steamboat Landing—an open area of 10 acres, with 1,000 feet front—and is intersected at right angles by 14 leading streets, named First, Second, Third, Fourth, and so on. Pearl street, parallel with the river, is the great jobbing mart. Fourth street is the "Fifth Avenue" of the town, a long, wide, elegant and fashionable promenade upon the crown of the First Terrace, following the course of the river, and overlooking its waters and windings. Fifth street contains the markets, and displays a scene of busy life through an extent of three or four miles.

PUBLIC BUILDINGS. The Cincinnati Observatory has a beautiful situation upon Mount Adams, in the eastern part of the city. It commands an extensive view of the Ohio, and of the surrounding country. It can be distinctly seen by the traveller from the steamboat, in passing up or down the river. It occupies four acres of land, the gift of Mr. Nicholas Longworth. It was built by the voluntary contributions of the citizens, who gave $25 each towards the erection of the building and the purchase of appropriate instruments. Much, however, is due to the energy and perseverance of Professor Mitchel, to whose unceasing labors they are principally indebted for the result. The corner-stone was laid on the 9th November, 1843, by the late John Quincy Adams, who called the edifice a "lighthouse of the skies." The telescope is of unsurpassed finish, accuracy, and power, made by Mentz & Mahler, of Munich, artists of the highest reputation. Its cost was $10,000.

The Masonic Hall stands on the north-east corner of Walnut and Third streets. It is an elegant structure, newly erected from designs by Hamilton and McLaughlin.

The Merchants' Exchange, or Cincinnati College, a beautiful new building, is situated in Walnut street, between Fourth and Fifth streets. It is of the Grecian Doric order, three stories high, exclusive of an attic, and 140 feet front, 100 deep, and 60 in height. The Exchange and Reading-room is 59 feet by 45, and one of the finest in the United States.

The Mercantile Library Association is in the same building as the Exchange, and on the same floor; it had, in 1853, no less than 2,300 members, and 13,000 volumes, besides a very large supply of American and foreign newspapers, periodicals, &c. This liberal supply of means is continually on the increase, and the Library promises to be one of the first in the land. The United States building for the accommodation of the Post Office, Custom House, and the U. S. Courts, is one of the most symmetrical edifices in the city, being a fine specimen of Corinthian architecture.

The Ohio Medical College is located in Sixth street, between Vine and Race; it contains a large lecture-room, library, &c., the latter having several thousand well-selected standard works, purchased by the State. The cabinet belonging to the anatomical department is amply furnished.

St. Peter's Cathedral is, perhaps, the finest building of its kind in the West; it is situated on Plum street, corner of Eighth, and is devoted to the services of the Roman Catholic Church. The building is 200 feet long by 80 broad, and 60 feet high. The roof is principally supported upon 18 freestone pillars, formed of a fluted shaft, with Corinthian tops, three and a half feet in diameter, and 35 feet in height. The ceiling is of stucco-work, of a rich and expensive character. The roof is composed of iron plates, whose seams are coated with a composition of coal, tar, and sand, which renders it impervious to rain. The building cost $90,000, and the ground $24,000. At the west end of the church is an altar of the purest Carrara marble, made by Chiappri, of Gonoa; it is embellished with a centre-piece, encircled with rays, around which wreaths and flowers are beautifully carved. An immense organ occupies its opposite end, having 2,700 pipes and 44 stops. One of the pipes is 33 feet long, and weighs 400 pounds. The cost was $5,500. Several paintings occupy the walls, among which is a St. Peter, by Murillo, presented to Bishop Fenwick by Cardinal Fesch, uncle to Napoleon.

The Episcopal Church, corner of Seventh and Plum streets, and the First Presbyterian, corner of Main and Fourth, are notable edifices. Besides these, there are, all told, more than 100 churches of every shade of faith and doctrine in Cincinnati.

The City Hall is in Plum street, between Eighth and Ninth. It is a comparatively new structure.

Theatres. Pike's Opera House is a superb edifice, fronting on Fourth street, between Walnut and Vine, and running back to Baker. The National, Sycamore between Third and Fourth streets, is the oldest establishment in the city. Wood's Theatre, corner of Vine and Sixth streets, is a newer place of resort. There is also a Museum called the "Western."

The educational institutions of the city are abundant. The public schools, under the elected Board of Trustees, embrace seventeen District, four Intermediate, and two High Schools; while private establishments of excellent grade are numerous. The St. Xavier (Catholic) College, Lake Theological Seminary, Wesleyan Female College, and three Medical Colleges, are the chief educational establishments under corporate charge.

With the rapid growth of the city, its list of literary attractions is ever increasing in a ratio with the advance in all other resources and attractions.

The chief Benevolent Institutions are the Lunatic Asylum, the Commercial Hospital, four Orphan Asylums, the Widows' Home, Asylum for Indigent Females, the House of Refuge, and the Hotels for Invalids.

The Suspension Bridge is a magnificent structure, now in process of erection across the Ohio River. A correspondent of the New York Evening Post, writing of this great work (December 7th, 1856), says—

"The Ohio River is really to be bridged at Cincinnati; not as it was last winter, by ice, and in defiance of the constitution of the United States—but by a splendid structure, that will stand against all weathers and freshets.

"John A. Roebling, Esq., architect of the Niagara Suspension Bridge, is at work, 'hammer and tongs,' building the towers of a structure on the same general plan, though not adapted for the passage of a railroad, as it might be with greater cost. This project has been long talked of, and the charter, I believe, was granted some years ago by the legislatures of Kentucky and Ohio. It was not till a quite recent date that subscriptions of stock could be secured to make a beginning. Some enterprising men have procured $350,000, and will probably issue bonds for as much more, which will complete the bridge. The progress of the work is very interesting. The towers, the foundations of which are laid 86 by 52 feet at the base, will be 230 feet high, and 1,006 feet apart. The cables will be anchored 300 feet back on each side of the river, pass over the tops of the towers, and thus be made to sustain the weight of the bridge. The entire span will therefore be 1,606 feet—a little short of one third of a mile. The elevation of the floor at the middle, above low-water mark, will be 122 feet. The great flood of 1832—the highest on record—rose 62 feet above low water; and, making allowance even for this, there will remain 60 feet, which is considerably more than will be required for the highest steamboat pipes on the river. It will be a novel spectacle to look down on those splendid floating palaces passing under the magnificent span.

"The highest grade of ascent at either end will be 7 feet in 100, and the strength of the bridge will be equal to every thing but a railroad train. The foundations were begun on the 1st of September, and the structure will be completed in three years from that time."

The **Residence of Mr. Longworth**, at the foot of Mount Adams, north-east end of the city, is a charming seat, with its vineyards, gardens, and conservatories, and its art-treasures. Mr. Longworth's name is familiar abroad, in connection with the culture of the grape, for which Cincinnati is so distinguished; for everybody knows and esteems the "sparkling Catawba" of this neighborhood.

In early times (that is, 25 years ago) Deer Creek, a green-margined, pebbly stream, wound gayly along the base of Mount Adams; now it is an under-ground sewer, carrying off the blood and offal of the extensive pork-killing and packing establishments, for which Cincinnati is so greatly renowned.

Vicinage.—For *Covington and Newport*, cities of Kentucky across the Ohio, from Cincinnati, see chapter on Kentucky; also, for the *Latonia Springs*, near by.

A short distance from the city, in its north part, are two beautiful villages—*Mt. Auburn* and *Walnut Hills*—occupied chiefly as country seats, by persons whose business is in the city. The latter place is the seat of Lane Seminary.

Spring Grove Cemetery is situated in the valley of Mill Creek, about four miles north-west of the city. It has a beautiful location, and contains about 168 acres. The road thence is a famous equestrian route.

North Bend, the Home and Tomb of General Harrison, is 16 miles below the city, in full view from the river. The venerable homestead of the regretted chieftain and President (now occupied by his son-in-law, Col. Wm. H. Taylor), is a plain wooden structure, some portions of weather-boarded logs, all agreeably embowered in shading trees. It lies some 250 yards back of the river. The grave of the departed hero is upon a knoll, some 200 yards both from the water and from the house, its position marked by a single white shaft. In the rear, upon the hill-top, there is a romantic little lake.

Running along the base of the hills, on the west of the city, is Mill Creek, three or four miles up which is the *Mill Creek House*, a famous resort of jolly excursionists, bent upon "having a time."

"**Over the Rhine.**"—The Miami Canal divides Cincinnati north and south, the upper portion being known to the initiated as "Over the Rhine." It is the German quarter, and has a German theatre, with lager-bier, pipes and tobacco, Schiller and Goethe, daily, but Sundays especially.

The Race Course, lies two miles below Covington across the Ohio.

The Buckeye House, opposite the Race Course, on the upper side of the river, is four miles below the low. This popular excursion terminus is kept by "Old Joe Harrison," so called, a rosy-hazy, Indian summery Boniface. His guests come to him on fast horses, and tarry long beneath his ancient and hospitable roof. This neighborhood is a fine pigeon or trap-shooting ground. In the shallows and surfy ripples of the Ohio hereabouts, salmon are taken with rod and line heavily leaded.

Cleveland.—Hotels:—

The *Weddell House*, Bank and Superior streets; the *Angier House*, Bank and St. Clair streets.

ROUTES:—From New York, by the New York and Erie Railway, to Dunkirk, on Lake Erie, 459 miles. The lake steamers, or the Lake Shore route R. R. (via Erie.), 142 miles. Total New York to Cleveland, 596 miles. Or, from New York, by Hudson River and Central roads to Albany and Buffalo, thence as above, 627 miles.

From Philadelphia, Penn. R. R., 355 miles to Pittsburg, etc. From Baltimore, via Harrisburg, thence to Pittsburg, Pa., or by the Baltimore and Ohio route to Wheeling, etc.

From Cincinnati, see Cincinnati from Cleveland; from Chicago, see Chicago, from Cleveland.

Cleveland, after Cincinnati, is the chief city of Ohio, with a population in 1860 of some 87,000. The town proper lies on the high and commanding bluff of the Lake, and is laid out with broad, well-paved streets, occasionally varied with open squares, the general appearance being very pleasing. The social and municipal institutions of the city are in a highly creditable condition. The churches and schools especially are numerous and excellent. Visitors must not fail to see the superb Perry Monument, the Medical College, the Marine Hospital, the new Water Works, on the highest ground west of the river, and the great Railroad Depôt, at which almost as many passengers daily arrive and depart as at any other point in the land.

Cleveland was the first settlement within the limits of Cayuga County, in that part of Ohio which has long been known as the Western Reserve. It was laid out in October, 1796, and named in honor of General Moses Cleveland, a native of Connecticut. Originally the town was confined to the eastern shore of the Cuyahoga, but subsequently Brooklyn or Ohio City sprung up on the opposite side, and both parts are now united under one corporation distinguishable only by the bridge across the river.

The trade of the city amounted in the year 1859 to $363,438,051. Eighty steam and sail vessels were built and equipped at the ship-yards in 1859-'60.

Columbus.—Hotels :—*Neil House.*

To reach Columbus from New York, Philadelphia, and intermediate places, *see* Cincinnati for route thence to that city, as far as Columbus. From Cleveland (Lake Erie), south-west, 135 miles, by the Cleveland, Columbus, and Cincinnati road ; from Cincinnati, by the same route, north-east, 120 miles ; from Wheeling, Va., terminus of Baltimore and Ohio road, 141 miles west, by the Ohio Central ; from Pittsburg, on the Pennsylvania road, by the Steubenville and Indiana route, via Steubenville and Zanesville, Ohio.

Columbus is near the centre of the State, upon the banks of the Sciota River, 90 miles from its debouchure on the Ohio. It was founded in the wilderness in 1812, and in 1860 had a population of some 19,000. It is the centre of a rich country, which is daily adding to its extent and opulence. Some of the principal streets are 100 and 120 feet in width, and elegantly built. Many of the public edifices are of very striking character. The *Capitol*, which is constructed of a marble-like limestone, has a façade of more than 300 feet, and an elevation, to the top of the rotunda, of 157 feet. Then they are, besides, the *Ohio Lunatic Asylum*, the *Institution for the Blind*, the Asylum for the Deaf and Dumb, and the State Penitentiary, all fine buildings. The Starling Medical College, endowed by the late Lyne Starling, was established here some few years ago. It occupies a Gothic edifice of brick, capped with a whitish limestone.

At *Eastwood*, close by, the traveller may see the gardens of the Columbus Horticultural Society, and the grounds of the Franklin County Agricultural Society.

Dayton.—Hotels :—*Phillips House.*

Dayton is at the meeting of various railway lines ; from Cincinnati, 60 miles, by the Hamilton, Cincinnati, and Dayton road ; from Xenia, 16 miles ; from Columbus, 55 miles ; from Zanesville, 114 miles, and from Wheeling (Va.), 196 miles, on the direct route from Baltimore to St. Louis. By the same route (from the West), 108 miles from Indianapolis, 181 from Terre Haute, and 368 from St. Louis.

The Mad River enters the Great Miami at Dayton, and it is also upon the line of the Miami Canal. This is one of the most populous and enterprising cities in Ohio. The situation is pleasant, and the streets, which are of remarkable width, are built with more than wonted elegance and richness. Many of the public edifices and private mansions are constructed of excellent limestone and marble, which abound in the vicinage. In 1860, the population of Dayton amounted to 20,482.

Zanesville.—Hotels :—*Stacy House ; Kane House.*

Zanesville is upon the route from Baltimore to Columbus, Cincinnati, Indianapolis, and St. Louis (*see* those cities for routes thither) ; from Wheeling, Va., 82 miles by Central Ohio line ; from Columbus, by same road, 59 miles ; from Cincinnati (Cincinnati, Wilmington, and Zanesville road), 167 miles.

The position of Zanesville upon the Muskingum River, and in the midst of a rich and populous valley region, promises an indefinite continuation of its past success, which has been upon the scale common to the cities of the West. Settlements were first made here in 1799, and here was the seat of the State Government during the two years immediately preceding the selection of Columbus as the capital in 1812.

Chilicothe.—Hotels :—

Chilicothe is on the Sciota River and the Ohio and Erie Canal, 45 miles below Columbus, and the same distance from the Ohio at Portsmouth. It is upon the Cincinnati and Marietta Railway, extending from Parkersburg, on the Ohio, a terminus of the Baltimore and Ohio road, to Cincinnati. From Cincinnati, 96 miles.

The fine hill-slopes which enclose the valley site of Chilicothe contribute greatly to the unusually attractive aspect of the landscape here. To describe the topography of this pleasant city would be but to repeat what we have already said of many other places on the fruitful plains of Ohio and the neighboring States—to talk only of spacious and regular streets, substantial and elegant buildings, all telling eloquent tales of prosperity and progress.

This city was founded in 1796, and was the capital of the State between the years 1800 and 1810.

Springfield.—HOTELS :—*Willis House.*

Springfield is in the midst of railways, 84 miles above Cincinnati, on the direct route thence from Sandusky City on Lake Erie, and 120 miles below Sandusky; from Columbus, 45 miles.

The Mad River and the Lagonda Creek meet at Springfield. These rapid waters afford abundance of fine mill-sites, which are all well employed by the manufactories of the town. This city is regarded as one of the most beautiful in the State, both in its position and in its construction. It is interesting as the birth-place of the famous Indian warrior Tecumseh.

Steubenville.—HOTELS :— *United States.*

Steubenville is upon the Ohio River, on the eastern boundary of the State, and on the great railway route from Philadelphia, via Pittsburg, and from Baltimore, via Wheeling, Va., to Cincinnati, and all points in the West.

The history of Steubenville dates from 1798. Railroad communication with the great world has of late years given to it, no less than to its neighbors, a new and strong impetus forward. The position of the town is upon high terrace land, overlooking a smiling and happy country in all directions.

Sandusky City.—HOTELS :—*West House—Townsend House.*

Sandusky City is upon Lake Erie, on the line of the Lake Shore Railways, from Dunkirk and Buffalo (N. Y.) to Toledo, Chicago, Cincinnati, etc. From Cleveland, 61 miles; from Toledo, 52 miles; from Cincinnati, 213 miles; from Dunkirk (N. Y. and Erie road), 203 miles; from New York, 662 miles.

The first church in Sandusky was built as late as 1830, and now the city is one of the most populous and opulent in the State. Its eligible position on the busy waters of Lake Erie and its beautiful harbor ensure it continued growth and prosperity.

Portsmouth.—HOTELS :—

Portsmouth is upon the Ohio, in the south-east part of the State. A railway extends northward to the line of the road from Cincinnati to Marietta and Wheeling (Va.) *See* Chillicothe. The river steamers from all points call here.

Toledo.—HOTELS :—*American Hotel.*

Toledo is upon the Maumee River, four miles from its entrance into Lake Michigan, and upon the great railway route from the eastern States westward. It is 52 miles west of Sandusky City, 113 miles west of Cleveland; 255 miles from Dunkirk (Erie road); 714 miles from New York, and 243 miles east of Chicago, by the Michigan Southern route.

Toledo is the terminus of the Wabash and Erie Canal, the largest in the United States. Its history as a city dates only from 1836, but it is already one of the chief commercial stations of the commerce of the Great Lakes.

INDIANA.

INDIANA extends about 275 miles from North to South, and 135 from East to West; on the North is the Lake and State of Michigan; on the East, Ohio; on the South, Kentucky (across the Ohio River); and on the West, Illinois (across the Wabash).

Topographically, this State bears a great resemblance to its neighbor, Ohio. In the South, bordering on the Ohio, is the same hilly surface; and above, the same, undulating or level land, of a more marked prairie character sometimes, and perhaps more of barrens and marshes northward. In this direction a great pine tract abuts on Lake Michigan in sand-hills of 200 feet elevation. The river lands are almost always rich and fertile.

As in surface, so in soil and climate, Indiana is very like Ohio. In the production of Indian corn she is the fourth State in the Union, Ohio being the first. The other products are much the same as those we have credited to her great sister State. (*See* Ohio.)

Coal, iron, copper, marble, freestone, lime, and gypsum are found here.

INDIANA.

The **Ohio** forms the entire southern boundary of Indiana, and receives the waters of nearly all the other rivers of the State.

The **Wabash**, after the Ohio, the largest river of the region, flows 500 miles, crossing the State and separating it in the lower half from Illinois. It is the largest tributary—from the north—of the Ohio, which it enters 140 miles from the Mississippi. In its passage, it passes Huntington, Lafayette, Attica, Terre Haute, Covington and other towns. It is navigable at high water for nearly 400 miles. The Wabash and Erie Canal follows its course from Huntington to Terre Haute, 180 miles.

The **White River**, the principal tributary of the Wabash, is formed by the two branches called the East and West Fork, which unite near Petersburg. It enters the Wabash after a course of some 40 miles—nearly opposite Mount Carmel, Illinois. Upon the West Fork, the longest branch of the White River flows south-west nearly 300 miles through the centre of the State, passing, among other places, Muncie, Anderson, Indianapolis, Martinsville and Bloomfield. On the East Fork are New Castle, Shelbyville, Columbus and Rockford. This Fork is 200 miles in length. It is sometimes called Blue River, until it reaches Sugar Creek near Edinburg.

The **Maumee**, which is formed in Indiana by the St. Josephs and the St. Marys rivers, passes into Ohio, where we have already met it.

Besides these rivers, there are many other lesser waters. Lake Michigan washes the northern border of the State for 40 miles. In this region there are also a number of other small lakes and ponds.

The most interesting natural curiosities here, (the peculiar landscape features of the region, in prairie reaches and richly wooded river banks excepted,) are the numerous and remarkable caves.

The **Wyandotte Cave** in Crawford County, 11 miles from Corydon, is a wonderful place, thought by many to equal in its marvels the famous Mammoth Cave of Kentucky. It has been explored for a number of miles, and has been found rich in magnificent chambers and galleries, in stalactites and other calcareous concretions.

Epsom Salts Cave is another notable place. It is on the side of a hill, 400 feet in height. Among its wonders is a white column 30 feet high and 15 feet in diameter. It is regularly and beautifully fluted, and is surrounded by other formations of the same character. Epsom salts, nitre, gypsum and alluminous earth are found in the soil of the floor here. Another curious object is the picture of an Indian rudely painted on the rock.

Ancient Mounds and earth-works are scattered over this State, as through Ohio.

Railways.—In our peep at Ohio, we have alluded to the wonderful reticulation of railway tracks, which so marks this State and its neighbors both East and West. These iron roads link all parts of Indiana to each other, and unite it thoroughly with all the Union from the Atlantic to the Mississippi. The railways here, as in Ohio, on the one side, and Illinois on the other, are links of the great highways across the Republic westward. Half a dozen trains often start together from the same depot in Indianapolis, the Capital, radiating to all points of the compass.

Indiana has at present but few large cities, the most populous not numbering, perhaps, more than 20,000.

Indianapolis. HOTELS:—The *Bates House*, and the *American*.

Indianapolis, the capital of Indiana, is in the centre of the State, and is the radiating point of railways in every direction. To reach the city direct from New York, see route thence to Cleveland on Lake Erie; from Cleveland take the Cincinnati and Columbus road to Crestline, and the Bellefontaine and Indiana route thence to Indianapolis. Distance from Cleveland, 281 miles; from New York, 840 miles.

From *Philadelphia*, see route thence to Cincinnati as far as Columbus, Ohio—from Columbus, proceed by the Columbus and Xenia, the Dayton and Western, and the Central Indiana roads.

From *Baltimore*, by the Baltimore and Ohio Railway to Wheeling, the Ohio Central to Columbus, and thence as in preceding route from Philadelphia.

From *Cincinnati*, by the Cincinnati and Indianapolis Railway, direct.

From *Louisville, Ky.*, by the Jeffersonville road, 108 miles.

From *St. Louis*, by the Ohio and Mississippi and the Jeffersonville roads.

Indianapolis may be readily reached, also, by railway from Chicago, and nearly every other city of the West.

The locale of Indianapolis was selected for the State Capital in 1820, at which time the whole region was a dense forest. Five years later, the public offices were removed hither from Corydon, and now, broad and beautiful and populous streets, lined with costly and elegant edifices and dwellings, are every year spreading farther and farther over the great plain.

The *Railway Station* here is an edifice of magnificent proportions, with a frontage of 350 feet, and trains are momently leaving it for every point of the compass. Some of the very many *Churches* are imposing structures. The *State House* is a fine building, 180 feet in length, ornamented on each side with a grand Doric portico, and surrounded by a noble dome. The *Court House*, the *Masonic Hall*, and the *Bates Hotel* will attract the particular notice of the visitor here.

Indianapolis is the seat of the Indiana Medical College, founded in 1849; here, too, is the State Lunatic Asylum.

New Albany.—Hotels:— The *De Pau House*.

New Albany, one of the chief cities of the State, is upon the Ohio River, three miles below Louisville, and two miles below the Falls. From Cincinnati, 136 miles. See Cincinnati and Louisville for routes to those points. The New Albany and Salem Railway comes to Albany, 288 miles from Michigan City, on Lake Michigan, where it connects with the routes to Chicago and the north-west, and with the Michigan Railways for Detroit, Niagara, and the Canadas, and with the lake-shore lines to New York via Dunkirk and Buffalo. The Jeffersonville Railway from Indianapolis, the capital, 108 miles above, terminates at Jeffersonville, just above New Albany, and opposite Louisville. The lines intersect and communicate with others, for all the towns of the Western States. Steamboats arrive and depart continually for all landings on the Ohio and the Mississippi Rivers and their tributaries. New Albany is one of the two largest commercial depots of Indiana. In 1860 its population numbered some 13,000, and it has since very greatly increased. The aspect of this city is very like that of most towns on the level prairie lands of the West, broad, regular, well-built, and agreeably shaded streets, with a general air of life and prosperity.

Madison.—Hotels:— *Madison Hotel.*

Madison (population 13,000 in 1860), is upon the Ohio, 90 miles below Cincinnati; 40 miles above Louisville; and 86 miles south-east of Indianapolis. See Cincinnati and Louisville for routes thither. From Cincinnati take the steamers on the Ohio River, or the Mississippi and Ohio route for St. Louis to Vernon, and the Jeffersonville road from Indianapolis.

Madison is in a pleasant valley, of three miles' extent, shut in on the north by bold hills, 400 feet in height. It was first settled in 1808.

Fort Wayne.—Hotels:—

Fort Wayne, in the north-east part of the State, is a great railway centre, on the grand route from New York via Cleveland and Toledo on Lake Erie, and from Canada via Detroit, to Springfield (Illinois), and St. Louis. It is the western terminus of the Ohio and Indiana Road, which connects at Crestline with the Ohio and Pennsylvania, for Philadelphia. The St. Josephs and St. Marys Rivers form the Maumee at this point, and the Wabash and Erie Canal comes in 122 miles from La Fayette, and 112 miles from Indianapolis. Fort Wayne was the ancient site of the Twight-wee village of the Miami Indians. The fort which gives name to the town, was erected here in 1794, by the command of General Wayne. It continued to be a military post until 1819. The Miamis were not removed westward until 1841.

Terre Haute.—Hotels:— *Terre Haute House.*

Terre Haute is on the western boundary of the State, upon the Wabash River; the Wabash and Erie Canal, and the great line of railways crossing Ohio, Indiana, and Illinois; from Wheeling, Va., through Zanesville, Columbus, Xenia, and Dayton, Ohio, Indianapolis, Indiana, and extending westward to St. Louis; communicating with Cincinnati, Louisville, Chicago, Cleveland, and all other points.

The town is most pleasantly situated upon a bank 60 feet above the Wabash. Fort Harrison Prairie, which sweeps away to the eastward, is famous for its charming landscape.

La Fayette.—Hotels:—*Bramble House.*

La Fayette is upon the Wabash River, and the Wabash and Erie Canal, and at the intersection of the New Albany and Salem Road from Michigan City on Lake Michigan, to New Albany on the Ohio; the La Fayette and Indianapolis Road from Cincinnati, via Indianapolis and the Toledo, Wabash, and Western road from Toledo on Lake Erie; from Toledo (railway always) 203 miles; from Indianapolis, 64 miles; from New Albany (Ohio river), 197 miles; from Michigan City (Lake Michigan), 91 miles.

Evansville.—Hotels:—*Pavilion Hotel.*

Evansville is upon a high bank of the Ohio, near the south-west extremity of the State, 200 miles from the Mississippi, and the same distance below Louisville in Kentucky. A railway extends north to Vincennes on the Ohio and Mississippi route (between Cincinnati and St.

Louis) to Vincennes, 51 miles, and to Terre Haute 109 miles (from Evansville).

Richmond is situated on a fork of Whitewater River, and four miles from the Ohio State line, 69 miles from Indianapolis, the capital, and 64 north-west of Cincinnati. It is a growing town, and has several flourishing manufactories of cotton, wool, iron, paper, and flour. The river furnishes abundant water-power, which is very generally taken advantage of by the inhabitants, for it has become the chief manufacturing town in the State. Richmond has ten or twelve churches, a public library, a Branch of the State Bank of Indiana, two fire companies, and a large number of retail stores. It is the centre of a rich and populous agricultural district, with which it does an active trade. The population is estimated at about 6,000. The Central Railway connects here, and there is a Railway to Newcastle, commencing at Richmond.

ILLINOIS.

ILLINOIS extends northward 380 miles, and westward (at the extremest point) 200 miles. Wisconsin lies on the north, Lake Michigan and Indiana on the east, Kentucky on the south (the Ohio between), and Missouri and Iowa on the west, the Mississippi intervening.

The general surface of the country here, as in Indiana and Ohio, is that of elevated table lands inclined southward, though it is more level than the neighboring States. In the lower portions there is a small stretch of hilly land, and some broken tracts in the north-west; and upon the Illinois River there are lofty bluffs, and yet higher and bolder points on the Mississippi.

The Prairies. The great landscape feature of Illinois is the prairie country, this unique phase of Nature being seen here in its most marked and happiest humor. No matter what may be its character, every work of art or nature, earnestly and magnificently done, affects and interests the human mind and heart. The want of variety and caprice which are ordinarily essential to landscape attraction, are more than compensated in the prairie scenery, as in that of the boundless ocean, by the impressive qualities of immensity and power. Far as the most searching eye can reach, the great unvarying plain rolls on; its sublime grandeur softened, but not weakened by the occasional groups of trees in its midst, or by the forests on its verge, or by the countless flowers everywhere upon its surface; any more than is the sea, by the distant sail here and there, by the far-off surrounding hills, or by the glittering sparkles of its crested waves.

The **Grand Prairie**, here, is the most striking example in the country of this aspect of Nature. Its gently undulating plains, profusely decked with flowers of every hue, and skirted on all sides by woodland copse, rolls on through many long miles from Jackson County, northeast, to Iroquois County, with a width varying from one to a dozen or more miles. The uniform level of the prairie region is supposed to result from the deposits of waters by which the land was ages ago covered. The soil is entirely free from stones, and is extremely fertile. The most notable characteristic of the prairies, and their destitution of vegetation, excepting in the multitude of rank grasses and flowers, will gradually disappear, since nothing prevents the growth of trees but the continual fires which sweep over the plains. These prevented, a fine growth of timber soon springs up; and as the woodlands shall be thus assisted in encroaching upon, and occupying the plains, human settlements and habitations will follow, until the prairie tracts shall be overrun with cities and towns.

The **Agricultural Capabilities** of Illinois are not surpassed anywhere in the Union. The soil on her river bottoms is often 25 or 30 feet deep, and the upper prairie districts are hardly less productive. The richest tracts in the State are the great American Bottom, lying along the Mississippi, between the mouths of the Missouri and the Kaskaskia Rivers, a stretch of 80 miles, the country on the Rock River and its branches, and that around the Sangamon and other waters. Forty bushels of wheat, or 100 bushels of Indian corn to the acre, is a common product here. In the growth of Indian corn, Illinois ranks as the third State in the Union; and her population and the amount of land employed considered, she is the first. In respect to other agricultural staples and products, what we have said of the adjoining States of Ohio and

Indiana, may be repeated of Illinois; so of the forest trees of the country.

In **Mineral Resources** the State is well provided. She shares with the adjoining States of Iowa and Wisconsin, extensive supplies of lead. The trade in this mineral is the chief support of the prosperous town of Galena, in the north-west part of Illinois. More than thirteen millions of pounds of lead have been smelted here in one year. Bituminous coal exists everywhere, and may be procured in many places without excavation. The Bluffs, near the Great American Bottom, contain immense beds. In the south part of the State iron is said to be abundant; and in the north, copper, zinc, lime, fine marbles, freestone, gypsum, and quartz crystals. Some silver, too, has been said to exist in St. Clair County.

Medicinal Springs, sulphur and chalybeate, are found in various parts of the State. In Jefferson County there is a spring very much resorted to, and in the southern part of the State are some waters, which taste strongly of Epsom salts.

Excepting the speciality of the prairie, the most interesting landscape scenery of this State is that of the bold, acclivitous river shores of the Mississippi, the Ohio, and the Illinois, more particularly.

The Mississippi forms the entire western boundary, and many of the most remarkable pictures for which its upper waters are famous, occur in this region, the tall, eccentrically shaped bluffs rising at different points from 100 to 500 feet. The Fountain Bluff of the Mississippi, is in Jackson County; it is odd in form, is 6 miles in circuit, and 300 feet in height; the summit is full of sink holes. See "*Mississippi River,*" elsewhere.

The Illinois, the largest river, here, flows through the centre of the State, south-westerly into the Mississippi, 20 miles above Alton. Exclusive of its branches, the Des Plaines and the Kankakee, its length is about 320 miles. Its navigable waters extend at some seasons 206 miles, to Ottawa, at the mouth of the Fox River. Peoria is upon its banks, 200 miles up. Half a hundred steamboats ply upon this river.

The picturesque heights of the Illinois, called the Starved Rock, and the Lover's Leap, are a few miles only below Ottawa. Starved Rock, 8 miles down the river, is a grand perpendicular limestone cliff, 150 feet in height. It was named in memory of the fate of a party of Illinois Indians, who died on the rock from thirst, when besieged by the Pottawatomies. Lover's Leap is a precipitous ledge just above Starved Rock, and directly across the river is Buffalo Rock, a height of 100 feet. This eminence, though very acclivitous on the water side, slopes easily inland. The Indians were wont to drive the buffaloes in frightened herds to and over its fearful brink. Lake Peoria is an expansion of the Illinois, near the middle of the State. Above Vermilion River there are some rapids, which boats pass only in periods of high water.

The Ohio bounds the State on its southern extremity. It is in this part of Illinois (Hardin County) that the famous Cave in the Rock of the Ohio shore occurs. See *Ohio River.*

The Wabash, on the Eastern boundary, divides Illinois in the lower portion from Indiana. See *Indiana.*

Rock River flows 330 miles from the neighborhood of Lake Winnebago, in Wisconsin, to the Mississippi, a little below the town of Rock Island. It enters Illinois at Beloit, and afterwards passes Rockford and Dixon. The course of Rock River is through a rich valley or plain, remarkable for its pictorial interest. The navigation of its waters is much obstructed by rapids; for it is, unlike the sluggish Illinois, a bold, swift stream, with a will and way of its own. Small steamboats ascend sometimes, however, 225 miles, to Jefferson, in Wisconsin.

The **Des Plaines** flows 150 miles from the south-east corner of Wisconsin to Dresden, where it unites with the Kankakee and forms the Illinois.

The **Kankakee** comes from the northern part of Indiana, 100 miles to Dresden. Its course is sluggish and through a region chiefly occupied by prairies and marshes.

The **Sangamon** travels 200 miles to the Illinois, about 10 miles above Beardstown. Small steamers ascend it in high water.

The **Fox River** comes 200 miles from Wisconsin to the Illinois, at Ottawa.

The Vermilion, the Embarras and the Little Wabash, are tributaries of the Wabash from Illinois.

Lake Michigan forms 60 miles of the northern boundary of the State. Excepting the expansion of the Illinois River, called Lake Peoria, and the waters of Pishtaka, in the north-east, there are no other lakes of importance.

Railways abound in Illinois as in all parts of the West. The whole country is traversed in all directions by grand lines of iron road, which unite all its cities and towns to each other and to all the surrounding States. At a moderate count we may speak of the miles of railway here in thousands.

In no part of the Union have towns and cities sprung up so rapidly and in such wonderful growth as in Illinois—increasing so fast in population, that the census of one year is no standard by which to count the people of the next.

ILLINOIS. 233

Chicago.—HOTELS:—The *Tremont;* the *Briggs House;* the *Sherman House.*

ROUTES — From *New York* — To Buffalo or Niagara Falls by the New York and Erie, and the Hudson River and Central Railways; from Niagara, by the Great Western Railroad (Canada) to Detroit, and from Detroit through Michigan by the Michigan Central Railroad; or, from Buffalo by the Lake Erie steamers; or the railway on the Lake Shore, through Erie, Pa., Cleveland, and Sandusky to Toledo, and thence by the Southern Michigan Route.

From *Philadelphia*—Pennsylvania Railroad, 355 miles to Pittsburg; by Pittsburg, Fort Wayne and Chicago Railroad to Chicago.

From *Baltimore.*—See pp. 16, 17.

From *New Orleans*—Mississippi Central Railroad to Memphis; thence to Cairo; thence on Illinois Central Railroad through Urbana.

Chicago is the largest and most important city of Illinois, and, in its rapid growth, the most remarkable in the Union. In 1831 it was only an Indian trading post, and as late as 1840, its population did not number 5,000, while to-day, it can show an actual population of 160,000 inhabitants. Its rapid progress in population and wealth, and its unrivalled position on the lake, vindicate its claims as the commercial metropolis of the north-west. The site of the city is a gently inclined plane, the ground in the western part of the city, three miles from the lake, being from 15 to 18 feet above the level of the lake. The city is well drained. The streets are much wider than those of New York or Philadelphia. They cross each other at right angles, and are for the most part paved with stone, or with the Nicholson pavement. The Chicago river and its two branches run through the city, and are crossed by swinging bridges, which allow the passage of vessels. The river affords a harbor for the largest vessels for more than five miles, at the entrance of which is a new iron lighthouse. Lake street is the Broadway of Chicago; while Michigan avenue and Wabash avenue are distinguished by princely edifices, and adorned with rows of magnificent trees.

The principal public buildings are the Custom House, the Chamber of Commerce, Crosby's Opera House, the Court-House, and the Armory. The Depot of the Illinois Central Railroad is also a very fine building of immense extent. There are spacious and elegant churches; among which are the First Baptist, the St. James, Episcopal, the First and Second Presbyterian, the Cathedral of the Holy Name, St. Patrick's Catholic Church, the Universalist Church, and the Church of the Holy Family.

The collections of books in the Chicago Historical Library, and the Young Men's Association Library, are both very large.

McVicker's Chicago Theatre is one of the largest and best appointed in the United States. Chicago is noted for the great extent of its grain and lumber trade. The receipts and shipments of grain are at present about 50,000,000 bushels each, annually.

Galena.—HOTELS:—*De Soto House.*

Galena is in the extreme north-west corner of the State, upon the Fevre River, six miles from its entrance into the Mississippi. It is 450 miles above St. Louis, and 171 from Chicago by the Galena and Chicago Union Railway. (See Chicago for routes to that place from the Atlantic cities.) Springfield, the capital of the State, is 250 miles below. The Fevre River, upon a rocky ledge of which Galena is built, may be considered as an arm or bay of the Mississippi. The situation is bold and picturesque, amidst lofty bluffs. The streets rise in terraces, one above the other, communicating by stairway or steps, making a very novel and striking picture. Below, on the levée, are the business depôts. Next come the churches and other public edifices; and yet above, rise, file on file, elegant private dwellings and villas.

Galena is famous as the centre of the great lead-mining districts of the Upper Mississippi. Copper, too, is found in considerable quantity in the country around.

With its railroad and steamboat access to all points of the vast territory on the north-west, it promises soon to double and treble its present population of perhaps 12,000.

Springfield.—HOTELS:—*The American,* an excellent house.

Springfield, the capital of Illinois. From New York, via Buffalo or Dunkirk, N. Y.; Cleveland, Crestline, and Bellefontaine, Ohio; and Indianapolis, Indiana. (See *Cleveland* and *Indianapolis.*)

From Cincinnati, via Indianapolis.

From St. Louis, *up* 97 miles, and from Chicago *down* 188 miles, on the Chicago, Alton, and St. Louis Railway.

Springfield lies south-west of the centre of the State, near the Sangamon River, upon the confines of a beautiful prairie district. In the centre of the city is a square, occupied by the State Capitol and other public edifices, and compassed by spacious and elegant streets. Railways diverge towards all points. Springfield was the residence, and is now the burial-place of Abraham Lincoln, the late President of the United States.

Peoria.—Hotels:—*Peoria House.*

Peoria lies northwest of the centre of the State, upon the Illinois River, at the meeting of the Peoria Branches of the Chicago and Rock Island Railways. By these routes it is connected, more or less directly, with all the other towns of Illinois and neighboring States. From Chicago, 160 miles; from Rock Island, 114 miles; from Springfield 70 miles north; from St. Louis, 167 miles.

Peoria is the most populous place upon the Illinois; and, commercially, the most important in the State. "It is situated upon rising ground," says a traveller; "a broad plateau, extending back from the bluff—and the river expanding into a broad, deep lake. This lake is the most beautiful feature in the scenery of Peoria, and as useful as it is beautiful, for it supplies the inhabitants with ample stores of fish; and in winter, with abundance of the purest ice. It is often frozen to such a thickness that heavy teams can pass securely over it. A substantial drawbridge connects the town with the opposite shores of the river. The city is laid out in rectangular blocks, the streets being wide and well graded. The schools and churches are prosperous, and the society good. A public square has been reserved near the centre. Back of the town extends one of the finest rolling prairies in the State, which already furnishes to Peoria its supplies, and much of its business." A post was established here by La Salle as early as 1684, but the history of the present town dates only from 1819. The population, in 1860, amounted to 14,425.

Alton.—Hotels:—*The Alton House.*

Alton is upon the Mississippi, 25 miles above St. Louis, on the Terre Haute and Alton Railway, direct route from Philadelphia, by the Pennsylvania Road to Pittsburg, Pa.; thence to Columbia and Dayton, Ohio, and to Indianapolis and Terre Haute, Indiana. It is 260 miles below Chicago, and 185 below Springfield, by the Chicago, Alton, and St. Louis line.

The Missouri enters the Mississippi three miles below Alton, contributing greatly to the commercial value of its position. It is, besides, one of the best landings on the great river. The present city, of about 7,000 people, has grown up since 1832, at which time the Penitentiary was established here. Upper Alton is the seat of the Shurtleff (Baptist) College.

Quincy.—Hotels:—

Quincy (population in 1860, about 14,000) is upon the Mississippi, 170 miles above St. Louis, and 104 miles west of Springfield; 263 miles from Chicago, by the Chicago and Burlington Road, 168 miles to Galesburg; and thence, 100 miles, by the Northern Cross Railway. By these lines Quincy is connected also with Galena, Rock Island, Peoria, and other cities. The town is built upon a limestone bluff, 125 feet above the river, in the vicinity of a fertile rolling prairie.

Rock Island.—Hotels:—*The Fanshaw House; The Fuller House.*

Rock Island is two miles above the mouth of the Rock Island River, on the Mississippi, at the foot of the Upper rapids, which extend 15 miles. It is upon the great highway from the Eastern States to Iowa and the north-west; from Chicago, 182 miles; from Iowa City (westward), 54 miles. This city is named after a large island near by. It is a picturesque and most thriving place.

Peru.—Hotels:—*Moore's Hotel.*

Peru is upon the Illinois River, and the Chicago and Rock Island Railway, at the intersection of the Illinois Central Road. From Chicago, 100 miles; from Rock Island, 82 miles. The Illinois and Michigan Canal terminates near Peru. The town is very advantageously situated, with such ready and general railway access, and being, too, as it is, at the head of natural navigation on the Illinois River. The population of some thousands is rapidly increasing.

Fulton, the western terminus of the "Air Line" from Chicago, west, is upon the Mississippi, 136 miles from Chicago. It gives fair promise to take a prominent position among the river towns. A superb railway bridge across the Mississippi is now in process of building.

Nauvoo is on the Mississippi River, at the second and last rapids below the Falls of St. Anthony, which extend up the river about 12 miles. It may be easily reached from Quincy, below, or from Burlington, the western terminus of the Chicago and Burlington Railway.

This is the site of the famous Mormon City, which was founded in 1840, by "Joe Smith" and his followers, and once contained a population of 18,000. It is located on a bluff, but is distinguished from every thing on the river bearing that name, by an easy, graceful slope, of very great extent, rising to an unusual height, and containing a smooth, regular surface, which, with the plain at its summit, is sufficient for the erection of an immense city. Nauvoo was laid out on a very extensive plan, and many of the houses were handsome structures. The great Mormon Temple, an object of attraction, and seen very distinctly from the river, was 128 feet long, 88 feet wide, and 65 feet high to the top of the cornice, and 163 feet to the top of the cupola. It would accommodate an assemblage of 3,000 persons. It was built of compact, polished limestone, obtained on the spot, resembling marble. The architecture, although of a mixed order, in its main features resembled Doric. In the basement of the Temple was a large stone basin, supported by 12 oxen of colossal size; it was about 15 feet high altogether, all of white stone, and well carved. In this font the Mormons were baptized. This building, without an equal in the West, and worth half a million of dollars, was fired by an incendiary, on the 9th of October, 1848, and reduced to a heap of ruins. Joe Smith and a number of his followers were arrested and confined in the county prison where, in June, 1844, they were put to death by a mob, disguised and armed. Expelled from Illinois, by force of arms, the Mormon community removed to their present settlements, in Utah. Since then, a company of French socialists, led by M. Cabet, have established themselves here.

MICHIGAN.

The unique character of the scenery of the upper peninsula of Michigan, and the present easy means of access, promise to make the region one of the most popular summer resorts in the Union. Excepting in portions of its southern boundary, this State is everywhere surrounded by the waters of the Great Lakes, insomuch that it has a coast of much more than a thousand miles. The country is, in shape, something after the position of a reversed letter Ӆ, divided into two peninsulas. All the northern shore of the upper portion or top stroke of the Ӆ, is washed by the waters of Lake Superior, with Lake Huron, Green Bay, and Wisconsin on the south. Of the lower peninsula, Lake Michigan forms the entire western boundary, while on the east there are Lakes Huron, St. Clair and Erie, and Michigan and Huron on the north. Of this immense lake-coast, 350 miles belong to Lake Superior, as much more to Lake Michigan, 300 to Lake Huron, 40 to Lake Erie and 30 to St. Clair. Besides these grand waters which encompass the State about like a girdle, there are many beautiful ponds scattered over the interior, and bearing thither the picturesque beauty of the shores.

The southern peninsula is more interesting in an agricultural than in a pictorial point of view. It is in surface notably unvaried—a vast plain, indeed, undulating but not broken by any elevations worthy of mention. It has, however, peculiar features which will interest the traveller, in its great prairie lands and that special characteristic of the western landscape—the Oak Openings—a species of natural park meagerly covered with trees.

The shores, however, even of this part of Michigan, are often picturesquely varied, with steep banks and bluffs, and shifting sand-hills, reaching, sometimes, a height of 200 or more feet.

The romantic lands of the "Lake State" are in the upper peninsula, which is rich in all the loveliness of rugged, rocky coast, of the most fantastic and striking character, in beautiful streams, rapids, and cascades. Here, making a part of the scenery of Lake Superior, which we have elsewhere visited, are the Wisconsin, or Porcupine Mountains, 2,000 feet in height, and those strange huge castellated masses of sandstone, celebrated as the Pictured Rocks. The famous straits of Mackinaw unite the converging floods of Lakes Huron and Michigan at the extreme northern apex of the lower peninsula, and the beautiful Sault de St. Marie conducts the wondering tourist from Lake Huron to Lake Superior on the north. The St. Mary separates the upper peninsula at its north-eastern extremity from Canada. The Pictured Rocks are about 60 miles west of this passage. Here the famous white-fish and other finny game may reward the patience of the angler.

The rivers of Michigan are chiefly small streams, but many of them, especially those in the mountain districts of the north, are replete with pleasant themes for the pencil of the artist.

MICHIGAN.

The history of this State has more points of interest than we are apt to find in this section of the Union, recording as it does some memorable incidents of Indian adventure and important exploits in the American and English war of 1812. After England had dispossessed the French, who first settled the country in the latter part of the seventeenth century, there arose among the Indian tribes the famous chieftain Pontiac, who availed himself of the opportunity afforded by the outburst of the Revolution, to attempt the entire expulsion of the white invaders of his ancestral lands. The chief planned a general attack upon all the English forts on the lakes; massacred the garrison at Mackinaw, and laid siege, for some months, to Detroit.

From its contiguity to Canada, Michigan was called early into the field in the war of 1812. Detroit was surrendered to the enemy by General Hull, August 15th, the fort at Mackinaw having already been captured. A number of American prisoners of war were butchered by the Indians at Frenchtown on the 22d of January, 1813. The State suffered at this period many trials, until General Harrison at length drove the British into Canada, carrying the war into their own country. Detroit was not surrendered to the United States until 1796. Michigan came into the Union as an independent State in the year 1837.

Railways.—The Michigan Central railway extends 280 miles from Detroit to Chicago, Ill.

The Michigan Southern, in connection with the Northern Indiana, traverses the Southern line of Michigan and the upper line of Indiana, 243 miles from Toledo to Chicago.

The Detroit and Milwaukee road crosses the State, 186 miles, from Detroit to Grand Haven, on Lake Michigan, opposite Milwaukee.

The Great Western (Canada) railway has its western terminus opposite Detroit. The Grand Trunk connects at Detroit with the Michigan Central, Detroit and Milwaukee, and Mich. So. railway.

The Amboy and Lansing road, 25 miles from Owasso to Lansing, the State Capital.

The Flint and Holly railway, from Holly on the Detroit and Milwaukee road, north to Saginaw.

The Detroit, Monroe and Toledo railway, from Detroit south to Toledo, Ohio.

The Jackson Branch of the Michigan So. railway, from Adrian north to Jackson on the Mich. Central.

Detroit.—HOTELS :—The *Russell House;* the *Biddle House.*

ROUTES :—From *New York.* By the Hudson River or Harlem Railway to Albany, thence by the Central Railway to Buffalo or Niagara Falls, or to Buffalo and Niagara by the New York and Erie Railway. (See these routes elsewhere.) From Buffalo or Niagara, take the Great Western Railway (Canada) to Windsor, opposite Detroit. Total distance from New York about 673 miles. Detroit may also be pleasantly reached from Buffalo or Dunkirk, via Cleveland, Sandusky City and Toledo, Ohio, by the railways on the southern shore of Lake Erie, or by the Lake Erie steamers. From Chicago to Detroit, by the Michigan Central or by the Michigan Southern and Northern Indiana railways; distance, 282 miles. Several trains daily on all these routes.

Detroit is one of the great commercial depots of the West, and the chief city of Michigan. It is pleasantly situated upon the Detroit River, a link in the chain of waters which unite Lake Huron and Lake Erie. This strait, for such it is, gives the city its French name—*dé-troit.* It is here about half a mile in width, and is charmingly dotted with beautiful islands. Detroit occupies a position admirable for business activity, being directly in the way of the flood of travel and transportation from the Atlantic to the Mississippi, and great railways and steamers, with their freights, necessarily paying it tribute. Not only is the city thus commercially alive, but it is distinguished also for its manufactories of many kinds.

Jefferson and *Woodward* avenues, and *Congress street,* are fine thoroughfares. The Campus Martius is a good example of its public squares. There is a fine open area called the *Grand Circus,* towards which the avenues of that part of the city lying back of the river, converge. The *Old State House* (Detroit was the capital of Michigan at one time) is a noteworthy edifice, with its dome and its tall steeple overlooking the town and its environs, Lake St. Clair above, and the Canadian shores. The City Hall is a brick structure of a hundred feet façade. The city possesses, also, a fine Custom House, a Marine Hospital, and other public edifices, and many elegant private residences.

Detroit was founded by the French in 1670. The Capital of the State was here until it was removed to Lansing. The present population is 45,619.

Lansing.—Hotels :—*Lansing House.*

Lansing, the capital of Michigan, is upon the Grand River, 110 miles north-west of Detroit. The Detroit and Milwaukee railway approaches within 25 miles at Owasso. From that station, a pleasant ride on the Amboy and Lansing railway speedily bears the traveller to the little capital. Lansing became the seat of the State government in 1847, at which period it was almost a wilderness. Its population is now about 3,000.

Ann Arbor is a flourishing place of 4,000 or more people, upon the line of the Michigan Central railway, 37 miles west of Detroit. It is the seat of the State University, founded in 1837. This Institution is liberally endowed, and has about 300 students.

Ypsilanti.—Hotels :—The *Hawkins House.*

Ypsilanti (with a population of about 3,000) is upon the line of the Central railway, 30 miles from Detroit.

Michigan towns and stations on the Central railway, and their distances from Detroit : Dearborn, 10 miles ; Wayne, 18 do. ; Ypsilanti, 30 do. ; Ann Arbor, 37 do. ; Dexter, 47 do. ; Chelsea, 54 do. ; Grass Lake, 65 do. ; Jackson, 75 do. ; Parma, 86 do. ; Albion, 95 do. ; Marshall, 101 do. ; Battle Creek, 120 do. ; Galesburg, 134 do. ; Kalamazoo, 143 do. ; Mattawan, — ; Paw Paw, 159 do. ; Decatur, 167 do. ; Dowagiac, 178 do. ; Niles, 191 do. ; Buchanan, 196 do. ; Terre Coupee, 202 do. ; New Buffalo, 218 do. ; Michigan City, 228 do.

Monroe City.—Hotels :—*Strong's Hotel.*

Monroe City, one of the principal towns of Michigan, (population about 8,000,) is upon the Raisin River, two miles from Lake Erie, 40 miles below Detroit, and at the eastern terminus of the Michigan Southern railway for Chicago and the West.

Monroe was settled by the French about 1776.

Grand Haven, on Lake Michigan, is the western terminus of the Detroit and Milwaukee railway. It is situated on a fine site at the mouth of the Grand River, directly opposite the beautiful and flourishing city of Milwaukee. It has an excellent harbor formed by the river and bays, extending some fifteen miles, with a depth of water of from thirty to fifty feet, sufficient for vessels of the largest size. The entrance to the harbor is six hundred and fifty feet wide. The distance from Grand Haven to Milwaukee is eighty miles. A line of transit steamers, fitted up in the most costly manner, with every regard for the safety and comfort of passengers, plies twice daily between the two ports, in connection with the regular trains of cars.

MISSOURI.

Missouri formed part of the ancient territory of Louisiana, purchased by the United States from France. A settlement called Fort Orleans was made within its limits by the French in 1719. The oldest town in the State, St. Genevieve, was founded in 1755. St. Louis was commenced in 1764. The State was visited in 1811 and in 1812 by a memorable series of earthquakes, which occurred in the vicinage of New Madrid. The face of the country was greatly altered by these events—hills entirely disappeared ; lakes were obliterated and new ones formed. The waters of the Mississippi River were turned back with such accumulations, that they overran the levees built to hem them in, and inundated whole regions, leaving it in its present marshy state. Missouri is the first of the States formed wholly westward of the Mississippi.

Landscape of Missouri. The surface of this great State is in many parts level or but slightly undulating. A wide marshy tract occupies an area of 3,000 square miles in the south-eastern part, near the Mississippi. In other sections are vast reaches of prairie lands, extending to the Rocky Mountains. The Ozark Mountains, which we have seen traversing also the State of Arkansas, extend through Missouri, centrally, from north to south. The Ozark hills are elevated table-lands. The rich alluvial tracts of the Mississippi lie east of this district, and westward are boundless deserts, and treeless plains, sweeping away to the base of the Rocky Mountain ranges.

Mineral Resources of Missouri. The State is remarkably rich in iron ore, lead and copper and coal mines, and in nearly all the mineral products. It possesses, too, a great variety of marbles, some of them beautifully variegated, and other valuable building stones.

Agricultural Products. The chief staples of Missouri are Indian corn, hemp, tobacco, flax, and all the varieties of grains, fruits, vegetables, and grasses, for the successful growth of which the soil is admirably adapted.

The Missouri River. The restless, turbid waters of this magnificent river flow fretfully, 3,096 miles from their sources in the remote west, to their *debouchure* in the Mississippi, not far above the city of St. Louis. The entire length of the river, including its course to the Gulf of Mexico by the Mississippi (1253 miles more), is 4,349 miles.

The head-waters of the Missouri are very near the springs, which find their way to the Pacific through the channels of the Columbia River. Their course is northward for 600 miles, until they reach the remarkable cataracts known as the *Great Falls*. Before their arrival here, however, and at a distance of 411 miles from their source, the waters make the grand passage of the bold chasms called the

Gates of the Rocky Mountains.

"Here, through a length of six miles, the giant rocks rise perpendicularly to an elevation of 1,200 feet. The dark waters, in their narrow bed, wash the base of these huge walls so closely, that not a foothold is anywhere to be found. It is a ghostly gorge on the sunniest day, but when its habitual gloom is deepened by the shadow of a stormy sky, its sentiment of solitude grows painfully impressive. Let a thunder peal reverberate, as often happens, in a thousand wailing voices through the rocky windings of this glen, and let the blackness of darkness be increased by the vanished gleams of the lightning flash, and you think you have left this fair world far behind you.

"We were once, with some friends, traversing this passage at such a fearful moment as we have described, when we became aware that we were pursued by a party of Indians. Noiselessly and breathlessly we urged on our canoes, pausing at intervals only, to ascertain the progress of our foes, hope and despair alternately filling our breasts, as we seemed at one moment to be gaining and at another losing ground. It was only now and then that we caught a glimpse of the savages, and the sound of their unceasing and unearthly yells came to our ears with such uncertainty, that it gave us no clue to their position. The excitement of the struggle was intense, as their random arrows flew about our ears, and as the deadly effect of our fatal shots was told to us by the death-cries from their own ranks.

"We took fresh courage as the increasing light spoke our approach to the terminus of the glen, and gave us hope, once on terra firma, of distancing our foes. New fears, though, seized upon us, lest our scanty supply of ammunition should be exhausted before we reached the prayed-for sanctuary. Happily the dread vanished, as the arrows of the savages sensibly decreased in numbers, and the chorus of their infernal shrieks died away.

"When we at last leaped, panting, upon the open shore, not a sound of pursuit was to be heard, leaving us the glad hope that we had slain them all, or so many as to secure us from further danger. But not stopping to verify this supposition, we made all possible haste to reach the camp, which we had so gaily left a few hours before. Once safe among our companions, we mentally vowed to be wary henceforth how we ventured within the Gates of the Rocky Mountains."

The Great Falls of the Missouri.

The descent of the swift river at this point is 357 feet in 16¼ miles. First comes a cascade of 26 feet, next one of 27 feet, then a third of 19 feet, and a fourth of 87 feet, the upper and highest. Between and below these cataracts there are stretches of angry rapids. This passage is one of extreme beauty and grandeur, and at some day, not very distant, perhaps, when these western wilds shall be covered with cities and towns and peaceful hamlets, this spot will be one of no less eager and numerous pilgrimage than many far less imposing scenes are now. The Falls of the Missouri are esteemed by the few tourists whose good fortune it has been to look upon these wonders, as holding rank scarcely below the cataracts of Niagara.

The upper waters of the Missouri flow through a wild, sterile country, and below, pass vast prairie stretches.

Above the river Platte, the open and prairie character of the country begins to develop, extending quite to the banks of the river, and stretching from it indefinitely, in naked grass plains, where the traveller may wander for days without seeing either wood or water. Beyond the "Council Bluffs," which are situated about 600 miles up the Missouri, commences a country of great interest and grandeur, denominated the Upper Missouri. It is composed of vast and almost boundless grass plains, through which runs the Platte, the Yellow Stone, and the other rivers of this ocean of grass. Buffaloes, elk, antelopes, and mountain sheep abound. Lewis and Clark, and other respectable travellers, relate having found here large and singular petrifactions, both animal and vegetable. On the top of a hill they found the petrified skeleton of a huge fish, 45 feet in length. The herds of gregarious animals, particularly of the buffalo, are innumerable.

The Missouri is navigable for steamboats, except during periods of extreme drought, 2,575 miles, from its mouth to the foot of the Great Falls.

The Yellow Stone, one of the principal tributaries of the Missouri, rises in the same range of mountains with the main stream. It enters from the south by a mouth 850 yards wide, and is a broad and deep river, having a course of about 1,600 miles. The Platte, another of its great tributaries, rises in the same range of mountains with the parent stream, and, measured by its meanders, is supposed to have a course of about 2,000 miles before it joins that river. At its mouth it is nearly a mile wide, but it is very shallow, and is not boatable except at its highest floods. The Kansas is a very large tributary, having a course of about 1,200 miles, and is boatable for most of the distance. The Osage is a large and important branch of the Missouri; it is boatable for 200 miles, and interlocks with the waters of the Arkansas. The Gasconade, boatable for 66 miles, is important from having on its banks extensive pine forests, from which the great supply of plank and timber of that kind is brought to St. Louis.

Railways in Missouri. This State is destined, as a chief depot for the products of the Great West, to be the centre of an interminable radiation of railways. Several magnificent routes are already opened towards the Pacific, and others are in progress.

The Pacific Railway is completed at this time from St. Louis westward 196 miles to Dresden, 71 miles beyond Jefferson City, the capital of the State. At Jefferson City the road leaves the Missouri River, not to touch it again until it reaches its western terminus at Kansas, on the Kansas River. West of Dresden, the line is in rapid progress. Its entire length from St. Louis to Kansas will be 283 miles. The South-West Branch of the Pacific road commences at Franklin, 37 miles west of St. Louis, and is now in successful operation as far as Rolla, 76 miles from Franklin, where it deflects from the Pacific road, and of 118 miles from the city of St. Louis.

The Hannibal and St. Joseph road extends 206 miles westerly from Hannibal, on the Mississippi, to St. Joseph, on the Missouri. The Great Salt Lake Mail, the Pike's Peak Express, and the N. Y. Pony Express, taking despatches to San Francisco in eight days, all start from St. Joseph. A road is in progress from St. Joseph to Council Bluffs and Omaha.

The North Missouri Railway extends 304 miles from St. Louis to St. Joseph. Mails leave St. Joseph, for Kansas, Leavenworth, Omaha City, Lawrence, Topeka, Lecompton, Great Salt Lake City, and for California.

The St. Louis and Iron Mountain Railway extends from St. Louis 86 miles south to Pilot Knob.

Besides the great roads leading from St. Louis westward, numerous lines from the east terminate here, as the Ohio and Mississippi, 340 miles, direct from Cincinnati. The Illinois Central (or its connections) from Chicago. The St. Louis, Alton, and Terre Haute, 189 miles. The great railways northward from New Orleans, Mobile, etc.

St. Louis.—HOTELS:—*Lindell House; Barnum's St. Louis; Everett House;* the *Planter's,* and the *Olive St. House.*

ROUTES.—From New York, via Chicago (see Chicago), and thence by the Chicago, Alton and St. Louis road; or to Cleveland, Ohio, 596 miles (see Cleveland); from Cleveland to Crestline by the Cleveland, Cincinnati and Columbus Railroad, 75 miles; Crestline to Indianapolis (Bellefontaine Line), 206 miles; Indianapolis to Terre Haute (Terre Haute and Richmond route), 73 miles; Terre Haute to St. Louis (Terre Haute and Alton road), 187 miles; total, 1137 miles. Or from Cincinnati in a direct line, by the new route of the Ohio and Mississippi Railway.

From Philadelphia: to Pittsburg by the Pennsylvania Railway, 355 miles; Pittsburg to Crestline (Pittsburg, Fort Wayne and Chicago road), 187 miles; Crestline to Indianapolis (Bellefontaine road), 206 miles; Indianapolis to Terre Haute (Terre Haute and Richmond road); 73 miles; Terre Haute to St. Louis (Terre Haute and Alton line), 187 miles; total, 1,008 miles.

From Baltimore: Baltimore and Ohio road to Wheeling, 397 miles; to Zanesville, 82 miles; Columbus, 141 miles; Dayton, 71 miles; Indianapolis, 108 miles; Terre Haute, 73 miles; St. Louis, 187 miles; total, 1,059.

History. The present site of the great city of St. Louis was chosen by Laclede, on the 15th of February, 1764. It was settled as a trading station for the trappers of the West. The average annual value of furs, brought here during the fifteen successive years ending with 1804, was $203,750. The number of deer skins was 158,000; beaver, 36,900; otter, 8,000; bear, 5,100; buffalo, 850, and so on. At this period of wild life, the population of St. Louis was between 1,500 and 2,000, half of whom were always away as *voyageurs* and trappers. Up to 1820, the number of the people had not reached 5,000.

In 1768 (August 11th), Rious and his band of Spanish troops took possession of the place, in behalf of Her Catholic Majesty, who kept possession until it was transferred to the United States, March 26th, 1804. The first brick house was built in 1813. The first steamboat arrived in 1817. The history of St. Louis as a city began in 1822, with the name bestowed upon it by Laclede, in honor of Louis XV. of France. Between 1825 and 1830, emigration began to flow in from Illinois, and the place thrived. The population in 1830 had reached 6,694; in 1840 it had swelled to 16,469; in 1850 it was 77,850; in 1852 it was (the suburbs included) over 100,000, to which a census to-day would add another hundred thousand. This is the magnificent way in which cities grow in the Great West.

St. Louis lies upon the right bank of the Mississippi River, 20 miles below the entrance of the Missouri, and 174 miles above the mouth of the Ohio. It is 744 miles below the Falls of St. Anthony, and 1,194 miles above the city of New Orleans. It is built upon two limestone plateaus, one 20 and the other 60 feet above the waters of the Mississippi. From the plain, into which the upper terrace widens, fine views of the city and its surroundings are presented. The entire extent of St. Louis along the curves of the river is about 7 miles, and back some 3 miles, though the densely occupied area spreads over a space of only 5 miles riverward and 2¼ miles inland. The streets are of good width, and regular. Front street, stretching along the levee, is 100 feet in breadth. This highway, and Main and Second streets, back of and parallel with it, are the great commercial streets.

Lafayette Square is almost the only public park of importance which the city yet possesses.

The *public edifices* of St. Louis, in its municipal buildings, churches, hotels, market-houses, and charitable institutions, are in every way creditable to the taste, munificence, and enterprise of the people. The *City Hall*, the *Custom House*, and the *Court House*, are worthy of a metropolitan fame. Of the churches, which perhaps much exceed 60 in number, many are very imposing: as the *Catholic Cathedral*, on Walnut street, between Second and Third; the *St. George* (Episcopal), at the corner of Locust and Seventh streets, and the *Church of the Messiah* (Unitarian), at the corner of Olive and Ninth streets. The *United States Arsenal* is a grand structure in the south-east part of the city; and 13 miles below, on the river banks, are the *Jefferson Barracks*.

If the visitor at St. Louis should chance to be benevolent, or literary, or educational, he will, perhaps, like to look at the *City Hospital*, the *Marine Hospital* (3 miles below the city), the *Home for the Friendless*, the *Sisters' Hospital*, and the *Orphan Asylums;* or the *University of St. Louis*, founded in 1832 by Roman Catholic patronage; *Pope's Medical College;* the *Washington University;* the *Carcudin College* of the Germans; the *Missouri University;* the *Mercantile Library Association*, established in 1846, and at the numerous excellent private schools and convents. Let the stranger, in this progressive metropolis of the great valley of the Mississippi, seek his pleasure as he will, here are the opportunities to find it.

Many railways bend their vast iron tracks towards St. Louis from all directions eastward, and soon they will be converging hither from the Rocky Mountain side. Already the great Pacific road comes in 282 miles from Kansas; the Hannibal and St. Joseph route comes from a distance of 206 miles towards the Rocky Mountains; and another line is traversing the State in a south-westerly direction.

St. Louis is the great starting point from civilization for savagedom—the place where adventurers for Kansas and Nebraska, and Utah, and for the wild traverse of the Rocky Mountains to the Pacific States and Territories, begin their rude forest journey.

Jefferson City.—HOTELS :—

Jefferson City, the capital of Missouri, is upon the Missouri River, 125 miles west of St. Louis, by the Pacific Railway, or 155 miles by steamboats up the river. The situation is bold and beautiful, overlooking the turbid waters of the Missouri and their cliff-bound shores. The population in 1853 amounted to about 3,000. Jefferson City is on the great route to Kansas, Nebraska, Utah, California, and all the Rocky Mountain region.

St. Joseph.—HOTELS :—*Patee House.*

St. Joseph is upon the Missouri River, 340 miles above Jefferson City, and 496 miles, by water, from St. Louis. It is the most important place in the western part of the State, and a great point of departure for the western emigrants. Population, 5,000. See *Hannibal*.

Hannibal.—Hotels:—*Planter's House.*

Hannibal is upon the Mississippi, 153 miles above St. Louis, and 15 miles below Quincy, Illinois. A railway over 200 miles long now extends entirely across the State, thus connecting Hannibal with St. Joseph, on the western boundary.

Lexington.—Hotels:—*City Hotel — Virginia Hotel.*

Lexington is upon the Missouri River, 120 miles, by land, from Jefferson City. The town has prospered by its trade with the Santa Fé and Great Salt Lake caravans. Population in 1860, about 4,000.

St. Charles City.—Hotels:—*The Virginia House—The City Hotel.*

St. Charles City is upon the Missouri, 22 miles from its mouth. By land it is 6 miles below the Mississippi. Population, between 3,000 and 4,000.

Cape Girardeau.—Hotels:—*St. Charles.*

Cape Girardeau is upon the Mississippi, 45 miles above the mouth of the Ohio. The *St. Vincent College* is located here.

Weston.—Hotels:—*The St. George.*

Weston is upon the Missouri, 200 miles by railroad beyond Jefferson City, and 5 miles above Fort Leavenworth. With the exception of St. Louis, Jefferson is the most active business town in the State. It drives a busy trade with the western emigrants, and supplies the garrison at Fort Leavenworth.

Palmyra is 6 miles from Marion City, its landing place on the Mississippi River. The railway from Hannibal across the State to St. Joseph now calls at Palmyra.

Carondelet is 6 miles below St. Louis, on the Mississippi.

St. Genevieve is 61 miles below St. Louis, on the Mississippi. It is the shipping point for the products of the Iron works at Iron Mountain.

New Madrid was formerly a noted place, but, owing to the dreadful earthquakes it experienced in 1811 and 1812, it has sunk into comparative insignificance. It is situated on a great curve or bend of the river, the land being extremely low, and the trees along the bank presenting a great uniformity of appearance. The view is most monotonous—a feature, indeed, characteristic of much of the scenery of the Lower Mississippi. On this side there is scarcely a dozen feet elevation for the distance of 100 miles. By the earthquake thousands of acres were sunk, and multitudes of lakes and ponds were created. The churchyard of this village, with its sleeping tenants, was precipitated into the river. The earth burst in what are called sand-blows. Earth, sand, and water were thrown up to great heights in the air. The river was dammed up, and flowed backwards. Birds descended from the air, and took shelter in the bosoms of people that were passing. The whole country was inundated. A great number of boats passing on the river were sunk. One or two that were fastened to islands, went down with the islands. The country was but sparsely peopled, and most of the buildings were cabins, or of logs; and it was from these circumstances that but few people perished.

IOWA.

IOWA is one of the new States, admitted into the Union in 1846. It lies wholly beyond the Mississippi, which washes all its eastern boundary. On this side, its neighbors are Wisconsin and Illinois. On the north is Minnesota; on the west, Minnesota and Nebraska, and upon the south, Missouri. The State has no very notable history, beyond the usual adventure and hardship of a lone forest life, among savage tribes. The settlement of the region was seriously began (first at Burlington) in the year 1833.

The landscape of Iowa is marked by the features which we have traced in our visit to neighboring portions of the north-west. The surface is, for the most part, one of undulating prairie, varied with ridges or plateaus, whose extra elevations impel the diverse course of the rivers and streams. The Coteau des Prairies enters the State from Minnesota, and forms its highest ground. On the Mississippi, in the north-east, the landscape assumes a bolder aspect, and pictures of rugged rocky height and bluff are seen. A few miles above Dubuque, Table Mound will interest the traveller. It is a conical hill, perhaps 500 feet high, flattened at the summit.

11

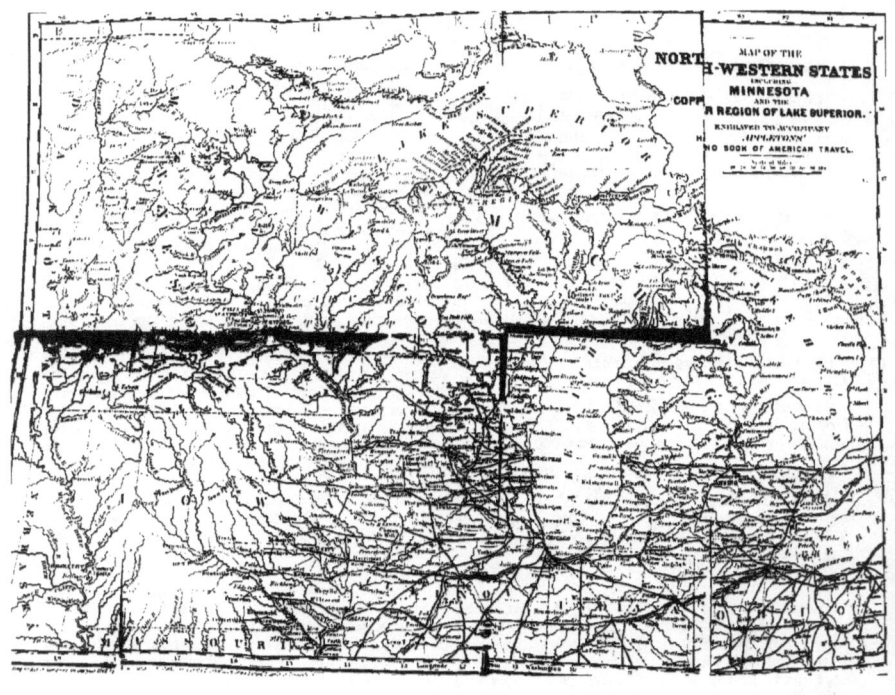

The **Prairies**, which are sometimes 20 miles across, present many scenes of interest, in their way—and it is a way not ungrateful to the unaccustomed eyes of the visitor from the Atlantic States—monotonous as it may, possibly, grow in time. The rivers in some parts of the State wind through ravines of magnesian limestone, amidst which they have gradually worked their way, leaving the rocks in every grotesque form of imagery.

Sink Holes.—The depressions in the ground, called sinks, are curious objects. These singular places, which are numerous, are circular holes, 10 and sometimes 20 feet across. They abound more particularly on Turkey River, in the upper part of the State. Near the mouth of this stream there are also to be seen many small mounds, sometimes rows of them, varying in height from 4 to 6 feet.

Minerals.—Iowa has many mineral products, among which is an abundant supply of lead. Copper and zinc are also freely found, and plenty of coal.

The **Des Moines River**, the most important stream in Iowa, rises in Minnesota and flows 450 miles through the State, to its southeast extremity, where it enters the Mississippi, 4 miles below Keokuk. It is navigable for small steamers 250 miles, or may be made so, with some practicable improvements.

The **Iowa River** is 300 miles in length, and is navigable from the Mississippi upwards, 80 miles, to Iowa City. The Skunk River, 200 miles, the Cedar, the Makoqueta, and the Wapsipinicon, are all tributaries of the Mississippi.

The **Missouri** and the Great Sioux rivers form the entire western boundary of Iowa.

Railways. The State of Iowa is like all the North-west, being rapidly covered by an endless network of rails. At Davenport and Rock Island, on the Mississippi, a great route comes in from the Atlantic cities, via Chicago, and is in progress westward, via Des Moines, to Council Bluffs and Omaha City, on the far-off shores of the Missouri. The far western portion extends now to Iowa City, which is (by rail) 238 miles from Chicago. The Dubuque and Pacific road is open 111 miles to Cedar Falls, and in progress to Fort Dodge.

The Chicago, Iowa, and Nebraska Railway, passing the Mississippi at Fulton and Clinton, extends, now, 82 miles beyond these towns to Cedar Rapids. The Keokuk, Fort Des Moines, and Minnesota, beginning at Warsaw and Keokuk, on the Mississippi, will intersect the road from Chicago to Council Bluffs and Omaha at Des Moines. It is open now from Keokuk 38 miles to Bentonsport.

Dubuque.—HOTELS :—*Washington House*; *Julien House.*

ROUTES. From the Atlantic cities or the South, *see* Chicago at St. Louis.

From Chicago, proceed by the Galena and Chicago Union Railway, through upper Illinois, *via* Galena, 184 miles, to Dubuque. From St. Louis, take the steamer up the Mississippi River, 450 miles.

Dubuque is upon the banks of the Upper Mississippi, in the midst of a very picturesque country. It is, indeed, if we possibly except Iowa City, the most beautiful town in the State. It occupies a broad, elevated terrace, which stretches along the great river for miles. Many fine buildings are to be seen here. Numerous railways of the West find their way to this point. Dubuque is the oldest town in Iowa, having been settled by the French as long ago as 1786.

Iowa City.—HOTELS :—The *Clinton House.*

Iowa City is charmingly situated upon some oval bluffs of the Iowa River about 80 miles from the meeting of that water with the Mississippi. It may be reached from Chicago (see Chicago, for routes to that point from New York and other places) by the Chicago and Rock Island Railway, 182 miles from Chicago to Rock Island, on the Mississippi ; and thence, 54 miles, by the Mississippi and Missouri Railway. From St. Louis, by the Illinois Railway, or by the Mississippi River, to Davenport ; and thence, 54 miles, by railway.

The site of Iowa City was a wilderness, in 1839, when it was selected as the seat of government of the then prospective State. Within one short year it had a population of 600 or 700 people ; and now, more than as many thousands. The town is delightfully embosomed in shady groves, and surrounded by fertile prairies. At the intersection of the chief streets, the Iowa Avenue and Capitol Street, which are each 100 feet wide, stands the former State House, a handsome Doric building, 120 feet in length. It is constructed of ringed and spotted stone, called "bird's-eye marble," which was quarried in the neighborhood. This edifice, with all its extensive grounds, have been granted to the State University, Fort Des Moines having been selected as the future capitol. The State Asylums for the deaf, the blind, and the dumb, are located at Iowa City. The Iowa River is at all times navigable to the city.

Burlington.—HOTELS :—*Basset House.*

Burlington, formerly the capital of Iowa, and one of the most populous and important places in the State, is upon the Mississippi River, 250 miles above St. Louis, 45 miles above Keokuk, and 88 miles east-south-east of Iowa City. See Chicago, for routes thither from the Atlantic cities, and from that point proceed by the Chicago and Burlington Railway, 210 miles south-westerly, across the State of Illinois to Burlington. Burlington is partly built upon the bluffs which characterize the shores of the Mississippi in this the most picturesque portion of its endless journey, from the Great Lakes to the Gulf of Mexico. In 1854, its population was about 7,000. At this time it is very much more, as it is growing rapidly, like all the cities of the West.

The famous Indian chieftain, Black Hawk, once dwelt at Burlington, and here his bones lie buried.

One of the present routes to Nebraska, starts from Burlington by railway a few miles to Mount Pleasant; and thence, by stages, to Omaha City.

Davenport.—HOTELS :—*Le Claire House.*

Davenport is on the Mississippi, at that foot of the Upper Rapids, opposite Rock Island, Illinois, and on the railways from Chicago to Iowa City. See route to Iowa City. The Iowa College was established here some 13 years ago. The landscape of this region is extremely attractive.

Keokuk.—HOTELS :—*The Billings House.*

Keokuk is at the foot of the Lower Rapids of the Mississippi, 205 miles above St. Louis, and 125 miles below Iowa City. See Burlington, for route, from Chicago and the eastern cites to that place. Keokuk is not far below, following the river.

This is the head of navigation for the largest steamers, and the outlet for the rich valley of the Des Moines, the most populous portion of the State. Fine steamers run daily between Keokuk and St. Louis, and a railway is in progress hence, 180 miles, to Dubuque, passing Burlington, Davenport, and Lyons, on the Iowa side of the Mississippi and Rock Island, Fulton, Galena, and other points upon the Illinois banks.

Muscatine is upon the Mississippi, 100 miles above Keokuk, and 32 miles from Iowa City. See Chicago, for routes to that city from the East, and take the Chicago and Rock Island Railway, 182 miles to Rock Island; cross the Mississippi to Davenport (opposite); thence, down the river, or, more expeditiously, by the Mississippi and Missouri Railway, Iowa Citywards 25 miles to Muscatine Junction, and 12 miles, by branch road, to Muscatine.

Muscatine is at the apex of a bend in the Mississippi, on the summit of a bold range of rocky bluffs, seen from the water 40 miles away. It was first settled by the whites, in 1836; before that period it was an Indian trading-post, called Manatheka. It is one of the most active and populous cities in the State.

Fort Madison is a growing town upon the Mississippi, 22 miles above Keokuk, and the same distance below Burlington. See routes to these places.

Fort Des Moines, selected as the capital of Iowa, in 1855, is at the junction of the Des Moines and Raccoon Rivers, 120 miles West of Iowa City. Steamboats ascend the Des Moines to this point from the Mississippi. The railway from Davenport to Council Bluffs (completed now to Iowa City) is to pass through Fort Des Moines. The place, as a United States military post, was evacuated in 1846.

WISCONSIN.

WISCONSIN came into the Union as a State as late as 1848, though the country was visited, as was all the wilderness of which it was so recently a part, by the French missionaries two centuries ago. Its growth has been, and continues to be, unexampled anywhere, excepting in the surrounding new States and Territories.

The topographical aspect of Wisconsin is very similar to that of other portions of the northwest section of the Union, presenting, for the most part, grand stretches of elevated prairie land, sometimes 1,000 feet higher than the level of the sea. Though there are no mountains in this State, there are the characteristic plateau ridges of the latitude, formed by depressions, which drain the waters, and afford beds for the rivers and lakes. The descent of the land towards Lake Superior is very sudden, and the streams are full of falls and rapids.

The floods of Lake Superior and Lake Michigan wash the northern and eastern boundaries of Wisconsin, and numberless lesser waters are scattered through the interior, and more abundantly over the north-western counties. The shores of these lakes are often most picturesque in rich forest growth, and in rocky precipice. The waters are clear and full of delicious fish.

Lake Winnebago, the largest of the interior waters of Wisconsin, lies south-east of the middle of the State. Its length is about 28 miles, with a width of 10 miles. The Fox or Neenah River unites it with Green Bay, an arm of Lake Michigan. A singular wall which might, in its regular formation, easily be supposed the work of art, instead of nature, follows the eastern shore of Winnebago for 15 miles. This wall rises through all its extent about 5 feet above the surface of the water, and sinks in places hundreds of feet below. Steamboats navigate the lake.

The **Mississippi River** forms much of the western boundary of Wisconsin, separating it from Iowa and Minnesota, with which States it thus shares the charming scenery of this portion of the great river—the noble expansion of Lake Pepin, with its bold precipices, and headland of the Maiden Rock, and the La Grange Mountain; Mount Trempleau in La Crosse County, with its perpendicular cliffs 500 feet in height; and many other striking scenes.

The **Wisconsin River**, the largest stream in the State, rises in a small lake called Vieux Desert, on the northern boundary, and flows south-westerly six hundred miles to the Mississippi at Prairie du Chien. Shifting sandbars obstruct the navigation very much, yet steamboats ascend as high as Portage City, 200 miles distant, by the windings of the River. At Portage City a ship-canal conducts small steamers to the water of the Neenah or Fox River (the outlet of Lake Winnebago), by which the navigation is continued through the State from the Mississippi to Lake Michigan. The Wisconsin River presents many beautiful pictures to the eye of the traveller. At the remarkable passage called the Grandfather Bull Falls, where the waters break through a bold gorge a mile and a half in length, and flanked on either hand with rugged walls 150 feet in height, some fine chalybeate springs add to the attractions of this charming spot, and promise to make it before long a favorite summer resort.

Petenwell Peak. On the Wisconsin, below Grandfather Bull Falls, some 60 miles, there is a singular oval mass of rock, 900 feet in length, and 300 wide, with an elevation above the surrounding country of 200 feet. This is Petenwell Peak. The summit for 70 feet is perpendicular, and the rocks in their fantastic groupings assume the most wonderful architectural appearances, almost persuading the voyager that he is transported back to feudal ages, and is passing through a barbaric land of castled and battlemented heights.

Fortification Rock is another interesting scene, a few miles below Petenwell Peak. The cliffs, here, have a vertical elevation of 100 feet.

The **Dalles of Wisconsin.** At the part of the river thus called, the water passes for half a dozen miles between hills of solid rock, in height from 30 to 100 feet. The narrowest width of the river here is 55 feet.

The **St. Louis River**, which forms part of the boundary between Minnesota and Wisconsin, is remarkable for a series of bold rapids, called the Falls of St. Louis. Of this scene we have spoken in our mention of the landscape of Minnesota. The St. Louis River is the original source of the St. Lawrence.

The Badaxe, Black, Chippewa, the Rock, the Des Plaines, the Fox, and other rivers of Wisconsin, are much broken by cataracts and rapids.

Mounds or Earth-works in Wisconsin. The antiquary, no less than the lover of natural beauty, may find here striking sources of pleasure, in objects scarcely less strange than the mystical relics of the old world. Scattered everywhere, over the plains of Wisconsin, are singular structures of earth, formed—who knows when, or by what people?—after the likeness of men and animals. At Prairieville, there is one of these weird works, 56 feet in length, which is in the similitude of a turtle; near the Blue Mounds is another, representing a man in a recumbent attitude, 120 feet in length; near Cassville yet another of these eccentric labors has been found, made in the image of the extinct Mastodon. At Aztalan, in Jefferson County, there is an old fortification 550 yards in length, and 275 wide. The walls are from 4 to 5 feet high, and more than 20 feet thick.

The **Blue Mounds** are in Dane County. The most elevated rises nearly 1,200 feet above the waters of the Wisconsin River.

The **Forest Scenery**, and the ever-welcome oak openings—the oasis of the prairie—will be among the gratifications of the nature-loving tourist in Wisconsin. The hunter may indulge his passion for the chase at will, whether he aspire to the wild game of the wilderness, or to the gentler sports by the brook-side.

WISCONSIN.

Railways in Wisconsin. Several hundred miles of railway are completed, and many other routes are in progress in Wisconsin.

The Chicago and Milwaukee extends along the western shore from Lake Michigan, 85 miles from Chicago to Milwaukee, connecting with various routes to other towns in the State.

The Milwaukee and Minnesota road extends the entire breadth of the State south, from Milwaukee to La Crosse.

The Milwaukee and Mississippi R. R. extends to Prairie du Chien, 200 miles, with branch to Janesville. The Milwaukee and Horicon, from Horicon to Stevens Point, on the Wisconsin. Racine is connected with Beloit by railway, 65 miles. Kenosha, Sheboygan, and Manitowoc are each building railways westward, and the Chicago, St. Paul, and Fond du Lac route is being urged forward northward from Fond du Lac to Lake Superior.

Besides these main lines, other lesser routes are in operation, connecting Beloit and Madison, Milwaukee and Watertown, and places on Lake Michigan, and in Illinois, with the more interior towns and villages of Wisconsin.

Milwaukee.—HOTELS:—*Newhall House.*

ROUTES. From Chicago take the railroad 85 miles to Milwaukee, along the western shore of Lake Michigan, or the steamers on the lake which arrive daily.

Milwaukee, the most populous city in Wisconsin, is built at the mouth of the Milwaukee River, on the western shore of Lake Michigan, 90 miles from Chicago, and 100 miles east of Madison. The town lies upon the river flats, and upon the bluffs which overlook the lake. The peculiar color of the "Milwaukee brick," of which many of the buildings are made, gives the city a very peculiar and pretty air. These famous bricks, which are much in fashion now all over the country, have a delicate cream or straw tint. In growth, this city of promise has kept pace with the rapid progress characteristic of the region. It was settled in 1835, incorporated in 1846, had a population in 1840, of 1,751 ; in 1850, of 20,061 ; in 1860, more than 45,000. It has increased greatly since, and, as the outlet of a large and rich country, will long continue to extend its borders. Several hundred miles of plank road once united the city with the interior, but now their office is better performed by good lines of railway. There are 40 or 50 churches of various denominations here, and numerous excellent literary institutions and schools.

Madison.—HOTELS:—*Capitol House.*

ROUTES. See Chicago and Milwaukee for routes from New York, Boston, etc. Thence by rail.

The town, in the centre of a broad valley enclosed by high grounds, occupies an isthmus between the Third and Fourth (Mendota) Lake. Mendota or Fourth Lake upon the upper side of the city, is about 6 miles by 4 miles in area. The Third Lake is somewhat smaller ; both are exceedingly picturesque waters, deep enough for steamboat navigation.

There was no building except a solitary log cabin, upon the site of Madison, when it was selected in 1836 for the Capital of the State ; yet in 1860 the population had reached nearly 7,000. The streets of this beautiful city of the wilderness, drop down pleasantly towards the shores of the surrounding lakes. "Madison," says a writer of the landscape here, "perhaps combines and overlooks more charming and diversified scenery than any other town in the West, or than any other *State Capital* in the Union. Its high lakes, fresh groves, rippling rivulets, shady dales, and flowery meadow lawns, are commingled in greater profusion, and disposed in more picturesque order than we have ever elsewhere beheld."

The Capitol, which has cost $150,000, is a lime stone edifice, in a public park 70 feet above the level of the lakes. The University of Wisconsin, founded in 1849, occupies an eminence a mile west of the Capitol, and 125 feet above the lakes. The State Historical Society and the State Lunatic Asylum are located here.

Kenosha.—HOTELS:—*City Hotel.*

Kenosha is upon Lake Michigan, just below Racine, 55 miles above Chicago (by railroad or by water), and 35 miles below Milwaukee.

Janesville.—HOTELS:— The *Hyatt House.*

Janesville, a populous and important city, steadily increasing in extent and population, is upon the Rock River, 45 miles below Madison. A branch of the railway between Milwaukee and Madison, extends to Janesville ; junction, at Milton.

Waukesha, once Prairieville, is upon the Milwaukee and Mississippi railway, 20 miles from Milwaukee and 78 from Madison. It is upon Pishtuka or Fox River, at the extremity of a pine prairie. Carroll College, founded 1846, is here.

Platteville, the centre of an extensive lead region, is about 22 miles north of Galena, Illinois, and 78 from Madison. See *route to Galena.*

Fond du Lac.—Hotels:—*Lewes House.*

Fond du Lac is a prosperous and populous town, at the south end of Winnebago Lake, 72 miles from Milwaukee and 90 miles from Madison, reached by railway from Milwaukee or from Chicago. Fond du Lac is remarkable, among other things, for its Artesian Wells, which are so numerous that nearly every household has its own. They vary in depth from 90 to 130 feet.

Beloit.—Hotels:—*Bushnell House.*

Beloit is upon the southern boundary of the State, on the line of railway from Chicago to Madison and to Dubuque, 98 miles from Chicago and 50 miles from Madison. From Milwaukee, by railway, 78 miles. Beloit is built on a beautiful plain, on the banks of Rock River. It is famous for elegant churches and fine streets. The Beloit College is located here.

Watertown is upon the Milwaukee and Watertown railway, 45 miles from Milwaukee.

Green Bay is at the mouth of the Neenah or Fox River, at the head of Green Bay. A railway from Milwaukee *via* Fond du Lac, is in process of building. Distance from Milwaukee, 114 miles; from Madison, 120 miles. The largest steamers of Lake Michigan stop here. The older part of Green Bay was formerly called Navarino. Fort Howard and Village is across the river.

Portage City is at the head of navigation on the Wisconsin River. Steamboats ply occasionally between this place and Galena, Illinois. It is upon the route of the La Crosse Railway, 95 miles from Milwaukee. The site of Portage City is at the famous Winnebago Portage, and at Old Fort Winnebago.

Sheboygan is at the entrance of Sheboygan River into Lake Michigan, 62 miles above Milwaukee.

Manitowoc is upon Lake Michigan, above Sheboygan, 93 miles from Milwaukee.

Racine—Hotels:—*Congress Hall.*

Racine is a beautiful city, situated on the west shore of Lake Michigan, at the mouth of Root River, twenty-five miles south by east of Milwaukee, and seventy miles north of Chicago. It is the second city of the State in population and commerce, and has one of the best harbors on the Lake, formed by the mouth of the river, which admits vessels drawing over twelve feet of water. It has a population of 12,000 inhabitants, and rapidly increasing. The public schools are among the best in the State. Over $80,000 have been expended by the citizens in the construction of a harbor; $30,000 for school buildings, and $350,000 for railways.

MINNESOTA.

ROMANTIC stories of the wonders of the land, which now forms the new State of Minnesota, were told two centuries ago by the zealous French missionaries, who had even at that remote period, pushed their adventures thither; still, only a very few years have elapsed, since emigration has earnestly set that way, calling up populous towns and cultivated farms along the rivers and valleys, before occupied by the canoe and the wigwam of the savage alone.

The magical development of Minnesota is in keeping with that marvellous spirit of progress so characteristic of the great Western sections of the United States. So rapid is this growth, and on such a sure and enlightened basis, that the church and the school-house spring up in the wilderness before there are inhabitants to occupy them. In Minnesota, one of the earliest foundations was that of a Historical Society, established almost before the history of the country had begun!

Area. Minnesota, before its organization as a State, included within its limits a vast region, extending from the Mississippi and the St. Croix Rivers, and from Lake Superior on the east, to the Missouri and the White-Earth Rivers on the west, a distance of more than 400 miles; and from the Iowa line on the south, to the British borders on the north—also 400 miles apart.

Surface and Soil.—"Almost the whole of this vast region," says Mr. Bond in his interesting volume about Minnesota and its resources, "is a fine rolling prairie of rich soil, a sandy loam adapted to the short summers of the climate, and which produces bounteously. The surface of the country, excepting the Missouri plains, is interspersed with numerous beautiful lakes of fresh water—all abounding in the finest fish, and their banks covered with a rich growth of woodland. The land is about equally divided between oak openings and prairies, the whole well watered by numerous streams navigable for steamers."

Forest Lands and Rivers.—In the eastern part, on the head waters of the Mississippi, Rum River, and the St. Croix, are extensive pine and hardwood forests, apparently inexhaustible for centuries; while from the mouth of *Crow-wing River*, a tributary of the Mississippi, an extensive forest of hard wood timber, fifty miles in width, extends south-westerly into the country watered by the Blue-Earth River, a tributary of the Minnesota River, emptying into it 150 miles above its mouth. The latter stream, rising near Lac Traverse, flows south-easterly a distance of 450 miles, and empties into the Mississippi at Fort Snelling, seven miles above St. Paul, and the same distance below St. Anthony. This is one of the finest streams in the valley of the Mississippi, and the country through which it flows is not excelled for salubrity of climate and fertility of soil by any part of the United States. In a good stage of water, steamboats can ascend it almost to its source. A portage of a mile or two then connects it from Big-Stone Lake with Lac Traverse; and the outlet of the latter, the Sioux Wood River (all of which are thirty miles in length), with the famous **Red River** of the North. This stream is navigable at all seasons for steamboats from the Bois de Sioux to Pembina, on the British line—to Selkirk settlements, 100 miles beyond—and even to Lake Winnipeg. The whole trade of these extensive regions will eventually seek this channel to a market, following down the Minnesota to the Mississippi at St. Paul, and thence to the States below. A railroad connection will eventually be made from the mouth of the Bois de Sioux to Fond du Lac; also from the same point to St. Anthony and St. Paul via Sauk rapids and the Mississippi. Another will connect the same point with Lac qui Parle, on account of the portage at Big-Stone Lake; thence down to the mouth of Blue-Earth; thence south-easterly through Iowa to some point, say Prairie du Chien or Dubuque, on the Lower Mississippi.

The only interruption to the navigation of the Lower Minnesota River in dry seasons, is what are called the "Rapids," some 40 miles above its mouth. This is a ledge of sandstone rock, extending across the stream, and will soon be removed.

The Mississippi above St. Anthony is navigable an almost indefinite distance to the north; and the steamer "Governor Ramsey" has already been running in the trade above the Falls for four years, as far as the Sauk Rapids (80 miles), which, with the Little Falls (40 miles beyond), are the main obstacles in a navigation of over 400 miles from St. Anthony to the Falls of the Pokegama. St. Croix Lake and River are navigable to the Falls, 60 miles above the junction of the lake and Mississippi; and the St. Louis River is navigable from Lake Superior 20 miles to Fond du Lac. Numerous other streams are navigable for light-draught steamers and flat-boats from 50 to 100 miles, penetrating into the interior to the pineries, and giving easy access into the country in all directions. These are the Blue-Earth, Rum, Elk, Sauk, Crow, Crow-wing, Vermilion, Cannon, and others.

On the north-eastern border of the territory is Lake Superior, with its valuable fisheries and its shores abounding in inexhaustible mines of copper, coal, iron, etc., besides affording the facility of that vast inland sea for immigration and commerce.

The Great Father of Waters, too, the mighty Mississippi—after rising in Itasca Lake, in the northern portion of the State, flows by a devious course for some 800 miles through the eastern part, and below the mouth of the St. Croix forms the dividing line between Minnesota and Wisconsin for some 200 more to the Iowa line. This mighty river gives the State the whole lower valley to the Gulf of Mexico for a never-ceasing market for its agricultural produce, lumber, and manufactures.

Various elevated ridges traverse the territory of Minnesota, though not of a mountain character. The plateau called the **Couteau des Prairies,** or the Prairie Heights, is one of these singular terraces. It extends 200 miles, with a breadth varying from 20 to 40 miles. The average elevation of this lofty plain is some 1,500 feet, and in some parts it rises nearly 2,000 feet above the level of the sea. In the north it is about 900 feet above the neighboring waters of Big-Stone Lake. There is another range of wooded heights, reaching 100 miles or more, called the "Coteau du Grand Bois." Then there are the "Hauteurs de Terre," which extend some 300 miles. These last-mentioned ridges form the dividing line of the rivers, which flow to Hudson's Bay on one side and to the Mississippi and Lake Superior on the other.

The Lakes of Minnesota are numberless and of extreme beauty. Sometimes there are little ponds a mile in circumference, and again, great waters 40 or 50 miles in extent. Their shores are charmingly wooded, and frequently present fine pictures of cliff and headland. The waters are pure and transparent, and are filled with whitefish, trout, pike, pickerel, sucker, perch, and other finny inhabitants. The largest of these lakes are the Minnetonka, the Osakis, White Bear, Kandiyohi, Otter-tail, and Mill Lac. The *Territory* of Minnesota has, in its curtailment to *State* limits, lost many noble natural points, among them Devil's Lake, which now lies in Dakotah.

Lake Pepin, a beautiful expansion of the Mississippi, is in this region. On the east bank is the famous Maiden's Rock, 400 feet high; and near the northern end, the La Grange Mountain rises in a bold headland, 230 feet above the water.

The Falls of St. Anthony.—HOTELS:—
The Winslow House.

ROUTES.—From the Atlantic cities to Chicago, and thence, to La Crosse or Dubuque, or Rock Island, by rail; thence by steamer on the Mississippi River.

The Falls of St. Anthony, in the Mississippi, lie within the territory of Minnesota, 8 miles from St. Paul. The river at this pass is divided by an island, as at Niagara, where it rushes over a bold and broad ledge of limestone.

"I visited the Falls of St. Anthony," said the Rev. Mr. Barnes, in a sermon of five years ago. "I know not how other men feel when standing there, nor how men will feel a century hence, when standing there—then, not in the *west*, but almost in the centre of our great nation. But when I stood there, and reflected on the distance between that and the place of my birth and my home; on the prairies over which I had passed; and the stream—the 'Father of Rivers'—up which I had sailed some 500 miles, into a new and unsettled land—where the children of the forest still live and roam—I had views of the greatness of my country, such as I have never had in the crowded capitals and the smiling villages of the East. Far in the distance did they then seem to be, and there came over the soul the idea of greatness and vastness, which no figures, no description, had ever conveyed to my mind. To an inexperienced traveller, too, how strange is the appearance of all that land! Those boundless prairies seem as if they had been cleared by the patient labor of another race of men, removing all the forests, and roots, and stumps, and brambles, and smoothing them down as if with mighty rollers, and sowing them with grass and flowers; a race which then passed away, having built no houses of their own, and made no fences, and set out no trees, and established no land-marks, to lay the foundation of any future claim. The mounds, which you here and there see, look, indeed, as if a portion of them had died and had been buried there; but those mounds and those boundless fields had been forsaken together. You ascend the Mississippi amid scenery unsurpassed in beauty probably in the world. You see the waters making their way along an interval of from two to four miles in width, between bluffs of from 1 to 500 feet in height. Now the river makes its way along the eastern range of bluffs, and now the western, and now in the centre, and now it divides itself into numerous channels, forming thousands of beautiful islands, covered with long grass ready for the scythe of the mower. Those bluffs, rounded with taste and skill, such as could be imitated by no art of man, and set out with trees here and there, gracefully arranged like orchards, seem to have been sown with grain to the summit, and are clothed with beautiful green. You look out instinctively for the house and barn; for flocks and herds; for men, and women, and children; but they are not there. A race that is gone seems to have cultivated those fields, and then to have silently disappeared—leaving them for the first man that should come from the older parts of our own country, or from foreign lands, to take possession of them. It is only by a process of reflection that you are convinced that it is not so. But it is not the work of man. It is God who has done it, when there was no man there save the wandering savage, alike ignorant and unconcerned as to the design of the great processes in the land where he roamed—God who did all this, that he might prepare it for the abode of a civilized and Christian people."

Fountain Cave is a remarkable spot two or three miles above St. Paul. A passage-way, 25 feet high, and nearly as wide, leads into a cavern of white sandstone, which has been penetrated for 1000 feet; first by a gallery 150 feet in length and 20 feet broad, and afterwards through narrow passes. A rivulet follows the course of this cave.

Fort Snelling is 5 miles from St. Paul, at the confluence of the Minnesota, or St. Peter's, and Mississippi Rivers, on the west side of the Mississippi. The buildings of the garrison are upon a high bluff, probably 200 feet above the level of the water in the rivers, and which stretches to the north and west in a gently undulating and very fertile prairie, interspersed here and there with groves of heavy timber. The steamboat landing of Fort Snelling is directly opposite the mouth of the Minnesota, from which a low island extends about two and a half miles down the Mississippi.

Mendota, which lies about half a mile below the mouth of the Minnesota, has been for many years a trading post of the American Fur Company, and is still a depôt of goods and provisions for the supply of the traders, who, at this time, have penetrated much farther into the Indian country. But it has, till lately, been included in the military reserve of Fort Snelling. It has not attained that degree of prosperity so remarkable in the villages of St. Paul and St. Anthony, and which its far more favorable position might justly have secured for it.

Pilot Knob. "From this summit, which lies back of Mendota," says Mr. Bond, whom we just quoted, "a view may be obtained of the surrounding country as far as the eye can grasp, affording to the spectator a sight of one of the most charming natural pictures to be found in this territory, so justly celebrated for scenic beauty. The view describes a circle of eight or nine miles, a grand spectacle of rolling prairie, extended plain and groves, the valley of the Minnesota with its meandering stream, a bird's-eye view of Fort Snelling, Lake Harriet in the distance—the town of St. Anthony just visible through the nooks of the intervening groves—and St. Paul, looking like a city set upon a hill, its buildings and spires distinctly visible, and presenting in appearance the distant view of a city containing a population of one hundred thousand human beings."

The St. Croix Falls, or Rapids, are in the St. Croix River, about 30 miles from its entrance into the Mississippi below St. Paul. The St. Croix continues the boundary line between Wisconsin and Minnesota, in the upper half of the territory, formed below by the waters of the Mississippi. The Falls in the St. Croix have a descent of 50 feet in 300 yards. The perpendicular walls of trap rock, between which the waters make their boisterous way, is a scene of remarkable picturesque interest. This wild pass is about half a mile below the Rapids. It is called the Dalles of the St. Croix.

The Sioux Rapids, in the Sioux River, is another striking point in the varied landscape of this region. The pass is through a grand quartz formation. The descent of the waters is 100 feet in 400 yards. There are three perpendicular falls of from 10 to 20 feet.

The Falls of the St. Louis River are a series of rapids extending 16 miles, the waters making, in that distance, a descent of 320 feet. These cataracts terminate about 20 miles from the mouth of the river.

In our enumeration of the landscape features and attractions of Minnesota, we have included only a few of the leading and most accessible scenes. There are, besides, the forest-hidden, laughing waters of Minnehaha, immortalized in the sweet song of Hiawatha, and a thousand cascades of beauty; gentle lakes and fertile flower-strewn prairies.

The Sportsman here will find plenty to do, whether it be with his gun in the woods, or with his line by the marge of the graceful waters. Immense herds of buffalo, deer, elk, antelope, and other noble denizens of the forest, still roam over the western plains, and the moose and the grizzly bear, the otter and the wolf, may all yet be found in Minnesota.

St. Paul.—HOTELS:—The *Winslow House;* the *Fuller House;* spacious and elegant establishments.

ROUTES.—Galena and Chicago Railway from Chicago to the Mississippi, and thence by steamer; or the Chicago, St. Paul, and Fond du Lac Railway from Chicago to Prairie du Chien, on the Mississippi; thence by steamer in summer, and stages in winter. The La Crosse Railway, from Milwaukee, is now in operation to the Mississippi at La Crosse, yet higher up.

This flourishing city of the Far West, the capital of Minnesota, is graphically described as perched on a high bluff overlooking the Mississippi, at the head of its navigable waters, 2070 miles from its mouth. "It is surrounded in the rear by a semi-circular plateau, elevated about 40 feet above the town, of easy grade, and commanding a magnificent view of the river above and below. Nature never planned a spot better adapted to build up a showy and delightful display of architecture and gardening than that natural terrace of hills. The town has sprang up like Minerva, full armed from the head of Jupiter, and now contains 10,401 inhabitants; its whole history of seven years forming an instance of Western enterprise and determined energy and resolution, hitherto unsurpassed in the story of any frontier settlement. The main street is fully a mile in length, with buildings running from shanties to five-story bricks. Its ten churches, with their lofty spires, show that the aspirations of the people of St. Paul are upward, and, though in the far-off West, they make the welkin ring. A travelling friend observed that he had, in Constantinople, where they have five Sabbaths a week, heard the Turkish Salims, the Catholic and Protestant, the Greek, Armenian, and Jew, each sending forth their summons for prayer to the faithful; but, measuring its religion by its bell-ringing, St. Paul far exceeds the Oriental capital."

The Falls of St. Anthony, Fort Snelling, and other points of interest to the tourist, are in the immediate vicinity of St. Paul. See mention of these places in preceding pages of Minnesota.

Stillwater.—HOTELS:—

Stillwater, upon the west bank of Lake St. Croix, 20 miles from St. Paul, was first settled in 1843, and is rapidly becoming a populous and

11*

important place. To be justly informed of the number of people in these cities and villages of the West, would require a monthly or weekly census.

St. Anthony.—HOTELS:—The *Winslow House*.

St. Anthony is a thriving town on the left bank of the Mississippi, at the famous Falls of St. Anthony, 8 miles above St. Paul (see St. Paul, and the Falls of St. Anthony in preceding pages.) The village is situated upon a lofty terrace overlooking the Falls. Its position at the head of navigation on the Father of Waters, is of immense commercial consideration, and the Falls afford incalculable water power for manufactures. This is the seat of the University of Minnesota.

Wabasha is upon the Mississippi, opposite the mouth of the Chippewa river. It is 90 miles below St. Paul.

Winona, Red Wing, Faribault, Rochester, Minneapolis, and **St. Cloud,** are large and growing towns, already containing populations varying from two to five thousand souls, and promising to become the most important places in the State, as some of them, indeed, already are

---

CALIFORNIA.

THE HISTORY OF CALIFORNIA must be of great interest to the traveller, especially as he surveys the astonishing progress which has been made within the last ten years.

The peninsula of Lower California was discovered by the expeditions of Cortez in 1534-5.

Upper California was seen by Cabrillo in 1542. Sir Francis Drake visited the coast and discovered Jack's Harbor, on the bay of Sir Francis Drake, a few miles to the northward of the bay of San Francisco, in 1579.

In 1769 the bay of San Francisco was discovered by the early Spanish missionaries, who established some 18 missions in the country; these continued to flourish until after the Mexican Revolution in 1822, falling into decay under the new government.

Capt. John Sutter established himself near the present site of Sacramento City in 1839.

In 1846 the war broke out between the United States and Mexico, which resulted in the conquest and purchase of California by the United States.

Gold was discovered in January, 1848, by James W. Marshall, in the employ of Capt. Sutter, at Sutter's Mill, on the South Fork of the American river, at the present town of Coloma.

From this date the unprecedented progress of the country commenced.

The State of California extends along the Pacific coast nearly seven hundred and fifty miles, from south-east to north-west, with an average breadth from east to west of two hundred and fifty miles, containing an area of 187,500 miles, or nearly twice the size of Great Britain. The whole country naturally falls into three great divisions:

First, The great valleys of the Sacramento and San Joaquin rivers, with all their lateral valleys; all of whose waters meet in the bay of San Francisco, passing through the Golden Gate to the Pacific Ocean.

Second, The portions of the coast range north and south of the bay of San Francisco, where the country is drained by streams falling directly into the Pacific, as the Klamath, Eel River, Russian River, the Salinas, San Pedro, and San Bernardino, with others of lesser magnitude.

Third, The country east of the Sierra Nevada Chain, the waters of which fall into the great basin, having no outlet to the ocean.

The ranges of mountains comprise the Sierra Nevada, which divides the State on the east from the Great Basin, and the Coast Range on the West.

Between the Sierra Nevada and the Coast Range, lies the extensive country of the First Division, a valley of some 500 miles in length, with an average breadth of 75 miles, with a rich soil and warm climate, producing all the fruits of the warm region with the products of the more temperate climes. The lateral valleys, with an elevation of from 1,000 to 5,000 feet above the level of the sea, producing the more hardy fruits and grains common to the more northern States of the Union.

A belt of gigantic timber, consisting of pines, firs, cedars, oaks, &c., &c., extends the entire length of the Sierra Nevada range, affording a supply of wood that can never be exhausted.

The mining region also stretches along this range, extending on the north into the Coast Mountains, passing into Oregon with an average breadth of 40 or 50 miles, at some points extending from the valley to near the summit of the Sierra, a distance of 100 miles in breadth.

Of the Second Division, located near the coast. This portion of California contains thousands of beautiful valleys, some of which are very extensive, as that of the Salinas, whose outlet is at the bay of Monterey and the country adjoining Los Angelos and San Diego. This portion has a cooler climate than the lower valleys of the First Division, owing to their proximity to the sea.

Every variety of product, from the orange and other fruits of the warm region at Los Angelos, to the more temperate clime and products of Humboldt Bay and Trinity River at the north. Gold is also found, and the richest quicksilver mines in the world. Oaks and gigantic red woods afford fine lumber.

Of the Third Division, the country east of the Sierra Nevada, but little is known, especially to the south-east, yet many fine valleys occur, as that of Carson's Valley, which now contains quite a population. Gold also is found along the eastern slope of the Sierra.

These, then, are the general features of the country. Much more might be said concerning the variety of climate incident to the location, the different natural productions, the mines of gold, quicksilver, coal, and iron, which are being daily discovered, with the many advantages of soil and climate adapted to grazing and agricultural purposes.

VOYAGE FROM NEW YORK TO SAN FRANCISCO.

The great route from New York to San Francisco is by the fine steamships continually leaving New York for Aspinwall; thence across the Isthmus, by the railway, to Panama, on the Pacific side, and from Panama by other steamers to California. The routes overland are becoming, by dint of travel, numerous and reasonably easy and safe. Continuous lines of railway will bear the tourist or the emigrant with rapidity from the Atlantic cities to St. Joseph, on the verge of civilization westward. From St. Joseph, the great mail sets out for Salt Lake City, so the Pike's Peak Express and the new Pony Express, which makes the journey in eight days.

Steamers ply during the entire summer season on the Missouri River, from St. Louis to Westport, Leavenworth City, Atcheson, or other towns above. The route from these points is the great track over which so many thousands have travelled within the past few years.

Emigrants starting from the extreme southwest travel the southern road, through Texas, along the Gulf coast to Powder Horn, on Matagorda Bay, reached by steamers from New Orleans. But to follow the great ocean route:—

Leaving New York on the 5th or 20th, we are soon upon the broad Atlantic. Crossing the Gulf Stream in about the latitude of Cape Hatteras, we sometimes catch a glimpse of the low, coral islands of the Bahamas. Five or six days bring us to the eastern extremity of Cuba, whose highlands, with those of the more distant mountains of San Domingo, look beautiful in the warm tropic haze as we pass between them. Soon after, the Blue Mountains of Jamaica loom up in the distance to our right—the last land seen until we arrive at Aspinwall, the Atlantic terminus of the Isthmus Railroad, to which we come, after a voyage of eight or ten days from New York.

Aspinwall.—HOTELS:— *City Hotel.*

Aspinwall is situated upon the Island of Manzanilla, at the north-east entrance to Navy Bay, and owes its importance, in fact, its existence, to the railroad. As it lies but a few inches above the waters of the sea, it is a perfect marsh, and is very unhealthy. The population consists of the employés of the railroad and steamers, together with a motley class of Jamaica negroes. There is nothing of interest in the place, and the traveller is glad to take his seat in the cars for Panama, leaving behind him a place with a population upon whose faces disease appears in its most pallid form.

For several miles the road passes through a deep marsh, reaching Gatun, on the Chagres River. Leaving the river a mile or two to the right, we traverse a dense tropical forest, with occasional clearings, and passing a few native huts, arrive at Barbacoas, crossing the Chagres River, upon a high wooden bridge. Beyond, the Cierro Gigante, the highest point upon the Isthmus, is seen on our right, from whose summit Balboa discovered the waters of the Pacific Ocean. Moving on, we get occasional views of the river, gleaming amid the rich verdure of gigantic trees and overhanging vines. Passing the little hamlet of Matachin, in seven miles we reach the summit, which is 250 feet above the tide level of the Pacific.

From this point we descend rapidly, a distance of 11 miles, shooting through the dense forests, and gliding over the level savannas, until, at last, we catch a gleam of the Pacific Ocean, and the spires of Panama.

Previous to the completion of the railroad, the steamers touched at Chagres, at which place the travel up the Chagres River was performed in native boats to Gorgona or Crucas, thence, by animals, to Panama; this generally occupied three or four days, and was attended with much exposure and discomfort, which very often resulted in an attack of the Isthmus fever, so fatal in its consequences.

In 1850, the survey of the railroad was commenced, under the superintendence of the well-known traveller, the late John L. Stephens.

In July, 1852, 23¼ miles of the road were completed, from Aspinwall to Barbacoas, on the Chagres River, and opened for travel. From that date Chagres sunk into utter neglect, as all the travel was diverted to Aspinwall, passing up the river from Barbacoas. December, 1854, saw the road completed to Culebra, on the summit. Panama was reached from the latter place by animals.

It was not until January 27th, 1855, that the first locomotive passed over the entire road, from ocean to ocean, a distance of 49 miles.

Nearly five years were thus consumed in the completion of this extraordinary American enterprise.

The cost of the road had been immense—some six millions of dollars, at the lowest estimate; while the sacrifice of life has been enormous.

The Isthmus afforded scarcely a material for its construction; not even food for the laborers. Every thing had to be imported from the United States or from Europe. A primeval forest was cut through, dense jungles were opened, deadly swamps were crossed, deep cuts were made, rivers spanned by bridges, whose timber was brought from afar; and, more than all, the pestilential climate swept thousands upon thousands into their graves ere the oceans were united.

But, to resume our travel—

As the small steamer is lying at the terminus of the railroad to convey the California passengers on board the Pacific steamer, which is waiting for us at the Island of Perico, some two miles distant, we shall not have an opportunity of visiting the City of Panama.

Yet we obtain a general view as we pass upon our transit to the steamer, its old towers and ramparts gleaming in the sun, overgrown with rank vegetation, presenting a time-worn and venerable appearance, finely relieved by the back-ground of hills, clothed in the richest green.

Panama.—HOTELS:—*Aspinwall House.*

The City of Panama contains many objects of interest; but, owing to the present arrangements, travellers, *en route* for California, have no opportunity of visiting or remaining here, unless they should lay over one steamer—a delay that might be fatal to the health, as the climate is usually pernicious to a northern constitution.

THE PACIFIC VOYAGE. On arriving at the steamer, we are soon under way for San Francisco, and, steering south, we pass the beautiful Islands of Toboga and Toboquilla, which are 12 miles from the city. Soon after, we pass the lovely Islands of Otoque and Bana, while, away to the south-east, a glimpse of the Pearl Island groups is obtained.

The next morning or evening finds us steaming past the Island of Quibo, with a distant view of the Mountains of the Isthmus.

We find the arrangements perfect on board the Mail Line, and our days pass pleasantly as we steam along the calm blue waters of the Pacific.

On the seventh day from Panama, we get a grand view of the Mountains of Mexico, and soon enter the fine harbor of Acapulco. Here the steamer takes in a supply of coal, which affords the stranger time to go on shore and take a look at this interesting place.

The **Harbor of Acapulco** is one of the most perfect in the world, protected on all sides by mountains, which rise almost from the water's edge.

We gaze with delight upon the fine groves of cocoas and palms, and look with interest upon the faded glories of this once important place. The motley population of Mexicans observed in the streets, which are alive with venders of all sorts of fruits and curiosities, is a study of itself, seen nowhere save in a Spanish city.

Resuming our voyage, we soon lose sight of the high mountain range of Mexico, the last land seen until, on the fourth day, we approach the southern extremity of Lower California, Cape St. Lucas.

CALIFORNIA. 253

From this point the weather suddenly becomes cold; and as we approach the port of our destination thick clothing comes into requisition. As we coast northward we sometimes see land, perhaps one of the barren islands off the coast of Lower California; but, as the atmosphere along the coast is generally very hazy, especially during the summer time, we find but little of interest until we approach the Golden Gate, the entrance to the noble Bay of San Francisco.

Upon the 14th day from Panama, or 22 or 24 days from New York, we see the mountains of the coast range, among which Tamul Pise stands pre-eminent, with Monte Diabolo in the distance, looming up from the waters; and soon after we near Point Lobos, with Point Boneta on our left, entering the Golden Gate; Fort Point is soon abreast, and we come into the Bay of San Francisco. Two miles further on we pass the Presidio, and catch a distant view of the western and northern portion of the metropolis of the Pacific. To our left, the Alcatraz Rock rises from the surface of the bay, bristling with cannon, and surmounted by a lighthouse; while beyond, Angel Island rises to the height of 900 feet.

Doubling Telegraph Hill, the city bursts upon our vision, rising picturesquely from the bay, which extends southward, like a vast inland sea.

We are soon along side of the wharf, and thus ends our pleasant voyage of 6,000 miles from New York.

San Francisco.—Hotels:—Among the principal Hotels, we may mention the *Oriental*, corner of Battery and Bush streets; *Metropolitan*, Bush and Sansome streets; *International*, Jackson street; *American Exchange*, Sansome street; *Railroad House; Tremont House.*

San Francisco, the principal city and seaport of the Pacific coast, is situated upon the Bay of San Francisco, near its entrance to the sea, and lies in lat. 37° 48' north, long. 122° 30' west, from Greenwich.

The Mission was founded and the Presidio established in 1776. The first house was erected by Capt. Richardson, in 1835, but up to January, 1847, bore the name of *Yerba Buena.*

At the time gold was discovered, in January, 1848, it contained but 200 buildings, of all kinds, and a population of 800 souls.

In the year 1860 the city covered an area of 8 or 9 square miles, with a population of some 65,000.

Among the principal buildings are the City Hall, fronting upon the Plaza or Portsmouth Square; the U. S. Custom House and Post Office, corner of Washington and Battery streets; the U. S. Marine Hospital, Rincon Point; Montgomery Block; Stevenson's and Tucker's Building, corner of Montgomery and California streets, and many fine structures of lesser note, as the Custom House Block, corner of Sansome and Sacramento streets; Mercantile Library Building, corner of Bush and Montgomery streets; U. S. District Court Building, corner of Washington and Battery streets; Masonic Hall, junction of Post, Market, and Montgomery streets; Odd Fellow's Hall, corner of Bush and Kearney streets, and others that would compare favorably with any buildings to be found in the large Eastern cities.

The principal churches are: First Presbyterian Church, Stockton Street, between Clay and Washington; Calvary Presbyterian Church, Bush Street, between Montgomery and Sansome; First Baptist Church, Washington Street, between Dupont and Stockton; First Congregational Church, corner of Dupont and California streets; Grace Church, corner of Stockton and California streets; Church of the Advent, Mission Street, below Second; First Unitarian Church, Stockton Street, between Clay and Sacramento; M. E. Church, North, Powell Street, between Washington and Jackson; St. Mary's Cathedral, corner of Dupont and California streets; St. Francis' Church, Vallejo Street, between Dupont and Stockton, and many other fine ones of less prominence.

The places of amusement are: The American Theatre, Sansome Street, between Sacramento and California streets; Maguire's Opera House, Washington Street, near Montgomery; Tucker's Academy of Music, in Tucker's Building, corner of Summer and Montgomery streets; Platt's New Music Hall, Montgomery Street, near the corner of Bush, and many others of less note.

A visit to the Mission, three miles southwest of the city, will interest the stranger. The Market Street railroad cars start for that point each half-hour in the day. Lines of omnibuses are running over the plank road to the same place, by a more circuitous route, passing through "The Willows," a pleasant suburban retreat, on their way. Many fine gardens are in the vicinity. The race course is a mile beyond. The Protestant Orphan Asylum is a fine building half a mile north. The Mission itself is an object of much interest. It is an adobe building of the old Spanish style, built in 1776. Adjoining is the Cemetery, with its well-worn paths and capricious monuments.

A branch of the Market Street railroad runs to Hayes Building, which contains the largest dancing hall in the city.

A line of omnibuses also runs to the Presidio, which is situated some three miles toward the Golden Gate; a mile further is Fort Point, so called from the fortification which protects the entrance to the harbor. Following the shore we pass Point Lobos and Seal Rock House, 8 miles, and the same distance further reach the Ocean House, situated on a little arm of the sea. From this point, returning to the city, 8 miles distant, the road winds through and over the San Bruno Hills, from whose peaks—1,200 feet above the level of the sea—a fine view of the Bay on one side, and of the ocean on the other, is to be had.

The view from Telegraph Hill, 290 feet high, at the northern extremity of the city, is unsurpassed. This view embraces the city, stretching along the semi-amphitheatre of hills, and overflowing the depressions toward the Presidio on the west, and the Mission on the south; both arms and the entrance to the bay, including the islands of Alcatraz, which is fortified, Angel Island, over 700 feet in height, and Yerba Buena; the Mountains of Marin County on the north, with the peak of Tamel Pais 2,600 feet high; and the Contra Costa Range on the east, with Monte Diablo rising in the back ground to a height of 3,790 feet.

The summits of Russian and Rincon hills also afford fine views.

A ferry boat connects the city with Oakland every hour. Oakland lies across the bay about 8 miles distant, and contains about 1,500 inhabitants and many fine residences. The college of California, a flourishing institution, is located in Oakland.

Steamers leave daily for various points on the bay—Petaluma, Sonoma, Napa, Suisun, &c., north, and for Alviso, connecting with San Jose by stage, south—all connecting with stages for the interior.

VISIT TO THE INTERIOR OF CALIFORNIA AND ALONG THE COAST.

From San Francisco *via* **Sacramento to Marysville**, up through the valley of the Sacramento to the Oregon line; thence through the mining regions of the eastern tributaries of the Sacramento and San Joaquin rivers, including a run over the Sierras into Carson Valley; returning from the Southern Mines *via* Stockton to San Francisco; thence proceeding *via* the Overland Mail Route through San Jose, Visalia, and Los Angeles, to the southern part of the State, returning by the ocean from San Diego, the most southern port, touching at Santa Barbara, Monterey, &c., and finally taking a glimpse at the more northern sections along the coast, and a voyage to Oregon.

By this arrangement all that is of interest can be seen, making an almost continuous tour of the State, without loss of time, or unnecessary travel or expense.

Many of the distances given are but an approximation to the exact, and the time and fares vary, as in other countries.

For the Sacramento Region.—Leaving San Francisco by the 4 o'clock afternoon steamer for Sacramento, we proceed northward toward Angel Island, in the bay, which we pass on the right some 6 miles from the wharf, soon after Red Rock Island, and enter the Bay of San Pablo, through the straits of the same name, at a distance of about 15 miles from San Francisco.

The Bay of San Pablo is a large and beautiful sheet of water, some 15 miles wide and 20 miles long, surrounded by picturesque ranges of mountains. The view looking westward is picturesquely fine; to the northward the fertile valleys of Petaluma, Sonoma, and Napa, bounded by the high mountains of the Coast Range, bathed in the warm summer haze so peculiar to California.

At the head of Napa Valley warm sulphur springs occur, which are a favorite place of resort, and offer fine accommodations to the visitor. Stages at Sonoma and Petaluma connect with Healdsburg, in the Russian River Valley, one of the most fertile sections of the State.

In the vicinity of Mt. Putas, or Geyser Peak, about 50 miles north of Petaluma, are the hot steam springs called the Geysers. The best means of communication is through Healdsburg. The picturesque scenery, deep ravines, towering mountains, and the springs—second only to the far-famed Geysers of Iceland—invest this section with an interest unsurpassed.

Still to the north the picturesque region of Clear Lake amply repays the tourist by its wild beauty and the fine hunting and fishing which the surrounding region and waters afford.

Resuming our voyage through San Pablo Bay, we pass Mare Island and Vallejo, where the U. S. Government have established a Dry Dock and naval station, and soon after enter the Straits of Carquinez, which connect the Bay of San Pablo with the Bay of Suisun.

The Straits of Carquinez.—These Straits are about 8 miles in length, and generally three-fourths of a mile wide. Benicia, the former capital of the State, is situated upon the north side, near the entrance to Suisun Bay, 30 miles from San Francisco. Vessels of the largest size can reach this point. The steamers of the Pacific Mail Steamship line are refitted at this place. Their extensive foundry and machine shop is the most important building in the place. The head-quarters of the U. S. Army are also located here.

The view as we approach Benicia is grand. Looking southeast, Monte Diablo, the most remarkable peak of the Coast Range, is seen rising to the height of 3,790 feet, while the little village of Martinez, with its groves of evergreen oaks, surrounded by hills, is a fine feature in the scene.

The Bay of Suisun.—We now enter the Bay of Suisun, another arm or continuation of the great Bay of San Francisco; here the waters of the Sacramento and San Joaquin unite, the former coming in from the north, and the latter from the south. The growing village of Suisun has become the outlet of a rich agricultural region, on account of being connected with the north side of this bay by means of a navigable slough.

Sacramento River.—Passing through the bay, we soon enter the mouth of the Sacramento River, about 45 miles from San Francisco. Much of the land adjoining this bay and the Sacramento and San Joaquin rivers is marshy, covered with tules, a kind of bulrush. Proceeding toward Sacramento, we pass a low range of hills to the left. Further on the banks are low and the country is marshy. Beyond, trees occur, and the river presents a more beautiful appearance. Sometimes in autumn the dry tules are on fire for miles, presenting a magnificent appearance to the passenger on the steamer.

In ten hours we arrive at Sacramento, the capital of the State, which is 125 miles from San Francisco.

Sacramento.—HOTELS:—*St. George*, corner of Fourth and J streets; *Orleans*, Second Street, between J and K; *What Cheer House*, corner of First and K streets.

Sacramento is situated at the confluence of the Sacramento and American rivers, and contains about 14,000 inhabitants, and is the centre of travel for the northern mines of California. It contains many fine buildings put up in the most substantial manner; and although it suffered, like San Francisco, from fires, in the early period of its growth, it has steadily improved, and is now the most important city in the interior.

The State Agricultural Pavilion is one of the finest buildings in California. In style it approaches the Romanesque. The main hall is said to be the largest *clear* chamber in the United States, being 100 by 120 feet. The central chandelier contains 56 burners.

The new State House is an ornament to the city not only, but a credit to the State.

Smith's Gardens, in the suburbs of the city, are well worthy a visit from the stranger.

Numerous lines of stages radiate from the city in every direction, connecting with all the interior mining localities.

The Sacramento Valley Railroad, running to Folsom, 22 miles east, adds to the facilities for getting to the interior of Placer and El Dorado counties. Other roads are in contemplation, one running east to connect Folsom with Placerville, another running north from Folsom to Auburn.

Steamers also leave for points on the Upper Sacramento and Feather River, such as Coluso, Tehama, and Red Bluff on the former, and Marysville and other points on the latter.

Marysville.—HOTELS:—*Merchants*; *St. Nicholas*; *Western*.

Proceeding on our journey through the great valley of the Sacramento, we reach Marysville by stage or steamer, distant by land 44 miles, and by river 20 miles further, arriving at the latter place by noon.

Marysville is next in importance to Sacramento among the northern places in the interior, and contains about 6,000 inhabitants. It is finely located near the confluence of the Feather and Yuba rivers; accessible at all times by steamer from either San Francisco or Sacramento. It commands much of the trade with the rich mining districts situated upon the Feather and Yuba rivers, with a rich agricultural region in the immediate vicinity.

The city is well built, giving the impression to the traveller who sees it for the first time from the steamboat landing, of a substantial city mostly built of brick.

There is a fine agricultural region around it, mostly comprised in the Yuba and Feather river bottoms. Briggs' fruit-ranches, the most extensive orchards in the State, perhaps in the world, are near this city.

The Marysville Buttes. From here a fine view is obtained of the isolated chain of mountains known as the Marysville Buttes. They rise from the plain of the Sacramento Valley to the height of 1,200 feet, and extend for some 8 miles in length, forming a remarkable feature in the valley of the Sacramento. There are three principal peaks and many subordinate ones, and from the central, elevated, broken, rocky mass, there run off spurs in all directions, forming valleys between them. It is about 30 miles around the Buttes. The view from the summit, which is easily accessible, is superb.

Daily lines of stages leave Marysville for all the mining localities to the north and east.

Oroville.—HOTELS:—*St. Nicholas; International.*

Journeying north we leave Marysville by stage for Oroville, distant 28 miles.

Oroville is situated at the base of the foot-hills upon the main Feather River, and is a mining town of considerable importance, while a rich agricultural region extends to the north and west. It is the county seat of Butte County, and contains about 2,500 inhabitants.

As the traveller journeys northward, many fine farms or ranches are passed, many fine views of the mountains of the Coast Range, some of whose peaks rival those of the Sierra Nevada, especially Mount St. Helen, Mount St. John, and Mount Linn, which are each from 7,000 to 9,000 feet above the level of the sea.

To the eastward on our right, the snow-capped peaks of the Sierra Nevada rise gleaming in the sunshine, bathed in snow; beyond Red Bluff we obtain a fine view of Lassen's Butte, a solitary peak rising to an altitude of 12,000 feet. Near Reading's Ranch, before reaching Shasta, we get a splendid view of Mount Shasta, the highest mountain in California, a vast cone of snow rising to the height of 16,000 feet above the level of the sea, the magnificent landmark at the head of the Sacramento Valley.

Red Bluff.—HOTELS:—*Lima House; Eagle Hotel.*

Travelling through a rich agricultural region over good roads, we cross the Sacramento River at Tehama, 50 miles north of Oroville, and proceeding up the western bank of that stream 14 miles we reach Red Bluff, a village of some 1,000 inhabitants, the county seat of Tehama County, and situated at the head of navigation on the Sacramento River.

This place is the centre of trade for the more northern mines of California, goods reaching here from San Francisco by the river to Colusa on the Sacramento, 75 miles below, or when the water is in a good stage, being shipped directly to this place.

The journey from Marysville to Red Bluff (90 miles) can be made in 13 hours for $15.

Shasta.—HOTELS:—*Empire; American.*

At a distance of 28 miles from Red Bluff we pass through Horsetown, a prosperous mining town of 500 inhabitants, and 9 miles further we arrive at Shasta. This town is situated in the foot-hills of the mountains stretching across the northern end of the State, connecting the northern Sierras with the Coast Range. It is a mining town of 800 inhabitants, at the northern extremity of what was once wagon navigation. Formerly all goods destined for mines further north, had to be packed on mules, but a good wagon road has recently been constructed over the Siskiyou Mountains, by the California Stage Company, for the purpose of transporting the U. S. Mail between Sacramento and Portland, Oregon.

Leaving Shasta for Yreka, we pass the Tower House, 12 miles, French Gulch, 15, Mountain House, 23, Gibb's Ferry, 35, Chadbourne, 43, Trinity Centre, 49, and Thompson's, 60 miles, arriving at New York House, 64 miles, at the base of Scott Mountain, which is now to be climbed. In a distance of 6 miles further we rise 2,600 feet. Every foot of the distance has been made into a superb roadway by cutting into the solid rock, bridging chasms, excavating the precipitous side of the mountain, walling up with stone, clearing away a dense growth of timber, and overcoming other obstacles. On the right rises the perpendicular embankment created in excavating for the road, while on the left the traveller looks down a fearful precipice its side bristling with sharp and jagged rocks. The summit reached, we are upward of 5,000 feet above the level of the sea. Here we obtain a glorious view of Mount Shasta on the right, covered with its snowy shroud.

A continuous descent of 7 miles brings us to the head of Scott Valley, and 3 miles beyond is Callahan's Ranch. Scott Valley is a level area 40 miles long and from 3 to 9 miles wide, a beautiful tract of country, hemmed in on all sides by bold and precipitous mountains. Passing through Fort Jones, 22 miles north, and crossing a lofty divide at the termination of the valley, we arrive at Yreka, 116 miles from Shasta.

Yreka.—HOTELS:—*Metropolitan; Yreka.*

Yreka, the county seat of Siskiyou County, is the most important mining town north of Oroville. The town contains about 1,500 inhabitants, is well laid out, has many fine buildings, and is lighted with gas. It is situated in the valley of Shasta Creek, is encompassed with mountains, and is distant from the State line 28 miles. The mines in the vicinity are very productive, giving the place a steady and rapid growth. A fine view of Mount Shasta, distant some 30 miles, is attained from the ridge east of the town.

Proceeding north we pass through Cottonwood, 20 miles, to Cole's, 28 miles, where there is a good wayside inn. Here we ascend the Siskiyou Mountain, 4 miles, and from its summit get the last glimpse of Mount Shasta. Descending the mountain 4 miles to its base, and traversing 20 miles of rolling country, we arrive at Jacksonville, 56 miles from Yreka.

Jacksonville is a prominent town of Southern Oregon, situated in the fertile Rogue River Valley, about 9 miles south of the river.

From Jacksonville, Althouse lies about 60 miles west, and Crescent City on the Pacific coast south of the boundary line, 125 miles, traversing an exceedingly rough country.

Having reached the northern limits of our journey, we return to Marysville *via* Yreka and Red Bluff.

Perhaps an excursion from Yreka west to the coast would be of interest. Scott's Bar, a rich mining locality, lies some 25 miles distant; the Forks of Salmon, a mining locality in Klamath County, about 90 miles distant; the mouth of the Trinity River 130 miles; and Trinidad on the coast about 100 miles.

From Shasta the rich mining localities in the vicinity of Weaverville, distant 38 miles, and Humboldt Bay on the Pacific coast, some 75 miles distant, can be visited on animals.

All these excursions can be accomplished on a riding animal. A mule is preferable, at an expense of about 6 dollars per day, which includes every thing. Good meals and sleeping accommodations are found along the routes, which is truly surprising, considering the almost impracticable nature of the country traversed.

The Sierra Nevada Mountains and Mines. Having returned to Marysville, we will now make an excursion among the Sierra Nevada Mountains, visiting the most celebrated mining regions which lie along their western slope.

Taking the Downieville road, we pass through the once rich localities of Long Bar, 14 miles, Camptonville, 35 miles, and Goodyear's Bar, 12 miles, arriving at Downieville, 4 miles, making the total distance from Marysville 65 miles. By this route we visit what were once the most important river and hill diggings in the State, finding good accommodations and fare at all points upon the road.

Downieville. — HOTELS :— *American Exchange; Andrews' Hotel.*

Downieville is the county seat of Sierra County, and contains about 1,500 inhabitants. It is situated at the junction of the east and south forks of the North Yuba River, high up in the mountains, its elevation being about 4,000 feet above the level of the sea. The scenery is bold and impressive, the mountains rising to a great height on either hand. The Sierra Buttes, 12 miles east, and 9,000 feet high, form a notable feature in the landscape.

An excursion from this point north would be of interest, passing through the rich and prosperous mining localities of Monte Cristo, 4 miles, Eureka, 7, Morristown, 11, Port Wine, 15, La Porte, 18, Gibsonville, 25, Newark, 27, Onion Valley, 30, Nelson's Point, 36, arriving at Quincy, 43 miles, over a very wild and rough country. To the northeast of Quincy, which is the county seat of Plumas County, lies Honey Lake, Honey Lake Valley, and Honey Lake Pass. Pilot Peak, near Onion Valley, is, next to Lassen's Butte, the highest peak of the northern Sierras.

Leaving Downieville for Nevada, we take a southerly direction through Forest City, 6 miles, Camptonville, 18, North San Juan, 26, arriving at Nevada, 38 miles, in 12 hours.

Nevada.—HOTELS :— *United States Hotel; National Exchange; Hotel de Paris.*

Nevada is the largest mining city in the State. It contains a population of about 4,500, and is the centre of a large, rich mining region. It is the county seat of Nevada County, situated well up among the foot-hills, and contains many fine hotels, stores, and churches. Having suffered like most of the cities of California from fires, the business portion has been rebuilt with substantial fire-proof structures.

In the vicinity occur many heavy mining operations, the hill diggings and tunnels. It has ever been the first in using the improved methods of hydraulic pressure, sluices, tunnels, &c. Immense outlays in water canals, for mining purposes, have been made in Nevada County.

A visit beyond, toward the head of the South Yuba, is of much interest. Besides the wild scenery, the stranger will be convinced of the exhaustless nature of the mining interests of California, as all the immense ridges dividing the Yuba River, extending for 20 or 30 miles in length, abound in the richest mines of gold, which will take centuries of labor to develop. The country also affords exhaustless quantities of the finest lumber, consisting of pine, cedar, fir, &c.

Nevada is the western terminus of an important route over the Sierras to Virginia City, in Washoe, called the Henness Pass route. A good road has been constructed over this route, which has an easier grade than any of those further south. The distance from Nevada to Virginia City is about 100 miles.

Grass Valley.—HOTELS :—*Exchange Hotel; Hotel de Paris.*

Grass Valley, 4 miles south of Nevada, is the great quartz mining centre of the State. It contains about 3,000 inhabitants, is finely located, and is noted for its fine hotels and beautiful residences, as well as numerous quartz mills. It is said to contain more wealth, including its mills and machinery, than any other mining town in the State.

Auburn.—HOTELS :—*Empire; American.*

Journeying southward from Grass Valley, we wind along through the mountains and forests, crossing Bear River, arriving at Auburn, 25 miles. This is a fine village of about 1,000 inhabitants, the county seat of Placer County, and the centre of a large farming region.

A rich mining region exists northeast of Auburn, between the north and middle forks of the American River, and many thriving mining towns have sprung up, among which are Illinoistown, Iowa Hill, Forest Hill, Yankee Jim's, Michigan Bluffs, Dutch Flats, and others, distant from 18 to 35 miles.

Dutch Flat, 32 miles distant, is the starting-point of what is called the Dutch Flat Wagon Road over the Sierras *via* the Truckee Pass to Virginia City. The latter is 120 miles from Auburn by this route.

Folsom.—Hotels:—*Patterson's; Central; Fremont.*

Folsom, a bustling place of 1,200 inhabitants, 18 miles south of Auburn, is the eastern terminus of the Sacramento Valley Railroad. It is situated at the base of the foot-hills, and is surrounded by both a mining and an agricultural region. Nine stages leave every morning, on the arrival of the cars from Sacramento, for various points in the mountains.

The Alabaster Cave.—As we shall not have a better opportunity, we will step into one of these stages, destined for the Alabaster Cave, 13 miles to the northeast in El Dorado County, and pay a visit to one of the great wonders and curiosities of California.

This cave was discovered in April, 1860, by two men who were quarrying limestone. It was immediately taken in charge by Mr. Wm. Gwynn, who arranged a convenient entrance, and keeps the interior constantly lighted.

(It has been called by a variety of names since it was discovered, but it is now known as the Alabaster Cave.)

This cave is situated on the Whisky Bar road, five miles from Centerville, in El Dorado County. A single step takes you from the street into the hall of the silent mansion. This entrance is not the one first discovered, but has been cut through the solid rock from another chamber to the outer world. Passing through this, the visitor is ushered into an irregular apartment two hundred feet in length by perhaps seventy-five in width, and of various heights, with numerous elevations, depressions, recesses, galleries, etc. A scene of wonderful magnificence is before him. Millions of jewels appear to be glittering from the walls. Shining pendants, some large, some small, some short, some very long, some reaching from ceiling to floor, some thick, some slender, some tapering, some uniform, some tubular, some solid, some clear as crystal, some of a bluish tinge, hang thickly from the marble roof. Here a little wrinkly stub of a stalagmite pushes itself up from the floor; and there stands Lot's wife turned into a pillar of salt, but of marble; and there, again, is Mt. Blanc rising with its snowy folds several feet above your head. Passing through this first chamber and descending a little, you turn to the left, through a cross section, from which shoot out several passages, some brilliantly lit, and beautiful to behold, and others, one at least, as yet unexplored. Turning still to the left, you enter the last chamber lying exactly parallel to the one first entered, but if any thing more beautiful. This is called the chapel, and has its belfry and pulpit as well. The pulpit especially is a thing of rare beauty, probably built in the olden time, as it is rather too near the ceiling to be of modern design. It has been formed by droppings from above, catching on a projection of rock, and then rising and spreading and folding over with the most graceful drapery underneath.

Placerville.—Hotels:—*Cary House; Orleans Hotel; Keokuk House.*

Placerville, the county seat of El Dorado County, is 28 miles east of Folsom. It is one of the oldest and one of the largest of the mountain towns, containing about 4,000 inhabitants. It is on the principal route of travel over the Sierra Nevada Mountains to the Washoe region in Carson Valley, and is a prominent fitting-out point.

The town of Coloma, 10 miles northwest of Placerville, was formerly the county seat of El Dorado County, and is distinguished as the place where gold was first discovered. The remains of the old saw mill of Captain Sutter are just below the present town, and will be looked upon with much interest by the stranger as a memento of the great event which has revolutionized the commerce of the world.

Carson Valley.—We will now take an excursion over the Sierras to Carson Valley, touching at other points on the eastern slope, where exist silver and gold mines of marvellous richness, which are attracting capitalists from the old and new world. There is a finely-graded road the whole distance, and it winds through some of the wildest and most beautiful scenery in California, as well as affording glimpses of the most grand and sublime.

The facilities of travel over the Placerville and Virginia City road, are of the best description. A line of mail stages runs daily between the two cities, bringing them within 30 hours of each other, allowing passengers time for meals and rest upon the road.

Leaving Placerville in the afternoon, we enter almost immediately upon the broad mountain road that by easy grades conducts us to the west summit of the Sierra Nevadas, a height of 7,000 feet above the level of the sea. As we approach the summit, the pines, firs, and cedars attain a gigantic size, and constitute a dense forest. At Crippen's, 26 miles from Placerville, we pass the night.

Renewing the journey at daybreak, we pass through Strawberry Valley, 50 miles from Placerville, where a good hotel affords the best of accommodations, and a few miles further reach the west summit. From this point of view we have a combination of mountain, lake, and valley scenery, unsurpassed in beauty and wild grandeur by any similar scene perhaps on the American Continent. At our feet lies Lake Valley, more than a thousand feet below. Granite ledges gleam through the dark pines that fringe its sides, which rise in places to snow-covered peaks. To the left and northeastward 7 miles, repose the deep blue waters of Lake Bigler, while beyond, the extremity of the valley loses itself in the distance. The effect of the whole scene is as charming as it is indescribable.

Descending into the valley by a roadway excavated from the side of the mountain, we pass over the east summit through Daggett's Pass, and 5 miles beyond find ourselves in the open, level, elevated plain of Carson Valley, 2 miles south of Genoa.

Carson Valley is a tract of nearly level land about 30 miles long and 10 wide, three-fourths of which is excellent land for agricultural purposes. Although shut in by high mountains on the east and on the west, it is itself an elevated plateau, more than 4,000 feet above the level of the sea. Carson River, fringed with willows and occasional cottonwoods, flows through it in a northerly direction. The eastern slope of the Sierras is very abrupt, rising at a sharp angle from the western limit of the valley, and is covered with pines, though none grow in the valley below.

Many of the emigrants to California, journeying across the plains, reach the country by the route which passes through Carson Valley, and terminates at Placerville. Though other routes across the mountains are much used, this has always been a favorite one.

The emigrant, leaving the frontier of Missouri at St. Joseph or Ft. Leavenworth, journeys up the valley of the Platte River, passing through the Rocky Mountains by the South Pass; thence *via* Great Salt Lake City, or *via* Bear River to the north of Great Salt Lake; thence to the head of the Humboldt River, and down the latter to the sink of the Humboldt, crossing the desert, and reaching Carson Valley. The distance travelled by this route across the plains is about 2,000 miles, and occupies wagon teams from three to four months. Mules can be ridden this distance in 50 days.

The Pony Express, which, by relays, is enabled to make the distance from the Missouri River to Sacramento in 12 days, passes over the same route, with the exception of taking Simpson's Cut-off between Great Salt Lake City and Carson Valley, which has of late become a favorite route for emigrants.

Genoa.—Hotels:—*Genoa; Nevada.*

This place is the present county seat of Carson County, in Nevada Ter., and contains about 600 inhabitants. It is the oldest town in this region, having been settled by the Mormons who formerly inhabited Carson Valley.

Carson City.—Hotels:—*Frisbie; Ormsby; St. Nicholas.*

Crossing a low divide we enter Eagle Valley, and arrive at Carson City, 15 miles north of Genoa. It is situated like the latter place at the base of the Sierras, and contains about 2,000 inhabitants. Notwithstanding its rapid growth, having attained this population in less than 18 months, it contains many substantial buildings and comfortable homes.

Virginia City.—Hotels:—*International; Barnum's; Bailey's.*

Passing through Silver City and Gold Hill, we arrive at Virginia City, 18 miles north of Carson City. This is the largest town in Nevada Ter., containing about 3,000 inhabitants. In its vicinity is the famous Comstock Lead—the richest vein of silver in the world. It is about 2,000 feet in length, 18 inches wide on the surface, expanding in some parts of the vein as it descends, and has been traced to a depth of more than 50 feet. It is divided into sections called claims, owned by different companies, as the "Mexican," "Ophir," "Central," "Gould & Curry," and others. The general direction of the "lead" is north and south. Other veins of more or less richness have been discovered parallel to the Comstock Lead, and also at points further north and south, which are supposed to be links in an extended system of silver deposits.

Gold quartz also exists in abundance throughout this region, and numerous quartz mills have been erected for crushing out the ore.

About 90 miles in a southeasterly direction from Genoa, is a region called the Esmeralda District, also rich in deposits of silver and gold. Still further south some 20 or 25 miles is the gold region of Mono Lake.

Of the many routes over the Sierra Nevada Mountains, we may mention, commencing with the most northerly, the Honey Lake Pass, the Beckwith Pass, the Henness Pass, the Truckee Pass, Daggett's Pass, or the Placerville Route, the Big Tree Route, and the Sonora Pass. The latter is the highest, being about 10,000 feet above the level of the sea.

Returning to Placerville *via* Genoa and Daggett's Pass, we proceed on our journey to the southern mines. Passing through Diamond Springs, Michigan Bar, and Drytown, to Jackson, the county seat of Amador County, 45 miles distant, we there take the stage from Sacramento, and crossing Mokelumne River on a fine bridge, reach Mokelumne Hill in 5 miles.

Mokelumne Hill. — Hotels: — *Union House; La Fayette House.*

Mokelumne Hill, the county seat of Calaveras

County, has a population of about 800, and contains many fine stone buildings, and other permanent structures. In the vicinity, some of the richest hill diggings in the State have been found. A canal for bringing water for mining purposes, a distance of 40 miles, has been in use several years; lumber is also floated down from the lumber region above.

In speaking of the canals of the mining region, it will be proper to state that millions of dollars are most profitably employed in their structure, and they are found traversing almost every ravine and flat; brought from far up the mountains at the sources of the streams; sometimes constructed of plank the entire distance, as the canal coming into Mokelumne Hill. These are among the greatest enterprises of the State, involving an immense outlay of capital and labor. In Calaveras County alone, there are 54 canals and water ditches, whose aggregate length is 550 miles, and the cost of construction $1,600,000.

San Andreas.—Hotels:—*Metropolitan; Kinderhook.*

Proceeding south by stage 9 miles, we arrive at San Andreas, a prosperous mining town of over 1,000 inhabitants. It is beautifully situated in close proximity to well-wooded hills. Northward and eastward on Murray Creek, are a number of orchards, gardens, and grazing ranches; to the west is the regular wall-formed Bear Mountain, covered with grass and oak timber.

Murphy's.—Hotel:—*Sperry's.*

A ride of 15 miles brings us to Murphy's, a village of six or eight hundred inhabitants, containing a fine hotel built of stone. In the immediate vicinity of the town rich deep diggings and hill diggings occur, and are worked on an extensive scale. A daily line of coaches reach Stockton from this point, and this is also the western terminus of the Big Tree Route over the Sierras into Carson Valley.

The Mammoth Tree Grove.—We are now within 15 miles of the celebrated grove of mammoth trees in Calaveras County.

Leaving Murphy's in the morning, we arrive at the grove in three hours, by carriages or on horseback, the road winding through a fine open forest, consisting of immense pines, firs, cedars, &c. At the grove a good hotel affords every accommodation to the visitor, and several days might be pleasantly spent at this point. The valley which contains these monster trees, is at the source of one of the branches of the Calaveras River, 81 miles from Stockton, and 203 miles from San Francisco. Here within an area of 50 acres some 90 odd mammoth trees are now standing. Full 20 of these exceed 25 feet in diameter at the base, and several of them are more than 300 feet in height. The one cut down in 1853 for the purpose of carrying a section of its trunk to the Atlantic States, stood near the house; the stump measures 96 feet in circumference, and the tree was 302 feet high. One, whose bark has been taken off 116 feet, is 327 feet high, and 90 feet in circumference at the base. One of the fallen trees measured 110 feet in circumference at the roots; it is 200 feet to the first branch, the whole trunk being hollow, and through which a person can walk erect. It is estimated that it was over 400 feet high.

Columbia.—Hotels:—*Broadway; Columbia; Munston House.*

Returning to Murphy's we take the stage *via* Douglas Flat, crossing the Stanislaus River at Abbeys Ferry, arriving at Columbia in 8 miles from Murphy's.

The scenery at the crossing of the Stanislaus is grand, and we find Columbia one of the largest and finest towns in the mining region, having a population of some 2,000, with fine brick stores, hotels, churches, &c.

In the vicinity many large mining operations are being carried on, which will interest the traveller.

Beyond, a little over a mile, is the thriving village of Springfield, and 2 miles further, lies Shaw's Flat, another important point.

Table Mountain, also, is well worth a visit. Many tunnels are found piercing the mountain for thousands of feet, yielding immense profits to the lucky owners. It is a formation of basaltic lava, and to the geologist its peculiar formation would be of much interest.

Sonora.—Hotels:—*City; Placer; United States.*

Hourly lines of stages connect Columbia with Sonora, the county seat of Tuolumne County, distant 4 miles.

Sonora, the most important mining town in the southern mines, contains about 2,700 inhabitants. A fine court house, several churches, three or four good hotels, and many fine stores, adorn the place. Daily lines of stages leave and arrive from Stockton and Sacramento, with many routes diverging north and south to the way places.

Coulterville.—Hotel:—*Coulter's.*

We will now take the stage for Coulterville, Mariposa County, passing through Jamestown, 5 miles. Montezuma, 9 miles, Chinese camp, 11 miles, crossing the Tuolumne River at Don Pedro's Bar, 25 miles, arriving at Coulterville the same day, distant from Sonora 40 miles.

Coulterville is a small mining town, containing a few stores which supply the miners in the vicinity: the traveller will find good accommodations at Coulter's Hotel.

The Valley of the Yo-Semite.—A trip to the celebrated valley of the Yo-Semite from this point would amply repay the tourist. The valley is about 45 miles east of here, and is reached upon animals; the trip can be made in 4 or 5 days, with ample time to view the different points in the valley.

Procuring animals at Coulterville for the trip, and providing ourselves with blankets and provisions, we start for the Yo-Semite Valley on the Mariposa Trail.

At a distance of 12 miles is Bower Cave, a singular, grotto-like formation, that lures the traveller aside for a few moments. Passing on to Deer Flat, 23 miles from Coulterville, we camp for the night. The next day we reach Crane Flat, 12 miles further, in time for an early lunch. Here the snow-clad Sierras begin to rise in serrated peaks above the horizon. To the right, about one mile, is a grove of mammoth trees, similar to those in Calaveras County, but fewer in number. One of these, consisting of two joined at their base, is called the Siamese Twins, and is 114 feet in circumference.

Two hours and 9 miles further, will bring us to Inspiration Point, whence we first look down into the wonderful cleft of the Sierras called the Yo-Semite. Descending into the valley, it is 7 miles to the foot of the trail, and 6 miles thence to the upper Hotel. After resting here over night, we will inspect the wonders of the valley.

The Yo-Semite Valley is about 10 miles long, rarely exceeding a mile in width, walled in by perpendicular rocks from 3,000 to 5,000 feet high on either hand. The Merced River winds through the grassy meadows at the bottom. It receives several tributaries, which pour over these granite walls at various points, forming water-falls on a magnificent scale. The most remarkable of these has been called the Yo-Semite Fall. It descends in two unbroken sheets, the upper one 1,450 feet in height, and the lower one 700 feet, while the rapids between the two have a fall of 400 feet, giving the total height of the whole fall 2,550 feet. Among the other falls are the Fall of the North Fork of the Merced, about 750 feet high, the Pi-wi-ack, or Vernal Fall, about 300 feet high, the Yo-wi-ye, or Nevada Fall, about 700 feet high, and numerous others of lesser note. Not the least remarkable objects of interest are the two domes, presenting nearly perpendicular faces on opposite sides of the valley.

The North Dome, or To-coy-ee, is about 3,000 feet high. The South Dome, or Tas-sa-ack, is 4,600 feet, the lower two-thirds of which is a sheer perpendicular rock, so that a stone tossed from its top would fall at its base. The volume of water pouring over the various falls varies according to the season of the year, being quite inconsiderable in the month of September; and one or two little lakes gem the valley.

Mariposa. — HOTELS: — *Union House; Franklin House.*

From Coulterville we can reach Mariposa by mules, crossing the Merced River, distance 20 miles; or by returning toward the plains and taking a circuitous route by stage, 50 miles.

Mariposa, the county seat of Mariposa County, is 91 miles from Stockton, with which it is connected by daily lines of stages occupying 25 hours. It is the most southerly of all the mining towns of importance in the State, and contains about 1,800 inhabitants. There are valuable quartz leads, and rich flat, gulch, and hill diggings, in the vicinity. It is here, in the valley of the Mariposa Creek, that the celebrated Fremont Grant is located. There is a good trail from Mariposa to the Yo-Semite Valley, distance 55 miles.

We now return *via* Stockton to San Francisco, whence we originally started. Passing through Hornitos, 16 miles, a brisk mining town of 800 inhabitants; Snelling, 12 miles, the county seat of Merced County; Loring's Bridge, 13 miles, where we cross the Stanislaus River on a fine suspension bridge; thence over an undulating country to Stockton, 50 miles.

Stockton.—HOTELS:—*Weber House; Magnolia; Main Street.*

This important place is situated upon a slough or arm of the San Joaquin River, and is 125 miles from San Francisco *via* the steamboat route, and 80 miles *via* the stage route through Alameda County to Oakland and across the bay.

Stockton is the centre of trade and travel for all the country south of the Cosumnes River, the district generally known as the Southern Mines.

The city contains a population of some 5,000 inhabitants, having several good hotels, fine churches, and a theatre. The streets are well graded and planked; many of the stores and other places of business are fine structures of brick. A daily line of steamers from San Francisco reach the place in the morning, and connect with the numerous lines of stages which leave every morning for the various mining towns in the interior. In the environs of Stockton, particularly toward the Calaveras River, many fine farms or ranches are located, and under good improvement. The State Asylum for the insane, a noble structure, is situated near the suburbs of the city.

The scenery is beautiful with evergreen oaks, while on the east the lofty Sierra Nevadas, with their snow-capped summits, are ever visible, and on the west the duplicate peaks of Mount Diablo rise in towering magnificence.

Leaving Stockton, on the return to San Francisco, we take the steamer at 4 o'clock P. M., and wending our way down the narrow crooked channel of the San Joaquin, we see but little to interest, as the country is almost a total marsh, covered with tules. Toward the setting sun, Mount Diablo rears his double summit to the height of 3,790 feet, presenting a grand outline, while the coast range stretches to the south as far as the eye can reach, forming the western boundary of the great valley which we have just traversed.

Morning finds us at our comfortable quarters in San Francisco, after the long tour through the great valleys of the Sacramento and San Joaquin, with the ride through the mountains of the Sierra Nevada, which forms the eastern boundary of these great valleys. The time occupied by such a trip would be in the neighborhood of two months, at an expense of about $500.

The Butterfield Overland Mail Route.—An interesting trip can be made to the southern part of the State, by the way of Visalia and Fort Tejon, by taking Butterfield's Overland Mail coaches to Los Angeles, thence returning up the coast, either by private conveyance along the shore, or by ocean steamers which connect San Francisco with the southern ports twice a month.

The overland stage leaves San Francisco at 12 M. each Monday and Friday. Proceeding southward we pass through Redwood City, 30 miles, the county seat of San Mateo County; and Santa Clara 20 miles beyond, a fine village of 1,000 inhabitants, containing the Pacific University, the Santa Clara College, and other flourishing institutions. Hourly omnibuses connect Santa Clara with San Jose, 4 miles distant.

San Jose.—HOTELS:—*Crandell's; Appleton's; Mansion House.*

Daily lines of stages leave San Francisco for San Jose; one line by the western side of the bay; the other is taken by crossing over to Oakland, 8 miles by steamer, and then by stage along the eastern side of the bay to San Jose. The distance by either line is about 54 miles; fare generally $3.00. A daily line of steamers runs to Alviso, situated on a slough at the southern extremity of the bay, and there connects by stages with San Jose, 7 miles distant.

San Jose, the garden city of California, is situated in the midst of a very fertile valley. It contains about 3,000 inhabitants, and has some fine public buildings and many elegant private residences. Besides the beauty of the valley of San Jose and its climate of perpetual spring, the gardens and Artesian wells, many places in the vicinity are worthy of a visit, especially the Quicksilver Mines and the Missions of Santa Clara and San Jose. The New Almaden Quicksilver Mines are about 16 miles south, and Monterey on the coast 126 miles.

Proceeding on our journey by the Overland Mail Route, we emerge from the valley of the Santa Clara River through the Pacheco Pass in the Coast Range, into the valley of the San Joaquin. Travelling up the western side of the valley, we first touch the river at Firebaugh's Ferry, 165 miles from San Francisco. Continuing on the same side we pass the great bend of the San Joaquin, and soon reach Fresno City, 184 miles from San Francisco. Crossing the little stream which seems to connect Tulare Lake with the San Joaquin, we cross King's River 40 miles beyond, and soon arrive at Visalia 248 miles from San Francisco.

Visalia.—HOTELS:—*Exchange; Warren's.*

This is the only town of importance between San Jose and Los Angeles on the Overland Route, containing about 1,200 inhabitants. It is located on the banks of the Kaweah River, about 18 miles from the mountains on the east and 20 from Tulare Lake on the west, in the centre of a large body of oak timber, and in the midst of a rich alluvial delta. The several creeks north and south of Visalia in its immediate vicinity, spread out on the large meadows and lose themselves and their channels before reaching the great Tulare Lake, which ordinarily has no well-defined outlet itself. The Coso silver mines lie about 100 miles east, and are reached by a trail over the Sierra Nevadas.

Proceeding southward we cross Tule River, 27 miles, Kern River, 87 miles, arriving at Tejon Canon, 128 miles from Visalia and 376 from San Francisco.

Fort Tejon.—Tejon Pass is at the head of the San Joaquin Valley. The Coast Range and Sierra Nevadas, gradually converging, join at this point. A fort has been built high up in this romantic pass, about 3,000 feet above the level of the sea, where there is a small spot of level land between the mountains, with fertile soil, grass, a pleasant brook, and fine oak trees.

From Tejon Canon we descend into and cross the arid plains of Palm Valley, a part of the great Basin, whose waters never find the sea. Then our road lies over the Coast Range through the San Francisquito Canon which opens into Santa Clara Valley, and crossing the Santa Clara River, our

CALIFORNIA. 263

way lies through the San Fernando Pass, over a spur of the Coast Range, whence we emerge into the vine-clad valley of the Rio Los Angeles, and in a few hours reach the city of the Angels, 491 miles from San Francisco.

Los Angeles.—Hotels:—*Bella Union; La Fayette.*

Los Angeles is situated near the foot of the Coast Range on the Los Angeles River. Most of the land in the valley which can be irrigated is planted with vines. The city contains about 5,000 inhabitants. The houses are many of them of the Spanish style, one story with flat roofs covered with asphaltum, which abounds in the vicinity. On the northwestern side of the town and very near the busiest part of it, is a hill about 60 feet high, whence an excellent view of the whole place may be obtained.

Along the banks of the river for miles are situated the vineyards and orange groves, the pride of Los Angeles.

Vast tracts of the fertile plains and river bottoms are irrigated by the waters of the river, producing every variety of fruit and vegetable common to the warm and temperate climes.

In the months of March and April, looking over these fertile plains, covered with the richest verdure, the snow-clad heights beyond contrast beautifully with the flowers at their feet.

To the east, Mount San Bernardino rises covered with snow, 80 miles distant. Its altitude is about 8,000 feet, and it marks the site of the pleasant valley in which the village of San Bernardino is situated.

Silver leads of more or less promise have been discovered in various parts of the neighboring mountains. A rich tin mine has been discovered at Temescal, about 60 miles distant on the overland route. The San Gabriel placer gold mines lie about 20 miles to the northeast. The sites of several old missions are in Los Angeles County.

The entire distance from San Francisco to St. Louis, by the Butterfield Overland Mail Route through Los Angeles, is 2,881 miles, the last 170 of which is performed by railroad. This route is now much travelled, each stage being capable of taking six through passengers. The distance is usually accomplished in 22 days. The traveller can obtain meals at way stations, which occur from 15 to 40 miles apart. He rides night and day without cessation, soon getting used to the motion of the stage so that he can get refreshing sleep at night, and arrives well and hearty at his journey's end.

The distance from Los Angeles to St. Louis is 2,890 miles, divided as follows:

From Los Angeles

To Fort Yuma..		288
Tuckson...	281	569
El Paso..	339	908
Fort Chadbourne...................................	428	1336
Red River...	384	1720
Fort Smith...	192	1912
Syracuse........................	308	2220
St. Louis...	170	2390

From Los Angeles a pleasant journey can be made to San Diego in two days. Setting out for Anaheim, 27 miles, we travel along the coast, passing through San Juan Capistrano, 55 miles, Los Flores, 78, San Luis Rey, 85, San Dieginto, 105, arriving at San Diego, 125 miles from Los Angeles.

San Diego.—Hotel:—*Franklin House.*

San Diego is a small town of 500 inhabitants, situated upon a harbor of the same name. San Diego harbor, next to that of San Francisco, is the best on the coast of California, being well protected, capacious, and having a fine depth of water. There is nothing remarkable about the town or surrounding scenery. A fine grazing country inland, abounds in large cattle ranches.

San Diego is connected with San Francisco by an ocean steamer, which makes two trips a month, touching at the intermediate ports of San Pedro, the port of Los Angeles, Santa Barbara, San Luis Obispo, and Monterey, and sometimes Santa Cruz.

Should the traveller prefer to return to San Francisco by land, he will find the route along the shore very pleasant, full of beautiful and romantic scenery; the mountains of the Coast Range and its spurs rising loftily on his right, and at times the waves of the sea dashing at his feet.

In proceeding northward then, the first point of interest is Santa Barbara, about 180 miles from San Diego by water, and 100 from Los Angeles by land.

Santa Barbara.—Hotel:—*City Hotel.*

The steamers afford a fine view of the coast, as they pass near the land; and approaching Santa Barbara, the view is very imposing. High ranges of mountains bound the view to the eastward, while the beautiful valley in which the town is situated, stretches far to the northward, finely relieved by a background of misty mountains, grand in outline.

This place, with the other ports along the coast, is famous for the hide business, formerly the staple product of California.

Santa Barbara has no protected harbor like San Pedro, and other places along the coast; it is only an open roadstead, dangerous during a south-easter, which, however, occurs only during the rainy season.

The town, like Monterey and the other old places in California, retains much of its old Spanish look—the buildings of adobes or sun-dried bricks, roofed with tiles, presenting a venerable appearance. A ride to the Mission of Santa Barbara, about 3 miles distant, would be of interest.

The climate below Point Conception (which lies between Monterey and Santa Barbara) is much milder. The northwest winds which prevail to the north, are not felt, and the climate is much warmer.

Continuing on our way up the coast, we touch the shore 110 miles above Santa Barbara, for the benefit of those who wish to go to San Luis Obispo, the county seat of San Luis Obispo County, and a small, unimportant Spanish town, in the midst of a beautiful grazing country.

Monterey.—HOTEL:—*Washington.*

The large open bay of Monterey lies about 120 miles north of San Luis Obispo by water. The town of Monterey is beautifully situated upon the southern extremity of this bay. It was formerly the seat of government, and principal port on the coast of California. But since the rise of San Francisco, its commerce and business have dwindled away, and now it is one of the most quiet places in the State, containing about 1,500 inhabitants.

The view of the town from the anchorage is very fine, especially if visited in the months of April or May. The green slopes upon which the town is built, contrast beautifully with the forest of pines which grow upon the ridges beyond.

Santa Cruz.—HOTEL:—*Exchange.*

It is 20 miles across the bay from Monterey to Santa Cruz, the county seat of Santa Cruz County. The town contains about 800 inhabitants, and is surrounded by a mountainous country covered with immense forests of redwood timber.

The distance from Santa Cruz to San Francisco by water is about 70 miles. To San Jose, crossing he Santa Cruz Mountains, it is about 35 miles.

Up the Coast.—Steamers leave San Francisco semi-monthly for Oregon, Washington Territory, and Victoria, in the British possessions, touching at Mendocino, Humboldt Bay, Trinidad, Crescent City, Port Orford, Portland, and sometimes Vancouver, on the Columbia River, the various points on Puget Sound, and Victoria on Vancouver's Island.

Sailing vessels also are constantly leaving San Francisco for Humboldt Bay, Port Orford, the Columbia River, Puget Sound, and Vancouver's Island.

Many of the northern mines near the coast, are easily accessible from Humboldt Bay, Trinidad, Crescent City, and Port Orford in Oregon, the gold range approaching the coast. Coal is also found in immense beds in the vicinity of Coose Bay, Oregon.

Embarking on one of the fine ocean steamers for a trip up the coast, we touch at Mendocino, or pass it 130 miles from San Francisco, Humboldt Bay upon which the thriving towns of Eureka and Arcata are situated, 230 miles, arriving at Crescent City, some 300 miles from San Francisco.

Crescent City.—HOTELS:—*Patchin House; American.*

Crescent City, the county seat of Del Norte County, is a thriving place of some 600 inhabitants. Most of the interior mining localities through a considerable range of country, obtain their supplies through this post. Extensive veins of copper have been discovered in the vicinity, some of which have been worked with profit. The surrounding region also abounds in gold and other minerals, but for want of systematic supplies of water have not yet been much worked.

Proceeding north we touch at Port Orford, 70 miles, a port from which much lumber is exported; Fort Umpqua, 140, near the mouth of Umpque River, which drains a fertile and productive valley; and arrive at Astoria on the Columbia River, some 300 miles north of Crescent City, and about 600 miles north of San Francisco.

The scenery of the Columbia River is wild and grand beyond description. Vessels of the largest size proceed up the river from Astoria at the mouth, to Vancouver, a distance of about 100 miles, and beyond to the falls of the river, where the Cascade Range of mountains cross.

Some of the mountain peaks of the Cascade Range, among which may be mentioned Mount Hood, Mount Jefferson, and Mount St. Helens, rival those of the Andes. They are covered with perpetual snow, and can be seen from various parts of the river.

Portland, the chief city of Oregon, is situated on the Willamette River near its mouth, and contains about 5,000 inhabitants.

The Willamette River flowing north between the coast and Cascade Ranges of mountains, empties into the Columbia about 100 miles from the ocean. The valley of the Willamette is the garden of Oregon, and contains a large population of permanent settlers, many of whom had located on farms some time before the Americans commenced settling in California.

A month's travel to the various places on the Columbia would amply repay the tourist, and can be made from San Francisco at an expense of about 175 dollars, including the fare each way.

Proceeding up the coast we find no other seaport till we reach Puget Sound, one of the most magnificent harbors in the world. While the Sound is so deep that vessels of the heaviest burthen can traverse any part of it with safety, it is nowhere too deep for convenient anchorage; and in many places vessels can ride boldly up to the shore, for purposes of loading, without the intervention of wharves. The lumber from some of the saw-mills on the Sound is shipped in this way.

The principal interests of the surrounding country, are lumber and agriculture.

Four thrifty towns have sprung up on different inlets of the Sound: viz.: Port Townsend, with 500 inhabitants and the Custom House; Olympia, with 700 inhabitants, formerly the capital of the territory, in the vicinity of the superb water power of Tum-water; Steilacoom with 800 inhabitants, and Seattle 500, are the termini of trails and military roads leading through the Cascade Range to the mineral regions beyond.

Whidby's Island, at the entrance of the Sound, contains many fine farms, and its verdant bluffs rising boldly from the water's edge, are very beautiful in spring and summer.

Several majestic mountain peaks are visible from the waters of the Sound, forming some of the most sublime scenes on the western coast of America. Among these are Mount Baker, Mount St. Helens, and Mount Rainier, whose summits are from 12,000 to 15,000 feet above the level of the sea, and covered with perpetual snow. Some of these have shown volcanic action within the last few years. Mount Baker, 14,000 feet high, was in active eruption in 1860.

From Port Townsend the traveller can reach the mouth of the Columbia, or indeed Sacramento in California, without returning by the ocean route. Proceeding by steamer to the head of the Sound at Olympia, or by stage on the west side of the Sound to the same point, he can proceed from thence through the cowlitz farms to Vancouver on the Columbia River.

Vancouver, the present capital, is one of the most promising places in Washington Territory, containing about 1,000 inhabitants besides the soldiers of the U. S. military post stationed near. The distance from Vancouver to Portland is 18 miles, and the entire distance from Port Townsend to Portland is about 230 miles.

From Portland the daily overland mail to Sacramento takes the traveller up the valley of the Willamette, across Umpqua and Rogue rivers to Jacksonville, and thence through Yreka, Shasta, and Marysville to Sacramento, 750 miles from Portland, making the longest stage route in the Union, with the exception of the overland routes.

The eastern slope of the Cascade Range in Washington Territory though but partially developed, gives indications of great mineral wealth. The Wenatchee, Samilkameen, and Rock Creek gold regions, have attracted many adventurers, and yielded their treasures bountifully. From Steilacoom a military wagon road leads through a pass in the Cascade Range to Walla Walla, 250 miles southeast on the Columbia River. Beyond Walla Walla lies the Nez Perces gold region.

Victoria, on Vancouver Island, the principal town of the British Possessions, contains about 3,000 inhabitants. It is the entrepot of goods for Fraser River. The gold diggings of the latter are still being successfully worked.

New Westminster, the capital of British Columbia, and next to Victoria the largest town in the British dominions on the Pacific, is situated on Fraser River near the head of navigation. The mines and inhabitants are protected from the depredations of Indians by the presence of soldiers at Fort Hope, Fort Yale, &c.

SUMMARY OF ROUTES FROM SAN FRANCISCO TO

PLACE.	DISTANCE.	TIME.	FARE.	REMARKS.
	Miles.			
Mendocino	128			Ocean Steamer.
Eureka	225	30 hours.	$30 00	"
Trinidad	240		30 00	"
Crescent City	280		40 00	"
Port Orford	340		40 00	"
Fort Umpqua	400	4 days.	50 00	"
Astoria	558		40 00	"
Portland	642	6 days.	40 00	"
Vancouver	632		40 00	"
Victoria	753		50 00	"
Port Townsend	773	} 8 days.	50 00	"
Seattle	810		50 00	"

CALIFORNIA.

SUMMARY OF ROUTES FROM SAN FRANCISCO TO

PLACE.	DISTANCE.	TIME.	FARE.	REMARKS.
	Miles.			
Stellacoom	836	} 8 days.	$50 00	Ocean Steamer.
Olympia	855		50 00	"
Santa Cruz	80		10 00	"
Monterey	92		12 00	"
San Luis Obispo	200		25 00	"
Santa Barbara	288		30 00	"
San Pedro	373		30 00	"
Los Angeles	395		40 00	"
San Diego	456		40 00	"
Acapulco	1840	8 days.	100 00	"
Panama	3280	16 "	150 00	"
New York		24 "	250 00	"
San Quentin	15	1¼ hours.	1 00	
Petaluma	48	4½ "	3 00	} Petaluma Steamer and Stage.
Healdsburg	32	5 "	3 50	
Geysers	50	12 "	Stage & Trail from Petaluma.	
Sonoma	45	4½ to 6 hrs.	3 50	Steamer.
Vallejo	25	2 to 3 "	2 00	
Napa	50	4 to 5 "	3 00	} Napa Steamer and Stage.
Sulphur Springs	18	2 to 3 "	3 00	
			From Napa.	
Suisun	60	5 to 7 "	4 00	Steamer.
Benicia	30	2 to 3 "	1 00 to 2 00	} "
Sacramento	120	6½ to 12 "	1 00 to 5 00	
Stockton	120	8 to 12 "	5 00	"
Alviso	45	2½ to 3 "	2 00	} Steamer and Stage.
San Jose	52	3½ to 5 "	2 50	
"	54	7½ hours.	3 50	
Redwood City	30	4½ "	2 00	
Visalia	248	36 "	20 00	
Fort Tejon	376	56 "	30 00	
Los Angeles	491	8 days.	80 00	} Butterfield's Overland Stage Route.
Fort Yuma	779	5½ "	70 00	
Tuckson	1060	8 "	100 00	
Mesilla	1353	9½ "	125 00	
El Paso	1399	10 "	125 00	
St. Louis	2881	21 "	200 00	
Monterey	130	16 "	9 00	Stage from San Jose.
Oakland	8	1 hour.	25	Steamer.

SUMMARY OF ROUTES FROM STOCKTON TO

PLACE.	DISTANCE.	TIME.	FARE.	REMARKS.
	Miles.			
Mokelumne Hill	50	6 hours.	$8 00	Stage.
San Andreas	45	6 "	6 00	} "
Murphy's	66	9 "	8 00 to 10.	
Big Trees	81	12 "		Mules from Murphy's $3 to 5 p. day.
Knight's Ferry	36	6 "	5 00	
Sonora	64	12 "	10 00	} Stage.
Columbia	68	12 "	10 00	
Coulterville	85	24 "	12 00 to 15.	"
Yo-Semite	130			Mules from C. $3 to $5 per day.
Mariposa	91	25 "	12 00 to 15.	Stage.

SUMMARY OF ROUTES FROM SACRAMENTO TO

PLACE.	DISTANCE.	TIME.	FARE.	REMARKS.
	Miles.			
Marysville {	44	6 hours.	$5 00	} Steamboat.
	64	7 to 12 hrs.	4 00	
Colusi	120	11 to 14 "	6 00 to 8 00	
Red Bluff	275	26 to indef.	18 00 to 20.	
Nevada	70	12 hours.	12 00	R. R. to Folsom, and thence Stage.
Auburn	40	4 "	6 00	"
Folsom	22	1 "	2 00	Railroad.
Alabaster Cave	35	3 "	3 50	R. R. to Folsom, and thence Stage.
Placerville	50	6 "	6 00	"
Lake Bigler	110		18 00	"
Carson City	145	26 "	26 00	"
Virginia City	163	30 "	30 00	"
Jackson	60	8 "	6 00	} Stage.
Mokelumne Hill	55	9 "	7 00	"
Sonora	80	12 "	12 00	"
Stockton	45	8 "	5 00	"
Napa	61	9 "	6 00	"

SUMMARY OF ROUTES FROM MARYSVILLE TO

PLACE.	DISTANCE.	TIME.	FARE.	REMARKS.
	Miles.			
Oroville	28	4 hours.	$3 00	
Red Bluff	92	13 "	15 00	} Stage.
Shasta	128	20 "	22 00	
Yreka	236	2 days.		
Downieville	76	15 hours.	16 00	Stage.
Nevada	40	7 "	6 00	"
Auburn	40	6 "	6 00	"
Colusa	23	5 "	3 00	

All the numerous mining towns in the counties of Calaveras, Tuolumne, Stanislaus, Merced, Mariposa, &c., can be reached by either of the above routes, or by lines of coaches in connection with the above, departing and arriving with excellent despatch.

OREGON.

OREGON was organized as a Territory August 16, 1848, and was admitted into the Union as a State February 16, 1859. It lies upon the Pacific, with an area of 95,274 square miles.

The first visit of the white race to Oregon was in 1775, when a Spanish voyager entered the Juan de Fuca Straits. Three years afterward (1778), the celebrated navigator, Captain Cook, sailed along its shores. In 1791 the waters of the Columbia River were discovered by Captain Gray, of Boston. An expedition, or exploring party, was sent out in the year 1804 by the United States, commanded by Lewis and Clark, who wintered in 1805-'6 at the mouth of the Columbia. From this period the coast was a great resort of both English and American fur traders.

By the treaty with Great Britain in 1846, this great territory, which had up to that time been jointly occupied by English and American adventurers, was divided—the one taking the portion above the parallel of 49° north latitude, and the other all the country south of that line.

Emigration to Oregon was earnestly commenced in 1839. For some years the settlement of the country was retarded by the more brilliant attractions of California, though the ultimate result of this neighborhood will be a great means of development, as Oregon is an agricultural land, whose products will be required by the mining population of the lower State.

Washington Territory, on the north, was a part of Oregon until the year 1853, when it was erected into a distinct government.

The coast of Oregon, viewed from the sea, is, like that of California, stern and rockbound, excepting that while in the latter region the nearer mountains follow the line of the shore, in Oregon they approach the ocean at a great angle. The lower or Pacific country occupies an area of from 75 to 120 miles wide, in which lie the great valleys of Willamette, Umpqua, and Rogue rivers.

Though the valley lands of the Willamette and the adjacent regions are extremely fertile, yet the greater portion of Oregon is unfit for tillage, being, as it is, a country of untamed and untamable hills.

The climate here, as on all the Pacific coast, is milder than in corresponding latitudes near the Atlantic. The winters are comparatively brief, and the snows, when snow falls at all, are very light.

Gold is found in various parts of southern Oregon, and silver, lead, and copper in the Cascade Mountains. Coal is abundant at Coosa Bay and other points.

The Columbia River of Oregon, is the greatest on the Pacific slope of this Continent. It rises in a small lake among the western acclivities of the Rocky Mountains, and flows in a devious course 1,200 miles to the Pacific, forming a great portion of the dividing line between Oregon and Washington Territory on the north. Its earliest meanderings are northward along the base of its great native hills, and afterwards its way is due west to the sea, though very capriciously. It is a rapid river, pushing its way through mighty mountain passes, and in many a cataract of marvellous beauty. In its course through the Cascade Range, it falls into a series of charming rapids, which may be numbered among the chief natural attractions of the country. The tide sets up to this point, 140 miles. For 30 or 40 miles from its mouth, the Columbia spreads out into a chain of bay-like expansions, from 4 to 7 miles or more in width. The shores are lined with grand mountain heights, making the landscape everywhere extremely interesting and impressive. We should far exceed our present opportunity in attempting even the briefest catalogue of the pictures on these noble waters. Vessels of 200 or 300 tons burden may ascend to the foot of the cascades, of which we have already spoken. Above this point the river is navigable for small vessels only, and but at intervals in its course.

The Willamette River flows from the foot of the Cascade Range, 200 miles, first northwest and then north to the Columbia, 8 miles below Fort Vancouver. Its way is through the beautiful valley lands which bear its name, and upon its banks are Oregon City, Portland, Corvallis, Eugene City, and other thriving places. Ocean steamers ascend 15 miles, to Portland. Ten miles beyond this point, a series of fine falls occur in the passage of the river, above which the waters are again navigable, perhaps 60 miles, for small steamboats. The Falls of the Willamette is a famous place for the capture of the finest salmon. Among the tributaries of the Willamette are the Tualatin, Yamhill, La Creole, Luckamute, Long Tom, Marys rivers, coming from the base of the Callepoosa and Coast Range Mountains, and the McKenzie, Santiam, Pudding, and Clackamus from the Cascade Range.

The Valley of the Willamette is a most fertile region, and most attractive in its natural curiosities. Many remarkable examples are to be found here of those eccentric mountain formations known as Beetlers—huge, conical, insulated hills. Near the mouth of the Coupe River, there are two of these heights which tower up 1,000 feet, but half a mile removed from each other at their base. They are called Pisgah and Sinai. They stand in the midst of a plain of many miles in extent. At a point near the Rickreall River, in the Willamette Valley, no less than seven snow-capped peaks of the Cascade Range may be seen.

Between the Blue Mountains and the Cascade Range lie a number of small lakes.

Forest Trees.—Oregon, like California, is famous for its wonderful forest growth. The Lambert pine, a species of fir, sometimes reaches, in the lower part of the country, the magnificent height of 300 feet.

Salem.—HOTELS:—*Bennett House; Marion Hotel.*

Salem, the capital of Oregon, is on the Willamette River, 50 miles above Oregon City.

Oregon City.—HOTEL:—*United States Hotel.*

Oregon City, the former capital of the territory, is upon the Willamette, hidden in a narrow, high-walled valley or cañon. Falls on the river at this point afford great manufacturing facilities to the growing settlement.

Portland.—HOTELS:—*Metropolis; Pioneer; What Cheer.*

Portland, the largest and most important town in Oregon, with a population of about 2,000, is upon the Willamette, at the head of ship navigation, 15 miles from its entrance into the Columbia.

St. Hellens is upon the west side of the Columbia, 30 miles from Portland.

Astoria, named in honor of its founder, John Jacob Astor, is on the south side of the Columbia River, some ten miles from its mouth. This was at one time an important fur depot.

Routes.—Steamboats ply regularly between San Francisco and the landings on the Columbia, the Willamette, and other rivers. There are also good stage routes to all points.

From St. Louis to Oregon, through Pass in the Rocky Mountains.

BY STEAMBOAT.

To St. Charles		40
Gasconade River	74	114
Osage River	32	146
JEFFERSON CITY	10	156
Booneville	53	209
Lexington	100	309
INDEPENDENCE	61	370
Kansas River Landing	12	382

BY LAND.

To Kansas River Crossing	75	457
Platte River	220	677
Forks of River	15	692
Chimney Rock	155	847
Scott's Bluff	22	869
Fort Laramie	60	929
Red Butter	155	1084
Rock Independence	50	1113
South Pass (Fremont's)	110	1244
Green River	69	1313
Beer Springs	191	1504
Fort Hall	50	1554
American Falls	22	1576
Fishing Falls	125	1701
Lewis River Crossing	40	1741
Fort Boisse	130	1871
Burnt River	70	1941
Grand Ronde	68	2009
Fort Walla Walla	90	2099
Umatillah River	25	2124
John Day's River	70	2194
Falls River	20	2214
The Dalles	20	2234
Cascades	45	2279
Fort Vancouver	55	2334
ASTORIA	100	2434

Or, from St. Louis, by the Missouri River, to Fort Benton, and thence by the new Military road, 400 miles, to the navigable waters of the Columbia.

KANSAS.

KANSAS was admitted into the Union as a State on the 29th January, 1861. It has an area of 80,000 square miles. Capital, Topeka.

"The face of this country," says a traveller, "is beautiful beyond all comparison. The prairies, though broad and expansive, stretching away miles in many places, seem never lonely or wearisome, being gently undulating, or more abruptly rolling; and, at the ascent of each new roll of land, the traveller finds himself in the midst of new loveliness. There are also high bluffs, usually at some little distance from the rivers, running through the entire length of the country, while ravines run from them to the rivers. These are, at some points, quite deep and difficult to cross, and, to a traveller unacquainted with the country, somewhat vexatious, especially where the prairie grass is as high as a person's head, while seated in a carriage. There is little trouble, however, if travellers keep back from the water-courses, and near the high lands. These ravines are, in many instances, pictures of beauty, with tall, graceful trees, cotton-wood, black walnut, hickory, oak, elm, and linwood, standing near, while springs of pure cold water gush from the rock. The bluffs are a formation unknown, in form and appearance, in any other portion of the West. At a little distance, a person could scarcely realize that art had not added her finishing touches to a work, which nature had made singularly beautiful. Many of the bluffs appear like the cultivated grounds about fine old residences within the Eastern States, terrace rising above terrace, with great regularity; while others look like forts in the distance. In the eastern part of the State, most of the timber is upon the rivers and creeks.

though there are in some places most delightful spots; high hills, crowned with a heavy growth of trees, and deep vales, where rippling waters gush amid a dense shade of flowering shrubbery.

"Higher than the bluffs are natural mounds, which also have about them the look of art. They rise to such a height as to be seen at a great distance, and add peculiar beauty to the whole appearance of the country. From the summit of these the prospect is almost unlimited in extent, and unrivalled in beauty. The prairie for miles, with its gently undulating rolls, lies before the eye. Rivers, glistening in the sunlight, flow on between banks crowned with tall trees;—beyond these, other high points arise. Trees are scattered here and there, like old orchards, and cattle in large numbers are grazing upon the hill-side and in valleys, giving to all the look of cultivation and home-life. It is, indeed, difficult to realize that for thousands of years this country has been a waste, uncultivated and solitary, and that months only have elapsed since the white settler has sought here a home.

"The climate is exceedingly lovely. With a clear, dry atmosphere, and gentle, health-giving breezes, it cannot be otherwise. The peculiar clearness of the atmosphere cannot be imagined by a non-resident. For miles here a person can clearly distinguish objects, which at the same distance in any other part of this country he could not see at all. The summers are long, and winters short.

"The winters are usually very mild and open, with little snow,—none falling in the night, save what the morrow's sun will quickly cause to disappear. So mild are they that the cattle of the Indians, as those of the settlers in Western Missouri, feed the entire year in the prairies and river-bottoms. The Indians say that once in about seven years Kansas sees a cold and severe winter, with snows of a foot in depth. Two weeks of cold weather is called a severe winter. Then the spring-like weather comes in February; the earth begins to grow warm, and her fertile bosom ready to receive the care of the husbandman."

The Kansas River, the largest stream of this region, excepting the Missouri, which washes its northeastern boundary, is formed by the Republican and the Smoky Hills Forks, which rise in the Rocky Mountains, and unite their waters at Fort Riley. The length of the Kansas, including its branches, is nearly 1,000 miles. Its course is through a productive valley region or plain, covered with forest trees, and varied here and there with picturesque bluffs and hills. The Kansas River is a tributary of the Missouri, and steamboats ascend from its mouth 120 miles to Fort Riley.

The Platte River rises in the Rocky Mountains in two arms, called the North and the South Falls, and runs 1,200 miles into the Missouri. It is navigable at high water for hundreds of miles, though it is usually shallow, as its name implies. It abounds in islands, and in some places spreads over a breadth of three miles.

The Arkansas River has nearly half its course within the Borders of Kansas. The Osage River flows nearly eastward, 500 miles to the Missouri, ten miles below Jefferson City.

Kansas City.—Hotels:—

Kansas City is upon the Missouri River, about 280 miles west of St. Louis, by the Pacific Railway, which is already in operation some 168 miles. The town may be readily reached by rail to St. Joseph, on the Missouri, and thence by boats. Kansas City is the point at which all goods to go up the Kansas River are transshipped.

Leavenworth City.—Hotels:—

Leavenworth, one of the most important places in Kansas, is upon the Missouri River, about midway between St. Joseph and Kansas City.

Routes. From the East, by railway to St. Joseph, and thence down the river *via* Atchison, or by steamer, up the Missouri from St. Louis.

Wyandotte City is at the confluence of the Missouri and the Kansas. Kickapoo is 15 miles further north. Atchison is yet 20 miles above, at the mouth of Independence Creek, and Doniphan is 20 miles yet further up.

Lawrence City. Douglass, Tecumseh, and Whitfield are upon the Kansas River.

Elm Grove, Council City, and Council Grove are upon the Santa Fé Trail.

Routes to Kansas.—From St. Louis, by the Pacific Railway (following the course of the Missouri River), 168 miles to Syracuse; thence to Kansas City, on the Missouri. Total distance from New York, 1,272 miles. Distance from St. Louis, by steamer on the Missouri, 480 miles. Or, by the North Missouri Railway, 304 miles from St. Louis to St. Joseph, on the east bank of the Missouri River; or, from Hannibal, on the Mississippi, 206 miles to St. Joseph.

St. Joseph is now a great starting point for the

farthest West. The Great Salt Lake Mail, the new Pony Express, and the Pike's Peak Express, all start thence.

DISTANCES AND NAMES OF PLACES BETWEEN ST. LOUIS AND FORT LEAVENWORTH, AND ALSO THE MOUTH OF THE YELLOW STONE BY STEAMBOAT.

To Cabris Island		3
Chouteau's Island	7	10
Mouth of the Wood River	5	15
Missouri River	3	18
St. Charles	22	40
New Port	46	86
Pinkney	7	93
Mouth of Gasconade River	21	114
Portland	10	124
Mouth of Osage River	21	145
To JEFFERSON CITY	9	154
Marion	16	170
Nashville	10	180
Rocheport	14	194
Boonsville	10	204
Arrow Rock	15	219
Chariton	16	235
Mouth of Grand River	26	261
Lexington	50	311
Blayton	18	329
Fort Osage	13	342
Liberty	18	360
Mouth of Kansas River	15	375
Mouth of Little Platte River	12	287
LEAVENWORTH CITY	39	425
Rialto	3	428
Weston	7	435
St. Joseph	15	450
Fort Pierre	1010	1460
Mouth of Yellow Stone	403	1863

WASHINGTON.

THIS Territory was organized March 2, 1853, with an area of 71,300 square miles.

It occupies the extreme northwestern part of the national domain. The recent erection of the territory of Idaho has greatly reduced its original extent, taking from it much of the western section. One of the chief sources of wealth to the inhabitants is in the utilization of their immense forests of fir, and spruce, and cedar; and another in the working of the mineral fields which are found in the region.

The forests abound in elk, deer, and other noble game. Wild fowl also, of many varieties, are plentiful; and in no part of the world are finer fish to be had.

The rivers of Washington are rapid mountain streams, replete with picturesque beauty, in bold rocky cliffs and precipices and in charming cascades.

Olympia, the capital of the territory, is upon Tcnalquet's or Strule's River, at its entrance into Puget's Sound, in the extreme western or Pacific section.

Routes to the settlements in Washington, are by steamboats from San Francisco, along the coast and to points on the Columbia River.

NEBRASKA.

NEBRASKA was organized as a Territory May 30, 1854. Its present area is 63,300 square miles, though it formerly embraced a very large region—reduced now by the erection of newer Territories from portions of its original domain.

Its mineral resources are now being rapidly developed; and its fertile valley lands are being turned to such account, that the present growth of the country is extremely promising of future power and wealth.

Omaha, the capital of the Territory, is upon the Missouri River, opposite Council Bluffs City. Bellevue is upon the Nebraska, 6 miles above its mouth. Fort Calhoun is 18 miles north of Omaha. Florence is 6 miles north of Omaha. La Platte is on the Missouri, 14 miles below the Capital. Plattsmouth is the first town south of the mouth of the Nebraska. Still further south, and along the banks of the Missouri, are Bluff Rock, Kenosha, Nebraska City, Kearney City, and Brownsville; north of the Nebraska, and beyond the places which we have already mentioned, are De Soto, Tekama, and Black Bird.

The chief interior settlements are Archer and Pawneeville, upon affluents of the Great Nehama Saline, on the Big Blue River; Magaretta, near the south bend of the Nebraska; Iron Bluffs, Elkhorn City, Fontenelle, and Catherine, on Elkhorn River; Pawnee, on Loup Fork; Maniton and Hauton in the county north of the Nebraska River.

The population of Nebraska was estimated at 40,000 in 1863.

Routes.—The best route to Nebraska, at present, is from St. Louis, by railway to St. Joseph. See St. Louis for route thither from the Atlantic cities; and see Kansas for route and table of places and distances from St. Louis to Fort Leavenworth.

St. Joseph	60	49
Nodaway River	14	503
Wolf River	16	524
Great Nemahaw River	18	542
Nishnebotna River	25	567
Little Nemshaw River	12	579
Fair Sun Island	16	595
Lower Oven Island	12	607
Upper Oven Island	4	611
Five Barrel Island	19	693
Platte River	16	638
Bellevue Trading-house	12	650
Omaha—opposite Council Bluffs	40	690

From St. Louis to Omaha (opposite Council Bluffs) on Missouri River, by Steamboat.

To Fort Leavenworth, as in Route to Kansas from St. Louis, see Kansas 425
Weston 9 434

There is a route to Omaha City now in vogue from Burlington, on the Mississippi (see Burlington, Iowa), 31 miles, by railway, to Mount Pleasant, Iowa; and thence across by stages to Council Bluffs and Omaha, on the Missouri. Railway lines are in course of construction over this course.

UTAH.

UTAH formerly extended some 700 miles from east to west, and 347 miles from north to south; but this vast region has been greatly reduced by the recent formation of other territories within and around it. Its present area is 109,000 square miles, its population, 88,193.

It is a country of elevated, sterile table-lands, divided in unequal parts by the Sierra Madre Mountains. The Great Basin, or Fremont's Basin, as it is otherwise called, extends over the western part, 500 miles from east to west, and 350 from north to south. This vast tract lies at an elevation of nearly 5,000 feet above the sea. Some portion of it is covered by a yielding mass, composed of sand, salt, and clay, and others with a crust of alkaline and saline substance. Great hills surround it on all sides, and detached groups cross its whole area. Near the centre it is traversed by the Humboldt River Mountains, which rise from 5,000 to 7,000 feet above the adjacent country. There are other great valley stretches in Utah, more sterile even than the Great Basin, as that lying between the Rocky and the Wahsatch Mountains. Only a small portion of this wide region can be turned to account in agricultural uses. The little fertile land it possesses, is that which skirts the streams and narrow tracts at the base of the mountain ranges. The most productive portion is that probably of the valleys extending north and south, west of the Wahsatch Mountains, and which is occupied by the Mormon settlements.

The Climate of Utah is said to resemble that of the great Tartar plains of Asia, the days in summer time being exceedingly hot and the nights cool. The winters are mild, and but little accompanied with snow. The temperature is liable to great and quick transitions from the changing currents of the winds.

The Great Salt Lake is perhaps the most remarkable of all the many natural wonders of these rude and desolate wilds. This singular body of water lies northeast of the centre of the territory. It is some 70 miles long and 30 wide. It is so highly impregnated with salt that no life is found in it, and a thick saline incrustation is deposited upon its banks by evaporation in hot weather; and yet all its tributary waters are fresh. In some of its features, as in the wild and weird aspect of much of the surrounding scenery, it has been compared to the Dead Sea of Palestine.

Utah Lake is a body of fresh water some 35 miles in length. It lies south of the Great Salt Lake, to which it is tributary, by the channel of the connecting river called the Jordan. Like its saline neighbor, the Utah Lake is elevated about 450 feet above the level of the sea. It is abundantly supplied with fine trout and other fish.

The Pyramid Lake lies on the slope of the Sierra Nevada Mountains, 700 feet yet above the Great Salt Lake. It is enclosed everywhere by giant rocky precipices, which rise vertically to the sublime height of 3,000 feet. From the bosom of the translucent waters of this wonderful lake, there springs a strange pyramidal rock 600 feet in air.

In the interior of the territory there are other smaller ponds, as Nicollet Lake, near the centre, and 70 miles yet southward, Lake Ashley. Mud, Pyramid, Walker's and Carson's Lakes are near the eastern base of the Sierra Nevada Mountains; Humbolt's Lake, formed by the waters of the Humbolt River, is about 50 miles east of Pyramid Lake.

The Boiling Springs is a scene of curious interest. The principal basin is described by Col. Fremont as having a circumference of several hundred feet, with a circular space at one extremity 45 feet in circuit, filled with boiling water. The temperature near the edge was found to be 206°.

Canons. Near Brown's Hole, in the vicinity of Green River, there are many of those singular ravines of the Great West, known as Canons. They are sudden depressions in the surface of the earth, sometimes of a vertical depth of 1,500 feet. Nothing can be more surprising and more grand than the pictures presented in these strange passages; the effect, too, is always heightened by the unexpected manner in which the traveller comes upon them, as no previous intimation is afforded, by the topography of the land, of their proximity.

Utah is famous as the home of the Mormons. This extraordinary people pitched their tents here in 1847, after they were driven out of Illinois and Missouri. They are the sole occupants of the region, excepting the native Indian tribes. They seem to be a prosperous and increasing community; for an enumeration of their numbers made in 1863, exhibits a population of over 83,000, exclusive of the Indians. At present new accessions are being made, and new settlers are daily wending their way thither from all quarters of the world.

The chief town of Utah is **Great Salt Lake City**, on the shores of its strange namesake waters. The population here is perhaps 8,000 or 9,000. A magnificent temple is to be erected for the celebration of the rites of the Mormon worship.

Besides Salt Lake City, the other principal settlements are Brownsville, Provo, Ogden, Manti, and Fillmore cities and Parovan.

Fillmore City is the capital of the territory. It is situated on the shores of the Nuquin, a branch of the Nicollet river. It is 1,200 miles west of St. Louis and 600 miles east by north of San Francisco.

Brownsville is on the east side of the Great Salt Lake.

Provo City is about 60 miles south-southeast of Salt Lake City.

Ogden City is 185 miles north of Fillmor. City, the capital of the territory.

Manti is 40 miles east-southeast of the capital.

Parovan is 110 miles south-southwest of Fillmore City.

Utah will no doubt soon seek admission into the Union as a State, and then will come under particular and universal consideration the institution of Polygamy, by the assertion and practice of which as a religious and political tenet, the people are more especially distinguished from those of all other parts of the Republic.

NEW MEXICO.

NEW MEXICO is a portion of the territory ceded to the United States by the treaty with Mexico of 1848 and of 1854. It was organized as a Territory September 9, 1850. Its area is, at present (as reduced by the subsequent formation of new territories), 124,450 square miles. It lies west of Northern Texas, with Colorado on the north.

Like the adjacent country, it is a region of high table-lands, crossed by mountain ranges, and barren to the last degree.

In the eastern part of this Territory are the valleys of the Rio Grande and its tributary waters skirting the base of various chains of the Rocky Mountains; as the Sierra Madre range, the Jumanes, and the Del Cabello. Mount Taylor, among the Sierra Madre, is said to rise 10,000 feet above the valley of the Rio Grande, which is itself a table-land of many thousand feet elevation.

Valuable mineral deposits exist in New Mexico—gold, silver, and other metals—though the resources of the mines have not yet been very much developed.

New Mexico is full of wonderful natural curiosities and beauties, though but a few of its many surprising scenes have been yet explored. Immense canons exist among the mountains of the Sierra Nevada; deep ravines, where rivers flow in darkness hundreds of feet down below the surface of the valleys. Red and white sandstone bluffs, too, abound; grand and lofty perpendicular precipices of rocks, wearing every varying semblance of cliff-lodged castle and fortress.

Waterfalls of surprising beauty are scattered through the mountain fastnesses. The Cascade Grotto is described as a series of falls, which, coming from a mineral spring in the hills, leap from cliff to cliff, a thousand feet down to the Gila below. A wonderful cavern, in which are some curious petrefactions, may be entered beneath the first of these cascades.

Two marvellous falls have been discovered in the Rio Virgen, one of which, 200 miles from its mouth, has a perpendicular descent of 1,000 feet.

The present inhabitants of New Mexico consist chiefly of domesticated nomad Indians, with a sprinkling of Mexicans and Americans. Emigration from the States has not yet turned much in this direction.

The chief towns are Santa Fé, with a population of about 5,000, La Cuesta, St. Miguel, Las Vejas, Zuni, Tuckelata and Mesilla.

Santa Fe.—Hotels:—

Santa Fé is the capital of the Territory. It is situated on the Rio Chicito, or the Santa Fé River, 20 miles from its entrance into the Rio Grande. It is the great depot of the overland trade, which has been carried on for 30 or 40 years past with Missouri. The town is built on a plateau elevated 7,000 feet above the sea, and surrounded by snow-capped mountains, 5,000 feet yet higher. The people are but a miserable set, and their home recommends itself to the stranger scarcely more than they do themselves. The houses here, as elsewhere in the region, are built of dark adobes or unburnt bricks. Each building usually forms a square, in the interior of which is a court, upon which all the apartments open. The only entrance is made of sufficient size to admit animals with their burdens.

Route from Independence City, in Missouri, to Santa Fé.

From Independence City, in Missouri, to the Kansas Boundary	22	
To Lone Elm	7	29
Round Grove	6	35
The Narrows	30	65
Black Jack	3	68
One-hundred-and-ten-mile Creek	32	100
Switzler's Creek	9	109
Dragoon Creek	5	114
Several creeks are then crossed, after which		
Big John Spring	34	148
Council Grove	1	149
Kaw Village and Placeto, in Council Grove	1	150
Sylvan Camp, in Council Grove	2	152
Willow Spring	6	158
Diamond Spring	13	171
Lost Spring	16	187
Cottonwood Fork of Grand River	12	199
Turkey Creek	29	228
Mud Creek	19	247
Little Arkansas	3	250
Cow Creek	20	270
Plum Buttes	14	284
Great Bend of the Arkansas	2	286
The trail then ascends the northern bank of the Arkansas River for 130 miles.		
Walnut Creek	7	293
Pawnee Rock	14	307
Ash Creek	6	313
Pawnee Fork of the Arkansas	6	319
Coon Creek	33	352
Caches	36	388
Old Fort Mann	14	402
Fort Sumner	4	406
Ford of the Arkansas	10	416
Jornado to Sand Creek	49	465
Lower Spring, on the Cimarron	11	476
Middle Spring, on the Cimarron	37	513
Willow Bar	28	543
Upper Spring, on the Cimarron	17	560
Cold Springs	6	566
McNees' Creek	26	592
Rabbit-ear	19	611
Round Mound	8	619
Rock Creek	13	632
Point of Rocks	17	649
Rio Colorado	20	669
Ocate	7	676
Wagon Mound	19	695
Santa Clara Spring	2	697
Fort Barclay, on Rio Mora	22	719
Los Vegas, on Rio Gallinas	19	738
Natural Gate	6	744
Ojo de Bernal	11	755
San Miguel	8	763
Pecos Ruins	26	737
Santa Fe	25	812

COLORADO.

COLORADO was organized as a **Territory March 2, 1861**. Its area is 106,475 square miles. It lies directly west of Kansas, and comprises the western part of the old Territory of Kansas, and portions of the former Territories of Nebraska, New Mexico, and Utah.

The estimated population of Colorado, in 1863, was 55,000, exclusive of 15,000 tribal Indians. This number is continually increasing, as is the population of all the newly-organized territories, by the unwonted emigration which is everywhere following the brilliant attractions of the great gold and silver mines of the region, of which new and more wonderful revelations are being made from week to week.

Golden City, the territorial capital of Colorado, has a population of about 1,000. It is situated at the base of the mountains, 15 miles from Denver.

Denver, the chief town of Colorado, is at the South Fork of the Platte River, 15 miles from the base of the Rocky Mountains. It is a busy and prosperous city, with a present population of about 6,000. It possesses a U. S. branch Mint, daily newspapers, churches, and schools, and other adjuncts of a much older settlement than is any one of the towns in this far-off portion of the national domain.

Central City, Nevada City, and Black Hawk, are settlements in the mountains 40 miles west of Denver. They are in the region of the chief lode mines, known as the Gregory mines. Each of these places is said to contain a population of 5,000.

Empire City, on the north clear creek, 15 miles from Central City, is in a rich lode mining region. Its population is 1,000.

Colorado City is near the base of Pike's Peak, on *Fontaine Que Bouille*, a tributary of the Arkansas. It is 100 miles south of Denver.

Hamilton, Montgomery, and Torry, are situated 100 miles west of Denver among the southwestern lode mines. They are all growing settlements.

Canon City is on the Arkansas, 120 miles south of Denver. Pueblo is 40 miles below Canon City, and 100 S. W. of Denver.

Other thriving settlements are found upon the western slope of the Snowy Range mountains, among the silver mines and the quartz lode mines of that region.

NEVADA.

NEVADA was organized as a Territory on the 2d of March, 1861. Its area is estimated to be 835,000 square miles, lying directly east of California and west of Utah. In 1862 it had a population of 40,000.

Carson City, the Territorial capital, has a population of 2,500.

Virginia, the chief business depot, is said to have had a population of 10,000 in November, 1863. The population of *Silver City* in 1863, was 1,000, and of *Gold Hill*, 1,500. The other principal towns and settlements are, *Washoe City, Ophir, Star City, Unionville, Humboldt, Dayton, Genoa, Aurora,* and *Austin.*

All the Territory of Nevada abounds in rich stores of mineral wealth, including gold, silver, quicksilver, lead, antimony, and other precious metals.

The finest silver deposits in the United States are said to exist in Storey County, in this Territory; and the silver mines of Lander County are reported to have increased the population there several thousands during the lapse of a single year.

The mining region of Nevada is described as an elevated semi-desert country; its surface a constant succession of longitudinal mountain ranges, with intervening valleys and plains, most of which are independent basins, hemmed in by mountains on all sides, and the whole system without drainage to the sea. The general elevation of these valleys is over 4,500 feet above the sea; and the mountains rise from 1,000 to 4,000 feet, and in some instances to 8,000 feet high.

The principal mining districts of Nevada are,

First. The Virginia District, including Virginia City and the eastern slope of Mount Davidson, Cedar Hill, and the upper part of six mile Cañon.

Second. The Gold Hill District south of Mount Davidson, including Gold Hill and the Gold Hill vein.

Third. The Devil's Gate District, in which is embraced Silver City and the mines on the sides of the lower part of Gold Cañon, and beyond the Carson River.

Besides these principal districts, there are the Florence District, the Argentine, the Sullivan, and the Galena. To the southward, the Esmeralda and the Mono regions are divided into two districts. Walker's River District and the Pacific Coal District are on Walker's River. The Prince Royal, the Castle, and the Antelope are names of districts in the Humboldt Mountains region, in the eastern portion of the Territory. New districts are continually being set off as the number of adventurers increase and extend over the land.

DAKOTAH.

DAKOTAH was organized as a territory March 2, 1861. It has an area of 152,000 square miles, lying directly west of Minnesota and of the north western part of Iowa. It has absorbed much of the western part of the old Territory of Minnesota, and of the eastern part of Nebraska.

Yankton, the territorial capital, is on the Missouri River, 65 miles from the Iowa line, and nearly due west from Chicago. The other principal settlements are Big Sioux Point, Elk Point, Maley Creek, Vermillion, Bonhomme, Greenwood, and Fort Randall.

Large quantities of valuable furs and peltries are obtained from this region; and recently, indications have appeared of the existence, in the Black Hills, 300 miles west from Yankton, of good supplies of gold, iron, and coal.

The climate of Dakotah is healthful and genial, and the soil is well suited to agricultural and grazing purposes, being rich in the yield of grain, fruits, and vegetables.

The Yankton and the Ponca Indians, also the Winebago, the Sioux, and the Santee tribes (recently removed from Minnesota), have extensive reservations on the Missouri River and Niobrara above Yankton. They are reported to have become domesticated, and to be devoting themselves to agriculture and stock-raising.

IDAHO.

IDAHO was organized as a Territory March 3, 1863. Its area is estimated to be 333,200 square miles. It is formed from the eastern halves of the old Washington and Oregon Territories, the western half of Nebraska, and a small part of northern Utah. It extends from Utah and Colorado on the south to the British possessions on the north.

It is said that the Indian word Idaho is, in English, "a star;" and again, that it stands for "the gem of the mountains." The Idaho region includes the rich gold fields of Salmon River, a stream of remarkably picturesque beauty, flowing, here and there, between grand perpendicular walls varying in height from 500 to 2,000 feet.

The very recent and rapid settlement of Idaho, commenced within two or three years past, has grown out of the gold discoveries in the neighboring British possessions. These discoveries attracted thousands of adventurers from California, who soon afterwards pushed their explorations toward eastern Oregon and western Idaho. From that period to the present, a steady and increasing torrent of emigration has set thitherward, and the resources of the land are being daily revealed and utilized, both in its mineral stores and its capacities in soil, climate, &c. Settlements are rapidly growing up, roads are being constructed, the waters are navigated, schools and churches are appearing, with all other adjuncts of permanent and progressive civilization.

The mineral resources of Idaho, even as at present known, compare well with the other great mining portions of the great Rocky Mountain region. Gold is found in most of the tributaries of the Missouri and the Yellow Stone. Platina, too, has been obtained in small quantities, while extensive deposits of this valuable metal are supposed to exist. Copper, and iron, and salt are abundant; and coal is found upon the Pacific slope of the Rocky Mountains, and on the upper Missouri and Yellow Stone rivers.

The climate of the Idaho region is bleak in the mountain ranges, but mild and agreeable in the valley districts.

ARIZONA.

ARIZONA was organized as a Territory February 24, 1863. Its area is estimated to be 130,800 square miles. It is formed from a portion of the old Territory of New Mexico, and is bounded by Mexico on the south, on the west by the Colorado River and California, and on the north by Utah and Nevada.

The Commissioner of the General Land Office, in his Report to Congress in December, 1863, says of this region, that it "is believed to be stocked with mineral wealth beyond that of any other Territory of equal extent in the great plateau between the Rocky Mountains and the Sierra Nevada."

No census of the population of Arizona has yet been made.

Montana is the name of a new Territory which is being organized as we now write.

INDEX.

A

Aberdeen, Miss., 199.
Acapulco, Harbor of, (Voyage to San Francisco from New York,) 252.
Adams, Mass., 60.
Adirondack Mountains, N Y., 115.
Alabama, State of, 194.
Alabama, the Hill Region, 195.
Alabama River, 194.
Alabama, Mineral Springs, 195.
Alabama, Railways, 195.
Alabaster Mountain, Ark., 213.
Albany, N. Y., 94.
Alexandria, Va., 148.
Allentown, Pa., 129.
Alton, Ill., 234.
Almanac, The Traveller's, 23.
Amherst, Mass., 56.
Amicalolah Falls, Ga., 192.
Andover, 54.
Androscoggin River, Me., 45.
Annapolis River, N. S., 40.
Annapolis, Md., 138.
Ann Arbor, Mich., 237.
Appalachicola, Fla., 183
Arkansas, State of, 212.
        Rivers, 213.
        Towns, 213.
Arkansas River, 213.
Arizona Ter., 276.
"Ashland," Home of Clay, Lexington, Ky., 219.
Aspinwall, 251.
Astoria, L. I., 81.
Astoria, Or., 265.
Athens, Ga., 189.
Atlanta, Ga., 189.
Atlantic City, N. J., 122.
Auburn, N. Y. Central Railway, 109.
Auburn, Cal., 257.
Augusta, Me., 45.
Augusta, Ga., Routes thence, 188.
Austin, Tex., 212.
Avon Springs, N. Y., 117.

B

Ballston Springs, 103.
Baltimore and Vicinity, 136.
        Cemeteries, 137.
        Churches, 137.
Baltimore Hotels, 136.
        Monuments, 136.
        Public Edifices, 137.
        Theatres, 137.
Baltimore and Ohio Railway, 140.
Baltimore to Philadelphia, 123.
Baltimore to Washington, 140.
Bangor, Me., 44.
Batesville, Ark., 214.
Bath, Me., 45.
Baton Rouge, La., Home of General Taylor, 204.
Battle of the Brandywine, 123.
Battle of Guilford Court House, N. C., 167.
Battle of Long Island, 118.
Battle of Trenton, 121.
Battle of Wyoming, 134.
Beaufort, S. C., 172.
Belfast, Me., 45.
Bellows Falls, Vt., 57.
Beloit, Wis., 246.
Bethlehem, Pa., 130.
Bennington, Vt., Battle of Bennington, 74.
Beverly, Mass., 53.
Beverly, on the Delaware, 122.
Big Black River, Ala., 198.
Binghamton, N. Y. and Erie Railroad, 98.
Birthplace of Henry Clay, 154.
Birthplace of Washington, 153.
Black Mountain, N. C., 167.
Black Warrior River, Ala., 195.
Blowing Cave, Va., 163.
Blue Lick Springs, Ky., 220.
Boston, Routes from New York, 48.
Boston, City of, 50.
        Cambridge and Harvard, 52.
        Churches, 52.
        The Common, 51.
        East Boston, 50.
        Faneuil Hall, 51.
        History, 50.
        Hotels, 50.
        Public Edifices, 51.
        South Boston, 50.
        State House, 51.
        Theatres, 52.
        The Vicinage, 52.
Bordentown, N. J., 121.
Brandon, Vt., 74.
Brattleborough, Vt., 57.
Brazos River, Tex., 210.
Bridgeport, Conn., 48.

INDEX.

Brighton, near Boston, 54.
Bristol, 121.
British America, 25.
Brooklyn, N. Y., 82.
    Atlantic Dock, 82.
    Churches, 83.
    Ferries from New York, 83.
    Hotels, 82.
    Public Buildings, 83.
    U. S. Navy Yard, 82.
Brown University, Providence, R. I., 61.
Brownsville, Tex., 212.
Brunswick, Me., 45.
Budd's Lake, N. J., 122.
Buffalo, N. Y., 110.
Buffalo Hunting, Nebraska Terr., 271.
Bull's Ferry, Hudson River, 85.
Bunker's Hill, Boston, 52.
Burlington, Vt., 107.
Burlington, N. J., 122.
Burlington, Io., 242.

C.

Caledonia Springs, Canada, 29.
California, History and Topography, 250.
California, Voyage up the Coast, 264.
California, Tables of Routes from San Francisco, Sacramento City, Stockton, and Marysville, to all other points in the State, 265.
Camden and Amboy Railway, 121.
Camden, opposite Philadelphia, 122.
Camden, S. C., 175.
Camden, Ark., 214.
Camel's Hump Mountain, Vt., 72.
Campton and West Campton, N. H., 67.
Canandaigua Lake, N. Y., 110.
Canandaigua, N. Y. Central Railway, 110.
Canada—Its Geography and Area: Discovery, Settlement, and Rulers; Government; Religion; Landscape; Mountains; and Rivers, 26.
Canada Railways, 29.
Cape Cod and the Sea Islands, 55.
Cape Girardeau, Mo., 241.
Cape May, N. J., 122.
Carlisle, Pa., 134.
Carondelet, Mo., 241.
Carquinez, Straits of, Cal., 254.
Carson Valley, 258.
Cascade Bridge, N. Y. and Erie Railroad, 97.
Cascade Range, Oregon, 265.
Castine, Me., 45.
Catawissa, on the Susquehanna, Pa., 134.
Catskill Village, Hudson River, 93.
Catskill Mountains (the), Routes thither, 100.
    The High Falls, 101.
    The High Peak, 102.
    The Mountain House, 100.
    The Plauterkill Clove, 102.
    The Stony Clove, 102.
Catskill Mountains (the), The Two Lakes, 101.
Caudy's Castle, Va., 164.
Cayuga, N. Y. Central Railway, 109.
Cayuga Lake, N. Y., 109.
"Cedarlawn," Home of J. T. Headley, 91.
Cedarmere, L. I., Home of Bryant, 118.
Centre Harbor, N. H., 66.

Charleston, S. C., Routes, 170.
Charleston, S. C., Description of, 170.
Charlestown, N. H., 58.
Charlotte, N. C., 167.
Charlotte, N. Y., 85.
Charlottesville, Va., 156.
Chattahoochee River, Ga., 186.
Chattanooga, Tenn., 216.
Chaudiere Falls, Quebec, 33.
Chelsea Beach, 58.
Cherry Valley, N. Y., 117.
Cheraw, S. C., 175.
Chesapeake Bay, Md., 138.
Chicago, Ill., 233.
Chilicothe, O., 227.
Cincinnati, O., Routes thence, and Hotels, 224.
    "North Bend," Home of General Harrison, 226.
    "Over the Rhine," 226.
    Public Buildings, Churches, Theatres, 224.
    Residence of Mr. Longworth, 226.
    Suspension Bridge, 225.
    The Vicinage, 226.
Claremont, N. H., 58.
Clarendon Springs, Vt., 73.
Clarksville, Ga., 190.
Cleveland, O., 226.
Coburg, Canada, 35.
Cohassett, Mass., 54.
Colorado Territory and Towns, 274.
Columbia Springs, near Hudson, N. Y., 93 and 117.
Columbia, Pa., 134.
Columbia, S. C., 175.
Columbia, Tenn., 216.
Columbia, Cal., 60.
Columbia River, Oregon Territory, 268.
Cold Spring, Hudson River, 91.
Columbus, Ga., 189.
Columbus, Miss., 199.
Columbus, O., 227.
Coney Island, 81.
Concord, Mass., 54.
Concord, N. H., 65.
Connecticut, State of, 59.
Connecticut River and Railways, 55.
Conway Valley, N. H., 66.
Cooperstown, N. Y., 117.
Cooper's Well, Miss., 199.
Cornwall Landing, Hudson River, 91.
Corning, N. Y., 99.
Covington, Ky., 219.
Coultersville, Cal., 260.
Cowpens, Battle-field, S. C., 178.
Cozzens' Hotel, Hudson River, 89.
Crescent City, Cal., 264.
Croton Falls, Harlem Railway, 103.
Cumberland, Md., 138.
Cumberland River, Ky., 217.
Cumberland Gap, Ky., 221.
Currahee Mountain, Ga., 192.

D

Dartmouth College, Hanover, N. H., 58.
Davenport, Io., 243.

INDEX. 279

Dayton, O., 227.
Dacotah Territory, 275.
Denver, Cal., 275.
Deerfield Mountain, Mass., 57.
Delaware, State of, 123.
Delaware River, 135.
Delaware Water Gap, Pa., 135.
Downieville, Cal., 257.
Detroit, Mich., 236.
Des Plaines River, Ill., 232.
Devil's Pulpit, Ky., 229.
Diamond Cave, Ky., 222.
District of Columbia, 166.
Dobbs' Ferry, Hudson River, 85.
Dover Plains, Harlem Railway, 103.
Drennon Springs, Ky., 220.
Dubuque, Io., 242.
Dunkirk, N. Y. & Erie R. R., 99.

E

Eastern Shore of Maryland and Virginia, 138.
Eastport, Me., 45.
Eastatoia Falls, Ga., 191.
Elgin Springs, Vt., 73.
Elizabeth, N. J., 120.
Ellicott's Mills, Md., 138.
Elmira, N. York & Erie R. R., 98.
Enterprise, Fla., 181.
Epsom Salts Cave, Ind., 229.
Erie (N. Y. &) Railway, 95.
Esculapia Springs, Ky., 220.
Evansville, Ind., 230.
Eutaw Springs, S. C., 175.

F

Fall River Route from New York to Providence and Boston, 49.
Fall River, Mass., 50.
Falls Village, Conn., 59.
Falls of the Passaic, N. J., 122.
Falls of St. Anthony, Mississippi River, 248.
Fayetteville, N. C., 167.
Fillmore City, Utah, 273.
Fishkill, Hudson River, 92.
Flatbush, L. I., 81.
Flatlands, L. I., 81.
Flint River, Ga., 186.
Florida, 179.
Flushing, L. I., 81.
Fond du Lac, Wis., 246.
Folsom, Cal., 258.
Fort Hamilton, N. Y., 81.
Fort Hill, S. C., Home of Calhoun, 177.
Fort Des Moines, Io., 243.
Fort Lee, Hudson River, 85.
Fort Madison, Io., 243.
Fort Moultrie, S. C., 170.
Fort Plain, Central Railway, N. Y., 109.
Fort Smith, Ark, 214.
Fort Snelling, Min., 248.
Fort Tejon, Cal., 262.
Fort Washington, Hudson River, 85.
Fort Wayne, Ind., 230.

Fortification Rock, Wis., 244.
Fountain Cave, Min., 248.
Fox River, Ill., 232.
Fox and Phillips Springs, Ky., 220.
Frankfort, Ky., 219.
Frederickton, N. B., 39.
Frederick, Md., 138.
Fredericksburg, Va., 153.
French Broad River, N. C., 168.
Fulton, Ill., 234.

G

Galena, Ill., 233.
Galveston, Tex., 211.
Gardiner, Me., 45.
Gates of the Rocky Mountains, Missouri River 238.
Genesee Falls, N. Y., 110.
Geneva, N. Y. Central Railway, 110.
Genoa, Nev. Ter., 259.
Georgia, State of, 184.
Georgia Railway Routes, 186.
Georgia, the Mountain Region, 189.
Georgia, Mountain Accommodations, 191.
Georgetown, D. C., 148.
Georgetown, S. C., 175.
Ginger Cake Rock, N. C., 167.
Glenn's Falls, N. Y., 104.
Grand Haven, Mich., 237.
Grand Trunk Railway—from Portland north, 46.
Gravesend, L. I., 81.
Great Barrington, Mass., 59.
Great Lakes, the, 36.
Great Bend, N. Y. & Erie R. R., 98.
Great Salt Lake, Utah, 272.
Great Salt Lake City, Utah, 272.
Great Falls of the Missouri, 238.
Grass Valley, Cal., 257.
Green Bay, Wis., 246.
Greenfield, Mass., 57.
Greenwood Lake, N. Y. & N. J., 122.
Green River, Ky., 17.
Greenville, S. C., 176.
Greenwood Cemetery, N. Y., 81.

H

Hagerstown, Md., 138.
Halifax, N. S., 41.
Hallowell, Me., 45.
Hamilton, Canada, 36.
Hampton, 53.
Hancock, N. Y. & Erie R. R., 97.
Hanging Rocks, Va., 164.
Hanover, Va., 154.
Hanover, N. H., 58.
Hannibal, Mo, 241.
Harlem Railway, 102.
Harrodsburg, Ky., 220.
Harrodsburg Springs, Ky., 220.
Harper's Ferry, Va., 141 and 153.
Harrisburg, Pa., 130.
Hartford, Ct., 49.
Harvard University, 52.
Hastings, Hudson River, 86.

280 INDEX.

Haverstraw, Hudson River, 87.
Havre-de-Grace, Md., 124.
Hawk's Bill, N. C., 167.
Hawk's Nest, Va., 163.
Hermitage, Home of General Jackson, 216.
Hiawassee Falls, Ga., 192.
Hickory Nut Gap, N. C., 167.
Highlands of the Hudson, 89.
Hingham, Mass., 54.
Hoboken, N. J., 81.
Holly Springs, Miss., 199.
Hornellsville, N. Y. & Erie R. R., 99.
Housatonic River, Valley, and Railway, 59.
Houston, Tex., 211.
Hudson River, description of, 84.
Hudson River Railway, 84.
Hudson River Steamboat Routes, 84.
Hudson, Hudson River, 93.
Huntsville, Ala., 195.
Hyde Park, on the Hudson, 92.

I

Ice Mountain, Va., 164.
Idaho Territory, 276.
"Idlewild," Home of N. P. Willis, 91.
Illinois, State of, 231.
Illinois Prairie, Grand Prairie, 231.
Illinois River, 232.
Indian Springs, Ga., 193.
Indiana, Rivers, Railways, Towns, etc., 228.
Indianapolis, Ind., 229.
Indian Territory—extent, character, and inhabitants, 276.
Introduction, Advice to Travellers, 5.
Iowa, State of, 241.
Iowa River, 242.
Iowa City, 242.
Irvington and "Sunnyside," on the Hudson, 86.
Isle of Shoals, Portsmouth, N. H., 53.

J

Jacksonville, Fla., 181.
Jackson, Miss., 198.
Jackson, Tenn., 216.
Jamaica, L. I., 82.
Jamestown, Va., 153.
Janesville, Wis., 245.
Jefferson City, Mo., 240.
Jersey City, N. J., 82.
Jocassee Valley, S. C., 177.
Juniata River, Pa., 132.

K

Kansas—character and settlement, 269.
Routes thither, 270.
Kansas River, 270.
Kansas Towns and Villages, 270.
Kankakee River, Ill., 232.
Katahdin Mt., Maine, 44.
Keesville, N. Y., 116.
Kentucky, State of, 217.

Kentucky Caves, 221.
Kentucky Railways, 218.
Kentucky Rivers, 217.
Kentucky Sink Holes, 221.
Kentucky Springs, 220.
Kentucky Towns, 218.
Kentucky River, 317.
Kenosha, Wis., 245.
Kennebec River, Me., 45.
Keokuk, Io., 243.
Keowee River, S. C., 177.
Key West, Fla., 184.
Killington Peak, Vt., 73.
Kinderhook on the Hudson, 93.
Kingston, Canada, 35.
Kingston, on the Hudson, 92.
King's Mountain, Battle-field, S. C., 178
Knob Lick, Ky., 220.
Knoxville, Tenn., 216.

L

La Fayette, Ind., 230.
La Chine, Canada, 34.
Lake Castleton, Vt., 74.
Lake Champlain, N. Y., 100.
Lake Dunmore, Vt., 73.
Lake George; Routes: Caldwell, Bolton, The Narrows, Sabbath Day Point, Rogers' Slide, Ticonderoga, 104.
Lake Mahopac, 103 and 116.
Lake Pleasant, Region of, N. Y., 116.
Lake Ontario, 35.
Lancaster, Pa., 130.
Lansing, Mich., 237.
Lauderdale Springs, Miss., 199.
Lebanon Springs, N. Y., 60 and 117.
Lebanon, Tenn., 216.
Lehigh River, 125.
Lettonian Springs, Ky., 220.
Lewiston, Me., 47.
Lexington, Mass., 54.
Lexington, Va., 155.
Lexington, Ky., 219.
Lexington, Mo., 241.
Licking River, Ky., 317.
Little Falls, Central Railway, N. Y., 109.
Little Rock, Ark., 213.
"Lindenwold," Home of Martin Van Buren, 93.
London, Canada, 34.
Long Branch, N. J., 81.
Long Island, N. Y., and Battle of, 118.
Look-out Mountain, Ga., 192.
Los Angeles, Cal., 268.
Louisiana, State of, 199.
  Railways, 200.
Louisville, Ky., 218.
Lowell, 54.
Lynn, Mass., 53.
Lynchburg, Va., 155.

M

Mackinac, the Straits of, 37.
Macon, Ga., 189.

INDEX.

Madison's Cave, Va., 163.
Madison, Ind., 230.
Madison, Wis., 245.
Madison Springs, Ga., 193.
Mahopac, Lake, 116.
Maine, State of, 43.
Manitowoc, Wis., 246.
Mammoth Cave, Ky., 221.
Mansfield Mountain, Vt., 73.
Maryland, State of, 185.
Marshfield, Mass., 54.
Martha's Vineyard, 54.
Mariposa, Cal., 261.
Massachusetts, State of, 47.
Mauch Chunk, Pa., 130.
Maumee River, Ind., 229.
Maysville, Ky., 220.
Memphremagog, Lake, 74.
Memphis, Tenn., 216.
Mendota, Minn., 248.
Miami River, O., 223.
Michigan, State of, 285.
Middleborough, Mass., 54.
Middlebury, Vt., 74.
Middleburg, Fla., 181.
Milwaukee, Wis., 245.
Milledgeville, Ga., 189.
Minnesota—Its Area, Surface, Soil, Forest Land, Rivers, &c., 246.
Mississippi, State of, 197.
  Railways, 198.
  Rivers, 198.
  Towns, 198.
  Watering Places, 199.
Mississippi River—Description of, and Table of Distances, 204.
Missouri, State of, 237.
  River, 238.
Mobile, Ala., Routes thence, 195.
Mokelumne Hill, Cal., 259.
Montana Territory, 276.
Montreal, Canada, Routes thither, Hotels, Public Buildings, Vicinity, 30.
Monticello, Home of Jefferson, 156.
Montgomery, Ala., 196.
Monterey, Cal., 264.
Montmorenci Falls, Canada, 33.
Montpelier, Vt., 73.
Moosehead Lake, 44.
Mount Ascutney, Vt., 58.
Mount Auburn Cemetery, 52.
Mount Desert Island, 45.
Mount Hope, Narraganset Bay, R. I., 62.
Mount Holyoke, Mass., 55.
Mount Independence, Lake Champlain, 106.
Mount St. Vincent, Hudson River, 85.
Mount Tom, Mass., 56.
Mount Toby, Mass., 57.
Mount Vernon—Home and Tomb of Washington, 143.
Mount Warner, 57.
Munroe City, Mich., 237.
Murfreesborough, Tenn., 216.
Murphy's, Cal., 260.
Muscle Shoals, Ala., 195.
Muscatine, Io., 243.
Muskingum River, O., 223.

N

Nacoochee Valley, Ga., 191.
Nahant, 52.
Nanticoke, Pa., 133.
Nantucket, 54.
Nantasket Beach, 53.
Napoleon, Ark., 214.
Narraganset Bay, R. I., 62.
Narrowsburg, N. Y. and Erie R. R., 97.
Nashville, Tenn., 215.
Natural Bridge, Va., 160.
Natural Bridge, Ala., 195.
Natural Bridge, Ky., 221.
Natural Bridge, Ark., 213.
Natchez, Miss., 199.
Nauvoo, Ill., 234.
Nazareth, Pa., 130.
Nebraska Territory, 271.
Nevada, Cal., 257.
Nevada Territory, 275.
New Albany, Ind., 230.
Newark, N. J., 119.
New Bedford, Mass., 54.
Newbern, N. C., 167.
Newburgh, on the Hudson, 91.
New Brunswick, General Description of, 38.
New Brunswick, N. J., 120.
Newburyport, Mass., 53.
Newcastle and Frenchtown Railway, 123.
New Hampshire, State of, 63.
New Haven, Ct., 49.
New Haven, Hartford, and Springfield Route from New York to Boston and Providence, 48.
New Jersey, State of, 119.
New Jersey Railway, 119.
New Madrid, Mo., 241.
New Mexico, Territory of, and Routes to, 273.
New Orleans, La., 200.
  Battle of, 203.
  Cemeteries, 202.
  Churches, Public Edifices, &c., 201.
  Hotels, 200.
  Theatres, &c., 201.
  The Levee, 203.
  The Markets, 203.
  Panorama of the City, 204.
  Pere Antoine's Date Palm, 240.
Newport, R. I., 62.
Newport, Ky., 219.
New Windsor, 91.
New York, State of, 74.
New York City, 75.
  The Arsenal, 80.
  Art Societies, 79.
  Artists' Studios, 79.
  Bloomingdale, 81.
  Churches, 80.
  Croton Aqueduct, 80.
  First-class Business Houses, 80.
  Fort Hamilton, 81.
  Greenwood Cemetery, 81.
  Harlem, 81.
  High Bridge, 80.
  Hotels, 75.
  Literary Institutions and Libraries, 78.

INDEX.

New York City, Manhattanville, 81,
    N. Y. Bay Cemetery, 82.
    Panorama from Trinity Church, 76.
    Public Buildings, 78.
    Public Parks and Squares, 77.
    Theatres and Places of Amusement, 80.
New York to Albany and Troy, 83.
New York to Boston—Route, 43.
New York to Buffalo, N. Y., 109.
New York to Buffalo and Niagara Falls, 90.
New York to the Catskill Mountains, 100.
New York and Erie Railway, 95.
New York to Lake Erie, 95.
New York to Montreal, via Lake Champlain, 106.
New York to Philadelphia, 119.
New York to Trenton Falls, 107.
Niagara Falls. Routes—Goat Island, the Rapids, Chapin's Island, the Toll Gate, the Cave of the Winds, Luna Island, Sam Patch's Leap, Biddle's Stairs, Prospect Tower, the Horse-shoe Fall, Gull Island, Grand Island, the Whirlpool, the Devil's Hole, Chasm Tower, the Maid of the Mist, the Great Suspension Bridge, Bender's Cave, the Clifton House, Table Rock, Termination Rock, 110.
Nickajack Cave, Ga. and Ala., 195.
Norfolk, Va., 152.
North Carolina, General Remarks, Railways, &c., 164.
North Carolina, Mountain Region, 167.
Northampton, Mass., 55.
North Point, and Battle of, 138.
Northumberland, on the Susquehanna, Pa., 134.
Norwalk, Ct., 48.
Norwich Route, from New York to Boston, &c., 48.
Nova Scotia, Description of, 40.
Nyack, Hudson River, 86.

O

Ockmulgee River, Ga., 186.
Oconee River. Ga., 186.
Oglethorpe University, Ga., 189.
Ohio, State of, 222.
    Rivers, 223.
    Railways. 223.
    Cities and Towns, 224.
Ohio River, Description and Distances, 207.
Olympia, Wash. Ter., 271.
Omaha City, Nebraska Ter., 271.
Orangeburg, S. C., 175.
Oregon, History, Topography, Rivers, Mountains, Towns, &c., 267.
Oregon City, 268.
Oroville, Cal., 255.
Oregon, Routes from St. Louis, 265.
Otsego Lake, N. Y., 117.
Ottawa River, Canada, Description of, 26.
Otter Creek Falls, Vt., 73.
Oswego, N. Y., 85.
Owego, N. Y. and Erie R. R., 98.

P

Paducah, Ky., 220.
Palatine Bridge, Central Railway, N. Y., 109.

Palisades, on the Hudson, 84.
Palmyra, Mo., 241.
Panama, 252.
Paroquet Springs, Ky., 220.
Passamaquoddy Bay, 46.
Patapsco River, 136.
Peaks of Otter, Va., 161.
Pearl River, Ala., 198.
Peekskill, Hudson River, 88.
Pendleton, S. C. 177.
Pennsylvania, State of, 125.
Pennsylvania Coal Region, 132.
Pennsylvania Railway, 131.
Penobscot River, Me., 44.
Pensacola, Fla., 183.
Peoria, Ill., 234.
Peru, Ill., 234.
Petenwell Peak, Wis., 244.
Petersburg, Va., 153.
Phillip's Beach, 53.
Piccolata, Fla., 181.
Pickens Court House, S. C., 177.
Piermont, Hudson River, 86.
Pilatka, Fla., 181.
Pilot Mountain, N. C., 167.
Pilot Mountain, Ga., 192.
Pilot Knob, Mo., 249.
Pittsfield, Mass., 60.
Pittsburg, Pa., 132.
Philadelphia and Vicinity, 126.
    Art Societies, 128.
    Benevolent Institutions, 128.
    Cemeteries, Laurel Hill, &c., 129.
    Churches, 127.
    Fairmount Waterworks, 129.
    Falls of the Schuylkill, 129.
    Germantown, 129.
    Hotels, 126.
    Literary Institutions, 128.
    Manayunk, 129.
    Medical Colleges, 128.
    Places of Amusement, 129.
    Prisons, 129.
    Public Edifices, 127.
    Public Squares, 126.
    The Schuylkill Viaduct, 129.
    Wissahickon Creek, 129.
Philadelphia to Baltimore, 123.
Philadelphia, Wilmington and Baltimore R. R., 123.
Philadelphia to Pittsburg and the West, 131.
Phillips' Beach, 53.
"Placentia," home of J. K. Paulding, 92.
Placerville, Cal., 258.
Plattsburg, N. Y., and Battle of Lake Champlain, 107.
Plantagenet Springs, Canada, 29.
Platteville, Wis., 245.
Plymouth, Mass., 54.
Plymouth, N. H., 66.
Popular Mountain Springs, Ky., 220.
Portage City, Wis., 246.
Port Clinton, Pa., 129.
Port Kent, Lake Champlain, 107.
Port Jervis, N. Y. and Erie R. R., 96.
Portland, Me., 46.
Portland, Oregon, 269.

INDEX. 283

Portsmouth, N. H., 58.
Portsmouth, Va., 152.
Portsmouth, O., 228.
Portage, N. Y. and Erie R. R., 99.
Potomac River, 135.
Pottstown, Pa., 129.
Pottsville, Pa., 129.
Poughkeepsie, Hudson River, 92.
Powder Springs, Ga., 194.
Preface, 5.
Prescott, Canada, 34.
Princeton, N. J., 120.
Princeton College, N. J., 120.
Providence and Vicinity, 61.

Q.

Quebec, Canada, Description of, Routes, Hotels, Objects of Interest in the City and Vicinity, 32.
Quincy, near Boston, 54.
Quincy, Ill., 234.

R.

Racine, Wis., 246.
Rahway, N. J., 120.
Raleigh, N. C., 166.
Ramapo Valley, N. Y. and Erie R. R., 95.
Reading, Pa., 129.
Red Bank, N. J., 81.
Red Bluff, Cal., 256.
Red Sulphur Springs, Ga., 193.
Rhode Island, Historical and Topographical mention of, 61.
Richfield Springs, N. Y., 118.
Richmond, Va., 151.
Richmond, Ind., 231.
Rideau Falls, Canada, 27.
Rio Grande, Texas, 210.
Rivers of Alabama, 194.
Rochester, N. Y. Central Railway, 110.
Rock Mountain, Ga., 192.
Rocky Point, Narraganset Bay, R. I., 62.
Rockaway, L. I., 81.
Rock Island, Ill., 234.
Rockland Lake, Hudson River, 87.
Rock River, Ill., 232.
Rondout on the Hudson, 92.
Rouse's Point, Lake Champlain, 107.
Rowland Springs, Ga., 193.
Rutland, Va., 73.

S.

St. Anne's Falls, Canada, 33.
St. Andrews, N. B., 39.
St. Anthony, Min., and the Falls of St. Anthony, 248 and 250.
San Antonio, Tex., 212.
St. Augustine, Fla., 182.
Santa Barbara, Cal., 263.
St. Catherines, Canada, 29 and 36.
St. Charles City, Mo., 241.

St. Croix Falls and Rapids, Min., 249.
San Diego, Cal., 263.
Santa Fé, New Mex., 274.
St. Francis, on Red River, Ark., 213.
San Francisco, Cal., from New York, 253.
San Francisco, Cal., 253.
St. Genevieve, Mo., 241.
San Jose, Cal., 262.
San Andreas, Cal., 260.
Santa Cruz, Cal., 264.
St. Joseph, Mo., 240.
St. Johns River, N. B., 38.
St. Johns, N. B., 39.
St. Johns River, Fla., 181.
St. Lawrence River; Thousand Islands; 26 and 34.
St. Leon Springs, Canada, 29.
St. Louis, Mo., 239.
St. Louis River, Wis., 244.
St. Mary, Straits of, 37.
St. Marys, Ga., 183.
St. Marks, Fla., 183.
San Pablo, Straits of, Cal., 254.
St. Paul, Min., 249.
Sacketts Harbor, N. Y., 35.
Sacramento River, Cal., 255.
Sacramento City, Cal., 255.
Saddle Mountain, Mass., 47.
Saguenay River, Canada, Routes thither and description of, 28.
Salem, Mass., 53.
Salisbury Beach, 53.
Salisbury Lakes, Ct., 59.
Salt Pond, Va., 164.
Salt River, 317.
Sandusky City, O., 228.
Sangamon River, Ill., 232.
Saranac Lakes, (the,) N. Y., 115.
Saratoga Springs, 103.
Savannah, City of, 187.
    Cemetery of Bonaventure, 187.
    Jasper's Spring, 188.
Savannah River, Scene of the invention of the Cotton Gin, Alligators, etc., 185.
Schooley's Mountain, N. J., 122.
Schenectady, N. Y., 109.
Schuylkill River, 126.
Schuylkill Haven, Pa., 129.
Scioto River, O., 223.
Sebago Pond, Maine, 44.
Seneca Lake, N. Y., 110.
Sierra Nevada Mountains, Cal., 257.
Sing Sing, Hudson River, 87.
Sioux River and Rapids, Min., 249.
Shaker Village, N. Y., 117.
Sharon Springs, 117.
Shasta City, Cal., 256.
Sheboygan, Wis., 246.
Sheffield, 59.
Shrewsbury, N. J., 81.
Shickshinney on the Susquehanna, Pa., 134.
Skaneateles Lake, N. Y., 109.
Skaneateles, N. Y. Central Railway, 109.
Skeleton Tours, 9.
Slicking Falls, S. C., 177.
Sonora, Cal., 260.
South Amboy, 121.
South Carolina, 168.

284   INDEX.

South Carolina, Railway Routes, 169.
          Seaboard and Lowlands, 172.
          Mountain villages and scenery, 176.
Spartanburg, S. C., 177.
Springs in New York, 118.
Springfield, Mass., 49.
Springfield, O., 228.
Springfield, Ill., 233.
Springs in Canada:—The Caledonia, the Plantagenet, the St. Leon, and St. Catherines. Routes thither, 29.
Squam Lake, N. H., 66.
Staten Island, N. Y., 77 and 81.
Starrucca Viaduct, N. Y. & Erie R. R., 97.
Staunton, Va., 156.
Steubenville, O., 228.
Stillwater, Min., 249.
Stone Mountain, Ga., 192.
Stockbridge, Mass., 50.
Stockton, Cal., 261.
Stonington Route from New York to Providence and Boston, 49.
Stonington, Conn., 49.
Stony Point, Hudson River, 86.
Sugar Loaf Mountain, Mass., 56.
Sugar Loaf Mountain, Maine, 44.
Sulphur Springs, Ga., 193.
Sunbury, Pa., 134.
Superior, Lake, 37.
Susquehanna Depot, N. Y. & Erie R. R., 98.
Susquehanna River, 133.
Swannanoa Gap, N. C., 167.
Syracuse, N. Y. Central Railway, 109.

T.

Table Mountain, S. C., 176.
Table Rock, N. C., 167.
Tallulah Falls, Ga., 191.
Tallahassee, Fla., 183.
Tampa, Fla., 184.
Tarrytown, Hudson River, 86.
Tar & Breckenridge White Sulphur Springs, Ky., 220.
Tar and Sulphur Springs, Ky., 220.
Taugkanic Mountains, Mass., 59.
Taunton, Mass., 54.
Tea Table, Va., 164.
Tennessee, State of, 214.
Tennessee Railways, 215.
Tennessee River, Tenn., 214.
Tennessee Towns, 215.
Tennessee Caves and Mounds, 216.
Terre Haute, Ind., 230.
Texas, State of, 209.
Texas, Landscape of, 210.
Texas Rivers, 210.
Texas Railways, 211.
Texas Towns, 211.
Texas Wild Animals and Birds, 211.
Thundering Springs, Ga., 193.
Ticonderoga, Fort, 105.
Tinton Falls, N. J., 81.
Toccoa Falls, Ga., 190.
Toledo, O., 228.
Tomb of the Mother of Washington, 154.
Tombigbee River, Ala., 194.

Toronto, Canada, 85.
Tour up the St. Lawrence, 34.
Tour to the great Lakes, Toronto to Collingwood, Mackinac, Sault St. Marie, Lake Superior, 36.
Tuwalaga Falls, Ga., 193.
Track Rock, Ga., 192.
Trenton Falls, N. Y., 107.
Trenton, N. J., 120.
Troy, N. Y., 94.
Tuscaloosa, Ala., 196.

U.

"Undercliff," Home of George P. Morris, 91.
United States—Extent and Population, 42.
Union College, Schenectady, N. Y., 109.
University of Virginia, 156.
Umbagog Lake, 44.
Utah Territory—Character, Climate, Lakes, Natural Wonders, Settlements, Mormons, 272.
Utica, N. Y. Central Railway, 109.

V.

Van Buren, Ark., 213.
Valley Forge, Pa., 129.
Valley of Wyoming, Pa., 133.
Vermont, State of, 72.
Vermont Central Railway, 72.
Vernon, Vt., 57.
Verplanck's Point, Hudson River, 87.
Vicksburg, Miss., 199.
Virginia—Historical and Poetical Memories, Eminent Men, Natural Beauties, Mineral Springs, Railway Routes, etc., 149.
Virginia Springs, 157.
    Alleghany Springs, 160.
    Bath Alum Springs, 159.
    Berkeley Springs, 159.
    The Blue Sulphur, 158.
    The Red Sulphur, 158.
    Capon Springs, 159.
    Dibrell's Springs, 160.
    Fauquier White Sulphur, 159.
    Grayson's Sulphur, 160.
    Healing Springs, 159.
    Hot Springs, 159.
    Huguenot Springs, 160.
    New London Alum, 160.
    Pulaski Alum Springs, 160.
    Rawley's Springs, 160.
    Red Sweet Springs, 159.
    Rockbridge Alum, 159.
    Routes, 157.
    Salt Sulphur, 158.
    Shannondale Springs, 159.
    Sweet Springs, 159.
    Warm Springs, 159.
    White Sulphur, 157.
    City, Nevada, Ter., 259.
Visalia, Cal., 262.

W.

Wabash River, Ind., 229.
Walled Banks of the Ausable, N. Y., 107.
Walhalla, S. C., 177.

Wapwollopen, on the Susquehanna, Pa.., 134.
Washington City—Description, National Edifices, Municipal Buildings, Hotels, Mount Vernon, etc., 146.
Washington Territory—Physical Aspect, Settlements and Towns, 271.
Warm Springs, N. C., 168.
Warm Springs, Ga., 193.
Water Falls in Canada—Niagara, Montmorenci, Chaudiere, on the Ottowa, the Chaudiere near Quebec, Rideau, Shawanegan, St. Anne's, 29.
Watering Places in Georgia, 193.
Watering Places in Alabama, 195.
Waterville, Me., 45.
Watertown, Wis., 246.
Waukesha, Wis., 245.
Weehawken, N. J., 81.
Weir's Cave, Va., 162.
Welaka, Fla., 181.
Wells River, 58, 67.
Wenham, Mass., 53.
West Point, Hudson River, 90.
Weston, Mo., 241.
Wheatland, near Lancaster, Pa.—Home of James Buchanan, 130.
Wheeling, Va., 157.
White Mountains, N. H., Routes thither, 64-68.
    The Ammanoosuc River, 70.
    The Basin, 71.
    Cannon Mountain, 71.
    The Crystal Falls, 70.
    The Devil's Den, 70.
    Description of routes, 64.
    Dixville Notch, 72.
    Eagle Cliff, 71.
    Echo Lake, 71.
    The Flume, 71.
    The Franconia Hills, 70.
    Glen House, 67.
    Great Notch, 68.
    Hotels, 68.
    Mount Lafayette, 71.
    Mt. Washington, 69.
    Oakes' Gulf, 70.
    The "Old Man of the Mountain," or Profile Rock, 71.
    The Pool, 71.
    The Profile Lake, 71.
    Scenes and Incidents, 68.

White Mountains, The Silver Cascade, 70.
    The Summit House, 70.
    Tuckerman's Ravine, 69.
    The Willey House, 70.
White River Junction, 58.
White Water Cataracts, S. C., 177.
Whitehall, N. Y., 106.
White Plains, Harlem Railway, 103.
White River, Ark., 213.
White Sulphur Springs, Ky., 220.
Wild Fowl of the Chesapeake, 139.
Williamsburg, Va., 154.
Wilkesbarre, Pa., 133.
Williamsport, Pa., 134.
Wilmington, N. C., 167.
Wilmington, Del., 124.
Willoughby Lake, Vt., 74.
William's College, Williamstown, Mass., 60.
Williamstown, Mass., 60.
Willamette River, Or., 268.
    Valley, Or., 268.
Winchester, Va., 153.
Windsor, Vt., 58.
Winnebago Lake, 244.
Winnipiseogee Lake, 65.
Winooski Valley and River, Vt., 72.
Wisconsin, State of, 243.
Wisconsin River, 244.
"Woodlands," S. C.—Home of W. Gilmore Simms, 172.
Worcester, Mass., 49.
Wyandotte Cave, Ind., 229.

Y

Yazoo River, Ala., 198
Yonah Mount, Ga., 191.
Yonkers, Hudson River, 85.
York, Pa., 134.
Yorkville, S. C., 178.
Yorktown, and Battle of, Va., 155.
Yo-Semite Valley, Cal., 261.
Ypsilanti, Mich., 237.
Yreka, Cal., 256.

Z

Zanesville, O., 227.

www.ingramcontent.com/pod-product-compliance
Lightning Source LLC
Chambersburg PA
CBHW032054220426
43664CB00008B/995